Principal Rice Areas of Asia

(The darker the shading, the more rice is grown)

THE RICE ECONOMY OF ASIA

THE RICE ECONOMY OF ASIA

Randolph Barker
Robert W. Herdt
with Beth Rose

RESOURCES FOR THE FUTURE / WASHINGTON, D.C.

Published by Resources for the Future, Inc., 1616 P Street, N.W.,
Washington, D.C. 20036
Resources for the Future books are distributed worldwide by
The Johns Hopkins University Press

Library of Congress Cataloging in Publication Data

Barker, Randolph.
The rice economy of Asia.

Bibliography: p.
Includes index.
1. Rice trade—Asia. I. Herdt, Robert W.
II. Rose, Beth. III. Title.
HD9066.A7B37 1985 338.1'7318'095 85-2382
ISBN 0-915707-14-4
ISBN 0-915707-15-2 (pbk.)

Resources for the Future is a nonprofit organization for research and education in the development, conservation, and use of natural resources, including the quality of the environment. It was established in 1952 with the cooperation of the Ford Foundation. Grants for research are accepted from government and private sources only on the condition that RFF shall be solely responsible for the conduct of the research and free to make its results available to the public. Most of the work of Resources for the Future is carried out by its resident staff; part is supported by grants to universities and other nonprofit organizations. Unless otherwise stated, interpretations and conclusions in RFF publications are those of the authors; the organization takes responsibility for the selection of significant subjects for study, the competence of the researchers, and their freedom of inquiry.

This book is a product of RFF's small grants program. Randolph Barker is Professor of Agricultural Economics at Cornell University. Robert W. Herdt is Scientific Advisor to the Consultative Group on International Agricultural Research at the World Bank. Beth Rose is an Asian scholar at Cornell University. It was edited by Ruth B. Haas and Sally A. Skillings. The book was designed by Elsa B. Williams and the figures drawn by Applegate Graphics. Beth Rose prepared the index.

Contents

List of Tables

List of Figures

Foreword

It is indeed an honor to write a foreword to this book. In a time when the world news is presented in 30 minutes on television, and when research often tends to focus on smaller and smaller problems, this book represents "old fashioned" classic scholarship. It gives comprehensive treatment to a single subject of great importance to humankind.

The growth of population on the globe will increase significantly the need for food in the next several decades. A great deal has been written about our ability to feed ourselves during this period. Whether one is an optimist or a pessimist in this respect, one fact is abundantly clear—expanded food production will have a significant impact on natural resource use. This impact will be reflected in problems of environmental quality and energy as well as in other ways. We at RFF are very conscious of these interrelationships and have shaped our program to better understand them. This book is squarely in that arena.

Hunger in Africa is in the headlines today. But, less than a decade ago, there was great concern about food supplies in much of Asia. While food is still a problem in Asia, enormous progress has been made. For example, a decade ago there was much concern about India's capacity to feed its growing population, but today India is self-sufficient in food. How has this occurred? Much of the answer revolves around rice because it is the major food for millions of Asian people. Thus, this book will be of special interest to policymakers and scholars concerned with Asia, but it also will be of use in understanding food problems elsewhere.

Resources for the Future has published many books with an international focus and this book continues that tradition. We expect that such efforts will accelerate in the period ahead as our world becomes more interdependent and fewer and fewer problems of resources and the environment can be solved within national boundaries and with domestic policies. Our publications reflect this increased emphasis on international work.

The book represents several decades of professional work by Barker and Herdt. Resources for the Future is fortunate in being able to make it available in book form. Resources for the Future has long had an external grants program, but because of financial constraints we have limited the amount available to any one project. As a result, such grants ordinarily are used to support either a promising idea (the initiation of research) or to help bring a major piece of work to conclusion. The support of Barker and Herdt was in the latter category. It provides dramatic evidence of how productive a small amount of strategically placed financial support can be.

March 1985

Emery N. Castle
President
Resources for the Future

Preface

The purpose of this book is to present a comprehensive picture of the role of rice in the food and agricultural sectors of Asian nations. Rice is a dominant consumption good, an important employer of resources, a strong political focal point, and the beneficiary of considerable research attention. To millions, rice is the center of existence, and more than 90 percent of the world's rice is produced in Asia.

This book is not the first of its kind. In 1941, Wickizer and Bennett provided an excellent picture of the prewar rice economy of monsoon Asia, and in 1952 Efferson described the prewar and early postwar world rice economy. Since then there have been enormous changes in the production and distribution of rice. In comparison with the earlier period, the statistical data and data collection methods are much improved. Although a vast body of literature dealing with the Asian rice economy already exists, there is no single comprehensive treatment.

This book is concerned principally with the changes that have occurred in the Asian rice economy since World War II, and in particular since the introduction of the modern varieties and technology in the mid-1960s. It is written for decision makers at national and international levels, for professionals whether they be social or agricultural scientists, and for students of development. The objective is not only to describe the many facets of the rice economy, but to present a clear picture of some of the critical issues concerning both productivity and equity.

The book consists of three major parts—an overview chapter, the main text divided into seventeen chapters, and a data appendix, which is published separately. The volume is organized to accommodate the varying degrees of interest that readers may have in the subject, with the central issues and conclusions presented in the overview chapter. Each of these issues is explored in more depth in the main text, which is divided into two parts. Part I contains nine chapters which deal with rice production and the development of modern rice technology. Part II contains eight chapters which deal largely with marketing, trade, and policy issues. The appendix includes annual estimates for each of the major rice growing countries of rice area, yield, price, and a range of other factors related to the rice economy. Where possible, the time series are carried back to the early part of this century.

A book such as this has a history of its own. Randolph Barker first went to Asia in the summer of 1963 as a consultant for the Chinese-American Joint Commission on Rural Reconstruction in Taiwan. He returned again in 1965 to serve for two years with the Cornell program as a visiting professor at the University of the Philippines College of Agriculture in Los Banos. In 1967, he joined the Rockefeller Foundation as a member of the staff of the International Rice Research Institute (IRRI), also located in Los Banos. He headed the Department of Agricultural Economics of IRRI until 1978 when he returned to Cornell University.

Robert Herdt's Asian experience began in India in 1962 where, for two years under Ford Foundation auspices, he conducted research in the Intensive Agricultural Districts Program. In 1967, he returned to India for a year and a half with Rockefeller Foundation support to conduct research for his doctoral dissertation on the aggregate supply response in Punjab agriculture and to teach at the Indian Agricultural Research Institute in Delhi. He joined the staff of the International Rice Research Institute in 1973, heading the Department of Agricultural Economics from 1978 until his departure in 1983. He is currently a scientific advisor with the Consultative Group on International Agricultural Research at the World Bank in Washington, D.C.

The authors served under three directors at the International Rice Research Institute: Robert F. Chandler, Nyle S. Brady, and M. S. Swaminathan. This book owes much to all three, each of whom strongly endorsed the agricultural economics program at IRRI. As social scientists in a primarily biological science institute, we could not have asked for better support.

The first economist at IRRI was Vernon W. Ruttan (1963–65). The research program that he established and his continued friendship and counsel through the years have provided a guideline and inspiration for our own work.

Yujiro Hayami, a long-term friend and a coworker at IRRI for more than two years, insisted that we should write this book. He said that it was a "book for the next century." Although we began the project in 1978, there were times when we thought that it would take at least a century to complete.

We were fortunate to have Beth Rose join the team in 1979 when she undertook the demanding task of assembling the appendix, which is based on individual country data sources. An Asian scholar in her own right, she critically edited each chapter and meticulously checked for consistency and accuracy. Her contribution has added immeasurably to the quality of the book.

Most of the manuscript was written either in Ithaca, New York, or in Los Banos in the Philippines. It was written in the midst of a word processing revolution. Rhonda Blaine deserves the credit for successfully orchestrating this task. Merla Domingo at IRRI and Judy Wiiki and Marjorie Peech at Cornell also deserve thanks for their help. Joe Baldwin is responsible for the competent art work in all the figures and graphs.

Emery Castle, president of Resources for the Future, provided funding and encouragement for the initial research required to complete the manuscript. The Henry Luce Foundation and the Rockefeller Foundation provided financial support for editing and publication.

We (Randolph Barker and Robert Herdt) shared a unique experience in Asia. We had the opportunity over almost two decades to interact with and learn from scientists not only at the International Rice Research Institute, but throughout the rice growing world. We have benefited from the comments of many of these colleagues on chapter drafts of this book. We were aided in our research at IRRI by an extraordinarily able group of Filipino research assistants. We traveled extensively in every major rice growing country in Asia and were involved in many aspects of research on rice with both social and agricultural scientists. However, this required two to three months of travel a year, and we were often away from home for a month or two at a time. We both were fortunate to have been supported in our work by our families.

It is only fitting that we should dedicate this book to our families and to our friends and colleagues in Asia with whom we have shared some of the best years of our lives.

March 1985 Randolph Barker
Robert Herdt

Editor's Note

- Complete citations for the sources of tables and figures in this volume can be found in the bibliography at the back of the book.

- A companion volume by Beth Rose, *Appendix to* The Rice Economy of Asia: *Rice Statistics by Country, Tables with Notes*, is available. For more information, please write to Book Order Department, Resources for the Future, 1616 P Street, N.W., Washington, D.C.

1

Overview

Rice occupies a position of overwhelming importance in the global food system. Over a third of the world's population, predominantly in Asia, depends on rice as a primary dietary staple. Many of these people live in densely populated countries on an average annual income of less than $US 100, of which a third or more is typically spent on rice. Millions more grow their own rice and are dependent on sales of surplus rice to provide them with cash to purchase other necessities.

Only wheat exceeds rice in terms of volume of production. Together, these cereals occupy over one-quarter of the arable land of the world. However, the two grains are produced in very different environments using sharply distinctive production systems. Wheat is principally grown in temperate zones over a wide geographic area, generally on large mechanized farms. Close to 24 percent of the crop is traded internationally, the four major exporting countries being the United States, Canada, Australia, and Argentina. By contrast, 90 percent of all rice is produced and consumed within Asia. India and China together account for approximately half of world production and acreage.

Most Asian rice farms are small by Western standards (3 hectares or less) and employ intensive labor practices in place of mechanization. Only about half of the crop enters commercial marketing channels, and less than 5 percent of the crop is internationally traded (half of that figure comes from outside Asia, primarily the United States).

The Asian rice growing region is bounded by Japan on the east and Pakistan on the west. It extends from 50 degrees north latitude in northern China to those parts of Indonesia that lie at 10 degrees south latitude. Wickizer and Bennett used the same regional boundaries in their seminal work, *The Rice Economy of Monsoon Asia*, after which this book is modeled.

Practically all of the territory within this region is characterized by a summer maximum and winter minimum of rainfall, with the wind blowing from the ocean toward the Asiatic highlands in the summer (summer monsoon) and reversing its course in the winter. In the tropics, rainfall is the dominant climatic variable, and the rice crop is normally limited to the rainy season unless irrigation water is available. In the temperate zone, temperature predominates, and rice is a summer crop.

The region consists of three geopolitical subregions—East Asia, Southeast Asia, and South Asia. East Asia lies mostly in the temperate zone north of the Tropic of Cancer. It includes three major rice growing countries—China, Japan, and Korea. Parts of China (including Taiwan) fall below the Tropic of Cancer and are semitropical. Southeast Asia includes eight major rice growing countries—Burma, Indonesia, Laos, Kampuchea, Malaysia, Philippines, Thailand, and Vietnam. South Asia includes five major rice growing countries—Bangladesh, Sri Lanka, India, Nepal, and Pakistan.

Within the region, rice dominates not only production and consumption patterns, but is also inextricably woven into the social and economic fabric of life. More farmers are engaged in rice production than in any other single activity, with rice absorbing more than half of the farm labor force in many countries.

Rice is also of paramount importance in the political arena. Adequate supplies of rice and relatively low rice prices benefit urban consumers and help to maintain competitive prices for consumer and export goods. Conversely, short supplies and rising rice prices produce inflationary pressures on wages and tend to create political instability. The importance of rice in both rural and urban economic development subjects the rice economy to political pressures that are often detrimental to efficient growth. To protect against uncertainty, developing Asian countries have instituted some or all of the following policies: striven for self-sufficiency; protected internal markets against the price fluctuations of the international market; controlled domestic marketing of rice; set price floors and ceilings; imported large volumes of rice in election years; and provided rice rations to politically powerful interest groups such as civil servants.

With the exception of the more developed countries of East Asia, until a little more than a decade ago, the Asian rice economies lacked the capacity for technical change that would permit rapid growth in rice production to create the food surpluses needed for economic development.

Development of the Modern Rice Production System

A well-known agricultural scientist, Richard Bradfield, once said at the height of a Philippine typhoon, that if rice did not exist we would have had to invent it—for a unique feature of the rice plant is its ability to thrive under the flooded conditions created by the monsoons. The rice plant was "improved," not in a matter of years using modern breeding practices, but over centuries through selection by cultivators.

Rice, *Orizya sativa*, originated in the remote foothills lying along the border regions that divide South and Southeast Asia. From there, it spread into all corners of Asia, supplanting taro and yams as the staple crops of much of the region. The lowland rice cultivation practices still followed in most of the region today—transplanting into bunded (dyked) and puddled (wet mud) paddy fields—were developed in the Yangtze River Delta of China in the eighth to the twelfth centuries. By following these practices, a farm family of six or seven members can produce enough rice year after year from a single crop on 1 hectare of land without modern inputs to meet basic household consumption requirements.

Economic development requires a food surplus, that is, an excess of output above producers' consumption needs. In contrast to the Malthusian doctrine of stagnation caused by continued population growth on a stable or declining land base, Boserup suggests that the development of agriculture and creation of a food surplus are stimulated by population pressure on the land. The process of searching for ways to increase land productivity not only raises input requirements, but also helps fuel overall improvements in agriculture. Accepting this, it is not surprising that the transformation of the Asian rice economies through the employment of inputs other than land and labor began in East Asia where population pressure historically has been most severe. By the turn of the century, however, only Japan had developed the preconditions for agricultural development: (1) a formal agricultural research system; (2) an industrial sector capable of producing inputs such as chemical fertilizer; and (3) a transportation and communications network to ensure that inputs and information could be supplied to farmers.

Prior to this, Japan and China (and some exceptionally population-dense spots in South and Southeast Asia, such as Java in Indonesia and Ilocos in the Philippines) relied heavily on irrigation and increases in multiple cropping to feed the expanding population. However, the cause and effect relationship between population growth and growth in potential food supply is not well understood. Most of the rest of Asia was relatively "land surplus" and experienced little population pressure. That is to say, increases in rice production were achieved by using the traditional cultural practices on newly developed lands. Production held its own with population, but the rice surpluses to supply burgeoning urban demand came increasingly from the three major export regions, the deltas of the Irrawaddy River in Burma, the Chao Phraya in Thailand, and the Mekong in Vietnam.

Elements of the Modern Technology

The conventional concept of "modern farming" is based on Western agricultural systems where farmers strive for both a high yield per hectare and a high output per man hour. Modern technology is normally visualized as including both land-saving and labor-saving elements. In labor-surplus Asia, however, an economically rational definition of modern farming does not necessarily include labor-saving technology. Asian rice farmers using water buffalo for land preparation and hand labor for other operations often can produce rice more cheaply than their fully mechanized Western counterparts.

In Asia today we can identify a range of rice farming systems extending from those using traditional practices handed down for generations to those using techniques such as chemicals and tractors, which their forbearers would not have recognized. Farmers typically use a combination of traditional and modern practices, emphasizing those that increase yields per hectare. We can find farmers at all stages in the modernization process. In South China and Java, 150 to 200 man days per hectare may be employed in combination with organic or chemical fertilizer, or both, to obtain yields of 4 metric tons per hectare (mt/ha) or more. In Taiwan and in the Punjab, mechanized farmers use less than 100 man days per hectare to obtain the same yield. However, in both cases the achievement of the 4 metric ton yield level requires the use of modern varieties and fertilizer.

Another key factor used to differentiate rice farming systems is the water environment. Five distinct water regimes include: irrigated, shallow rainfed, deepwater, floating, and dryland. Essentially all of the rice in East Asia is irrigated. In South and Southeast Asia, however, irrigation accounted for approximately one-third of the total area and 50 percent of the total production in the mid-1970s. Modernization has been confined largely to the irrigated areas for reasons that will be discussed subsequently.

In summary, modern rice farming in Asia today may or may not include mechanization or the use of chemicals to control weeds, insects, and diseases. A key first step in the process of modernization has been the adoption of the short, stiff-strawed, fertilizer-responsive varieties to raise yields. Thus, our discussion of the modern production systems begins with varietal improvement.

Varietal Improvement

As noted earlier, at the turn of the century Japan initiated a government program to increase rice production. Advances were based on the selection and dissemination of improved varieties by veteran farmers (*rono*), coupled with the development of commercial supplies of organic fertilizers. Rice production increased at about the same rate as population (1.2 percent per year) until the end of World War I, when increasing demand and a slackening of production growth forced Japan to turn to its colonies, Korea and Taiwan, for additional supplies.

Varieties suited to the semitropical conditions of Taiwan were developed and disseminated in the mid-1920s. The *japonica* varieties known as *ponlai* (heavenly rice) were not only higher yielding than the native *indica* varieties, but had a shorter growth duration that permitted a significant increase in double cropping.

The rapid closing of the land frontier, as a result of population expansion in South and Southeast Asia after World War II, drew attention to the need to increase rice yields in the tropics. The average rice yield in Asia was 1.5 metric tons per hectare, or about 60 percent of the yield in Japan at the time of the Meiji Restoration (1868). Of course, the rice lands in much of tropical Asia were unirrigated. Particularly in the irrigated areas, fields of lodged rice (with stalks bent over and panicles lying flat on the ground) were a familiar sight at harvest-time. Fertilizer was not used because the application of nitrogen to tall indica varieties weakened the stalks, advancing the date of lodging and further reducing yields.

In the 1960s, rice scientists in tropical Asia, particularly at the International Rice Research Institute (IRRI) in the Philippines, began to focus on breeding a new plant type—a shorter, stiffer-stemmed variety that would respond to nitrogen fertilizer. The same kind of plant type had already markedly improved yields of Japanese japonica varieties developed in Japan and of wheat and other small grains (barley and oats). The Chinese initiated a similar breeding program at the Academy of Agricultural Sciences in Guangdong Province in 1956. Working independently (in those days there was no information exchange between Chinese and non-Chinese rice research centers), the two groups successfully developed short-stemmed varieties using Chinese indica varieties with a single recessive gene for dwarfness. Scientists at IRRI obtained this dwarfing gene from native Taiwan varieties that were crossed with tall indicas. Not until the late 1970s was it learned that the gene used as a source of dwarfing in the two programs was identical.

The release of the first of the IRRI varieties, IR8 (a cross betwen the Taiwan Dee-geo-woo-gen and the Indonesian variety Peta), established a maximum yield of about 10 metric tons per hectare for a single crop under ideal growing conditions. This yield ceiling was not broken until the 1980s. Because such rapid strides were made in yield improvement, most subsequent research has been devoted to developing varieties and production strategies that would allow the researcher to transfer high experimental yields to the heterogeneous farm environments found throughout Asia. This has proved to be an enormous challenge.

Under the growing conditions found in monsoon Asia, pests and diseases are a major constraint. The increased application of fertilizer and intensification of rice production through double cropping further exacerbates the problem. Tungro virus, blast (a fungus), stem borer, brown planthopper, and rats are among the major enemies of the rice plant. A unique feature of rice varietal improvement in Asia has been the initiation of breeding programs for insect as well as disease resistance. IRRI has played a leading role in the development of resistant germplasm, which it has made freely available to national programs around the world.

After the realization in the 1970s that the new varieties performed best under irrigated conditions, research was undertaken to develop varieties better suited to the less than ideal soil and water conditions that typify most of the Asian rice growing environment. The major achievement of the 1970s, however, was the development of the 100 day varieties, which not only made it possible to escape drought in the areas of short water supply, but to increase cropping intensity in areas where water was more plentiful.

Because the new varieties have typically been associated with fertilizer response and high yield, the impact that shortening the growth duration has had upon production is often overlooked. Traditional indica varieties take five months or more to reach maturity. The first of the modern varieties (such as IR8) reduced the growth duration by a month to 130 days. Varieties developed at IRRI (IR36) and in national programs, such as in Sri Lanka (BG series), reduced the time from planting (transplanting) to harvest to 3 to 3.5 months. Parenthetically, under more temperate conditions, the Chinese successfully introduced 100 day indica varieties (so-called Champa varieties from Vietnam) in the Yangtze River Valley in the eleventh century (Song Dynasty). Unfortunately, the impact of short-season varieties on rice production cannot be readily identified in national rice statistics in the manner, for example, that one can note the progress in yield per hectare. However, it is perhaps fair to say that shorter growth duration has been as important a factor as high yield in the contribution of varietal improvement to increased production over the past two decades.

As noted previously, a new yield ceiling was established with the introduction of the modern varieties in the 1960s. It is becoming apparent now that further increases in yield potential will depend on advances in biotechnology. An indication of this is seen in the work of the Chinese in the development of the F_1 hybrids. The hybrid rice breeding program was initiated in 1971 in Hunan Province. By 1974, the first true hybrid rice was grown, and by the late 1970s, 15 percent of Chinese rice land was planted to these new hybrids. Although the hybrid rice varieties can increase yields by as much as 20 percent, problems in producing and distributing hybrid seeds suggest that their impact on most of Asia will be minimal in the foreseeable future.

The Spread of Modern Technology

Modern varieties initially released in the 1960s spread rapidly throughout South and Southeast Asia, covering 9 million hectares or 12 percent of the rice area by 1970, 23 million hectares or 28 percent in 1975, and 33 million hectares or 40 percent in 1980 (excluding Kampuchea, Laos, and Vietnam). The spread was even more dramatic in China and in South Korea, where a program crossing indica and japonica varieties was initiated in the early 1970s.

Closely linked to the spread of modern varieties has been the rapid rise in fertilizer use. When the word "fertilizer" is mentioned today, most people think of chemical fertilizers. However, organic fertilizers, particularly animal manures, composted garbage, and plant refuse, have historically been important in maintaining soil fertility, are still important in some countries today, and may increase in importance at some future date when fossil-fuel-based supplies are exhausted.

Nitrogen is the most important fertilizer nutrient, representing two-thirds or more of total nutrient consumption in Asia. The yield response of rice to phosphorus and potassium is much less predictable. However, there are many soils in Asia that are deficient in phosphorus and other trace elements such as zinc. Intensification of production will further deplete these nutrients, leading to an increased demand in the future.

The rise in prominence of nitrogen-based fertilizers, such as ammonium sulfate and urea, was caused by technological advances in production and marketing that allowed nitrogen prices to fall steadily from the early part of the century up to the early 1970s. Even so, in the 1950s, approximately one-third of the Japanese and most of the Chinese fertilizer nutrients came from organic sources. Intense population pressure and hence the availability of vast amounts of labor in East Asian countries made it profitable to use organic fertilizer materials.

The consumption of chemical fertilizer has doubled and redoubled in many South and Southeast Asian countries in the short period since the introduction of modern varieties. Fertilizer nutrient application rose to about 20 kilograms (kg) in the mid-1970s, and to about 30 kg by 1980, but still remains a small fraction of the 200 kg/ha level applied in East Asia. The wide variation in the growth of fertilizer consumption by country is largely a function of the suitability of the environment for the use of fertilizer-responsive varieties, but also reflects differences in government policies across countries, including price, credit, and other measures to encourage (or discourage) fertilizer use.

The 1973–75 fertilizer crisis illustrates the importance that many Asian countries attach to adequate fertilizer supplies. With the sharp upward movement in oil prices, the historic downward trend in nitrogen prices came to an abrupt end. Fearing a fertilizer shortage, most Asian countries overreacted, importing larger than normal quantities and further driving up the prices. The price of urea, which had been $US 60 to $US 70 per metric ton in 1970, rose to almost $US 400 per metric ton in late 1974, but fell back to $US 125 per metric ton a year later.

Many Asian countries expanded domestic fertilizer production in the 1970s, but South and Southeast Asia still rely on imports for about half of their fertilizer needs. Thus, countries that are moving toward self-sufficiency in rice have become increasingly dependent on fertilizer imports. This led to support for research in biological nitrogen fixation and other alternative sources of plant nutrients. Nevertheless, research findings indicate that a significant increase in production could be achieved through more efficient application of chemical fertilizers that are already used. Chemical fertilizers will continue to be the major source of plant nutrients until more progress is made in developing economical biological sources.

Farm chemicals other than fertilizers include insecticides and herbicides. Along with fertilizer, the use of insecticides has grown rapidly with the spread of modern varieties. However, herbicide use has been largely confined to East Asia (excluding China) and to pockets of South and Southeast Asia with high wages, since herbicides tend to substitute for labor.

In general, cash expenses by farmers for insecticides and herbicides are only a fraction of the amount spent for fertilizer. This is perhaps a blessing in disguise since the overuse of insecticides and herbicides creates serious environmental problems. Toxic waste problems are less serious in flooded soil conditions, but elimination of natural predators, the spread of hitherto unimportant insects, diseases and weeds, and the capacity of insects to develop immunity to chemicals can aggravate pest control problems and raise production costs. The development of insect- and disease-resistant varieties, while itself not offering a single long-term solution (for example, varietal resistance can break down), does offer the opportunity for South and Southeast Asia to avoid dependence on costly chemicals found in Japan, Korea, and Taiwan. Furthermore, recommending a resistant variety to farmers is much simpler than instructing them in the appropriate use of insecticides. Research conducted in a number of Asian countries following prophylactic or predetermined levels of insecticide impact has shown that the probability that benefits will exceed costs is frequently 50 percent or less. Research on more cost-efficient strategies is needed to improve the level and probability of benefits from insecticide use and to determine how insecticide can best be employed in combination with resistant varieties.

Irrigation Sets a Limit to Adoption

The enthusiasm that accompanied the initial release of the new varieties led many to assume that it was just a matter of time before the new seed fertilizer technology spread to all parts of Asia. But the importance of irrigation, noted by some observers of the earlier East Asian experience, soon became even more evident in tropical Asia.

The confinement of the adoption of early modern varieties largely to the irrigated areas can best be understood by examining the yield response of rice to nitrogen under different environments. Compared to the strong yield increases and consistent year-to-year performance under dry-season irrigated conditions when solar energy is at a peak, the response is lower and highly variable in the wet season. Under rainfed conditions, the uncertainty of adequate moisture in many areas and the flooded conditions in others discourage the adoption of modern varieties and the use of fertilizer. In a large area of eastern India and Bangladesh, for example, farmers adopted modern varieties and applied high levels of fertilizer in the dry season, but these same farmers frequently did not use modern inputs in the wet season. A variety that has gained popularity in this area, *Mahsuri* (called *Pajam* in Bangladesh), is neither traditional nor modern. It was developed in Malaysia by Japanese breeders in the 1950s and seems to have performed extremely well under poor water control conditions and at low levels of fertilizer input. In recent years, new varieties have outperformed the old under rainfed conditions, but the major gains in yield per hectare have been made in the well-irrigated areas.

South and Southeast Asia can be divided into two major geographical regions based on irrigation characteristics. A line drawn from Saigon in the lower Mekong River Delta to Kathmandu in Nepal will include Vietnam, Kampuchea, Laos, Thailand, Burma, Bangladesh, Eastern India (the states of Assam, Bihar, Uttar Pradesh, West Bengal), and Nepal. This region encompasses all four of the major river deltas (Mekong, Chao Phraya, Irrawaddy, and Ganges/Brahmaputra) and the traditional exporting countries, Burma, Thailand, and Vietnam. Less than 20 percent of the rice growing area in this region is irrigated. This region accounted for 63 percent of South and Southeast Asian rice area and production in 1960. By the late 1970s, however, it accounted for 62 percent of the area but only 55 percent of production. Twenty percent of the area was planted to modern varieties.

The remainder of South and Southeast Asia includes the island peninsula countries, Indonesia, Malaysia, Philippines, and Sri Lanka representing traditional importers, plus Pakistan and India (with the exception of the eastern states). Over half of this area is irrigated, and 58 percent of the area was planted to modern varieties in the late 1970s. The introduction of modern varieties shifted the comparative advantage in rice production in favor of the countries and states represented in this region.

The introduction of modern varieties increased the profitability of investment in irrigation, particularly for dry-season crops. Many major storage schemes were undertaken with government and foreign donor financing. Tubewells became an important source of irrigation water in Pakistan and parts of India and Bangladesh. Government expenditure for irrigation as a percentage of total agricultural investment rose sharply, particularly in the island-peninsula countries. However, irrigation investments were heavily subsidized, with the water charges paid by farmers seldom covering the cost of operation and maintenance of the systems.

More recently, planners have begun to consider small, local systems. Not only are such systems able to overcome many of the organizational and accountability problems that plague large systems, but they also offer planners the opportunity to use community resources in place of state funds. As an added bonus, small communal systems already account for a major portion (perhaps half) of irrigated area, thus removing the need to train people in rudimentary irrigation management techniques. Further, the high degree of centralization of investment and management decisions in the large government-operated systems has led to a lack of communication between the system operators and the farmer users.

Only East Asia seems to have achieved a balance between national government support and local control in investment and management decisions. South and Southeast Asian countries will have to move toward a more decentralized system placing more decision-making responsibility in the hands of users if they are to achieve significant improvements in water use efficiency.

Mechanization and Employment

There is considerable controversy about the desirability of agricultural mechanization in Asia. For many, tractors and mechanization have become the symbol of modern agriculture because they are the most visible difference between farmers in the developing and the developed world. But those who oppose mechanization argue that almost any form of agricultural mechanization represents a straightforward substitution of capital for labor, and that under the labor-surplus conditions existing in most Asian countries, any such substitution is socially undesirable.

Mechanization, of course, means different things to different people. In this discussion, we refer principally to power equipment, particularly tractors (power tillers) and power threshers. In this section, we consider the effect of mechanization on labor, wages, and productivity.

An examination of the pattern of mechanical adoption shows considerable variability among countries. Commonly, however, mechanization begins with irrigation, then threshing, then land preparation, and finally proceeds to other cultural practices. The particular pattern of adoption reflects the technical efficiency of substituting mechanical power for human labor.

The demand for mechanization in the free market economies has been generally shaped by market forces, particularly wage rates and rice prices, although government rice price supports and subsidies and taxes on imported machinery have encouraged or discouraged mechanization in specific countries. East Asia (excluding China), with wage rates and rice prices well above the rest of Asia, is highly mechanized. The relatively high level of mechanization of China, on the other hand, has been more a reflection of government policy than of market forces.

Advocates of mechanization argue production is increased and, in some cases, even employment opportunities are improved. Production can be increased through higher yields or through greater cropping intensity. However, there is virtually no research evidence to support the claim that power tillers and power threshers have increased either yields or cropping intensity. In fact, peak cropping intensities were achieved in Japan in 1957 and in Taiwan in 1972, prior to extensive mechanization and declined thereafter as mechanization proceeded rapidly.

Even before the introduction of power machinery,

labor input per hectare in irrigated rice production seems to have shown a historic decline in some locations, such as Japan and Java, from over 200 man days per hectare to 150 man days or less. Labor input has been reduced through the introduction of animal power and the adoption of simple tools like the sickle in place of the *ani ani* knife (used to harvest one stalk at a time). The introduction of modern varieties at least temporarily reversed this trend since it led to an increase in labor demand, particularly for weeding. These gains in employment have been offset in some regions by the mechanization of land preparation and threshing. But in other areas, the expansion of irrigation in the dry season has greatly increased employment. The indirect employment effects of irrigation development (jobs created by expanded irrigation) are extremely large, much larger than the indirect employment effects of mechanization.

While hired labor benefited from the increased demand for labor associated with modern varieties, the introduction of machinery for land preparation and threshing resulted in a transfer of income to the capital owners. Since there is no significant gain in output through mechanization, the decision to promote mechanization is a decision to support transfer of income from labor to machinery without an offsetting benefit to labor.

Against this setting, a discouraging employment situation prevails in Asia, not as a result of the introduction of modern rice technology, which has increased employment, but rather because of rapid population growth and lack of alternative employment opportunities. Increased productivity in the rice sector in East Asia was accompanied by rising real wages and rising (supported) rice prices. By contrast, increased productivity in the rice sector in much of South and Southeast Asia has been accompanied by stagnant or falling real wages and a stable rice price. Thus, it is likely that most South and Southeast Asian countries will not undergo rapid mechanization in the 1980s.

Real wages in the rural areas appear to be very sensitive to shifts in labor demand. Employment gains have been made in the irrigated areas, but slow economic development in the rainfed areas is placing increasing employment pressure on the irrigated areas. The rapidly growing population of landless and near landless, some of them migrating from the uplands to irrigated rice areas, is a reflection of the problem. In the heavily populated regions, changes in the institutional arrangements by which the rice harvest is shared provide the mechanism through which the wage for hired labor is lowered. In a social context, the situation appears to be worse in South Asia, where the outlawed caste system remains a dominant force, than in Southeast Asia, where the traditional patron-client relationship offers the poor and disadvantaged some protection against falling wages. If real wages are to rise, continued technological development must be accompanied by other government policies and social reforms that will help on the one hand to increase the demand for labor and access to resources among the rural poor, and on the other hand, to slow the rate of population growth.

The Beneficiaries

Those scientists who initially developed the modern rice varieties gave little thought to the complexities of the physical and socioeconomic environment and the effect that they might have on the pattern of adoption. Nor had the experience with the adoption of rice technology in the more homogeneous environment of East Asia brought this issue to the forefront. The research objective was seen largely as one of raising rice yields in the tropics by the most expedient method. Given the uncertainty as to how long it might take to develop fertilizer-responsive varieties or to continue to convince "conservative" peasant farmers to accept the varieties and to purchase chemical fertilizer, it seemed reasonable to select the most promising environment for this research.

The rapid and almost simultaneous spread of the modern rice and wheat varieties in Asia led to an extensive debate about the distribution of the socioeconomic benefits of the new technology. The early literature on the Green Revolution reflected a polarization of views among supporters and detractors. Scholars studying the same events, and sometimes the same data sets, drew opposite conclusions. At one extreme were those who believed the technology was widening the gap between the rich and the poor, leading, in the Marxist interpretation, to an inevitable conflict between classes. At the other extreme were those who saw technological change as a necessary component of development. In the neoclassical tradition of the right, they divorced the issue of economic growth from distribution. Whether the new technology promoted equity or reinforced inequity was determined not by the nature of the technology itself, but by the pattern of ownership of resources and the institutional setting, which was an internal political affair.

This debate has led to a much better understanding of the equity issue among social and agricultural scientists alike. One consequence has been a greater effort to develop technologies suitable for unfavorable environments and resource-poor farmers. These factors notwithstanding, much confusion and disagreement still remain about the socioeconomic impact of the rice technology.

As pointed out earlier, the process of modernization has centered largely around the adoption of modern varieties and increases in the use of fertilizer. There is ample evidence to show that the adoption of the new seed-fertilizer technology led to higher yields. The development of rapid-maturing varieties also had a major impact on crop intensification, although in contrast to yields, there are few data to substantiate the magnitude of this impact. While there has been growth in the use of labor-saving technology, such as tractors and herbicides, there is no evidence to show that the adoption of this technology has been fostered by the introduction of modern varieties or has led to higher yields and production.

Major beneficiaries of the introduction of new technology have been the consumers who benefit from lower rice prices. Those for whom rice represents a major portion of the diet, that is to say the poorest families, clearly benefited the most. National governments also recognized a major political advantage in maintaining low and stable rice prices and were quick to promote and support the adoption of the new technology. However, countries tended to benefit in proportion to the amount of rice land under irrigation. As with countries, so also with farmers, those with irrigation benefited relative to those without.

The main controversy regarding beneficiaries has centered, not around the above issues, but around the issue of farm size. Much of the Green Revolution literature paints a picture of two distinct sectors in the rural village, the big farmers who monopolize technology, fertilizers, and credit, and the small farmers who have no access to inputs and cannot afford the cash expenses or the risk associated with the new technology. That such a situation should develop is almost intuitively obvious to those familiar with the rural areas of Asia. The fact that this bimodel pattern of adoption has not emerged, therefore, needs careful documentation and explanation. The large surveys conducted in India in the 1960s and 1970s are perhaps the best source of evidence. They show conclusively that there is no significant difference in the level of modern variety (MV) adoption, yield, and fertilizer use among farm-size classes. Significant differences do, however, exist in the level of these three factors across states, and this is closely associated with the proportion of area irrigated.

We have not studied the evidence for other crops, but irrigated Asian rice is perhaps somewhat unusual. Rice is a crop that for centuries has responded to intensive management and labor input. It is not uncommon to find that the yields of small farms are higher than those of large farms. This factor, coupled with the neutrality of the new technology in terms of economies of size, appears to have offset the institutional bias favoring large farms. The majority of those in Asia who have tried to consolidate rice holdings into large management units and to apply modern inputs, including mechanization, have failed. It is no accident that most Asian rice continues to be grown on farms of 3 hectares or less.

Although we observe that the new technology has been adopted by large and small farms alike and has generally resulted in an increase in the use of labor in production, these tendencies are subject to many exceptions. One cannot say unequivocally how the benefits have been spread among various groups in society without a more holistic analysis. The village rice economy is frequently made up of several groups—landlords, tenants, owner-operators, and landless laborers. A handful of village-level studies has been conducted to examine the distribution of income among these groups before and after the introduction of new technology. The results are not uniform and not easily summarized. But it is clear that, in relative terms, hired labor stands to gain as much if not more than farm operators from the introduction of the new technology. Alternatively, where yields have stagnated, no gains accrued to hired labor.

In absolute terms, however, there is no question that the distribution of benefits from technological change and increased income earnings is determined largely by the ownership of resources. It would be difficult to argue that the introduction of new rice technology has had any significant impact on the pattern of resource ownership, which is much more skewed in South and Southeast Asia than in East Asia. What does seem clear is that the new technology and perhaps the very nature of the rice crop itself have afforded little opportunity for the large and affluent to strengthen their position of wealth at the expense of the poor.

Rice Markets and Trade

The development of a modern rice production system in Asia, as well as other factors affecting supply and demand, have resulted in significant changes in the patterns of consumption, marketing, and trade. As in the case of production, so also in consumption and marketing, one has to distinguish between the more advanced economies, Japan, South Korea, and Taiwan, and the rest of Asia. The former economies have reached a stage of development where population growth has slowed, the society is becoming increasingly urban, and a smaller portion of the household budget is spent on food. The per capita consumption of rice and other cereal grains has leveled off or begun to decline as higher quality and more costly food items are substituted in the diet. This change is also taking place in city states such as Hong Kong and Singapore.

In most of Asia, however, growing population and rising incomes are continuing to contribute significantly to the increasing demand for rice. The growth in marketed surplus has been much more rapid than growth in total rice consumption, as an increasing percentage of the population has been absorbed by urban centers. This has placed pressure on the procurement and processing sector to expand and modernize facilities, and on national governments to maintain sufficient stocks to stabilize urban prices.

Because rice is widely grown and consumed in Asia, the domestic rice markets are relatively well developed and economically efficient in comparison with markets for many other commodities. By contrast, only a small portion of rice is traded internationally, less than 5 percent of total world production. Because of the thinness of the international market and the domestic policies of the trading countries, world prices are very volatile and the market is not efficient.

Prior to World War II, Asia accounted for 93 percent of rice exports and 75 percent of rice imports. By 1980, Asian export and import shares had dropped to 60 and 39 percent respectively. This changing pattern of trade is a result of a complex set of interactions among technological, demographic, and economic factors and the policies of trading nations. Although in absolute terms, Asian rice imports have remained fairly constant, Asian exports, particularly to the Middle East and Africa, have risen sharply since the late 1970s; in the meantime, Asian wheat imports have increased dramatically. These changes in consumption, marketing, and trade outlined above are described in more detail in the sections that follow.

Trends in Consumption

Based on rice consumption patterns, Asian countries can be divided into three major groups: (1) rice and wheat producers, (2) rice-dependent economies, and (3) rice producers and rice and wheat importers. The first group of countries includes China and India, the world's two largest rice producers, which also grow large amounts of wheat and other cereal grains, and Pakistan, the only country in the region where the level of consumption of wheat products greatly exceeds that of rice.

The rice-dependent countries have the highest level of per capita rice consumption. They include the traditional rice exporters, Burma, Thailand, and Vietnam, plus Bangladesh. The level of rice consumption is about 200 kg per capita per year, 50 percent above the level of other Asian countries. This same high level of consumption exists in portions of South China and eastern India. The dependency on rice in these regions stems from the fact that crops other than rice cannot be grown in the delta areas, which are flooded throughout much of the main growing season.

The final group of countries, the rice producers and importers, includes the island peninsula economies stretching from South Korea and Japan in East Asia, to Sri Lanka in South Asia. These countries produce little if any wheat and are traditionally rice importers. However, wheat consumption and thus wheat imports have increased, particularly in the urban areas.

Although average food consumption in most Asian countries is close to 2,000 kilocalories per capita per day, this does not take into account the highly skewed distribution of income and hence ability to purchase food. The insular countries as a group have shown an upward trend in per capita consumption of food in the past two decades, reflecting the stronger growth rate in their economies. Among the other two groups of countries, the pattern is very mixed. There is little evidence to suggest that increases in rice production have done much more than keep pace with population growth. The proportion of the population that cannot afford an adequate diet is large, by some estimates as high as one-third to one-half of the total. In absolute terms, the numbers of malnourished appear to be rising in many parts of Asia.

Rice and wheat are the preferred cereal grains in all countries of Asia, and, with the exception of Japan, their per capita consumption is either static or rising. Wheat products continue to substitute for rice in the diets of many higher income urban dwellers. In rural areas, however, a rise in income generally leads to a substitution of rice for inferior staples such as corn, millets, and root crops.

Rice will continue to be the dominant staple in Asia. To a large degree, the nutritional status of a particular Asian population can still be judged by per capita rice consumption. However, despite a growing amount of information on overall consumption patterns, our knowledge of the nutritional status of the poorer and more vulnerable segments of the population is still very inadequate. In many countries, it is still difficult to assess the degree to which the nutritional situation is improving or deteriorating.

Modernization in Marketing

The areas of most rapid growth in marketed surplus also have been those of fastest growth in rice production. Some areas have experienced severe although usually fairly temporary shortages of transportation, storage, and processing facilities. As in the case of production, capital intensity in marketing is often seen as a symbol of progress, but much of the modern equipment has proved to be inappropriate given local resources and economic conditions.

There are three major marketing channels in most countries: the local channel, which is largely private and serves the rural consumer; the private channel serving the urban consumer; and the government market channel. Because most of the population of Asia still lives in rural areas, the local channel is by far the largest. Governments normally acquire about 10 percent or less of the rice crops, but control both imports and exports.

Because rice is widely produced and consumed throughout Asia, the rice market is relatively well developed. The conventional image of the middleman notwithstanding, the domestic rice markets of Asia are, for the most part, efficient in an economic sense, being both competitive, and well integrated. That is to say, differences between farm and retail prices have tended to reflect the cost of marketing, and differences in price across regions have tended to reflect the cost of transportation. The private sector has dominated marketing activities in most countries, with the notable exceptions of China and Sri Lanka where, in the past, the government tightly controlled farm rice prices while subsidizing urban prices without generating sufficient revenue to cover marketing and handling costs.

A major marketing cost is milling, or the removal of the rice husk, bran, and other thin layers. Even today much of the rice that remains in the rural areas of Asia is milled by a traditional and labor-intensive process called hand pounding using a simple wooden mallet and bowl. Essentially all of the marketed rice is now processed in single huller mills, disc shellers, or in modern mills. Single hullers are found in the rural areas and normally handle small volumes of rice, while the larger mills are found in or near urban areas. As one moves from huller to disc sheller to modern mill, the percentage of whole grains, recovery rate, and appearance are improved.

While the large modern mills are technically more efficient, their appropriateness in a particular setting in Asia must be judged on the basis of economic efficiency. For example, available evidence suggests that, for most of Asia, it is economically more efficient to sun-dry rice rather than use mechanical driers, to employ disc shellers rather than rubber rollers, and to use bags rather than bulk storage. Thus, while there has been considerable expansion in milling and storage facilities in the past three decades, the basic technology has changed very little.

Changing Patterns in Rice Trade

Burma, Thailand, and Vietnam emerged as major rice exporters in the latter part of the nineteenth century and maintained their position of dominance until World War II. After the war, Thailand alone among the three retained its position as a leading rice exporter. Asia's share of world rice exports fell from 93 percent in 1935 to 70 percent from 1960 onward. The decline in Asian rice imports as a percentage of world imports, however, was even more dramatic, dropping from 75 percent in 1935, to 65 in 1975, and 39 in 1980.

Over the past three decades, the volume of Asian rice imports has remained stable at about 4 to 5 million metric tons per year. However, exports increased sharply in the late 1970s as a result of a rise in import demand in the Middle East and Africa, particularly among the Organization of Petroleum Exporting Countries (OPEC) countries. World rice trade increased from 8 million metric tons in the mid-1970s to 12 million metric tons in the early 1980s. What effect recent changes in the world economy, such as rising interest rates and falling oil prices, will have on trade remains to be seen.

While Asian rice imports have been stable, wheat imports grew at 3.7 percent per annum for the region as a whole from 1960 to 1980. The growth was most rapid in the non–wheat producing countries that, with the exception of the traditional rice exporters, Burma and Thailand, increased their wheat imports at a rate exceeding 8 percent per annum. We have already noted the tendency to substitute wheat for rice, particularly in the diets of the wealthier urban Asian population. But wheat consumption also rose in cities such as Dacca and Saigon because Bangladesh and Vietnam received most of their foreign grain aid in the form of wheat.

In attempting to understand the changes that have occurred in the post–World War II period, it is useful to separate the factors affecting market function in the short run from those affecting long-run trends.

While the world price has been relatively stable in the long run, it has been very volatile in the short run. This can be traced to national government policies that have adversely affected market performance. National governments have controlled, either directly or indirectly, the volume of imports and exports with the objective of increasing domestic price stability. Most Asian countries have determined the volume of rice to be traded (either imported or exported) on the basis of the adequacy of domestic production and supplies and have tended to be unresponsive to changes in world price. As a result, a major portion of price instability has been shifted to the world market.

The world market is integrated in the sense that price movements of most traded rice grades are highly correlated. But at any given point in time, there is no market-determined average world price as there is in the case of other major cereal grains such as wheat or corn. In fact, over half of the traded volume is handled through government-to-government contracts. The thinness of the market, the volatility in year-to-year demand among importers, and the fact that there is no central market clearing price greatly adds to the cost of sellers locating buyers and to the risk of expanding rice exports.

The changing patterns of long-term trade have been influenced primarily by technological change, by the relationship of rice to wheat prices, and by domestic rice-pricing policies. The introduction of modern rice varieties favored the importing over the exporting countries of Asia as has been noted previously. The new rice technology offered Asian importers the opportunity to reduce and sometimes eliminate their dependence on a volatile world market.

Perhaps the major factor encouraging wheat imports has been the low price of wheat relative to rice. If the additional processing and preparation costs for wheat are considered, on a per calorie basis wheat is similar in cost to rice. However, since most countries already have wheat-processing facilities, wheat imports result in a considerable savings of foreign exchange.

The fact that rice prices rose relative to wheat prices after World War II appears to be caused by both supply and demand factors. Technological changes led to a more rapid growth in yield and production of wheat than rice. At the same time, the demand for rice grew more quickly than for wheat because of more rapid population growth in rice-producing regions of the world, and because of the higher income elasticity of demand among the relatively poorer rice-consuming population.

In terms of domestic pricing policies, the Asian countries can be divided into three groups. The more affluent areas, which include Japan, South Korea, Taiwan, and Malaysia, have domestic rice prices well above the world price. At the other extreme, the exporters, Burma, Pakistan, and Thailand, have held domestic prices well below the world price, discriminating against producers. The remaining countries have held prices somewhat below the world level, but have mounted major campaigns to increase rice yields and production by introducing modern technology and expanding irrigation.

As a result of these policies, the pattern of world trade that has emerged does not reflect long-term comparative advantage. Nor have such policies led to efficient market performance in the short run. While one can blame the poor performance of the market on the policies of the trading nations, the cause and effect relationships are not clear. Nor do we know what measures can realistically be implemented to improve market efficiency.

Rice Policies

In the final five chapters of the book, we examine rice production from a more macro perspective. An underlying question is whether the growth in rice production achieved over the past two decades can be sustained for the foreseeable future. The ability to sustain growth in production depends on the one hand on the research establishment, and on the other hand, on supportive government policies. First, we examine the rice research and extension system as it has developed in Asia. Then we review the policy objectives and programs to achieve these objectives. Finally, we summarize the findings of our projections study, which illustrates in quantitative terms the continuing dependence of the rice sector on governmental policy support to sustain future growth in production.

Rice Research and Extension

Within the context of research, scientists must first continue to improve the potential of the new, high-yielding rice varieties to perform at a high level in the more disadvantaged physical and socioeconomic environments. Basic rice research is also needed to improve the biological yield potential of the rice plant. Throughout this century, the level of investment in rice research in the developed countries of East Asia has been several fold that in South and Southeast Asia. Furthermore, while East Asia has favored investment in research over extension by a ratio of two or three to one, just the opposite has been the case in South and Southeast Asia. However, the rice research system in South and Southeast Asia has matured rapidly since World War II. It now consists of international, national, and in some countries, regional institutions, all of which are linked to varying degrees. In fact, it might be fair to say that horizontal communication among rice scientists in Asia has advanced more rapidly than vertical communication among researchers, extension workers, and farmers. The result is that the Asian rice research system is heavily "top down," with relatively little feedback from farmers.

Although the overwhelming yield superiority of the modern varieties in the well-irrigated areas led to widespread farm adoption despite the weak researcher-farmer linkages, farm yields were still disappointingly short of experiment station yields. Extensive research at IRRI in cooperation with several Asian countries has definitively shown that the gap between farm yields and the economic optimum in that particular environment is very low. Furthermore, the research showed that cultural practices appropriate for achieving high yields under experiment station conditions were not consistently appropriate for farm use.

As more research attention has shifted to the development of appropriate technology for the more unstable rainfed environments, the gap between farm and experiment station has become more apparent. As a result, greater emphasis is being placed on efforts to understand the complexities of the farming system by conducting research in farmers' fields, and on a more participatory (farmer-involved) approach to research. This so-called cropping systems, or farming systems, style of research offers promise for improving research efficiency by combining the experiential knowledge of farmers with the scientific knowledge of researchers.

Under the most favorable environments, the yield ceiling established with the release of the new high-yielding varieties in the mid-1960s remains unchanged. A breakthrough is likely to require further advances in genetic engineering. For the immediate future, raising yield potential in the less favorable environments seems to offer the greatest potential benefits from the standpoint of both productivity and equity. But progress is likely to be slow given the complexity of these environments and the traditional top-down approach of Asian research and extension.

Policy

The overriding aim of any society is survival, and this requires a minimum level of political stability. In keeping with this goal, an important objective of rice policy in most Asian countries is to maintain adequate rice supplies, particularly in the urban areas, at relatively low and stable prices. Other objectives of rice policy include reduced dependency on foreign supplies, saving of scarce foreign exchange or increasing government revenues, increasing producer incentives and farm income, and achieving greater equity in income and more adequate nutrition among the poor. Of course, there is a conflict among some of these objectives, and at various stages in their development governments typically choose to emphasize some objectives over others.

There is a wide range of policy instruments used by governments to affect either the supply of or the demand for rice, although price manipulation is one of the most common approaches. Policies may be designed to adjust the prices of rice or other inputs; to increase investments in irrigation systems, roads, market and credit facilities; to expand research and extension; and to increase government revenues through taxation. Prices play an important part in rice policy because prices act on both ends of the marketing–consumption continuum by limiting consumption levels and providing an incentive to producers. In the long run, governments may attempt to seek all of the objectives outlined above. In the short run, many governments try to maintain low and stable prices to consumers and other privileged groups, often at the expense of long-run incentives for production.

Asian governments can be classified in three major groups according to price policy. The medium-to-high income countries support the domestic rice price well above world market prices. In these countries, the political appeal of rice self-sufficiency continues to be so strong as to encourage what many observers regard as unreasonable protectionist measures. The Japanese paddy rice price, for example, was several times the world market price in 1980.

At the other extreme, low-income traditional exporters, such as Burma and Thailand, have set domestic rice prices well below the world market. Until recently, traditional exporters have not greatly benefited from modern technology, since most of their rice is produced under rainfed conditions. Furthermore, because of the relative uncertainty of future world demand for rice, there has been little incentive for them to expand production beyond current levels.

The bulk of Asian countries fall into the category of low-income traditional importers, even though in recent years some of these countries have achieved a degree of self-sufficiency. These countries as a group have tended to maintain domestic rice prices somewhat below world market levels. At the same time, they have used a wide range of other policy instruments to take advantage of the opportunity offered by the new rice technology.

The countries in the latter group include the largest and most densely populated in Asia—China, India, and Indonesia. Most of them have undertaken programs or campaigns designed to promote the new technology and stimulate production. Extension and credit programs have been designed to get the package of inputs into the hands of farmers. Investment in irrigation and other infrastructure has been greatly expanded, and fertilizer is frequently subsidized. And in some countries, agrarian reform has been enacted in an attempt to provide more secure property and tenure rights, although such programs in South and Southeast Asia have had limited success. The major crises that have occurred to create sudden shortfalls in production, either regionally or locally, have provided further impetus to these production campaigns.

In summary, the pragmatic balance of low rice prices, government subsidies for irrigation, investment in research and extension, and subsidized credit has proved very successful in a number of Asian countries in achieving policy objectives and in increasing rice production. However, it would be wrong to conclude that these policies achieved economic efficiency either for specific countries or for the region as a whole. Rice production increases, whether encouraged through price supports in the medium-to-high income countries or through other forms of subsidy in the low-income countries, have been achieved at considerable cost. However, the key to a more rational policy that takes into consideration regional comparative advantages probably rests with efforts to persuade the more well-to-do countries to reduce their protectionism.

At the domestic level, the foundations of a rational rice policy (and a national development policy) must be built on two programs: technical change in agricultural production and redistribution of assets in favor of the poor. The historical underinvestment in rice research has been discussed previously. Redistribution of assets in favor of the poor can raise the employment and income potential and hence stimulate the demand for rice and other goods in the economy.

Projecting the Asian Rice Situation

Projecting future levels of rice supply and demand is at best a dubious art. It must always be remembered that at any given point in time, supply will be equated with demand. Nevertheless, projections can be useful in suggesting whether or not there will be strong upward pressures on price or, alternatively, in suggesting the level of investment commitments that will be needed to sustain production growth.

Our projections were based on simplified models of rice sectors of eight major rice-producing and consuming countries. Projections were made to the year 2000. The future demand for rice was based on assumptions regarding future trends in population and income growth. Projections of supply were not based on past trends, as is commonly the practice, but on projections of input use and output based on existing technical relationships. The current level of adoption of modern varieties and fertilizer was used to determine the potential for additional growth from further adoption. The current proportion of irrigated rice land was used to determine the future potential for conversion of rainfed to irrigated land. The availability of unused land was used to determine the potential for new land in rice.

Projections were made to compare production and per capita consumption in the year 2000 at constant prices. The results suggest that a substantial increase in imports would likely be required if the level of technology remains as it was in the 1970s. The largest possible investment in irrigation, complete fertilizer availability, and the full spread of modern varieties will not be sufficient to meet rice demand except in a few countries. Thus, the ability to meet future demands for rice without substantial increases in either price or imports depends on further modest productivity improvements. In short, these findings corroborate our previous assertion that a national rice policy must have as its foundation a strong research program.

The combination of continued growth in populations and rising incomes will cause the demand for rice to grow at 3 percent or more per annum in most of Asia throughout the remainder of this century. There will be very little expansion in cultivable land suitable for rice production. In our view, the technical capacity for meeting future demands does exist. Failure to sustain investments in irrigation and in technology development could dampen what currently appears to be a very favorable outlook for rice. However, sustaining such investments to meet market demand provides no guarantee that Asian countries will be able to raise the level of consumption of the less advantaged segments of society. Progress in this area will depend on reforms and economic developments that extend well beyond the rice sector.

2

The Origins, Classification, and Dissemination of Rice Cultivation

Rice is a plant of such antiquity that it seems unlikely that we will ever know its exact place of origin. The location of the earliest centers of domestication are still a matter of conjecture, as is almost every other aspect of the prehistoric culture and spread of rice. Wild species of rice are distributed over a broad area extending from India and South China southward and eastward across continental and insular southeast Asia.

Asian rice, *Oryza sativa*, is believed to have evolved from an annual progenitor in a broad belt extending from the Gangetic plain below the foothills of the Himalayas, across upper Burma, northern Thailand, to North Vietnam and South China.[1] Domestication could have occurred independently and concurrently at many sites within this area. Man took annual wild types, subjected them to the selection pressures of cultivation, harvesting, and sowing, and this gave rise to the *O. sativa* cultivars in Asia.

Numerous archaeological investigations throughout Asia have established that rice was domesticated as early as the fifth millennium B.C.[2] There seems to be general agreement that domestic rice varieties originated in the flood plains rather than in upland areas,[3] but whether rice was first grown as a shallow or deepwater crop is a matter of conjecture. It has been suggested that both shallow and wet-field cultivation occurred in the upland valleys, and that the subsequent introduction of terracing and water control into this same region permitted the development of a more productive wet-field system.[4]

Since primitive man had little, if any, control over the environment, the climate in the center of origin must have been particularly suitable for the domestication of rice. This area would have to have been very warm and humid, with a strong monsoon rhythm in the rainfall pattern and limited solar radiation.[5]

There are two cultivated rice species, *Oryza sativa L.*, and *Oryza glaberrima*, the former having its origin in tropical Asia and the latter in West Africa. The cultivation of *O. glaberrima* is limited to the high rainfall zone of West Africa while *O. sativa* is grown throughout the world.

Because of the wide range of environmental factors that influence the evolution of rice types, classification is a complex task, and a number of systems have developed. Rice is commonly divided into subspecies—*sinica* or japonica, indica, and *javanica*—or into cultural types—lowland, upland, and deepwater (floating).[6] Classification according to cultural types is based on soil and water conditions, while classification according to subspecies is based on growth response of the rice plant to temperature and light. These basic classifications are widely used in the literature and are particularly helpful in understanding modern rice production and consumption patterns from a socioeconomic as well as an agronomic perspective.

Cultural Types and Their Dissemination

Rice cultural types are often poorly defined in the literature and cause considerable confusion. The usual distinctions are between wetland and dryland culture or between lowland and upland culture. In wetland or lowland rice culture, the fields are flooded, leveled, and bunded (that is, enclosed with an earth levee or dyke called a bund to contain the water) before planting or transplanting. Water is supplied by natural rainfall, floodwater, runoff from higher ground, or irrigation, and the fields typically remain flooded throughout much of the growing season, depending on rainfall or water availability. In Asia, most rice is grown under wetland conditions.

The confusion in definitions arises partly because rice was first grown in relatively flat and low-lying sites (lowlands) and evolved through time from a dryland to a wetland culture. Rice, of course, does not need to be grown in a continuously flooded field. Upland cultural practices developed later and are typically found in areas of low rainfall or in higher elevations where the land is not terraced. Because fields are not leveled and flooded, water comes solely from rainfall, and therefore the supply is less certain. The upland rices generally have a shorter growing season and are more resistant to drought.

Because of low levels of productivity, upland rice culture is no longer very important and today accounts for only 10 to 15 percent of the area planted to rice in South and Southeast Asia.[7] Agricultural practices used in upland culture range from very primitive to modern.

Table 2.1, summarizing the main types of rice culture by water regime, shows a classification system developed in the late 1970s by scientists at the International Rice Research Institute in collaboration with scientists in national programs throughout Asia.

In China, rice was planted in the low plain areas of Hunan and the central Shaanxi basin between 3000 and 2500 B.C. Fields were typically established a short distance from a river, where floods would not damage the crop.[8] As rice spread southward and eastward, a lowland rice culture evolved, allowing substantial gains in productivity. By the second to third century B.C., irrigation was widely practiced, and some large-scale works were in operation. The water buffalo, indigenous to Southeast Asia, was in common use, the iron plough, hoe, and sickle were primary implements of cultivation, and rice was the staple grain in the middle and lower Yangtze River Valley.

During the early Christian era, rice cultivation extended southward in China, and manuring and transplanting were adopted. In the period from A.D. 300 to A.D. 500, rice became a staple in central and east China south of the Yangtze River. Between the eighth and twelfth centuries, the spike-tooth harrow and roller compacter were introduced, greatly facilitating weed control and transplanting activities.

The cultural practices perfected during this period in China and still commonly practiced in Asia today gradually spread to Southeast Asia as more and more

Table 2.1. Classification of Rice Cultures by Water Regime

Water regime	Description of culture	Dominant varietal type	Typical yield mt/ha
Irrigated—wet season	Fields are bunded and puddled. Rice is transplanted. Water is added to the fields from canals, river diversion, pumps, tanks, etc. to supplement rains.	Modern semidwarf-to-medium (100–130 cm)	3.0
Irrigated—dry season	Similar to wet season, but water must be supplied from storage reservoirs or from pumps. Solar energy levels are normally much higher in wet season.	Modern semidwarf (100–120 cm)	3.5
Shallow[a] rainfed	Maximum water depth from tillering to flowering ranges from 0 to 30 cm. Fields are bunded and puddled. Rice is transplanted.	Modern and traditional semidwarf (100–130 cm)	2.0
Deepwater[b]	Maximum water depth from tillering to flowering ranges from 30 to 100 cm. Rice is either broadcast in dry fields or transplanted in bunded or puddled fields.	Traditional tall (150 cm)	1.5
Floating	The maximum water depth from tillering to flowering exceeds 1 m and may run as high as 6 m. Rice seeds are normally broadcast in dry, unbunded fields before the onset of rains.	Floating rice with elongating potential	1.0[c]
Dryland	Rice is grown on flatland, terraces, or slopes without leveling, bunding, and impounding standing water in fields.	Medium-to-tall traditional (130–150 cm)	1.0

[a]In 1981, IRRI identified four subtypes within this category: (1) shallow favorable, (2) shallow drought prone, (3) shallow drought and submergence prone, (4) shallow submergence prone. See IRRI, *A Plan for IRRI's Third Decade* (Los Banos, Philippines, 1982).

[b]Three subtypes have been identified: (1) medium-deep waterlogged, (2) medium-deep tidal swamp, (3) deepwater.

[c]Recent data show that yields in some floating areas are much higher.

of the lowland areas were cleared and developed for cultivation. Similar techniques spread in the Indian subcontinent, although the origin and dissemination of the cultural practices are not as well documented as in the case of China. In areas such as Burma, Thailand, and Indonesia, which have been exposed to both cultures, it is easy to identify agricultural implements of both Chinese and Indian origin today.

As rice moved into the lowlands, it gradually spread into the major river deltas of South and Southeast Asia. Traditional tall indica varieties are well adapted to a water regime between 50 and 100 cm in depth. However, in those areas of the deltas where water normally rises to a depth of 1 meter or more during the growing season, a distinctly different deepwater or floating rice culture developed. Today, a little over 5 percent of Asian rice land is planted to floating rice varieties. There is no mention in the literature of the exact origin of this culture. However, floating ability under steadily rising and deepwater conditions is apparently a primitive trait that enabled wild progenitors to adapt to haphazard water regimes.[9]

Subspecies or Varietal Types and Their Dissemination

In most of Asia, rice sown at the beginning of the rainy season is considered the main crop, and rice grown under irrigated conditions in the dry season is the secondary crop. The seeds can be sown directly by broadcasting or started in a nursery and transplanted.

There are three phases of rice plant development. The vegetative phase covers the period from germination to the initiation of the panicle (the floret). Next comes the reproductive phase, in which the plant grows taller and continues to develop to the flowering stage. During ripening, the starchy portion of the grain gradually turns into a hard dough and finally dries out. Early-maturing varieties of rice need 110–120 days to grow to maturity. Medium- and late-maturing varieties range from 140–150 and 150–180 days, respectively.

There are hundreds of rice varieties grown in South and Southeast Asia, and they differ in grain quality, yield, physical features, sensitivity to light, resistance to stress (bad weather, pests, and diseases), and responsiveness to fertilizers and cultivation methods.

The traditional indica varieties are native to the tropics and are tall, leafy, and high-tillering plants. The japonica varieties are most commonly found in subtropical and temperate zones. Since they were first grown in China and then spread to Japan, they are also known as sinica or *keng*. The javanica varieties are found mostly in Indonesia and in the mountain regions of northern Luzon, the Philippines, and in Taiwan. Table 2.2 shows some of the characteristics of these three subspecies.

Even before the advent of modern breeding programs, japonica rice possessed characteristics favoring fertilizer responsiveness (medium height, greater resistance to lodging). Furthermore, the nonphotoperiod sensitivity of japonica rices made dissemination of improved selections an easier task.

Grain type and quality are other differentiating characteristics and are important in consumer preference and marketing. Japonica rice grains are generally short and are stickier when cooked. However, among the diverse indica rices, there are also medium-grain glutinous types often referred to as "sticky or sweet rices." These are grown widely throughout Asia for use in speciality dishes, but they are also the staple grain throughout much of the Mekong Delta, including most of Laos, a large part of Kampuchea, and Northeast Thailand.

The close genetic affinity between the japonica varieties and the *aus* (autumn) varieties of northeastern India and Bangladesh has given rise to the theory that the early maturing, nonphotoperiod-sensitive aus rice may have differentiated from the indica, which may in turn have given rise to japonica

Table 2.2. Selected Characteristics of the Three Rice Subspecies

	Subspecies		
Characteristics	Indica[a]	Japonica	Javanica
Tillering	High	Low	Low
Height	Tall	Medium	Tall
Lodging	Easily	Not easily	Not easily
Photoperiod	Sensitive	Nonsensitive	Nonsensitive
Cool temperature	Sensitive	Tolerant	Tolerant
Shattering	Easily	Not easily	Not easily
Grain type	Long-to-medium	Short and round	Large and bold
Rice texture	Nonsticky	Sticky	Intermediate

[a]Traditional type prior to development of modern semidwarf varieties.

rices.[10] It has been suggested that the aus varieties be classified as an intermediate subspecies. Recent studies show that the upland rices of Southeast Asia and the javanica rices of Indonesia are closely related to each other and also to the aus varieties.[11]

The keng (sinica or japonica) rices of China could have differentiated on the northern slopes of the Himalayan ranges where, today, japonica is frequently found on the high slopes and indica in the valleys. From here japonica spread northward and eastward into the tributaries of the Yangtze and the Yellow rivers and hence into the North China Plain.[12] Archaeological discoveries place the beginning of rice culture in east-central China at 5000 B.C., in the Yangtze basin in about 4000 B.C., and in the Yellow River basin between 2500 and 3000 B.C. The earliest rice appeared in the Yangtze basin slightly later (4000 B.C.).[13] From China, japonica rices were introduced into Korea (date unknown) and Japan in the third century B.C.[14]

Early-maturing varieties were transported from the state of Champa in central Indochina to Fujian Province in South China in the eleventh century and subsequently extended from Fujian to the Yangtze and lower Huai areas in 1011.[15] Champa rice was drought resistant and matured in 60 to 100 days after transplanting, compared with 150 days or more for the traditional varieties.[16] The yield, however, was considerably lower than the late-ripening varieties.

The introduction of early-maturing varieties permitted the expansion of double-cropped area and of rice land to the hillier and more marginal areas. Within two centuries of the introduction of the Champa varieties, the landscape of the eastern half of China's rice area was substantially changed.[17] The success of the double-cropping system in the Yangtze River Valley permitted rapid population growth and assured the economic dominance of this region throughout the Southern Song, Yuan, and Ming dynasties (A.D. 1127 to 1662). It is estimated that rice accounted for more than two-thirds of Chinese foodgrain production during this period and that more than two-thirds of the population was located in the central and southern rice growing provinces. In the early Qing Dynasty (mid-seventeenth century), the introduction of foreign food plants such as maize, sweet potato, and peanuts into northern China began to redress the agricultural imbalance between north and south that resulted from the dissemination of the early-ripening rice.[18]

The early tropical and subtropical cultivars, except for a few very early-maturing aus types, could not have been grown beyond 36 degrees north latitude.[19] The northward expansion of the rice growing area in more recent times has been due to the creation of early-maturing varieties well adapted to the cool climate, and to the development of techniques to protect the seed beds from cold damage. Just prior to the nineteenth century, rice culture spread to Japan's northern island of Hokkaido (41 to 46 degrees north

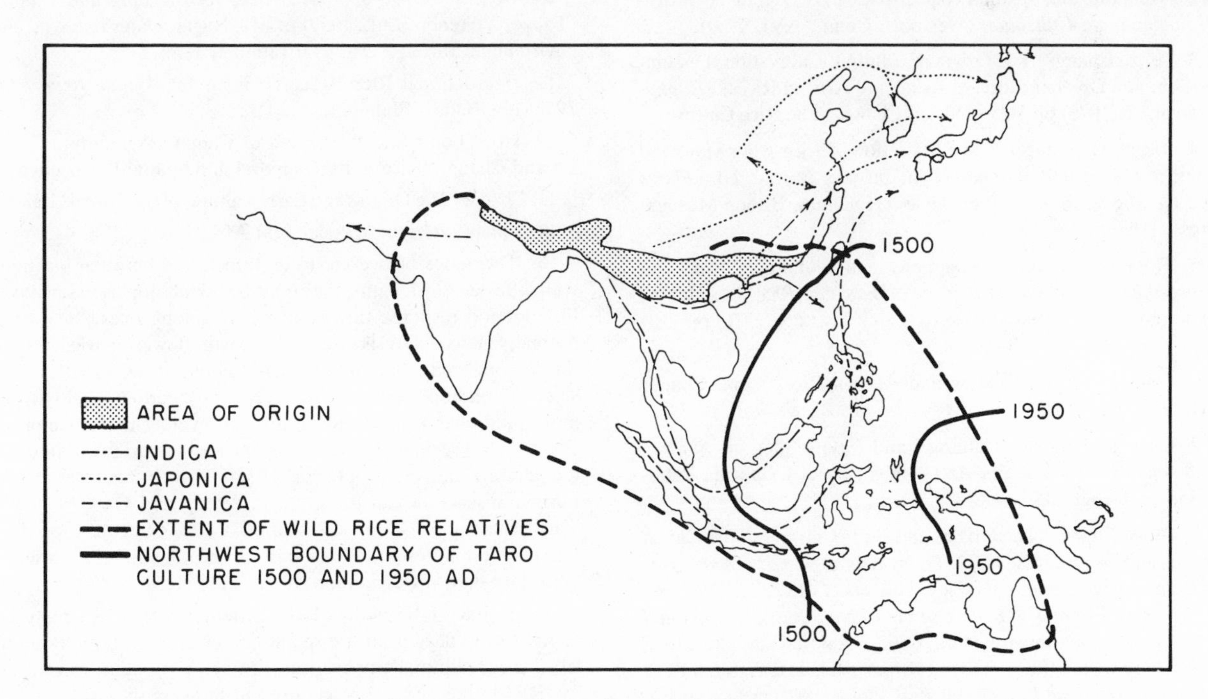

Figure 2.1. Origin and spread of rice culture and the retreat of taros (Sources: Adapted from T. T. Chang, "The Rice Cultures," p. 144; and Joseph E. Spencer, *Shifting Cultivation in Southeast Asia*, p. 113)

latitude).[20] In the first half of this century, the northernmost limits were extended by about 5 degrees north latitude.[21] Rice is currently grown as high as 53 degrees north latitude.

The indica cultivars spread southward through India to Sri Lanka, southeastward across the Malay Archipelago, northward into Central and South China, and westward to the Middle East, Africa, and Europe. Historical evidence suggests that taro and yams were initially the staple crops throughout much of the Indian subcontinent, upper Southeast Asia, and South China, covering an area roughly identical to that occupied by the wild varieties of *O. sativa* (figure 2.1). There was a gradual substitution of grain for root crops, with the taro-yam culture giving way first to pulses, millets, and beans and ultimately to rice as the staple crop.[22] As figure 2.1 shows, the frontier of the taro-yam culture retreated to the southwest until today only New Guinea remains as a major site. Permanent dry-field and shifting cultivation techniques were predominant, except in very localized areas of agricultural terracing and water control. In the Chao Phraya Delta of Thailand, for example, lowland rice area surpassed upland area only in the fifteenth century.[23] On the Malayan peninsula, in the Philippines, and in Indonesia, wet-field rice agriculture remained of minor importance until the coming of the Europeans.[24]

As we discuss the various aspects of the Asian rice economy in the chapters that follow, we will frequently refer to the basic cultural and varietal classifications described above. Prior to World War II, most work on modernization of Asian rice production occurred in the temperate zone and involved the improvement of japonica varieties. The Green Revolution marked the beginning of the development of high-yielding indica varieties in the irrigated areas of tropical Asia. Only in the past decade has attention been given to improving rice yields in the rainfed, deepwater (floating), and dryland rice areas.

Notes

1. Te-tzu Chang, "The Rice Cultures," in *The Early History of Agriculture* (London, Oxford University Press, 1977) p. 143.

2. For a discussion of archaeological and historical studies pertaining to rice, see Jonathan J. Lu and Te-tzu Chang, "Rice in Its Temporal and Spatial Perspectives," in B. S. Luh, ed., *Rice: Production and Utilization* (Westport, Conn., AVI, 1980).

3. Te-tzu Chang, "The Origin, Evolution, Cultivation, Dissemination, and Diversification of Asian and African Rices," *Euphytica* vol. 25 (1976) pp. 425–440 and Chang, "The Rice Cultures."

4. Joseph E. Spencer, "The Migration of Rice from Mainland Southeast Asia into Indonesia," in Jacques Barrau, ed., *Plants and the Migration of Pacific Peoples* (Honolulu, Bishop Museum Press, 1963) pp. 84–86.

5. Robert E. Huke, "Geography and Climate of Rice," in *Proceedings of the Symposium on Climate and Rice* (Los Banos, Philippines, International Rice Research Institute, 1976) pp. 31–50.

6. Chang, "Origin of Asian and African Rices;" and Chang, "The Rice Cultures."

7. Chang, "The Rice Cultures" and Joseph Spencer, *Shifting Cultivation in Southeastern Asia* (University of California Publications in Geography, 1966) vol. 19, p. 111.

8. Te-tzu Chang, "The Origins and Early Cultures of the Cereal Grains and Food Legumes," in D. N. Keightley, ed., *The Origins of Chinese Civilization* (Berkeley, University of California Press, 1982); Mark Elvin, *The Pattern of the Chinese Past: A Social and Economic Interpretation* (Stanford, Calif., Stanford University Press, 1973); Ping-ti Ho, "Loess and the Origin of Chinese Agriculture," *American Historical Review* vol. 75 (October 1969) pp. 1–35; and Kwang-chih Chang, *The Archaeology of Ancient China* (New Haven, Conn., Yale University Press, 1977).

9. Te-tzu Chang and H. I. Oka, "Genetic Variousness in the Climatic Adaptation of Rice Cultivars," in *Proceedings of the Symposium on Climate and Rice* (Los Banos, Philippines, International Rice Research Institute, 1976) p. 99.

10. Isaburo Nagai, *Japonica Rice, Its Breeding and Culture* (Tokyo, Yokendo, Ltd., 1959) p. 141. Nagai relates rice type and environment through characterization systems.

11. International Rice Research Institute, *Annual Report for 1982* (Los Banos, Philippines, IRRI, 1983).

12. Ho, "Loess and the Origin of Chinese Agriculture;" and Lu and Chang, "Rice in its Temporal and Spatial Perspectives."

13. Chang, "The Origins and Early Cultures of the Cereal Grains."

14. Chang, "Origin of Asian and African Rices," p. 432.

15. If varieties of rice are to be transferred between locations with different day lengths, they must be nonphotoperiod sensitive. Photoperiod response differs markedly among rices, but rice is normally viewed as a short-day plant, with flowering triggered by the seasonal reduction in day length among those varieties classified as photoperiod sensitive. The photoperiod sensitivity of most traditional, tropical indica varieties severely limits their diffusion to other regions. However, as one moves northward into the temperate zone, the proportion of sensitive varieties decreases. Further discussion can be found in Chang and Oka, "Genetic Variousness in the Climatic Adaption of Rice Cultivars," pp. 88–92. The nonphotoperiod sensitivity of the Champa rice varieties permitted their dissemination over a wide area.

16. The first of the modern IRRI varieties, such as IR8, matured in 120 to 130 days from transplanting, which was about a month less than traditional tropical rices. Modern varieties, developed by IRRI in the mid-1970s, matured in 110–115 days.

17. Ping-ti Ho, "Early Ripening Rice in Chinese History," *The Economic History Review* vol. 9 (1956) pp. 214–215.

18. Chang and Oka, "Genetic Variousness in Rice Cultures," pp. 216–217.

19. Nagai, *Japonica Rice, Its Breeding and Culture,* p. 132.

20. Matsuo Takane, *Rice Culture in Japan* (Government of Japan, Ministry of Agriculture and Forestry, 1954); and Chang, "The Rice Cultures."

21. Nagai, *Japonica Rice, Its Breeding and Culture,* pp. 133 and 200.

22. Spencer, *Shifting Cultivation in Southeast Asia,* p. 111.

23. Tadayo Watabe, "The Development of Rice Cultivation," in Ishii Yoneo, ed., *Thailand: A Rice Growing Society* (Honolulu, University of Hawaii Press, 1978) pp. 6–10.

24. Spencer, "The Migration of Rice," p. 86.

3

Asian Rice Farming Systems

The way in which rice is grown and the particular rice farming systems in a given location reflect the interaction between physical, environmental, and socioeconomic factors such as institutions, available technology, and government policies. Before describing the common rice farming systems, we characterize the variability in topography and climate, relating this to the rice classification system described in chapter 2.

Topography

The rice areas of South and Southeast Asia are classified by water regime in table 3.1 following the system defined in table 2.1. Topography is closely linked to water regime. The major topographic features of the landscape are shown in figure 3.1. The subsections that follow describe three broad topographic environments—the lowland plains, the hills and plateaus, and the river deltas.

Lowland Plains

The lowland plains represent the most important rice regions in terms of total area and total production. The lowland plains gained prominence with the coming of the colonial powers, which undertook substantial investments in irrigation (chapter 7). The large government-managed irrigation systems are typically found in these areas. Originally these rice lands were irrigated through river diversion systems, but today a number of storage systems (for example, Upper Pampanga in the Philippines, Muda River in Malaysia, Mahaweli in Sri Lanka) have permitted the irrigation of two rice crops in a single year in many locations. These rice bowls have been the center of expansion of the new rice technology, which performs exceptionally well under dry-season irrigated conditions. In other lowland plains, such as the Indus and Gangetic river plains of the Indian subcontinent (which includes the Punjab), physiographic conditions favor the use of tubewells and pumps to lift irrigation water from shallow aquifers. Here also the expansion of irrigation and the new rice technology have been very rapid.

Rainfed Terraces and Plateaus

The rainfed areas are so diverse that it is impossible to talk about a typical rainfed rice culture. It is also difficult to classify the rainfed areas in terms of water regime, although flooding depth is the principal criterion for classification. In this section and the section that follows, we distinguish between the rainfed terraces and plateaus and the river deltas. The rainfed terraces and plateaus are identified in figure 3.1 and form the bulk of the area classified as shallow rainfed and dryland in table 3.1.

Much rainfed rice is grown in gentle to steeply sloping terraces in areas that are difficult to irrigate. Different soil and water conditions on the upper and lower terraces affect the type of paddy grown and, therefore, the cultural practices and cropping patterns.

Table 3.1. Rice Area by Water Regime, South and Southeast Asia, 1970s
(thousand hectares)

Country	Irrigated		Shallow rainfed (0–30 cm)	Deepwater (30–100 cm)	Floating (100 cm+)	Dryland
	Wet season	Dry season				
South Asia						
Bangladesh	170	987	4,293	2,587	1,117	858
Bhutan	—	—	121	40	—	28
India	11,134	2,344	12,677	4,470	2,434	5,973
Nepal	261	—	678	230	53	40
Pakistan	1,710	—	—	—	—	—
Sri Lanka	294	182	210	22	—	52
Total South Asia	13,569	3,513	18,100	7,389	3,604	6,979
Percent of total	25	7	34	14	7	13
Southeast Asia						
Burma	780	115	2,291	1,165	173	793
Indonesia	3,274	1,920	1,084	534	258	1,134
Kampuchea	214	—	713	170	435	499
Laos	67	9	277	—	—	342
Malaysia	266	220	147	11	—	91
Philippines	892	622	1,207	379	—	415
Thailand	866	320	5,128	1,002	400	961
Vietnam	1,326	894	1,549	977	420	407
Total Southeast Asia	7,685	4,100	12,396	4,238	1,686	4,642
Percent of total	22	12	36	12	5	13
South and Southeast Asia						
Total	21,254	7,613	30,375	11,587	5,290	11,593
% of total area	24	9	35	13	6	13
% of total production[a]	34	15	33	9	3	6

Source: Robert E. Huke (1982)

[a]Assuming yields as follows: Irrigated, wet season, 3.0 mt/ha; irrigated dry season, 3.5 mt/ha; shallow rainfed, 2.0 mt/ha; deepwater, 1.5 mt/ha; floating dryland, 1 mt/ha (see table 2.1).

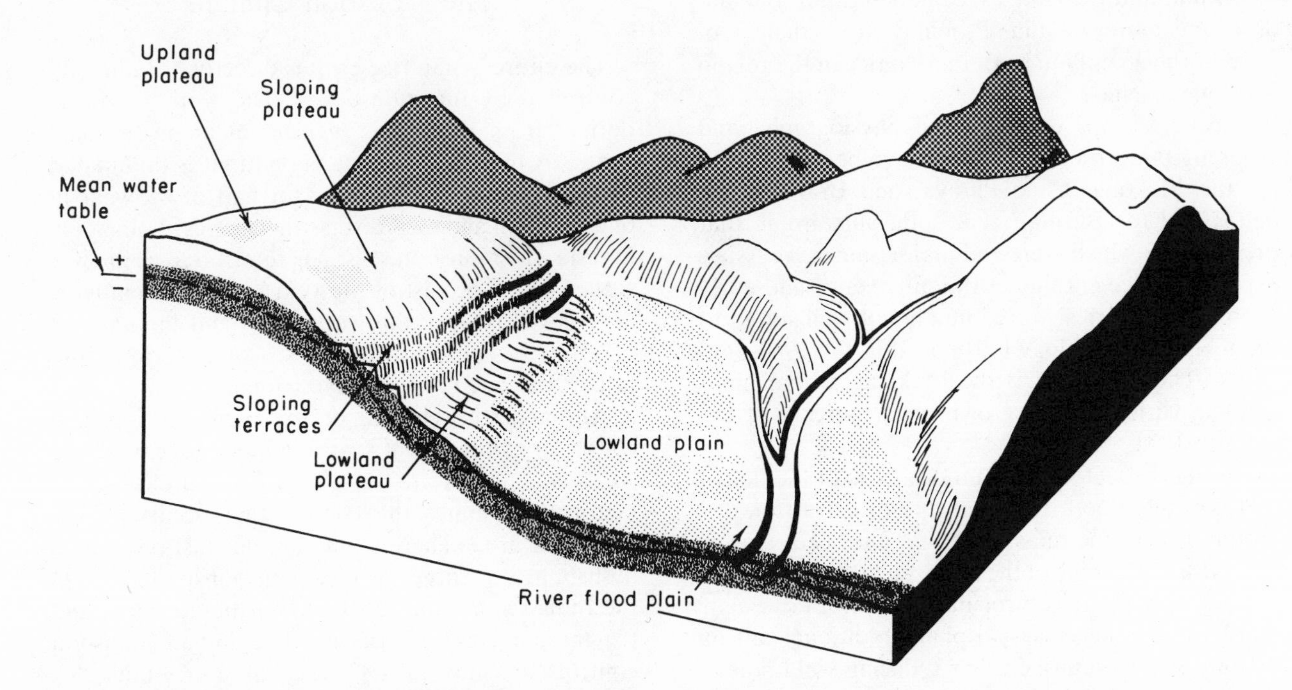

Figure 3.1. A concise view of the physiologic and hydrologic characteristics of rice cultural systems (Source: Adapted from J. C. O'Toole and T. T. Chang, "Drought and Rice Improvement in Perspective," p. 5)

Farmers are highly conscious of these differences. Many areas have local systems for classification of paddy position that depict, in microcosm, the system described in figure 3.1. For example, in the terraced areas of the Chotanagpur hilly region of eastern India (covering parts of Bihar, Orissa, and West Bengal), the rainfed terraces are divided into four categories: upland (coarse-textured, light, badly eroded soils), medium upland, medium lowland, and lowland (with intermittent flooding and drainage).[1] Soils are more fertile and yields higher in the medium lowland and lowlands, and a small fraction of the fields are planted to high-yielding varieties in the most fertile areas.

The single largest plateau area extensively planted to rice is the Khorat Plateau in northeastern Thailand. More than 3.5 million hectares are planted to rice—over 40 percent of Thailand's total rice area—but only 100,000 hectares are irrigated. The Khorat Plateau soils are infertile, and average rice yield is less than 1.5 mt/ha (metric tons per hectare), and very variable, compared with about 2 mt/ha for the rest of Thailand.

The River Deltas

The river deltas include most of the area classified as "deepwater" and "floating" rice in table 3.1. Four major river deltas in South and Southeast Asia are planted largely to rainfed rice—the Mekong in Vietnam, the Chao Phraya in Thailand, the Irrawaddy in Burma, and the Ganges–Brahmaputra in Bangladesh and eastern India. Probably more than one-third of the 35 million ha of rainfed rice in the region is in these deltas.

Excellent maps depicting the physiography and cropping pattern of delta areas have been prepared for the Mekong, Chao Phraya, and Brahmaputra deltas and for Burma.[2] It is somewhat ironic that these areas, which were the major source of Asia's exportable rice surplus in the nineteenth and early twentieth centuries, were initially not well suited to the modern technology introduced in the 1960s and 1970s. The exception is the Mekong, where high-yielding varieties were grown on 27 percent of the delta in 1973.

During the wet season in those parts of the deltas that typically flood, rice seeds are broadcast on dry ground before the rains. Although the seeds germinate with the coming of the rains, plant establishment is poor and weed control inadequate. Because of flooding, it is necessary to plant traditional tall or floating rice varieties, further reducing yield potential (see table 2.1).

Most of the rice grown in the rainfed delta areas is photoperiod sensitive, that is, flowering and harvest date are determined by day length, but there are more than 2 million ha of photoperiod-insensitive rice such as the aus rices in Bangladesh and eastern India. In some of the deeply flooded areas of the Ganges–Brahmaputra, Irrawaddy, and Mekong deltas, a single crop of photoperiod-insensitive rice is planted as the flood waters recede.

There are major soil problems in the delta areas. About 700,000 hectares in the South and Southeast Asian deltas are covered with acid sulfate soils. Some of these soils can be used for rice production if tolerant varieties are developed. Salinity is also a problem on more than 4 million ha in the Ganges–Brahmaputra and Irrawaddy deltas.[3]

Cropping patterns in the deltas are closely related to soils. Heavy clay soils severely limit the potential for growing a nonrice second crop following rice. A second crop is more common in the Ganges–Brahmaputra Delta where soils are sandier and the water table is shallower. About one-third to one-half of the area is double cropped. A detailed description of the Chao Phraya River basin illustrates the relationship between physiography, water conditions, soil fertility, farm size, and paddy production.[4] Generally speaking, farm size is larger where broadcasting is used, but yields are much higher where transplanting is used.

The Monsoon Climate

The entire Asian rice crop is directly or indirectly controlled by the monsoons.[5] The word monsoon derives from the arabic *mansin*, meaning seasonal wind.[6] A monsoon climate is one that is dominated by seasonal winds that blow for half of the year in one direction and then reverse themselves. Commonly in Asia, the monsoon season is referred to as the wet season since the rest of the year is usually relatively dry. The arrival of the wet season and the amount of water brought during this season determine the fate of the rice crop, even in irrigated areas.

Although the climate of the region is typically characterized by a wet season and a dry season, the weather patterns are varied, and a complex set of factors determines the climate that occurs in any particular area. There is, in fact, not just one Asian monsoon, but three distinct monsoon patterns: (1) the Indian monsoon, (2) the Malayan monsoon, and (3) the Japanese monsoon.[7] The Indian monsoon controls air movements over Pakistan, India, Sri Lanka, and the west coast of Burma; the Malayan monsoon is operative over Indonesia, the Philippines, most of mainland Southeast Asia, and most of China; the Japanese monsoon affects Korea, part of northern China, most of northeast China, and Japan. In general, during the winter months, the air currents move southward and eastward, while in the summer the prevailing air drift is from the south—the southwest in the case of the Indian monsoons, the southeast in the case of the Japanese monsoon.

The winter months bring cold dry air from continental Asia, and the summer months bring moist air from the ocean. However, the pattern is not entirely uniform. Calcutta, Rangoon, Bangkok, Saigon, Guangdong, and Manila have heavy rains in the summer months. Toward the equator (Singapore), the rainfall pattern is less distinct, and it is reversed south of the equator. Colombo experiences two peaks of precipitation, one coming in the late fall. In much of the temperate zone, the summer is less pronounced.

Typically in the tropics, there is a period of four months when average rainfall exceeds 200 mm (8 inches). However, parts of Thailand and Burma (for example, Mandalay) fall in a rain shadow, and the monsoon in southeast India (Madras) is exceptionally short.

Weather Threats to Crops

Droughts, floods, and typhoons can cause severe damage to the rice crop. Drought is by far the most serious threat. Regions subject to drought are also those with relatively high variability in precipitation—south India, eastern Indonesia, and other areas that fall in a rain shadow (figure 3.2). More adequate and reliable irrigation must be provided in these areas to ensure stable production, even for the main rice crop. Widespread and damaging droughts occurred in 1965, 1966, and 1972. In India, one of the most severely affected areas, rice production fell by 20 percent from 1964 to 1965.

Floods occur principally in the major river deltas where one-third of Asia's rice crop is produced. Although the geographical area of the deltas is smaller than the area subject to drought, rice is almost the only crop grown.

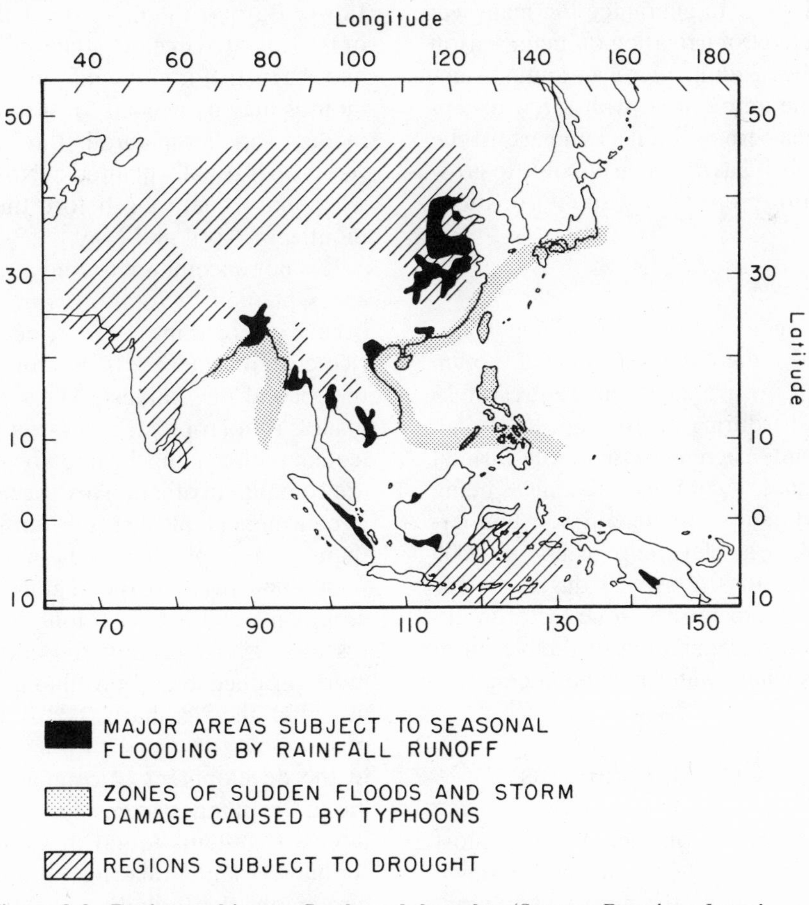

Figure 3.2. Regions subject to floods and droughts (Source: Based on Joseph Spencer and William Thomas, *Asia East by South*, p. 183)

Typhoons, known in the West as hurricanes and in India as cyclones, occur in two zones: the Western Pacific, extending fom the Philippines northward to Japan, and the Indian Ocean. Typhoons occur most frequently in the late summer and fall months. High winds damage the rice crop by causing lodging (bending over of the stalks). Sterility can occur in the flowering stage, rice shatters near harvest, and general flooding destroys infrastructure and causes delays in land preparation, planting, and harvesting. Although damage is more local than in the case of droughts, crop losses can be severe.

In summary, throughout most of monsoon Asia, rainfall is plentiful but highly seasonal, coming primarily in the summer months. In the "home" area of rice extending from eastern India throughout continental Southeast Asia and South China, rainfall generally is adequate for the main rice crop. Most of the rice in this area is still grown under rainfed conditions. Outside of this area, uncertainty is much greater with respect to both the onset of the monsoons and the quantity of rain in any season. The irrigation facilities that have been developed up to the 1970s in most of South and Southeast Asia can provide only supplemental water to guarantee the main wet season crop. The effect of irrigation on main season crop yield diminishes with increases in annual rainfall. Throughout the entire area, with a few exceptions such as southeastern Sri Lanka and parts of the southern Philippines, it has not been possible to grow a second or dry-season crop of rice without irrigation.

Other Threats to Crops

As one moves to higher elevations and higher latitudes in Asia, temperature rather than rainfall is often the critical factor. Low average temperatures of 15 to 18°C (60 to 65°F) during a growing season can damage the rice plant. There is considerable varietal variability in tolerance to cold, with japonicas being more resistant than indicas. In general, low temperatures are a problem at elevations of about 1,000 meters (m) and latitudes above 20 degrees. This vulnerability to low temperature is one factor inhibiting the more rapid dissemination of double cropping in some areas where water may be adequate.

Rice-Based Cropping Patterns

Cropping systems based on rice are the most common form of agriculture in Asia, and a single crop of rice per year may be the most widely practiced land use pattern in Asia.[8] In the tropics, where water for irrigation is available in adequate amounts, two crops of rice are grown. In the temperate areas, cold weather limits rice production to one crop per year, but in Korea and Japan some rice paddies are planted to barley in the winter months. In the warmer areas, other crops are planted after or before rice on some fraction of the land.

Figure 3.3 shows the diversity in timing and duration of rice crops in a number of locations throughout Asia.[9] The precise dates of planting and harvest vary from year to year, depending on weather and other environmental and cultural practices. Varietal development can shorten or lengthen the crop duration and may even permit the cultivation of an additional crop under appropriate circumstances.

Except for the high-latitude countries, the availability of adequate water is the main factor determining when rice is planted. Because of the pronounced monsoon and dry season, even the two-crop locations (figure 3.3) usually produce a second crop of rice only where irrigation is available. In most places, the cultivation season begins in May or June with the onset of the main monsoon showers. The first crop is broadcast in parts of Central Thailand, Lower Burma, Vietnam, Bangladesh, and Sri Lanka or transplanted in most other areas during June, July, and August. If it is a traditional variety, four or five months may be needed for it to mature, while some modern varieties mature in three months. The second rice crop is usually planted in November, December, or January, maturing before the hottest and driest months of April and May.

It is not uncommon to find a wide range of upland crops planted in rotation with rice. This normally occurs where rainfall is not adequate for a second rice crop or where a third crop can be grown after the second rice harvest. Maize is common in some places; root crops such as sweet potato, and various legumes like mung beans, cowpeas, and soybeans are common in others. However, in general, the area rice farmers plant to crops after rice is much smaller than the area planted to the main rice crop.

In some parts of the region, complex, intensive cropping systems have a long history. A substantial research effort was initiated in the 1970s to develop more productive and profitable cropping systems to use after the rice harvest.[10] A major thrust of the program was to work with and learn from farmers in the development of complex and intensive rice-based cropping systems for widespread use. Four intensive systems found in China, the Philippines, Indonesia, and India are broadly described in the paragraphs that follow.

We start with China because cropping systems in

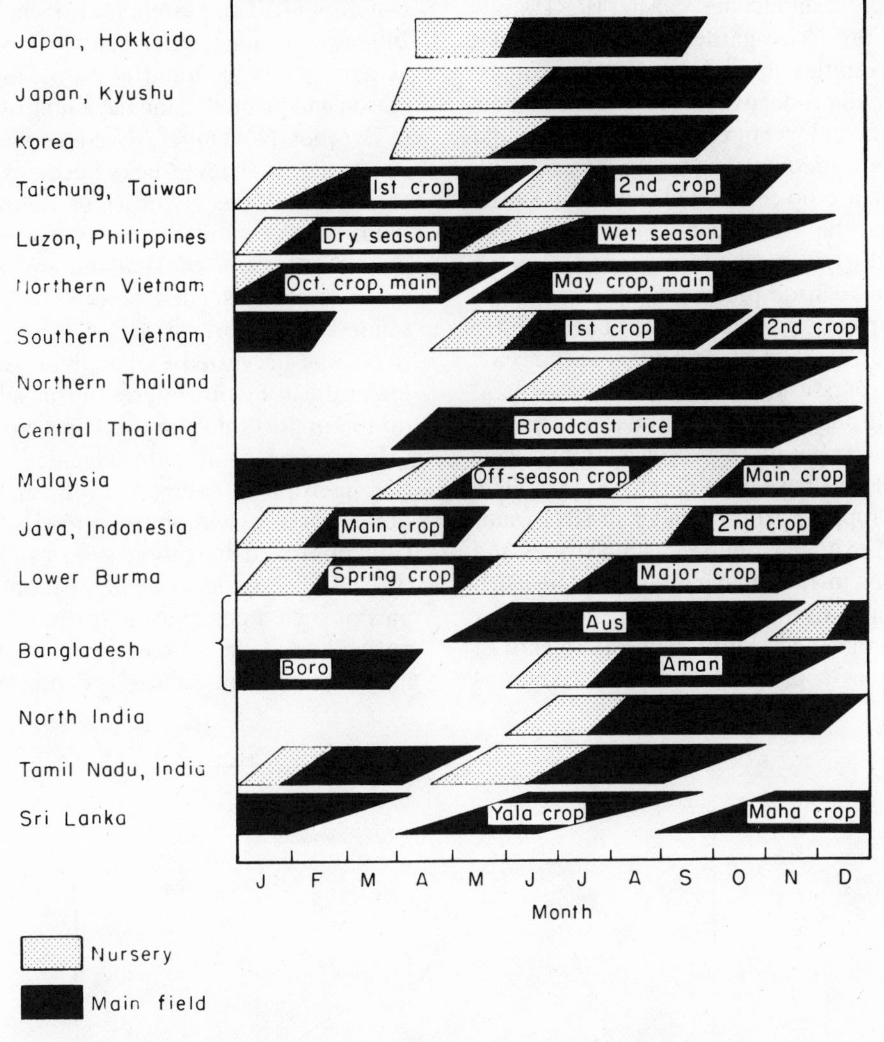

Figure 3.3. Seasonal duration of rice crops in Asia (Source: Adapted from A. Tanaka, "Comparisons of Rice Growth in Different Environments," pp. 438–440)

Taiwan and South China are the most intense found in Asia. The cropping index (number of crops harvested per year per unit area) reached a record high of 189 in Taiwan in 1964. The cropping index in South China today is 187, close to the peak achieved in Taiwan.

Cropping systems of rice-rice-sweet potatoes, rice-rice-maize, rice-rice-vegetables, rice-vegetables-rice-vegetables, and other highly intensive systems are used in Taiwan. A major factor permitting this level of intensity is the high degree to which the Chinese control water.

In Guangdong Province, South China, attempts have been made to grow as many as three crops of rice and one winter crop in a single year using relay planting. Under ideal conditions, close to 24 mt of grain, including 18 tons of paddy rice, can be produced from four crops in a single year. However, in terms of the added resources required, the addition of a third rice crop has not proven economical. Although this system was initially encouraged by the government, it has not been widely adopted.[11]

Another example of an extremely intensive rice production system can be found in the Philippines. In 1975, scientists at the International Rice Research Institute (IRRI) discovered a farmer in Central Luzon who was planting and harvesting rice continuously on a daily basis, one small plot at a time. In 1976, IRRI began experimenting with a similar system. A 1-hectare field was divided into forty plots of 250 m^2. A plot was transplanted every other day, and one was harvested on alternate days, six days a week. During 1977, about 23 tons of rice were harvested and three men were fully employed, with a relatively constant flow of cash income.[12] The IRRI study concluded that this "rice garden" system provides dramatic opportunities to increase labor earnings, family income, and productivity of small rice farmers if they have the high level of control over water that is required. With adequate management and with levels of inputs per crop comparable with those used by many farmers, this system can double output and productivity and triple or quadruple labor input. A major constraint to widespread adoption of such a system is that typical Asian rice farmers have neither access to nor control over water throughout the year.

As in China, pressures for intensive use of land are very high throughout most of Java, Indonesia, but the quality of irrigation varies considerably. Rice cropping technologies have been developed that allow more intensive cropping patterns under variable water conditions. For example, figure 3.4 illustrates the cropping patterns in Indramayu, East Java in 1973. Farmers with only five months of irrigation water plant seeds in a dry seed bed prior to the release of irrigation water in a production system known as *gogorantjah*. This system is also widely used in many Indonesian rainfed areas and was practiced at least as early as 1920. Land is prepared during the dry season and planted at the beginning of the rainy season in October–November. Rice grown as a dryland crop during the first few weeks becomes a wetland crop as soon as there is sufficient rainfall to flood the field.[13] This is similar to the practices followed in the river flood plains of Thailand and Vietnam, where rice is directly seeded before the floods. In many rainfed and poorly irrigated areas, this allows the rice to be harvested earlier than would be possible under the usual procedures of transplanting, opening up the opportunity for a second crop.

In tropical Asia, major changes have been occurring in cropping patterns as a result of the introduction of varieties with shorter growth duration coupled with improved irrigation systems. Punjab State in North India provides one of the more dramatic examples of such a change. In North India, rice has long been planted on a limited area of land during the monsoon season. Maize and millets were planted

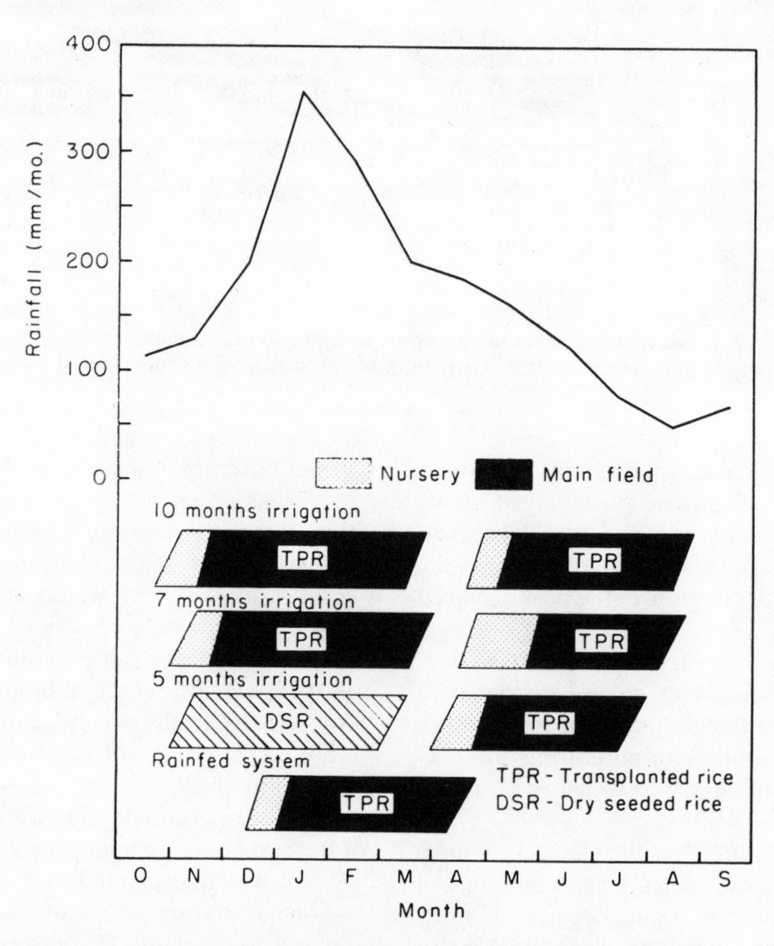

Figure 3.4. Calendar of four cropping patterns with different water availability, Indramayau, West Java, Indonesia

during the same season on lighter soils. Wheat was widely planted on most of the land beginning in November, but long-duration traditional varieties limited the area that could be planted to wheat following rice. Instead, a low-yielding, low-input crop of legumes was grown in many areas.

Short-duration wheat and rice varieties have increased the amount of land that can grow two grain crops per year, and, as a result, cropping systems are changing rapidly. Rice has traditionally not been an important food, but in the 1960s and 1970s strong demand from the other Indian states encouraged Punjab farmers to increase rice production. The availability of short-duration wheat and rice varieties permitted increased double cropping in a two-season farming system in which the monsoon *kharif* crops of rice, maize, oilseeds, and fodders are grown from July through November and the winter *rabi* crops of wheat, gram, and cotton are grown during the second season. Most farmers also have some sugarcane that occupies the land throughout the year.

This well-documented experience in North India has been repeated in many other parts of Asia. In the North China Plain, for example, higher-yielding varieties of rice, wheat, and maize have replaced other grains, soybeans, and oilseeds in many rotations. The growth duration of tropical rice varieties was shortened from 150 or more days to 125 days in the 1960s and subsequently to 100 days in the 1970s. The impact of reduced growth duration on crop production has probably been as significant as the impact of higher yields. However, because of the way in which crop statistics are collected, it is much more difficult to verify the impact of crop intensification compared with yield, and hence its significance is often overlooked.

Farming Systems in Asia

The traditional farming systems of Asia include both wetland (lowland) and dryland (upland) systems. The cultural practices and techniques developed in these systems have been practiced for centuries and remain dominant in many parts of Asia today. During the colonial period, the more productive systems spread throughout the lowlands of Asia, and the output from dryland culture gradually became a minor portion of total production.

The intensification of production through control of water dates back many centuries. However, the development of high-yielding, fertilizer-responsive varieties, the use of chemical fertilizers and other chemical inputs, and the introduction of farm power are twentieth century phenomena. Since water control is the *sina qua non* in the process of modernization, modern farming systems have begun to emerge primarily in the irrigated lowland areas. The most modern techniques can only be used where careful land shaping has made it possible to control the application and drainage of water to a high degree.

An important distinction in the process of modernization is the degree to which farming systems rely on labor-intensive or capital-intensive techniques. High labor-use practices are being followed in the more labor-abundant areas of Asia, where opportunities for nonfarm employment are extremely limited. By contrast, other parts of Asia are following a more capital-intensive path.

Table 3.2 outlines the characteristics of six generalized Asian rice production systems that are described in this section. They are differentiated by methods of production rather than size. None is typically large, as practically all of Asia's rice production is carried out on small farms. In fact, the farms using the most modern systems, such as those in Japan and Korea, are typically smaller than the farms using traditional systems, such as those in Burma and Thailand.

In addition to production techniques, the level of development of rice farming systems can be characterized by the intensity of land use throughout the year. In the simplest of traditional systems, only one crop of rice is harvested each year. The most highly developed rice farming systems in Taiwan may grow two wetland rice crops and one dryland crop on the same field during one year. Three crops of rice are grown in some areas of Southeast Asia where irrigation is well developed.

Cropping systems may be ranked in order of intensity, beginning with rice/fallow systems, followed by rice/dryland crop systems, rice/rice/systems, rice/rice/dryland systems, and finally rice/rice/rice and rice/upland/rice/upland systems. The more intensive systems are limited to very small areas, even in highly developed rice-producing countries, so that no country or major region averages over 1.9 crops harvested per field per year (the maximum reached in Taiwan and South China). Increasing land-use intensity is facilitated by the modern rice varieties, which are rather insensitive to day length and mature in a shorter time than traditional varieties. Many Asian rice farmers are using transitional technology and, at the same time, are beginning to grow two or more crops per year. A basic decision not only for Asian farmers but also for agricultural policy makers is *when* to shift from labor- to capital-intensive technology. Farmers and policy makers alike too frequently associate capital intensity with modernization.

Table 3.2. General Types of Rice Production Systems Used in Asia in the 1980s

Characteristic	Dryland rice	Wetland rice				
		Traditional	Transitional labor-using	Transitional capital-using	Modern labor-using	Modern capital-using
Rice area/farm (ha)	1–3	1.5	0.1–3	1–10	0.1–2	0.1–2
Crop duration (days)	150–250	150–250	110–150	110–150	80–120	80–120
Degree of water control	None	Little	Moderate	Moderate	High	High
Power for land preparation[a]	F/H	H/A	H/A	P/T	A	P/T
Method of stand establishment[b]	B	B/T	B/T	B/T	T	T
Nutrient supply[c]	F/O	O	O/C	O/C	O/C	C
Plant protection chemicals	None	None	None	Modest	Modest	High
Reaping[d]	F/H	F/H	H	H	H	H
Threshing[e]	H/A	H/A	H	T	H/T	T
Predominant disposal	Subsistence	Subsistence	Market	Market	Market	Market
Production per ha (metric tons)	0.8	1.5	2.5+	2.5+	4.0+	4.0+
Areas where important	Border hills of S and SE Asia	Lower Burma, NE Thailand, Bangladesh, Eastern India	Sri Lanka, Java, China	Central Luzon, W. Malaysia, C. Thailand, North India	China high & stable yield areas, Japan, and Taiwan prior to WWII	Taiwan, South Korea, Japan

[a]F = burning of forest cover; H = human labor; A = animal power; P = power tiller; T = 4-wheel tractor.
[b]B = broadcast; T = transplanted.
[c]F = burning of forest cover; O = organic manures; C = chemical fertilizer.
[d]F = cutting with finger knife; H = hand reaping with sickle.
[e]H = human; A = animal; T = power thresher.

Before describing the wetland systems, we first discuss dryland rice techniques. As noted in chapter 2, it is not known whether dryland or wetland culture was developed first. But today, dryland rice production systems are, in a sense, atypical, occupying less than 10 percent of Asia's rice land.

Dryland Rice Systems

The major characteristic of dryland rice is that the farmer has no expectation that standing water will remain in the field during the production cycle. Two main subtypes are continuous cropping systems and the swidden, or slash-and-burn, still in use in the border hills of mainland South and Southeast Asia and some mountainous areas of the Philippines and Indonesia.

In continuous dryland systems, the fields are not bunded and puddled as in wetland culture, but are prepared as for any other upland crop. With the exception of highly localized areas, most dryland rice systems produce one rice harvest per year. The varieties grown are traditional types that are better adapted to withstanding drought than modern wetland varieties. Little fertilizer is used, and yields are usually less than 1 metric ton/ha. The land is usually prepared with draft animals after the first rains of the season soften the soil. Several plowings and harrowings uproot the weeds and create a seedbed. Seeds are planted by digging a hole with a dibble stick rather than broadcasting seed or transplanting started seedlings. Weeds are a major problem, and farmers have devised various control methods. One method is to run a harrow through the crop 30–45 days after planting. This uproots some of the rice along with some of the weeds, but simplifies the task of hand weeding that follows. Rice varieties with tough, deep roots are naturally preferred by farmers who use this technique. Some dryland rice producers use large amounts of labor to control the weeds simply by hand pulling. Harvesting and threshing are carried out by hand.

Classical swidden agriculture has been the subject of numerous anthropological studies.[14] It is often practiced where there is a relative abundance of land with poor soil or steep slopes. Important characteristics include the shift to a new plot of land every year or so, the use of fire to clear the land and partially prepare the soil, and reliance on ashes as a fertilizer source. Careful timing is needed to cut and pile the forest cover so it will be dry and ready to burn at the end of the dry season. Following burning, seeds are planted after the first rains of the season. Crops of varying duration are planted in the swidden to provide food over an extended period. Rice may cover a major portion of the area, or may be a minor component, with maize or sweet potatoes dominating. After one, two, or three years of cropping, the plot is abandoned for five to twenty years while natural cover regrows.

If population pressure or lack of alternatives shorten the period of regrowth or extend the period of cropping beyond that for which a particular soil and topography are best suited, the system deteriorates. The forest fails to regenerate, and the land may be taken over by noxious vegetation not adaptable to the swidden system. In Southeast Asia, a tough, vigorous grass, *imperata*, covers much land that has been lost to the swidden system. Where the soil and topography are adapted and commercial demand for tree fruits is strong, gradual intensification of swidden agriculture may take place, with coconut, coffee, rubber, cacao, or fruit trees replacing natural forest; annual production of dryland rice or other field crops can be carried out beneath the canopy. Many variants of the tree crop-rice system exist, from land that aerially appears to be forest, but which also grows rice beneath the canopy, to fields that are solidly rice with tree crops planted along the borders.

Traditional Wetland Systems

Farmers who use traditional production techniques with traditional rice varieties are almost always limited to growing one crop of rice per year. This is because traditional varieties cover the entire annual period when rainfall is adequate for unirrigated crop production.

Traditional production techniques involve few inputs other than those available on the farm or created by the farmers' labor: seed saved from last year's harvest and hired labor to repair dikes and raise, pull, and transplant seedlings. Human and animal labor will also be involved in harvesting and threshing operations.

Rice is normally transplanted by hand, but in some places seasonal labor shortages encourage broadcasting. Nearly all rice is harvested by hand with a sickle, although a finger knife (ani-ani) is used to harvest one panicle at a time in some places. In many places, rice-growing techniques have not appreciably changed in the past 100 years. Typically, farmers spend little time on cultural practices such as weeding or pest control. The tall stature of traditional varieties shades out competing weeds, and most are at least somewhat resistant to insect pests and diseases. In some areas, considerable effort may be invested in water control. Yields of 1.3 to 1.5 mt/ha are common. Most traditional farms produce little income over and above subsistence levels. Table 3.3 presents some data taken from two studies of traditional systems in Burma and Northeast Thailand conducted in 1932 and 1969, respectively. Although the Burmese farms are considerably larger than the Thai farms, inputs and returns are very similar, illustrating how little

Table 3.3. Labor Use, Costs, and Returns with Traditional Rice Production Techniques in Burma, 1930s, and Northeast Thailand, 1960s

	Burma		Thailand	
Labor use (days of labor/ha[a])	Human	Animal	Human	Animal
Land preparation	9.9	20.0	12.8	12.4
Raise, pull, transplant seedlings	30.8	0.0	28.9	1.7
Fertilize, irrigate, weed, etc.	0.0	0.0	0.4	0.0
Harvest, thresh, winnow, haul	6.4	4.0	14.4	0.0
Total	47.1	24.0	56.5	14.1
Paid-out costs (kg of paddy/ha)				
Hired labor	598		0	
Purchased fertilizer	0		83	
Hired animal labor	136		0	
Value of seed	52		28	
Land rent	515		0	
Returns				
Farm area (ha)	10.1		2.9	
Paddy yield (kg/ha)	1,548		1,321	
Paid-out costs (paddy equivalent, kg/ha)	1,301		111	
Net income (paddy equivalent, kg/ha)	247		1,210	
Farm income (paddy equivalent, kg/farm)	2,497		3,509	

Sources: Burma: Agriculture Department of Burma, *The Rice Crop in Burma* (1932). Thailand: Land Policy Division, Land Development Department, *Cost-Return Information for Selected Crops by Soil-Series in Ubonrajthani for 1969* (1971).

[a]Thai data given in hours per rai, converted at 6 hr/day, 6.25 rai/ha. Burmese data given in months per farm, converted at 20 days/month or in Rs/acre, converted at Rs 0.4/day (wage implied in the data).

the traditional systems have changed in the past fifty years.

Both sets of data represent traditional rice farming systems even though they are somewhat different. The Burmese farms were substantially larger than the Thai farms and hired labor to assist in cultivation, while the Thai farms hired no labor. A small amount of fertilizer was used on the Thai farms, none on the Burmese. The Burmese farmers paid land rent while the Thais owned their land. Mainly because of the land rent of 515 kg/ha of paddy, Burmese family income per farm was 2,497 kg, about 30 percent below the Thai farm family income. The Thai farmers, because they supplied all their own labor and animal power, retained about 90 percent of the gross value of production as a return to land and their family-owned factors, while the Burmese farmers, using hired labor and renting their land, retained only about 16 percent.

Transitional Wetland Systems

The two most significant changes made by farmers who are moving from traditional to modern techniques are a switch to new varieties and the use of chemical fertilizer. These innovations were widespread throughout Asia by 1980, although they were certainly not universal. It is not uncommon to find farmers in South Asia using new varieties and high levels of fertilizer in the dry season and traditional varieties, with low yields, in the wet season.

Changing from the traditional photoperiod-sensitive varieties to nonsensitive varieties that ripen in a fixed number of days after planting may require substantial labor and capital adjustments on the part of farmers. In some locations, the rainy season may last three months, while in other areas it may last six to nine months. The switch to nonsensitive varieties may be difficult for farmers whose fields remain covered with water when the nonsensitive variety is ripe. Clearly, improved control over water is critical to the successful adoption of new varieties.

Where water control is not a problem, or where irrigation facilities make a dry-season crop possible, many farmers have changed from traditional to modern varieties. Because the modern varieties mature quickly, there is a chance for a second crop during the dry season, and if this second crop is rice, it is likely to be a nonphotoperiod-sensitive type. The opportunity for a second crop provides an impetus to improve irrigation and also encourages farmers to prepare fields as quickly as possible to take advantage of as much of the natural rainfall as possible. For these reasons, when modern varieties are introduced, farmers tend to use more inputs for irrigation and farm power. Two general transitional production techniques can be identified—one that is relatively labor intensive and one that is relatively capital intensive.

Rice crop budgets for two transitional farming systems are shown in table 3.4. The first budget is based on a survey made in Kurunegala District in Sri Lanka for the wet season of 1972/73 and the second on surveys in Central Luzon and Laguna Province in the Philippines for the 1974/75 wet season. Despite fairly similar yields, different inputs were used. Most striking is the level of labor input per hectare, which was twice as high on the Sri Lankan farms (162 vs. 82 days per hectare), with most of the extra time used for land preparation and in post-harvest activities. Fertilizer and chemical inputs were somewhat higher on the Philippine farms, but machinery rental for land preparation was ten times higher in Kurunegala. Thus, while the farms surveyed in Sri Lanka are labor intensive relative to the Philippine farms, they also use more capital. Yield per hectare is similar and significantly higher than for traditional farmers (table 3.3). The Sri Lankan farmers have substantially higher incomes because they have larger farms, even though their per hectare costs are higher than the Philippine farmers. In contrast to the traditional farms, most farmers who are in the transitional stage have adequate water supplies to grow two crops of rice on at least part of their land.

Insecticides are often part of the "package of inputs" that government programs and private industry persuade farmers to adopt. It is already evident that excessive use of chemicals in Japan and Taiwan has led to serious environmental degradation. Given their inherent complexity and the low level of understanding of this technology by farmers in much of Asia, in most cases it seems doubtful that the benefits from the use of insecticides will outweigh the hazards (including injury to humans, livestock, and the destruction of the ecological balance among insects). Resistant varieties have been more effective than chemicals in controlling insects, and further research on the appropriate combination of control measures in integrated pest management may lead to lower cost and lower risk methods of pest control.

Labor-Using Techniques

Some transitional systems are more labor intensive than those described in table 3.4. For example, in some areas of Java and the wet zone of Sri Lanka, human labor supplies much of the power used for land preparation as well as for carrying out most

Table 3.4. Labor Use, Costs, and Returns for Transitional Farms in Central Luzon/Laguna, Philippines, and Kurunegala District, Sri Lanka

	Kurunegala District 1972–73 wet season	Central Luzon/Laguna 1974–75 wet season
Labor use (days of labor/ha)		
Land preparation	27.7	10.7
Raise, pull, transplant seedlings	23.5	20.5
Fertilize, irrigate, weed, etc.	80.6	31.9
Harvest and post-harvest	70.6	18.5
Total	162.1	81.6
Paid-out costs (kg of paddy/ha)		
Hired labor	714	286
Purchased fertilizer	159	239
Insecticide	50	48
Herbicide	50	16
Machinery	621	61
Land rent	336	127
Returns		
Area in rice (ha)	4.8	2.5
Paddy yield (kg/ha)	2,917	2,449
Paid-out costs (paddy equivalent, kg/ha)	1,931	777
Net income (paddy equivalent, kg/ha/crop)	986	1,672
Farm income (paddy equivalent, kg/farm/crop)	4,732	4,180

Sources: Sri Lanka: K. Izumi and A. S. Rantaunga (1974). Philippines: R.W. Herdt (1978) pp. 63–80; R. Barker and V. G. Cordova (1978) pp. 113–136.

other tasks. After the monsoon begins, or when water is available in the irrigation system, the fields are flooded, and the land is allowed to soak for several weeks until it is saturated. Some plowing may be done with animals or two-wheeled tractors, but the most common power source is human labor. Further soaking precedes the secondary tillage operation, which is sometimes carried out by harrowing with animals, and sometimes by hoeing.

In most labor-intensive systems, considerable effort is expended in controlling water. Part of the effort is made prior to stand establishment, when fields are carefully leveled and bunds repaired. Later in the season, time is spent controlling the flow of water onto the fields and the drainage of water from the fields. In some places, farmers own pumps that allow them to supplement gravity-fed water supplies.

The high labor intensity observed in some locations has not been caused by sudden changes introduced in the past few years, but is a continuation of long trends. In the 1970s on Java, innovative farmers used 200 to 250 days of labor to produce 1 hectare of rice.[15] This is more than three times the level reported for Thailand and Burma and does not include harvest labor.

Harvesting operations are carried out by hand in most of Asia using a sickle to cut several stalks at one time. Harvesting practices on Java have been rapidly changing from a traditional highly labor-intensive panicle harvesting (that is, harvesting one panicle at a time with an ani-ani) to the use of a sickle. Whether this change has occurred as a result of the introduction of modern varieties is hotly debated.[16] In the second half of the twentieth century, panicle harvesting in Asia is the exception, rather than the rule. Even in otherwise highly labor-intensive areas, including the wet zone of Sri Lanka and Bangladesh, sickle-harvesting procedures are used.

Some of the same areas that use panicle harvesting also use highly labor intensive hand-pounding instead of machines for removing the outer husk from the paddy. In most of Asia this is achieved by using small village milling machines, but the traditional hand-pounding method is still used in some parts of Bangladesh and Java. Like changes in harvesting methods, the reasons for the decline of hand-pounding are an issue of some controversy.[17] Nevertheless, significant changes in the pattern of rural employment have resulted. Because hand-pounding was traditionally a female job, changes have also occurred in income distribution.

There are millions of farmers in Asia who are modernizing their rice production practices by using new varieties, fertilizer, and irrigation. Many are using additional labor inputs for weed control, stand establishment, and water regulation but, at the same time, using less labor for other operations. The net effect has been to increase labor use per hectare harvested but to reduce labor input per kilogram of rice produced (chapter 9).

Capital-Using Techniques

In some areas, additional capital may be substituted for increased labor inputs. The capital is primarily used in three production operations: tractors for land preparation, machines for threshing, and pumps for irrigation.

A number of alternatives to using animal power, which requires 15 to 25 days per ha for plowing followed by numerous harrowings over a period of several months, have become established in parts of Asia. Four-wheeled tractors can initially plow or rotavate 1 hectare in a matter of hours. Most farmers then complete the task by harrowing with animal power. In some areas of the Philippines and Thailand, two-wheeled tractors of 7–10 hp have become popular for land preparation. The entire land preparation operation can be completed in about 30 hours per ha. Two-wheeled tractors are preferred where irrigation allows two rice crops per year because the heavier tractors tend to become stuck in the mud. Aside from preparing land, tractors are used to pull trailers to transport supplies and products. They are used for little else, however, because other rice production operations are not adapted to mechanization.

The productivity of fertilizer-responsive varieties is greatly enhanced by controlling the depth and timing of water applications. Many governments and individual rice farmers are making substantial capital investments in irrigation. Individuals may install low-lift pumps capable of irrigating 2 to 5 ha where the water table is high throughout the year. In Thailand, some farmers have large-diameter low-lift pumps suitable for pumping from surface canals or for pumping water out of fields to reduce flooding. Many deep tubewells with the capacity to irrigate 50 to 100 ha have been installed in rice-producing regions of Bangladesh and India, and canal irrigation is being expanded in nearly all countries. Thus, irrigation systems are being improved in both capital- and labor-using systems.

Threshing machines are a third type of capital investment being introduced into some traditional systems in Thailand, the Philippines, and parts of India. Most of these are simple, stationary machines used to thresh grain that has been previously cut by hand.

Modern Systems

In Western systems of farm production, modern agriculture is typically capital intensive because of the high cost of labor relative to capital. In Asia, on the other hand, many countries have a surplus of labor, and rice farming systems that achieve high levels of land productivity with widely different levels of capital and labor inputs may all be regarded as modern. In this context, the fundamental requirements for modern rice farming systems are a high level of water control and the use of modern varieties and inputs that permit high production per hectare per year.

The recent history of rice production in Taiwan illustrates the shift from modern labor-using to capital-using technology. Labor use in rice production in Taiwan reached a peak of about 120 man-days per crop after World War II, and rice yields approached 4 mt per ha in the early 1960s (exceeded only by Japan and South Korea in Asia). Sample surveys in Central Taiwan (table 3.5) show that between 1961 and 1972, labor input per hectare fell by more than 40 percent while costs, principally for inputs such as insecticides and herbicides, increased 50 percent.[18] Despite much higher yields in 1972, income from the rice crop was only 23 percent higher than in 1961.

An even more vivid example of the contrast between labor- and capital-intensive systems can be seen by comparing collectives in China with Japan, Taiwan, and South Korea. Vast supplies of surplus labor have encouraged China to emphasize labor-intensive technology over the capital-intensive technology favored in the rest of East Asia. China's collective system facilitated the mobilization of labor for such activities as composting, irrigation development, and manual pest control. Although yields were lower than its neighbors, averaging 4 mt per ha in 1980, yield levels were still well above those in South and Southeast Asia. Even today a major portion of China's plant nutrients are supplied by organic fertilizers. One 1965 Chinese report indicated that the cost of compost containing 100 kg of nitrogen was 35 to 45 labor days and 20 to 25 animal labor days.[19]

Today, Japan has the most capital-intensive rice production system in Asia. Typical Japanese rice farmers in the 1970s had less than 1 ha of lowland paddy, with about 0.1 ha of upland area planted to vegetables or tree crops. The high government price support for rice (seven to eight times the world market price in 1980) has encouraged many rice farmers to continue to grow rice on a part-time basis, although many have given up less lucrative nonrice enterprises. Nearly all operations are mechanized—land preparation, transplanting, chemical application, harvesting, and threshing. Despite this, Japanese farmers still employ about 70 man-days of labor per hectare compared with 500 days in the most intensive areas of China.[20]

Table 3.5. Labor Use, Costs, and Returns with Modern Labor-Using Techniques in Taiwan, 1961, and with Modern Capital-Using Techniques in 1972

	1961 Central Taiwan (modern labor-using)		1972 Central Taiwan (modern capital-using)	
Labor use (days of labor/ha)	Human	Animal	Human	Animal
Land preparation	19.4	17.6	4.0	1.2
Raise, pull, transplant seedlings	14.7	0.1	15.4	0.1
Fertilize, irrigate, weed, spray, etc.	56.4	0.0	37.3	0.0
Harvest and post-harvest	30.0	0.0	14.9	0.5
Total	120.5	17.7	71.6	1.8
Paid-out costs (kg of paddy/ha)				
Hired labor	88		142	
Purchased fertilizer	666		685	
Insecticides	59		332	
Herbicides	00		73	
Value of seed	48		25	
Returns				
Area in rice (ha)	1.0		1.0	
Paddy yield (kg/ha)	4,071		5,229	
Paid-out costs (paddy equivalent, kg/ha/crop)	861		1,257	
Net income (paddy equivalent, kg/ha)	3,210		3,972	

Source: L. Tsai (1976).

Farm Size and Tenure in Asia

Farms in Asia are small, with half or more including less than 2 ha. Bangladesh, India, Indonesia, Korea, and Japan have an especially large number of holdings below 1 ha, while Thailand and the Philippines have relatively few extremely small holdings (table 3.6).[21] Small farm size is no absolute barrier to agricultural progress, however, as illustrated by Japan and Korea, which have the highest rice yields and fastest economic growth rates in Asia.

Small size is, in some sense, both a result of and a necessary condition for rice production, at least for rice production as traditionally practiced in Asia. Rice absorbs up to twice as much labor as other grain crops grown under similar economic and social conditions. For example, a recent study in Ferozepur, India showed that farmers used 405 hours of labor for 1 ha of wheat and 764 hours for 1 ha of rice.[22] A 1930s study in four mixed cropping areas in China showed that farmers used 180 days of labor/ha of rice, 90 days/ha of wheat, and 67 days/ha of barley.[23] Today, Japanese farmers use about 250 hours of labor to produce 1 ha of wheat, and about 800 hours of labor per hectare of rice.[24] Because farmers use more labor on rice than on other crops, this suggests that rice gives a relatively more attractive return per unit of labor used. The large amounts of labor required by rice have helped to keep the size of rice farms small. The typical farm family relying on its own labor resources would find it impractical to cultivate more than 1 or 2 ha of paddy.

Where newly opened land suitable to rice cultivation has become available for settlement, the average farmer seldom plants more than a few hectares of rice because of labor requirements.[25] As population grows and availability of hired labor increases, rice production can absorb astounding amounts of labor. This process is perhaps best exemplified on Java, where up to 300 days of labor have been used to grow 1 ha of rice. This contrasts with about 100 days/ha for all rice production activities in the Philippines.

Large farms are not unknown in Asia, however, and although only a relatively small proportion of farms are large, in some countries those few large farms control a disproportionate amount of land. As shown in table 3.6, the largest 10 percent of farms control about 40 percent of the land in Pakistan, 50 percent of the land in India, 35 percent of the rice land in the Philippines, and 35 percent of the rice land in Indonesia. In Korea and Japan, on the other hand, the largest 10 percent of farms (which are nearly all below 5 hectares) control only about 10 percent of the land.

The national distribution of farms by size does not, of course, reflect the distribution in any particular village or production area. At this level, the contrasts are even greater. In one Javanese village, for example, the size distribution of farms showed that 35

Table 3.6. Distribution of Farms and Farm Area by Size of Holdings in Asia
(percent)

	0–1 ha		1–3 ha		3–5 ha		5–10 ha		over 10 ha	
	Farms	Area	Farms	Area	Farms	Area	Farms	Area	Farms	Area
South Asia										
Bangladesh, 1974	66	24	29	53	3	13	2[a]	10[a]	—	—
India, 1971	51	9	29	22	9	16	8	22	4	31
Pakistan, 1972	—	—	44[b]	12[b]	25	18	21	27	11	43
Sri Lanka, 1960[c]	44	15	40	43	16[d]	42[d]	—	—	—	—
Southeast Asia										
Indonesia, 1973	70	29	24	38	3	13	2	10	1	10
West Malaysia, 1973	35	15	56	61	8	18	1[a]	7[a]		
Philippines, 1971[c]	15	3	54	34	22	28	7	17	3	19
Thailand, 1971[e]	13	3	36	18	31	32	16	32	4	16
East Asia										
South Korea, 1974	67	58	25	35	8[d]	7[d]	—	—	—	—
Japan, 1970	67	33	29	46	2	7	2[a]	14[a]	—	—

Source: A.C. Palacpac (1982) pp. 61–62.
[a] Includes all farms above 5 ha.
[b] Includes all farms below 3 ha.
[c] Includes only rice farms.
[d] Includes all farms above 3 ha.
[e] Size categories for Thailand are 0–0.96, 0.96–2.4, 2.4–4.8, 4.8–9.6, over 9.6.

percent of the households farmed no land, 19 percent operated tiny holdings of less than 0.2 ha, 28 percent had holdings between 0.2 and 0.5, and 11 percent had holdings between 0.5 and 1 ha. The top 7 percent had holdings in excess of 1 hectare.[26] This last group operated 54 percent of the total rice area and included those who cultivate "village land" in exchange for serving as village-level governmental functionaries, an arrangement that appears to outsiders to be highly inequitable. In an intensive rice-producing village in the Philippines, a 1974 survey showed that 43 percent of the households were landless, 9 percent had holdings smaller than 1 ha, 46 percent had holdings between 1 and 5 ha, and two had holdings in excess of 5 ha.[27]

In many areas land ownership is more highly concentrated than the distribution of operational units, with owners of large holdings renting their land to farmers in small units. Many varieties of rental arrangements exist and have various degrees of inequity and disincentive effects. The most common types are *share cropping* and *fixed rent*.

Share cropping is widely practiced in Bangladesh, India, Pakistan, and Indonesia. It was the predominant form of tenure on Philippine rice farms until 1973. Under this arrangement, the tenant produces the crop on the landlord's land and each gets a percentage of the output. Within this general type, there are many variations, with the terms of the land contract evolving as production conditions change.[28] In some sharing systems, the landlord shares in the cost of certain inputs in the same proportion as the output is shared. Labor costs may be shared in this way in a relatively labor-short location. The landlord may finance the production costs, and under many traditional arrangements, the landlord is the *patron* of the share tenant, providing him with assistance in time of need.[29] Sharing has other advantages for the tenant. If bad weather or pests cause a reduction in output, the rent is correspondingly lower. Thus, the degree of exposure to risk is lower than with a fixed rental.

Fixed rent systems exist in all countries of the region, but are less common than share rents. Fixed rents may be imposed as part of a land reform, with the intention of reducing the burden to the tenant and providing tenants with increased incentives for production. In a dynamic technological situation, fixed rents increase incentives because all of the increased output resulting from technological improvement is retained by the tenant, rather than a share of the increase going to the landlord. For this reason, fixed rents are often seen as being preferable to share rents. However, the net advantage of one system or the other to the tenants depends on all the terms and conditions under which the two alternatives would be available.

Land Reform

Obviously, farmers would have higher incomes if they paid no rent. This has led to numerous land reform movements. The experience of several countries suggests that credit needs and risk survival mechanisms are important components of land reform programs that seek to transform tenants into owner operators. Therefore, the government must provide the credit and the insurance against risk formerly provided by the landlord if land reform is to be successful.

In cases of more radical land reforms, such as in China and Burma, land has been expropriated by governments and held in public ownership. Farmers do not pay rent, but the price at which output may be legally sold is held low by forced sales to the state at fixed prices. In these cases, farm incomes may be low, but not as low as might have been true with high land rents.

The land tenure and farm structure systems in Japan and Taiwan are seen by many as models for other Asian countries to follow. Following World War II, a strong land reform program was implemented in Japan.[30] Prior to the reform, tenants made up 27 percent of the farm households and afterward only 5 percent.[31] Only 30 percent of households had owned all their land prior to the land reform; afterward 62 percent owned all their land. Landowners were given bonds for their land, and former tenants were deemed owners. The size distribution of holdings remained virtually unchanged because prior to the reform the operating units were very small.

In Taiwan, landowners were also compensated with bonds. Prior to land reform, cultivators retained 67 percent of the farm income to cover cultivating costs and domestic expenses. After reform, this rose to 82 percent of farm income, while the share of landlords and money lenders fell from 25 to 6 percent, and the government's share rose from 8 to 12 percent.[32]

Asian countries other than China, Burma, North Korea, and Vietnam have had less complete land reforms. Following India's independence, land reform was a priority program of Nehru's government,[33] but control over agricultural matters in India is vested in the individual state, not in the national government. Each state passed its own law and enforced it with varying degrees of effectiveness, but in most areas the largest land holdings were redistributed, leaving former landlords with holdings smaller than the "land ceiling."

In the Philippines, land reform was enacted on all rice and corn land by the Marcos martial law government in 1972. It was expected that most farmers would immediately become fixed rent tenants and, when the legal steps were completed for each farm, would purchase their land over a period of fifteen years from a government bank, which would pay the former landlord. The first step was implemented fairly rapidly, but because of difficulties with land valuation, the second step was never carried out. After more than a decade, relatively few Philippine rice farmers owned their land. In this context, the land reform has been regarded as a failure. From the perspective of the political leadership, however, the land reform was a success because it increased farm income and strengthened the position of larger tenant farmers who are leaders in many villages.

These examples illustrate some of the complexities in rice farming structure and some of the difficulties that countries have had in attempting to achieve a more equitable distribution of land resources. The difficulties go well beyond the realm of technical and environmental constraints. There is nothing in the process of rice production as practiced in Asia that places farms of 1 ha at any great disadvantage compared with those of 5 ha or more in terms of economic efficiency (that is, cost per unit of production). In fact, large holdings may be relatively less efficient because of the farmers' inability to provide the intensive care and management to which the rice plant responds. On the other hand, access to resources, not only land but credit and inputs that are needed to produce a rice crop, typically favors the large over the small and near landless, and the owner-operators over the tenants. With the modernization of the Asian rice economy, the greater reliance on cash inputs, and the growing population pressure on the land, these distributive issues have become of increasing concern.

Summary

Rice cultivation systems used in Asia today range from the simplest traditional system with one crop of monsoon rice to modern systems with two crops of rice and a dryland crop, all harvested from the same land in one year. The intensive systems involve more labor, more nutrients, more power, and more management. The more intensive systems also use more inputs for each crop of rice and give higher yields. Rice farming in the more advanced regions of Asia (Japan, South Korea, Taiwan, and Punjab State in India) reflects both forms of intensification—increased inputs per crop and increased crops per year (where the latter is climatically possible).

There is wide variation in the structure of farming—the number of farms of various sizes and the institutional and contractual arrangements under which rice production takes place. Technological and institutional changes have swept the rice farming sectors of many countries in the decades since World War

II, but these changes have been far from universal. Some farmers are using technology undreamed of by their forebears—modern varieties, chemical fertilizers, irrigation pumps, and tractors. Others are adopting innovative institutional changes that permit greater cooperation and more equitable utilization of credit and inputs such as fertilizers, pesticides, and water. However, for a large portion of Asian farmers, particularly those in nonirrigated areas and disadvantaged environments, rice farming continues to be based on centuries-old techniques and institutional arrangements.

Notes

1. Randolph Barker and Robert W. Herdt, "Rainfed Lowland Rice as a Research Priority—An Economist's View," in International Rice Research Institute, *Rainfed Lowland Rice: Selected Papers from the 1978 International Rice Research Conference* (Los Banos, Philippines, IRRI, 1979) pp. 12–13.

2. Three of the four maps are found in Yoshikazu Takaya, "Rice Cropping Patterns in Southeast Asian Deltas," *Southeast Asian Studies* vol. 13 (1975) pp. 256–281. A map for Burma is in Hazel Richter, "Burma's Rice Surpluses: Accounting for the Decline," The Australian National University Development Studies Working Paper No. 3 (Canberra, 1976).

3. F. N. Ponnamperuma and A. K. Banyopadhya, "Soil Salinity as a Constraint on Food Production in the Humid Tropics," in *Priorities for Alleviating Soil-Related Constraints to Food Production in the Tropics* (Los Banos, Philippines, IRRI, 1980) p. 205. In Southeast Asia there are an estimated 49.4 million ha of saline soil (about half of which lies in the Indus Basin), 12.4 million ha of alkali soils, 5.4 million ha of acid sulfate soils, and 20.9 million ha of peat soils.

4. H. Fukui, "Environmental Determinants Affecting the Potential Productivity of Rice—A Case Study of the Chao Phraya River Basin of Thailand" (Ph.D. dissertation, Kyoto University, 1973).

5. Isaburo Nagai, *Japonica Rice, Its Breeding and Culture* (Tokyo, Yokendo Ltd., 1959).

6. Harry Robinson, *Monsoon Asia: A Geographic Survey* (New York, Praeger, 1967) p. 20.

7. Joseph E. Spencer and William L. Thomas, *Asia East by South* (New York, Wiley, 1971) p. 175.

8. The World Bank, *The Philippines, Priorities and Prospects for Development* (Washington, D.C., 1976).

9. Documentation on the extent of single- and multiple-cropped land is scarce, especially for less developed Asian countries. The most comprehensive review of macro-level data on multiple cropping is Dana G. Dalrymple, *Survey of Multiple Cropping in Less Developed Nations*, Foreign Agricultural Economic Report No. 12 (Washington, D.C. U.S. Department of Agriculture [USDA], 1971).

10. International Rice Research Institute, *Cropping Systems Research and Development for the Asian Rice Farmer* (Los Banos, Philippines, IRRI, 1977).

11. In their eagerness to increase food production, the Chinese leadership promoted cropping intensity beyond its economic level. Thomas Wiens analyzed cropping systems in Suzhou Prefecture, Jiangsu Province. He showed that the rapid replacement of triple cropping with double cropping since 1978 is increasing profitability and reducing labor requirements. See Thomas B. Wiens, "The Limits of Agricultural Intensification: The Suzhou Experience," in Beth Rose and Randolph Barker, eds., *Agricultural and Rural Development in China Today* (Ithaca, N.Y., Cornell University 1983) pp. 54–77.

12. Y. Morooka, R. W. Herdt, and L. D. Haws, "An Analysis of a Labor Intensive Continuous Rice Crop Production System at IRRI," IRRI Research Paper Series 29 (Los Banos, Philippines, IRRI, May 1979).

13. S. K. De Datta, *Principles and Practices of Rice Production* (New York, Wiley, 1981) pp. 238–239.

14. See for example, H. C. Conklin, "Hanunoo Agriculture: A Report on the Systems of Shifting Cultivation in the Philippines" (Rome, FAO, 1957); J. D. Freeman, "Iban Agriculture: A Report on the Shifting Cultivation of Hill Rice by the Iban of Sarawak" (London, H. M. Stationary Office, 1955). There is a remarkable similarity between swidden rice systems in Asia and in Latin America. One case of the latter is described in R. H. Bernsten and R. W. Herdt, "Toward an Understanding of Milpa Agriculture: The Belize Case," *Journal of Developing Areas* vol. 11, no. 3 (April 1977).

15. Sajogyo and William L. Collier, "Adoption of New High Yielding Rice Varieties by Java's Farmers," in R. T. Shand, ed., *Technical Change in Asian Agriculture* (Canberra, Australian National University Press, 1973) pp. 80–107.

16. Yujiro Hayami and Anwar Hafid, "Rice Harvesting and Welfare in Rural Java," *Bulletin of Indonesian Economic Studies* (July 1979) pp. 94–112; William L. Collier, Soentoro, Gunawan Wiradi, and Makali, "Agricultural Technology and Institutional Change in Java," *Food Research Institute Studies* vol. 13, no. 2 (1974) pp. 169–194.

17. C. Peter Timmer, "Choice of Technique in Rice Milling on Java," *Bulletin of Indonesian Economic Studies* vol. 9 (1973) pp. 57–76; and W. L. Collier, J. Colter, Sinarhadi, and R. d'A. Shaw, "Choice of Technique in Rice Milling: A Comment," *Bulletin of Indonesian Economic Studies* vol. 10 (March 1974) pp. 106–120.

18. Lih-Yuh Tsai, "Production Costs and Returns for Rice Farms in Central Taiwan, 1895-1976: Analysis of Structural Changes" (M.A. dissertation, University of the Philippines, School of Economics, 1976).

19. Bruce Stone, "The Use of Agricultural Statistics: Some National Aggregate Examples and Current State of the Art," in Randolph Barker, Radha Sinha, and Beth Rose, eds. *The Agricultural Economy of China* (Boulder, Colo., Westview Press, 1982) p. 238.

20. Wiens, "The Limits to Agricultural Intensification: the Suzhou Experience."

21. The size distribution of farms growing rice shows a somewhat higher portion of very small farms, but data are not available for recent years in most countries.

22. A. S. Kahlon and Gurbachan Singh, "Social and Economic Implications of Large-Scale Introduction of High-Yielding Vari-

eties of Wheat with Reference to Ferozepur District, Punjab," (Ludhiana, India, Punjab Agricultural University, 1973).

23. Wei-sen Tong and Sin-chaw Tu, "A Study of the Farm Economy of China through an Analysis of Farm Accounts in Selected Districts," *Agricultural Sinica* vol. 1, no. 12 (1936) pp. 405–507. Similar results are reported in J. L. Buck, *Land Utilization in China* (New York, Agricultural Development Council, 1956).

24. Japan, Ministry of Agriculture, Forestry and Fisheries, *Abstract of Statistics on Agriculture, Forestry and Fisheries* (Tokyo, 1978).

25. Two studies that confirm this tendency are: W. E. James, "An Economic Analysis of Land Settlement Alternatives in the Philippines," IRRI Agricultural Economics Paper 78-30 (Los Banos, Philippines, IRRI, 1978); and Aman Djauhari, "Present Stage of Agricultural Development, Indramayu and Lampung," Annual Report 1975–76 Cropping Systems Research (Bogor, Indonesia, Central Research Institute for Agriculture, 1977).

26. Gillian Hart, "Labor Allocation Strategies in Rural Javanese Households" (Ph.D. dissertation, Cornell University, 1978).

27. Yujiro Hayami, *Anatomy of a Peasant Economy: A Rice Village in the Philippines* (Los Banos, Philippines, IRRI, 1978).

28. James Roumasset, "Land Tenure and Labor Arrangements in Philippine Agriculture: Some Lessons from the New Institutional Economics," paper presented at the Transition in Agricultural Organization Workshop (Los Banos, Philippines, January 1982).

29. Mahar Mangahas, Virginia A. Miralao, and Romana P. de los Reyes, *Tenants, Lessees, Owners: Welfare Implication of Tenure Change* (Quezon City, Philippines, Institute of Philippine Culture, 1974).

30. Toshihiko Isobe, "Land Reform's Achievement and Limits, the Case of Japan," Symposium on Institutional Innovation and Reform (Kyoto, The Ladejinsky Legacy, Kyoto International Center, October 1977).

31. Japan, Institute of Developing Economies, *One Hundred Years of Agricultural Statistics in Japan* (Tokyo, 1969); and Ministry of Agriculture, Forestry and Fisheries, Statistics and Information Department, *Statistical Yearbook of Ministry of Agriculture, Forestry and Fisheries, Japan, 1978-79* (Tokyo, 1980).

32. T. H. Lee and T. H. Shen, "Agriculture as a Base for Socio-Economic Development," in T. H. Shen, ed., *Agriculture's Place in the Strategy of Development: The Taiwan Experience* (Taipei, Joint Commission on Rural Reconstruction, 1974).

33. Daniel and Alice Thorner, *Land and Labor in India* (Bombay, Asia Publishing House, 1962).

4

Trends in Production and Sources of Growth

Asia dominates the world in rice production and consumption, and most imports and exports involve Asian countries, although other regions have increased their contribution to world rice output and trade in the past fifty years. Asia has the largest concentration of poor people in the world, two-thirds of whom rely on rice as a primary foodstuff. Many depend on rice for both consumption and income generation. How successful have Asian countries been in meeting the increasing demand for rice from their populations? What countries provide the rice that is traded internationally to enable deficit producers to meet their needs? What have been the major sources of growth in rice output over the period since World War II?

To gain a perspective on the importance of rice in Asia, this chapter examines changes in rice production from 1910 to 1980. The dominance of Asia in world production, trade, and consumption over the entire period is illustrated in the first section. The second section discusses the relative importance of countries within Asia regarding total regional production and the significance of rice consumption within each country. The third section analyzes the growth in output that has occurred in the Asian producing countries since World War II and the contribution of increased land and yield to that growth.

Asian Rice in World Perspective

The world rice scene was centered in Asia even more completely in the early part of the twentieth century than it presently is. Despite the absence of a number of countries from the reported data,[1] 95 percent of estimated rice output came from Asia in the 1911–30 period (table 4.1). The area planted was equally concentrated (table 4.2).

A detailed analysis of growth in production over this period could be somewhat misleading because of the fragmentary nature of the available information. Area and production for the 1911–30 period cannot be directly compared with subsequent years for many reasons. Primary among them is that fewer countries were included in the early period. A major gap is a time series on rice area for China. Hence, our world rice area data begin with 1930. China's production is estimated for only selected years before 1930 by the U.S. Department of Agriculture (USDA), so we do not attempt to include China's production or area in the time series in tables 4.1 and 4.2, but it is shown later in the chapter. Output of each country was reported as milled rice prior to 1930 instead of being reported as rough (unmilled) rice. Even when reconverted to rough rice, the pre-1930 production data show a large discontinuity with the post-1930 data. No data are available for many countries during World War II so the generalizations that follow should be interpreted with care.

1911 to 1930

Asia dominated the world rice scene during the first third of the century, with 95 percent of the reported world production (table 4.1). Nearly every country

Table 4.1. Rice Production by Continent, 1911–30

Area	Milled rice (thousand mt)				Average percent of world
	1911–15	1916–20	1921–25	1926–30	
Asia[a]	49,382	52,413	54,353	55,018	95
Africa	1,114	1,091	1,381	1,503	2
South America	145	428	529	674	1
North America	409	558	503	603	1
Europe	478	487	521	626	1
Oceania	9	9	4	7	—
World	51,537	54,986	57,291	58,431	100

Source: U.S. Department of Agriculture, *Yearbook of Agriculture*.
[a]Excluding China. Time series are not available in the source used. See table 4.8 in this volume for estimates of Chinese rice production.

Table 4.2. Rice Area by Continent, 1911–30

Area	Average area (thousand ha)				Average percent of world
	1911–15	1916–20	1921–25	1926–30	
Asia[a]	—	—	—	78,000	96.0
Africa	1,316	1,398	1,682	1,766	2.2
South America	185	306	503	821	1.0
North America	348	502	435	447	0.5
Europe	194	187	188	212	0.3
Oceania	5	6	5	5	—
World	—	—	—	81,251	100.0

Source: U.S. Department of Agriculture, *Yearbook of Agriculture*.
[a]Excluding China. Time series are not available in the sources used. See table 4.9 in this volume for estimates of Chinese rice area.

in Asia had substantial production. Of course, India and China had by far the largest output, although no specific estimates are available for the period. In the other regions of the world, one or two countries provided most of the rice output. Europe produced an average of 500,000 mt between 1911 and 1930, mostly in Italy and Spain. The United States produced 80 percent of the 500,000 mt in North America, with Mexico and three Central American nations providing the balance. Brazil dominated South American output, with 80 percent of the region's total production. In Africa, French Guinea, Madagascar, Egypt, and Sierra Leone each contributed 100,000 mt, and together produced 97 percent of that continent's output. Rice was an important commodity to farmers and national economies alike, but compared with Asia, any one of the countries named was a minor producer.

Asia was a net exporter of rice until World War II. Africa and North America were net importers until the end of the 1920s, but production gradually increased until, during the decade of the 1930s, those regions became net exporters (table 4.3). Shipments to South America increased over the early part of the century while shipments to Europe gradually decreased. Most of the trade in rice was among various Asian nations—they absorbed imports of about 3 million out of the 5 million mt exported by the region. Other regions were marginal in the total trade picture.

1930 to 1980

During the Great Depression and after World War II, rice production continued to be concentrated in Asia, although other regions of the world rapidly increased their production, especially after 1950 (table 4.4). In the 1931–40 decade, Asia produced 139 million mt of rice, 96 percent of the world's total. Africa produced 1.3 percent, South America 1.1 percent, and North America and Europe each produced less than 1 percent. By the 1971–80 period, Asia's production had more than doubled, but its share in world output had declined to 92 percent, while Africa's share had increased to 2.2 percent, South America's to 3.2 percent, and North America's to 2 percent. Production had increased somewhat in Europe, while it increased eightfold in Oceania because of Australia's entrance into rice production in the 1920s, but that area still contributed less than 0.2 percent of total world production in 1971–80.

Asia's rice production in the 1970s increased by 130 percent over the level of the 1930s, although some of the apparent increase was due to improved statistical coverage in a number of countries.[2]

Table 4.3. World Rice Imports, Exports, and Net Exports, by Continent, 1909–13 to 1921–30
(thousand mt milled rice)

Area	1909–13			1911–20			1921–30		
	Total		Net	Total		Net	Total		Net
	Imports	Exports	exports	Imports	Exports	exports	Imports	Exports	exports
Asia	3,211	5,142	1,932	2,583	4,098	1,515	3,493	5,427	1,934
Africa	44	24	−19	23	16	−7	33	71	38
South America	190	— [a]	−190	214	n.a.	n.a	344	18	−326
North America	110	8	−102	127	108	−18	45	138	93
Europe	1,633	594	−1,038	1,340	427	−912	1,007	464	−543
Oceania	—	—	—	—	—	—	15	—	—
World	5,188	5,769	583[b]	4,287	4,651	578[b]	4,937	6,119	1,196

Source: U.S. Department of Agriculture, *Yearbook of Agriculture*.
[a] Less than 500 mt reported.
[b] Calculated as the column total. Actual world net exports are, of course, zero, but trade statistics always show a positive or negative balance because of different data reporting systems in various countries, omission of certain countries, and other statistical abberations.

Production in South America increased by over 500 percent and by almost the same proportion in North America, while in Africa it increased by nearly 300 percent.

The distribution of land devoted to rice closely parallels the distribution of production and is the major factor determining production (table 4.5). In the early 1930s, nearly 80 million hectares of rice were grown in Asia. That amounted to 96 percent of the world's total. By the 1970s, there had been a 60 percent increase in Asia's harvested rice land, bringing it to 125 million ha. Area planted increased fivefold in South America and Australia, threefold in North America, and nearly that proportion in Africa. Asia's share of the world total fell to 91 percent by 1971–80.

Along with the significant increases in production that have occurred because of the growth in land devoted to rice, there has also been substantial improvement in yields over the half century since 1930. Average world rice yields increased 42 percent, Asia's rice yields increased by 43 percent, and North America's yields increased by over 90 percent (table 4.6).

Table 4.4. Rice Production by Continent, 1931–40 to 1971–80

Area	Average production (thousand mt, paddy)					Percent increase 1971–80 over 1931–40
	1931–40	1945–50	1951–60	1961–70	1971–80	
Asia	139,102	146,605	181,719	237,331	321,608	131
Africa	1,914	2,987	3,954	4,454	7,572	296
South America	1,641	3,389	4,984	8,074	10,874	563
North America	1,136	1,941	3,048	4,446	6,863	504
Europe	1,043	991	1,631	1,553	3,790	263
Oceania	49	101	127	202	479	878
World	144,885	156,014	195,463	256,060	351,186	142

Sources: Rossiter, Willahan, Cummings (1946); U.S. Department of Agriculture, *Rice Situation*.

Table 4.5. Rice Area by Continent, 1931–40 to 1971–80

Area	Average area (thousand ha)					Percent increase 1971–80 over 1931–40
	1931–40	1945–50	1951–60	1961–70	1971–80	
Asia	78,338	84,427	100,891	113,944	126,051	61
Africa	1,519	2,321	3,024	3,401	4,212	177
South America	1,127	2,055	2,987	4,973	6,326	461
North America	561	950	1,205	1,329	1,766	215
Europe	221	230	364	340	904	309
Oceania	14	35	40	33	83	493
World	81,780	90,018	108,511	124,020	139,342	70

Sources: Rossiter, Willahan, Cummings (1946); U.S. Department of Agriculture, *Rice Situation*.

Table 4.6. Rice Yields (paddy) by Continent, 1931–40 to 1971–80

Area	Average yield in paddy (mt/ha)					Percent increase 1971–80 over 1931–40
	1931–40	1945–50	1951–60	1961–70	1971–80	
Asia	1.78	1.74	1.80	2.08	2.55	43
Africa	1.26	1.29	1.31	1.31	1.80	43
South America	1.46	1.65	1.67	1.62	1.72	18
North America	2.03	2.04	2.53	3.35	3.89	92
Europe	4.72	4.30	4.48	4.57	4.19	−11
Oceania	3.50	2.89	3.18	6.12	5.77	65
World	1.77	1.73	1.80	2.06	2.52	42

Sources: Rossiter, Willahan, Cummings (1946); U.S. Department of Agriculture, *Rice Situation.*

Yields are substantially higher in the developed continents than in Asia, South America, and Africa. Australia, the main producing country in Oceania, has yields of over 5 mt of rough rice per hectare. Yields in Europe reached 4.7 mt/ha during the 1930s, but stagnated until the 1970s, when they fell by about 10 percent. North America showed the largest proportional increase in yields over that period, from 2 mt/ha to 3.9 mt/ha. Within North America, the United States, the largest producer, had yields of over 5 mt/ha by the late 1970s, but poor yields in other countries lowered the continent's average to 3.9 mt/ha.

Asia has the highest average yield of the developing areas of the world, and yields have been rising since the 1950s. These trends are partly attributable to Japan and Korea, the most developed countries of Asia. However, there have also been significant yield increases in India and China as well as in other developing Asian countries.

Average rough rice yields in Asia were nearly constant around 1.8 mt/ha through the 1930s, 1940s, and 1950s. Yields increased to 2.1 mt/ha in the 1960s and to 2.65 mt/ha in the 1970s. Yields in Africa and South America have been somewhat lower than in Asia, mainly because most of the crop is grown under dryland conditions, but in both areas, yields increased from the 1930s to the 1960s.

On a percentage basis, Asia does not claim as large a share of world rice imports and exports as it used to (table 4.7). In the 1950s, Asia contributed about 75 percent of the 5.5 million mt of world trade. By the 1970s, Asia's share of imports and exports had declined to about 60 percent. North America emerged as the second most important exporter while Europe's imports doubled. Imports to Africa also increased substantially. During the 1960s, Asia was a net importer, but by the 1980s the cumulative effect of improved technology and gradually slowing rates of population growth, along with the policy of exporting rice and importing wheat, made Asia a net rice exporter again.

Table 4.7. World Rice Imports, Exports, and Net Exports by Continent, 1951–60 to 1971–80[a]
(thousand mt milled rice)

Area	1951–60			1961–70			1971–80		
	Total			Total			Total		
	Imports	Exports	Net exports	Imports	Exports	Net exports	Imports	Exports	Net exports
Asia	3,960	4,170	210	4,800	4,472	−329	5,466	5,921	454
Africa	349	224	−125	648	530	−118	1,297	217	−1,080
South America	51	169	118	68	325	257	185	499	314
North America	321	721	400	300	1,519	1,219	373	2,215	1,842
Europe	767	368	−399	1,138	331	−807	1,547	717	−830
Oceania[b]	15	38	23	32	92	61	59	277	218
World[c]	5,463	5,690	227	6,987	7,270	283	8,928	9,846	918

Sources: 1950–59: U.S. Department of Agriculture, *Agricultural Statistics*; 1960–80: U.S. Department of Agriculture, FAS, *Grains*, FG-38-80.

[a]Refers to local marketing year trade data. Negative values indicate net imports.

[b]Data refer to Australia and Papua New Guinea only.

[c]Positive net exports arise from statistical discrepancy and drawdown of stocks.

Asian Rice in Historical Perspective

There are large data problems in any effort to provide an historical perspective on rice production in Asia. The changes brought by colonialism, war, independence, and the national realignments that swept across Asia during the twentieth century are difficult to quantify for most countries. The devastation of World War II pushed statistical activities to the background so that much of the data for the 1939–46 period are less reliable than for other periods. Independence arrived in most Asian countries in the fifteen years following World War II. For several years most countries were too busy establishing viable governmental organizations to put much emphasis on statistics. However, many countries gradually changed and improved statistical collection methods. Thus, some reported changes may be due to statistical modifications.

To develop the kind of comprehensive picture we desired, efforts were made to go back to original national statistical data sources to obtain the production, trade, and other data shown in the appendix tables.[3] Therefore, some of the totals for Asia in this section may differ from those in the previous one, which depended on USDA data. It seemed most appropriate to present the pre–World War II data separately from the postwar data in order to recognize the realities of national boundaries as they existed in each period. Many changes occurred during the 1940s and data are missing for many countries. We examine the 1901–40 data, omit 1941–45, and then consider the 1946–80 data.

1901 to 1940

China and India dominate Asian rice production, much as Asia dominates the rest of the world. Data for both are fragmentary and particularly unreliable during the first few decades of the century (table 4.8). Based on the available numbers for the twenty years from 1921 to 1940, China had an estimated 48 percent of Asia's rice production, and India had 21 percent of production. Japan, the next most important with

Table 4.8. Rice Production by Region and Country, Asia, 1900–40

Region & country	Average production (thousand mt, paddy) 1901–10	1911–20	1921–30	1931–40	Average annual growth rate 1911–20 to 1931–40 (percent)
East Asia	—	62,326	68,620	75,357	0.95
China	—	49,173[a]	54,317[a]	60,000[b]	1.00
Japan	8,734	10,429	10,973	11,415	0.45
Korea	—	1,846	2,178	2,283	1.07
Taiwan	749	878	1,152	1,659	3.23
Southeast Asia	—	17,286	20,802	21,996	1.21
Burma	6,762[c]	6,448	6,999	7,114	0.49
Indochina[d]	—	3,278	3,614	3,721[e]	0.64
Java and Madura	—	3,067[f]	3,470	4,072	1.43
Malaya	—	182[g]	213	327	2.97
Philippines	821[h]	1,106	2,058	2,216	3.54
Thailand	2,737[i]	3,248	4,448	4,546	1.70
South Asia	—	26,527	25,461	24,466	−0.40
India[j]	25,812	26,295	25,204	24,160	−0.42
Sri Lanka	—	219[f]	257	306	1.69
Total Asia	—	106,139	114,883	121,819	0.69

Source: Appendix tables.
[a]Derived by assuming a constant growth rate over the period.
[b]Interpolated back from 1946–55 data.
[c]1909–10 av.
[d]Includes Kampuchea, Laos, and Vietnam.
[e]1931–38 av.
[f]1920 data only.
[g]1910–20 data only.
[h]1908–10 av.
[i]1907–10 av.
[j]Includes India, Pakistan, and Bangladesh.

Table 4.9. Rice Area by Region and Country, Asia, 1900–40

Region & country	Average area (thousand ha) 1901–10	1911–20	1921–30	1931–40	Average annual growth rate 1911–20 to 1931–40 (percent)
East Asia	—	—	—	30,205	—
China	—	—	—	25,000[a]	—
Japan	2,862	3,025	3,137	3,172	0.24
Korea	—	1,464	1,555	1,378	−0.30
Taiwan	432	486	552	655	1.50
Southeast Asia	—	—	17,794	19,584	—
Burma	3,726	4,203	4,699	4,942	0.81
Indochina[b]	—	4,115	5,193	5,598	1.55
Java and Madura	—	3,225[c]	3,366	3,831	0.86
Malaya	—	—	268	305	—
Philippines	1,118[d]	1,287	1,753	1,996	2.22
Thailand	1,461[e]	1,906	2,515	2,912	2.14
South Asia	—	—	27,461	28,007	—
India[f]	26,516	27,414	27,134	27,667	0.05
Sri Lanka	—	—	327	340	—
Total Asia	71,508	72,851	73,870	77,796	0.33

Source: Appendix tables.
[a]Interpolated from 1946–55 data.
[b]Includes Kampuchea, Laos, and Vietnam.
[c]1916–20 av.
[d]1909–10 av.
[e]1907–10 av.
[f]Includes present-day India, Pakistan, and Bangladesh.

9 percent of production, was far below the first two. Burma had 6 percent of production while Indochina and Java-Madura each had about 3 percent. The contribution of each country to area (table 4.9) differed from its share in production because the countries of East Asia had more irrigated rice and relatively higher yields. For example, India accounted for 35 percent of the Asian rice area and China had 32 percent, while their contributions to production were 21 percent and 48 percent, respectively.

The period was not a dynamic one for rice production. The data in table 4.8 show that output grew very slowly over the period. Taiwan and the Philippines increased production at over 3 percent annually, and Malaya nearly reached that pace, but output in most countries grew at near or below 1 percent per year. Thailand and the Philippines were able to increase the area devoted to rice at over 2 percent per year, Malaya and Indochina increased their area at more than 1 percent per year, but in the other countries rice area increased more slowly (table 4.9). In Korea, area actually declined over the period.

India, Japan, Sri Lanka, Malaya, and Indonesia were the major rice importers in the prewar period (table 4.10). Major exporters included Burma, Thailand, Indochina, and Korea. On the whole, the region had net exports of about 2.5 million mt/year over the entire period from 1911 to 1940.

1946 to 1980

Figure 4.1 shows trends in area, production, and yield for the three main regions of Asia since 1960. Clearly, Southeast Asia shows the fastest growth, but there is also a steady growth in production in South and East Asia. Both area and yield have contributed to output growth.

Table 4.11 shows the average annual rice production for the Asian countries for the post–World War II decades. In contrast to the prewar era, the data show that most countries increased their rice production significantly during the period. Some of the reported increase, especially during the late 1940s and 1950s, was a result of improved statistics, but by no means was that the only source. Most countries continued the increases over decades. Those equaling or exceeding 3 percent annual growth in rice output included North Korea, Malaysia, Thailand, the Philippines, Indonesia, Pakistan, and Sri Lanka.

Table 4.10. Annual Average Net Exports of Rice by Region and Country, Asia, 1911–40
(thousand mt)

Region & country	1911–20	1921–30	1931–40
East Asia			
Japan	–263	–371	–1,178
Korea	245	641	1,178
Taiwan	120	194	546
Southeast Asia			
Burma	2,230	2,683	2,687
Indochina[a]	1,222	1,491	1,325
Indonesia	—	–559	–281
Malaya	–348	–211	–390
Philippines	–154	–66	–35
Thailand	853	1,307	1,551
South Asia			
India[b]	–1,450	–1,522	–2,615
Sri Lanka	–374	–438	–518
Total shown above	2,081	3,148	2,271

Note: Negative signs indicate imports.
Source: Appendix tables.
[a]Includes present-day Kampuchea, Laos, and Vietnam.
[b]Includes present-day India, Pakistan, and Bangladesh.

Vietnam, Burma, and India were not far behind. Only Japan, Bangladesh, Nepal, Vietnam, and Laos grew substantially more slowly.

Production in East Asia grew at over 2.5 percent annually, with all countries except Japan having significant growth. Japan had largely outgrown the need for increased rice production by 1970 because incomes had reached the level where consumption was being diversified (chapter 11). Southeast Asia grew at the most rapid rate, with all countries except Burma and Laos exceeding 2.5 percent per year. South Asia's rate of output growth was somewhat less than Southeast Asia's, with Pakistan and Sri Lanka growing faster than India, and Bangladesh and Nepal growing more slowly.

Growth in output after World War II came both from increased area planted and from increased yields as modern biological and chemical technologies were applied to the rice paddies of Asia. Area planted increased substantially in North Korea, Malaysia, Thailand, the Philippines, Pakistan, and Sri Lanka (table 4.12). In most other countries, rice area increased at 1 percent per year or less over the entire period, with the rest of the growth in output coming from increased yields.

Yields increased fairly rapidly after the end of World War II. Japan's rice yields, long the highest in Asia, increased from the prewar level of 3.6 mt/ha to over 5 mt/ha by the middle 1960s (table 4.13). Yields in North and South Korea also improved dramatically. Most of the other countries of Asia had yields around

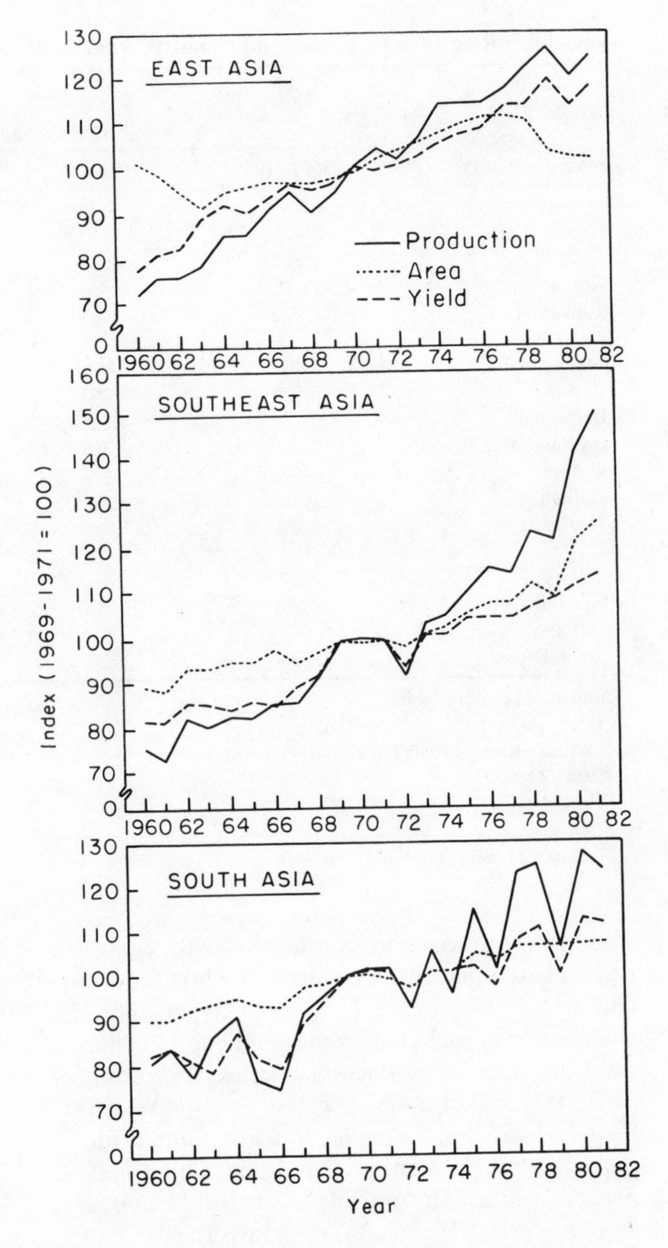

Figure 4.1. Trends in regional paddy production, area, and yield, 1960–81

1 mt/ha in the prewar period. These increased to around 2 mt/ha in the 1960s and approached 3 in many countries in the 1970s. During the 1920s, Burma and Thailand had the highest yields in Asia except for Japan, but yields stagnated at around 1.5 mt/ha until the late 1970s while they gradually increased in other countries in the region. Only in the 1970s have yields in Burma shown an increase. Data on production and yields in Nepal are not available for the prewar period, but in the 1950s Nepal had fairly high yields, although these did not measurably increase.

Table 4.11. Rice Production by Region and Country, Asia, 1946–80

Region & country	Average production (thousand mt, paddy)				Average annual growth rate 1946–55 to 1976–80 (percent)
	1946–55	1956–65	1966–75	1976–80	
East Asia	82,269	107,135	137,946	164,149	2.54
China	63,643[a]	82,993	110,164	134,400	2.76
Japan	12,157[b]	15,368	16,090	14,793	0.72
South Korea	3,514[b]	4,426	5,560	7,135	2.61
North Korea	1,173[b]	1,712	2,981	4,462	4.98
Taiwan	1,782	2,636	3,151	3,359[c]	2.33
Southeast Asia	31,223	41,711	57,445	72,545	3.11
Burma	4,995[d]	7,037	8,054	9,579	2.40
Indonesia	10,180[b]	12,159	18,738	25,695	3.42
Laos	506[b]	559	844	899	2.11
Malaysia	743[b]	971	1,679	1,862	3.40
Philippines	2,768	3,747	5,060	7,221	3.55
Thailand	6,546	8,177	13,182	16,400	3.40
Vietnam	5,485[b]	9,061	9,888	10,889	2.52
South Asia	50,667	67,298	84,274	101,321	2.55
Bangladesh	11,140[e]	13,735	16,700	19,230	2.01
India	35,955	49,164	60,773	73,475	2.63
Nepal	1,752[b]	1,937	2,268	2,306	1.00
Pakistan	1,258[b]	1,625	3,212	4,589	4.82
Sri Lanka	562[b]	837	1,321	1,721	4.15
Total Asia	164,159	216,144	279,665	338,015	2.66

Source: Appendix tables.
[a]1949–55 av.
[b]1950–55 av.
[c]1976–79 av.
[d]1946–51 av.
[e]1947–55 av.

At the same time, yields in the rest of South Asia showed a gradual increase from around 1 mt/ha in the early 1950s to around 2 in the late 1970s.

International trade in rice has been characterized by two interesting phenomena: (1) until the 1970s the major exporting countries were fairly unimportant producers and (2) international trade is a very small proportion of total output. The leading Asian exporters—Thailand, Burma, Indochina, Korea, and Taiwan—produced 25 percent of Asia's output but contributed 90 percent of Asia's exports in the 1911–30 period. Malaysia, India, Indonesia, Japan, and the Philippines produced 43 percent of Asia's output; each was a net importer during that period, and together they absorbed about 50 percent of world imports.

Compared with the other major grains, only a small fraction of world rice production enters the market. During the 1970s, only 4 percent of the 224 million-metric-ton rice crop (milled) was traded internationally.[4] Thus, rice is consumed mainly where it is produced. For those countries that buy or sell rice, however, trade is an important part of their total rice economy, with implications for domestic prices and government budgets, especially in urban areas. Because most rice is consumed directly by people, its demand is relatively inelastic with respect to price changes. Also, developing countries tend to have limited storage capacity. To avoid consumer distress, most countries insulate their domestic rice price from the world price. This generally involves subsidies by government and export taxes by some exporters. Both have important budgetary implications that are discussed in chapters 13 and 16, which cover international trade policy. Table 4.14 shows net exports of rice for the countries that either import or export appreciable quantities. Japan was a net importer of rice until the late 1960s, obtaining most of those imports from its prewar colonies, Taiwan and Korea. (The dramatic increase in those shipments between 1910 and 1940 is evident from the export data in table 4.10.)

The continuing dependence of Malaysia, the Philippines, Indonesia, Bangladesh, and Sri Lanka on

Table 4.12. Rice Area by Region and Country, Asia, 1946–80

Region & country	Average area (thousand ha)				Average annual growth rate 1946–55 to 1976–80 (percent)
	1946–55	1956–65	1966–75	1976–80	
East Asia	32,845	36,447	38,022	40,968	0.81
China	27,641[a]	30,735	32,420	35,640	0.93
Japan	3,054[b]	3,271	2,944	2,592	−0.59
South Korea	1,007[b]	1,139	1,203	1,222	0.71
North Korea	407[b]	527	683	753	2.26
Taiwan	736	775	772	761[c]	0.12
Southeast Asia	22,038	26,132	29,690	32,971	1.48
Burma	3,686[d]	4,403	4,755	4,939	1.07
Indonesia	6,213[b]	7,011	7,979	8,906	1.32
Laos	508[b]	677	722	681	1.07
Malaysia	431[b]	474	679	709	1.83
Philippines	2,370	3,149	3,287	3,524	1.45
Thailand	4,970	5,634	7,478	8,990	2.18
Vietnam	3,860[b]	4,784	4,790	5,222	1.11
South Asia	40,722	45,875	50,566	53,436	0.99
Bangladesh	8,185[e]	8,651	9,793	10,071	0.76
India	30,139	34,383	37,404	39,458	0.98
Nepal	1,008[b]	1,062	1,184	1,264	0.83
Pakistan	951[b]	1,202	1,527	1,941	2.63
Sri Lanka	439[b]	577	658	702	1.72
Total Asia	95,605	108,454	118,278	127,375	1.05

Source: Appendix tables.
[a]1949–55 av.
[b]1950–55 av.
[c]1976–79 av.
[d]1946–51 av.
[e]1947–55 av.

foreign rice is evident. Malaysia and Sri Lanka took between a quarter and a half million metric tons a year throughout the period, and continue to be large importers. Imports to Indonesia increased in the 1950s and 1960s and showed no sign of decreasing through the 1970s until 1980 when Indonesia's production increased dramatically. Total shipments to the countries previously making up India have declined, but Bangladesh continues to import.

China is no longer a rice importer, but now substitutes wheat imports. Because wheat can be purchased on the world market at a lower cost per calorie of food than rice, to increase total food grain availability, the Chinese have imported wheat, exported rice, and improved their foreign exchange position. Because of the war, Vietnam lost its surplus production capacity and became a net importer in the 1970s.

Thailand continues to play a dominant role in world rice exports, although shipments have dropped below 1 million metric tons in some years. Burma's exports fell off substantially after World War II. In the early part of the century, Burma exported twice as much as Thailand, but by the 1950s their exports were about equal and by the 1970s, Burma's were only one-third of Thailand's.

Sources of Growth in Production

Most Asian countries were able to increase their rice production significantly faster during the post–World War II era than in the early part of the century. In fact, during the decade from the early 1950s to the early 1960s, the average growth rate exceeded 3 percent annually, and from the early 1960s to the early 1970s was only slightly lower. Most individual countries increased their production fairly steadily, at rates up to 6 percent per year over the period. Some countries generated most of this growth by adding to their rice land, either through the construction of new irrigation systems to facilitate double cropping, or simply by opening new land to cultivation. In other cases, the growth was generated by increasing per hectare yields.

In most countries, expansion of area planted was a more important source of growth in the 1950s than

Table 4.13. Rice Yields by Region and Country, 1931–80

Region & country	Average paddy yield (mt/ha)				
	1931–40	1946–55	1956–65	1966–75	1976–80
East Asia		2.50	2.94	3.63	4.01
China	1.77	2.33[a]	2.71	3.39	3.77
Japan	3.60	3.97[b]	4.70	5.48	5.71
South Korea	2.44	3.47[b]	3.88	4.62	5.83
North Korea	2.44	2.89[b]	3.23	4.35	5.93
Taiwan	2.54	2.40	3.40	4.09	4.42[c]
Southeast Asia		1.41	1.64	1.94	2.20
Burma	1.44	1.36[d]	1.59	1.69	1.93
Indonesia	1.06[e]	1.63[b]	1.73	2.34	2.89
Laos	0.66[f]	1.00[b]	0.83	1.20	1.32
Malaysia	1.07	1.72[b]	2.03	2.46	2.62
Philippines	1.11	1.16	1.19	1.54	2.05
Thailand	1.56	1.32	1.44	1.76	1.82
Vietnam	(f)	1.42[b]	1.90	2.06	2.08
South Asia	—	1.24	1.47	1.67	1.90
Bangladesh	—	1.36[g]	1.58	1.71	1.91
India	1.30	1.19	1.43	1.62	1.86
Nepal	—	1.74[b]	1.82	1.91	1.82
Pakistan	—	1.32[b]	1.35	2.09	2.36
Sri Lanka	0.90	1.27[b]	1.48	2.00	2.45
Total Asia		1.71	2.00	2.37	2.66

Source: Appendix tables.
[a]1949–55 av.
[b]1950–55 av.
[c]1976–79 av.
[d]1946–51 av.
[e]Java and Madura only.
[f]Refers to Indochina.
[g]1947–55 av.

in the 1960s or 1970s. Of course, there was a good deal of variability in the rates of growth achieved by various countries, and in the rates needed by them to keep up with demand. Table 4.15 shows post–World War II growth rates in production and the contribution of area and yield to the growth in production.

Some countries experienced rather long periods of steady expansion while others alternated between periods of stagnation and growth. Weather-related fluctuations in output are important in nearly all countries. In the 1953–67 period, expansion of land area contributed about half or more of the growth in output in eleven of the seventeen countries shown in table 4.15.[5] In the 1967–72 period, area expansion contributed about half or more in only four countries, and in 1972–77, area expansion also dominated in four countries. Where yield increase dominated, production grew at an average rate of 2.7 percent while where area increase dominated, production grew at an average rate of 2.6 percent annually in the 1972–77 period.

In addition to effectively increasing land productivity, irrigation also improves the environment for plant growth. Chemical fertilizer, carefully applied at appropriate rates, can give notable yield increases even for traditional tropical rice varieties, and farmers in a number of countries began using fertilizer on rice during the 1950s and 1960s. Japan had fertilizer-responsive varieties even before World War II and continued to increase its use of fertilizer on rice after the war. The development of fertilizer-responsive varieties for the tropics enabled many farmers throughout Asia to use fertilizer, irrigation, and new varieties to raise their yields far beyond earlier levels. Better education and extension programs were instrumental in bringing the new technologies to farmers.

Two techniques have been used to separate the contribution of various factors to the growth in output. The *growth rate* technique divides the total growth in output into a portion attributable to increases in irrigated and nonirrigated land, a portion attributable to increases in yield directly due to fertilizer, and a residual. These attributions are based on calculations of the growth rates of production, irrigated area, nonirrigated area,[6] and the assumption that each kilogram of fertilizer produced 10 kg of unmilled rice.[7]

The residual may be the result of many factors. First, the above assumption made about fertilizer may be wrong. There may be a number of factors in addition to fertilizer that indirectly influence the effiency of the fertilizer input (that is, the conversion rate of plant nutrients into grain). Alternatively, there may be other inputs or management practices that operate independently of fertilizer to raise yields, which are reflected in the residual. The *production function* technique allocates total growth in output to the factors included in an estimated production function. Production function estimation requires data on the inputs used in production and so it is more data intensive than the growth rate technique, but the two approaches are complementary and both have been used here.

The Contribution of Land, Irrigation, and Fertilizer

This section describes the relative contributions of irrigated and nonirrigated land expansion and yield

Table 4.14. Average Net Rice Exports, 1951–80
(thousand mt, milled)

Region & country	1951–60	1961–70	1971–80
East Asia			
China	—	946	1,529
Japan	—	−166	285
South Korea	—	−295	−475
North Korea	—	59	249
Taiwan	120	103	104
Southeast Asia			
Burma	1,551	898	480
Kampuchea	536	230	−145
Indonesia	−600	−567	−1,319
Laos	—	−61	−79
Malaysia	—	−384	−268
Philippines	−63	−161	−109
Thailand	1,258	1,420	1,817
Vietnam		−420	−518
South Asia			
Bangladesh	−201	−325	−297
India	−348	−488	99
Nepal	—	234	146
Pakistan	—	145	760
Sri Lanka	—	−485	−334
Total of above	—	683	1,925

Note: Negative signs indicate imports.
Source: Appendix tables.

Table 4.15. Rice Production Growth Rates, and Relative Contribution of Area and Yield to Output Growth, 1953–77[a]

	1953–67			1967–72			1972–77		
	Output growth rate	Percent contributed by		Output growth rate	Percent contributed by		Output growth rate	Percent contributed by	
Region & country	(percent)	Area	Yield	(percent)	Area	Yield	(percent)	Area	Yield
East Asia									
China	2.50	19	81	4.00	53	47	2.50	54	46
Japan	2.40	19	81	−2.50	—	—	0.90	—[c]	—
South Korea	2.80	52	48	2.40	—[b]	—	5.40	10	90
North Korea	5.60	60	40	5.00	35	65	6.20	23	77
Taiwan	2.90	3	97	−0.10	—	—	1.40	23	77
Southeast Asia									
Burma	2.30[c]	69	31	1.10	28	72	2.70	26	74
Indonesia	2.60	49	51	6.40	24	76	3.30	46	54
Laos	3.50	100	0	1.70	—[c]	—	0.70	60	40
W. Malaysia	4.60	59	41	6.90	59	41	−0.20	—[c]	—
Philippines	2.60	53	47	3.10	20	80	5.80	27	73
Thailand	3.90	49	51	2.00	75	25	3.80	86	14
Vietnam	3.40	38	62	3.70	21	79	1.10	—	—[c]
South Asia									
Bangladesh	2.80	36	64	−0.10	—	—	2.80	27	73
India	2.60	48	52	3.10	23	77	2.90	30	70
Nepal	1.40	62	38	1.40	72	28	0.30	—	—[c]
Pakistan	5.30	62	38	5.90	4	96	5.20	87	13
Sri Lanka	4.80	46	54	5.30	44	56	1.50	13	87

Source: Derived from data in appendix tables.
[a]Based on 5-year averages centered on the years shown.
[b]Not calculated because area or yield declined over the period.
[c]Based on 5-year averages centered on 1958 and 1967.

increases attributable to fertilizer in each country for which data are available. Unfortunately, data are not available to separate the contribution of irrigated from unirrigated land before the mid-1960s, so the detailed analysis reported in table 4.16 contrasts the growth achieved between 1967 and 1972 with growth achieved between 1972 and 1977. Five-year averages were used to minimize weather effects.

East Asia. Japan presents a unique picture of production growth during the post–World War II period. Fertilizer use was high, fertilizer-responsive varieties were disseminated, and rice production was completely irrigated by the early 1950s. Increasing fertilizer use and continued intensification occurred during the 1950s, when the demand for rice increased and output grew at 2.4 percent annually, mainly from yield increases. During the 1960s, Japan's domestic demand slowed abruptly as skyrocketing incomes changed consumption patterns away from cereals toward fruits, vegetables, and protein-rich products. With surpluses building up during the 1960s, acreage restrictions were imposed on rice to limit the government's financial burden from storage of stocks. As a result of acreage controls, area planted to rice fell at about 2 percent per year, and output decreased slightly during the 1960s. However, because the Japanese government wanted to protect its farmers (and because it could afford to do so), it maintained rice price supports at a very high level. With this price incentive, farmers continued to increase yields.

North Korea has had one of the most rapid sustained rates of output growth, but lack of information on fertilizer use and irrigated area prevents us from disaggregating the sources beyond that shown in table 4.15. South Korea also had a rapid rate of output growth, especially in the 1972–77 period. As shown in table 4.16, growth in fertilizer contributed the bulk of the yield increases achieved between 1967 and 1972, but in the 1972–77 period, fertilizer use increased less rapidly, and most of the increase in total output as well as yield was unaccounted for.

Just over half of China's output growth since 1967 resulted from increases in area planted to rice. Much of this came from increased double and triple cropping owing to irrigation, but the data available are inadequate to permit a detailed accounting. In Taiwan, rice production grew at nearly 38 percent annually from 1953 to 1967, basically because of yield increases. By 1967, rice farmers in Taiwan were applying over 200 kg of fertilizer nutrients per hectare. However, because of changing incentives, during 1967 to 1972, fertilizer use fell to about 165 kg/ha, and nonirrigated land in rice was also reduced. There were evidently offsetting gains in productivity because production was about constant between 1967 and 1972. After 1972, fertilizer use picked up to move above its earlier level, and production increased to over 1 percent per year.

Table 4.16. Annual Percentage Rates of Growth Attributed to Growth in Irrigated Land, Nonirrigated Land, Fertilizer, and the Residual, 1967–77[a]

	1967–72					1972–77				
Region & country	Total	Irrig. area	Non-irrig. area	Fer-tilizer	Resid-ual	Total	Irrig. area	Non-irrig. area	Fer-tilizer	Resid-ual
East Asia										
Japan	−2.50	−3.60	0.0	1.20	−0.10	0.90	−0.40	0.0	0.30	1.00
South Korea	2.40	1.40	−1.70	2.00	0.70	5.40	0.50[b]	—[c]	1.30	3.60
Taiwan	−0.10	−0.30	−0.50	−2.60	3.30	1.40	−1.20	1.50	4.40	−3.40
Southeast Asia										
Burma	1.10	0.40	−0.10	0.20	0.60	2.70	0.30	0.40	0.80	1.20
Indonesia	6.40	1.50[b]	—	2.30	2.60	3.30	1.50[b]	—	2.90	−1.10
W. Malaysia	7.20	5.40	−0.20	0.90	1.00	1.50	0.90[b]	—	—	—
Philippines	3.10	0.80	−0.20	1.80	0.70	5.80	0.60	0.90	1.00	3.20
Thailand	2.00	0.20	1.30	0.20	0.40	3.80	—	3.20	1.00	−0.50
South Asia										
Bangladesh	−0.10	0.90	−0.80	0.30	−0.40	2.80	0.0	0.70	0.60	1.50
India	3.10	0.50	0.20	2.00	0.50	2.90	0.50	0.40	3.00	−1.10
Pakistan	5.90	0.40	0.0	1.50	4.00	5.20	4.50	0.0	1.80	−1.10
Sri Lanka	5.30	2.50	−0.10	0.70	2.20	1.50	0.40	−0.20	0.60	0.70

Source: Derived from data in appendix tables.
[a]5-year averages centered on the years shown.
[b]Data not available to separate contribution of irrigated from nonirrigated area.
[c]Data not available.

Southeast Asia Burma's rice output grew at 2.3 percent in the 1953–67 period, with about 70 percent of the growth coming from increased area. In the late 1960s, output growth slowed substantially and then picked up again during the 1970s. In the latter period, irrigated land and nonirrigated land contributed roughly equally to growth. Fertilizer was important, but the unexplained portion of growth was surprisingly high, even higher than in Indonesia.

Substantial growth in Indonesian rice production was due to increases in irrigated land from 1967 to 1972. Growth in irrigated land slowed somewhat during the 1970s. Fertilizer and the unexplained residual each contributed substantially to output growth during both the 1960s and the 1970s.

Malaysia had one of the most variable situations. Until 1972, output grew rapidly, largely on the basis of increases in irrigated land. After 1972, output grew much more slowly, in spite of increases in fertilizer use and continued slow growth in irrigated area.

Output growth in the Philippines accelerated from 2.6 to 3.1 to 5.8 percent annually over the three periods shown in table 4.15. During 1953–67, about half of this growth came from area increases, but thereafter, yields were the dominant source. Increased fertilizer was responsible for 1.8 percent annual growth in 1967 and 1972 and for 1 percent during 1972–77. Other factors increasing productivity (the residual) were the most important source of rice production growth for the Philippines in the 1972–77 period.

Growth in Thailand's output was caused by nonirrigated land increases. Fertilizer has been of little importance. Throughout the period since 1967, area increases contributed over two-thirds of Thailand's growth in output.

South Asia Bangladesh achieved a 2.8 percent rate of output growth during the 1950s and 1970s, but in the 1967–72 period when revolution and civil strife affected the country, rice production stagnated, then picked up again in the late 1970s. Unlike many countries, irrigation played a rather minor role in explaining Bangladesh's area increases of the 1970s. Fertilizer contributed in a modest way, but it was the unexplained residual that seemed to be most important in the 1970s.

Being the largest among the South Asian countries, India's successes and failures seem to dominate thinking about the region. India had an average rate of rice output growth in excess of 2.9 percent per year over both periods. In the 1950s and early 1960s, that growth was about equally derived from area and yield increases (table 4.15). In the latter half of the 1960s and 1970s, yield increases dominated the picture, with fertilizer providing a 2 percent rate of output growth in the late 1960s and a 3 percent rate of growth in the 1970s.

All rice in Pakistan is irrigated, so the entire growth in rice area is attributed to gains in irrigated land. During the 1967–72 period, yields grew very rapidly, with only a relatively small proportion of the growth of yields attributable to increased fertilizer. Instead, the unexplained residual was more important. From 1972 to 1977, area expansion again became dominant and, along with fertilizer, resulted in a better than 5 percent annual growth rate.

Sri Lanka had a rapid growth in production through the 1950s and 1960s, but it slowed significantly during the 1970s. Increasing area was nearly as important as yields during the first two periods, and while fertilizer was important in the 1967–72 period, other productivity gains were also important. During the 1970s, the importance of the residual diminished somewhat.

An Aggregate Production Function Estimate of Productivity

The above identification of the sources of output growth can be supplemented using a production function analysis. Rice grows by using solar energy, water, and soil nutrients in an environment that is controlled by human labor and mechanical power. A production function is a quantification of the relationship between those factors and the rice produced. The measurement and analysis of the effect of climatic factors is beyond the present exercise because we are mainly interested in measuring the effect of factors under human control and in characterizing the reasons for the broad differences in production across countries. Six variables were hypothesized to determine the level of rice production: land, irrigation, fertilizer input level, fertilizer productivity, labor, and capital.

Land was measured as the area of rice harvested in each country. Irrigated rice area was used to reflect water, although it is recognized that the quality of irrigation varies widely across the countries of the region. For example, farmers in Japan and China have excellent control over the application and drainage of water. Farmers in Bangladesh and Thailand, by contrast, have almost no control over water. Unfortunately, other than general impressions, there are no indicators of the quality of rice irrigation at the national level. However, there is a strong correlation between the proportion of rice area irrigated and the quality of irrigation. As more land becomes irrigated, there is a demand for improving the control over water in existing systems and so the two factors move together (chapter 7). Therefore, the area of rice irrigated was used as a reflection of water control in the production function estimation.

Good measures of inherent soil fertility do not exist, but the quantity of chemical fertilizer applied reflects efforts to increase soil nutrient availability to rice. Fertilizer use data are reported by most countries, and estimates of the quantity of fertilizer used on rice have been developed (chapter 6).

The area of land planted to modern, fertilizer-responsive varieties was included because it is commonly believed that fertilizer applied on these varieties is more productive than on old varieties (chapters 5 and 6). A positive coefficient for modern-variety area would reflect the relative size of that shift in technology.

Labor and capital are alternative inputs to accomplish the physical tasks of land preparation, transplanting, weeding, harvesting, and threshing. No data are available on the labor used in rice production, but most countries have census data on the agricultural labor force that were used in this analysis. The number of agricultural tractors was used as a proxy variable for capital. The amount of labor and tractor power used on rice was approximated by assuming that those inputs were allocated to rice in the same proportion as land was allocated to rice.

Every mathematical form used for a production function has certain restrictive characteristics. The log-log or Cobb-Douglas form used in this analysis assumes that a given percentage change in an input results in a constant percentage change in output at all levels of the input. Five-year averages of the variables for 1951–55, 1961–65, and 1971–75 were used in this analysis.

The first estimated equation explained 94 percent of the variation in rice output in the observed data set. Irrigated land area and labor were the most important explanatory variables. Nonirrigated land area, modern varieties, and capital were not significant at the 5 percent confidence level, but had reasonable magnitudes and signs. The coefficient of fertilizer was negative, the most unexpected result. Because fertilizer was expected to be a major factor generating growth in output over time, an explanation for its negative sign was sought.

There are three possible explanations: (1) the negative sign may arise because of intercorrelation between fertilizer and other factors in the equation; (2) the negative sign may arise because the extremely wide range of input levels existing among Asian countries cannot be adequately captured by a single equation because of limitations imposed by its mathematical form (fertilizer use ranged from zero to over 300 kg/ha and irrigation from zero to 100 percent of rice area); and (3) it may be an accurate reflection of the effect of fertilizer, that is, fertilizer reduces production.

The third explanation is rejected on the grounds that too many scientists and farmers have found convincing evidence to the contrary. Modern-variety area was highly correlated with fertilizer, so the first explanation has some basis. In subsequent equations only the fertilizer term was retained. Also, to reduce the extreme range of input levels and conditions, three classes of countries were formed: those where less than 25 percent of the area was irrigated (low irrigation), those where more than 80 percent of the area was irrigated (high irrigation), and those that were intermediate. Analysis of covariance (regression with dummy variables) was used to estimate constant terms and fertilizer response terms for each group separately. A number of variations on the equations were tried, with the final results shown in table 4.17.

Modern variety area and nonirrigated area were dropped from the final equation because their effects were not significantly different from zero, even at

Table 4.17. Estimated Coefficients (and corresponding *t*-values) of a Log-Log Production Function Fit to Rice Production Data for Thirteen Asian Countries, 1951–55, 1961–65, 1971–75

Irrigation class	Constant	Irrigation	Fertilizer, variety	Labor	Capital	R^2
Low	0.49	0.37	−0.16	0.49	0.05	—
	(5.8)		(3.2)			
Medium	0.49	0.37	0.09	0.49	0.05	0.97
	(5.2)	(5.5)	(3.3)	(8.9)	(2.7)	
High	−0.23	0.37	0.15	0.49	0.05	—
	(2.3)		(4.1)			

Note: The equation is: $\ln Q = a + b_1 \ln I + b_2 \ln F + b_3 \ln L + b_4 \ln C + b_5 D_M + b_6 D_H + b_7 D_M F + b_8 D_H F$ where Q, I, F, L, C stand for quantity produced, irrigation, fertilizer, labor, and capital, respectively. D_M is a dummy variable that takes the value 1 for countries in the medium irrigation group and zero otherwise. D_H is a similar dummy variable for countries in the high irrigation group.

the 90 percent level of confidence. However, this does not mean they had no effect. Modern variety area is highly correlated with fertilizer use (r = .76), therefore their separate effects cannot both be estimated. The effect of modern varieties is included in the fertilizer coefficient.

The results of this exercise are extremely interesting. The statistics indicate that the coefficients are estimated more closely and that the equation has a higher total explanatory power compared with alternatives without dummy variables. The measured productivity of fertilizer increases, as expected, as the proportion of irrigated rice land increases. The group of countries with less than 25 percent of land irrigated have a negative coefficient of fertilizer, but the level of fertilizer in those cases only ranges from 0 to 6 kg/ha so the basis for estimating the coefficient is very limited. The productivity of fertilizer with high irrigation is substantially higher than with medium irrigation.

Figure 4.2 illustrates the productivity of fertilizer for the three groups, low, medium, and high irrigation. The increase in productivity of fertilizer across the three groups is remarkable. The increase is only partly due to irrigation, even though the three groups were identified on the basis of percentage of rice land irrigated. All other factors affecting the productivity of fertilizer—modern varieties, labor, and capital—are higher on a per hectare basis for the high group and lower on a per hectare basis for the low group (table 4.18). This suggests that there is some additional factor, perhaps education or quality of irrigation, that has a part in raising rice output in those countries where it has increased, hence the functions in figure 4.2 are labeled according to technology level, not irrigation.

The growth in output accounted for by each of the factors was calculated by multiplying the change in the level of each input by its estimated coefficient. Table 4.19 shows the average proportion of increased output accounted for by each of the inputs in those countries that grew at a sustained rate. Countries whose output declined were not included in the averages shown in the table because the declining output figures complicate the interpretation.

Forty-four percent of the growth in output in the high irrigation countries was attributed to fertilizer and the associated change from traditional to modern varieties. Increases in irrigated area provided 25 percent of the growth. Growth in countries with moderate levels of irrigation was obtained about equally from additional irrigation, fertilizer, and labor. In both groups of countries, capital provided a relatively small source of growth. Subsequent chapters examine how each of the primary inputs directly affects rice production.

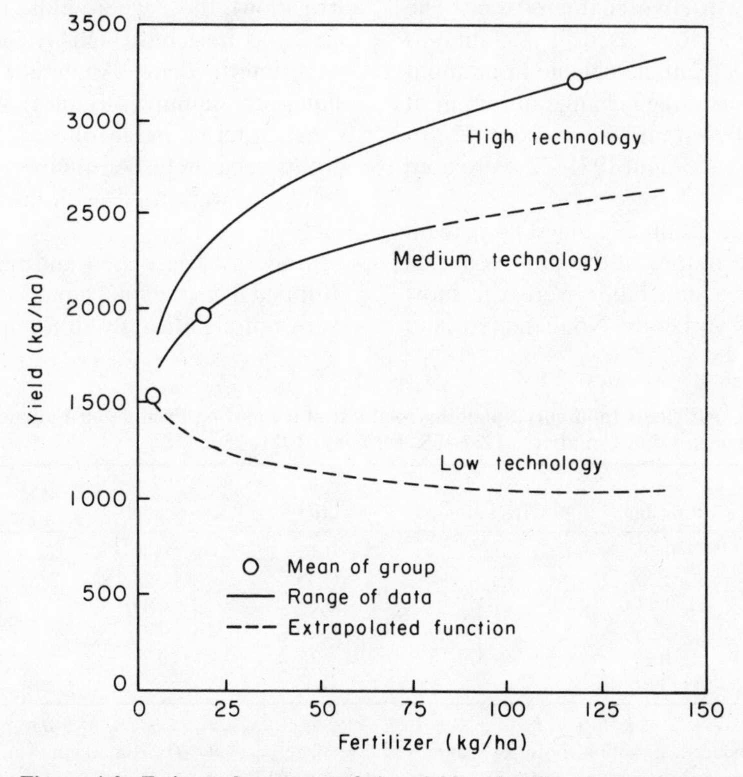

Figure 4.2. Estimated response of rice yield to fertilizer based on log-log production function fit to data for thirteen Asian countries, 1951–55, 1961–65, and 1971–75

Table 4.18. Average Values of Production Variables[a]

Irrigation class	Percent of rice irrigated	Percent of rice in MV	Fertilizer applied/ha	Labor force/ ha	Tractors/ thousand ha
Low	15.7	5.6	4	1.3	0.46
Medium	54.2	8.8	17	1.5	1.13
High	92.0	48.1	114	1.9	6.06

Note: MV = modern varieties.
[a]From the three groups of observations used in the rice production analysis, table 4.17 and figure 4.2.

Table 4.19. Average Contribution to Growth of Rice Output of Four Productive Factors, 1950–70[a]

Group of countries	Percent of output growth attributed to			
	Irrigation	Fertilizer variety	Labor	Capital
High irrigation[b]	26	44	12	16
Medium irrigation[c]	30	30	25	15

[a]As estimated from a cross-country production function in countries with sustained output growth.
[b]South Korea, Pakistan, Indonesia.
[c]Malaysia, Sri Lanka, Philippines, China, India.

Notes

1. The data used in this section are those reported by the U.S. Department of Agriculture (USDA) in various publications, mainly *Agricultural Statistics* and *Rice Situation*. Prior to 1936, the data in *Agricultural Statistics* were published in the *Yearbook of Agriculture.* These were collected by USDA representatives in various parts of the world and from publications of the International Institute of Agriculture, the predecessor to the Food and Agriculture Organization of the United Nations. We used these sources for the data series in this section that compare rice production and trade among continents.

2. For example, USDA reported that the 1930s and 1940s data for India cover approximately 95 percent of the total rice area.

3. Detailed explanation of the data sources and peculiarities for each country are in the appendix tables, which are printed separately and are available from Resources for the Future, Washington, D.C.

4. Food and Agriculture Organization of the United Nations, "Food Outlook 1981 Statistical Supplement," January 1982.

5. The contributions of area (A) and yield (Y) increases to output (Q) are calculated as follows: By definition, $Q = A\,Y$ and if Q is growing at the rate r, $Q_t = Q_0\,(1 + r)^t = A_0(1 + a)^t\,Y_0\,(1 + y)^t$. Taking logarithms of this equation and simplifying, then taking antilogs, one can write $r = a + y + ay$, that is, the rate of growth of output equals the rate of growth of area plus the rate of growth of yield, plus an interaction. The growth rates are calculated from the relationship: $(Q_t/Q_0)^{1/t} - 1$, where Q_t is the average during the second of two periods, Q_0 is the value during the first period, and t is the number of years between the two periods.

6. The attribution to irrigated and nonirrigated area follows a method similar to that used in computing the contribution of area and yield. Because the growth rates of irrigated and nonirrigated area are rates, and because they have different initial absolute values in most cases, they do not sum to the growth rate of total area. They do indicate the relative contribution of each type of land, however. Therefore each is divided by their sum, and the results in turn multiplied by the growth rate of total area to give the growth rate attributed to each.

7. The change in yield that would be obtained from the increased fertilizer used, at a 10:1 ratio, was computed and subtracted from the ending period yield. The growth rate of that adjusted yield over the initial period is the residual.

5

Varietal Improvement

A discussion of modern rice technology must begin with varietal improvement, since to a large extent the production potential of a variety sets an upper ceiling on rice production potential. The production ceiling is raised not only through the development of higher yielding varieties, but also by creating varieties with shorter growth time in order to increase the cropping intensity. The degree to which the potential of the new variety is realized will depend, of course, on a wide range of complementary research activities as well as other socioeconomic factors.

Farmers have been selecting for improved varieties since the beginning of agriculture. The formalization of this process in plant breeding research led initially to pure-line selection in which superior individuals within a population are identified and purified. The potential for increasing production through this method is limited, but the procedure is simple and quick.

The possibility of making desirable genetic combinations through crossbreeding or hybridization represented a significant advance.[1] The rediscovery of Mendel's findings in 1900 initiated a considerable amount of research on the hybridization of rice. The Japanese appear to have been the first to adopt hybridization in rice breeding, achieving a crossing in 1898.[2] By 1913, twenty varieties developed by hybridization were being grown by Japanese farmers.

For hybridization to be successful, it is necessary to identify the specific breeding objectives. With the exception of Japan, before World War II most of Asia lacked the professional expertise and technical skills to use this technology effectively. Even in Japan, the lag between the initial cross and final distribution of a new variety was twelve to thirteen years.[3] Pure-line selection remained popular in tropical Asia until after World War II when improved techniques in plant breeding greatly reduced the time required to develop varieties through hybridization. This factor, coupled with the extensive exchange of genetic materials and research information since the founding of the International Rice Research Institute (IRRI) in the early 1960s, resulted in rapid advances in production potential.

Research Prior to World War II

At the turn of the century, agricultural and economic development as well as natural and human resource endowments varied widely among the countries and regions of Asia. These differences have influenced the patterns of agricultural growth in this century.

An obvious contrast is between the temperate and tropical regions. In China and Japan, a growing population and a severely limited land base created a chronic problem of finding ways to increase agricultural production. In most of Tropical Asia on the other hand, rice production continued to grow because of migration and the development of new rice lands. The contrast between Japan and the rest of Asia is particularly sharp. Economic development was considerably more advanced in Japan than in either China or tropical Asia. Much has been written elsewhere about the emergence of Japan as an industrial nation.[4] Japan alone among the Asian countries had the technical skills needed to breed fertilizer-responsive varieties, the industrial capacity to support a fertilizer industry, and the transportation network needed to deliver chemical fertilizer to the farmers.

We can distinguish three patterns of growth in rice production during the first half of the twentieth century: (1) China's continuing emphasis on intensification of land use through multiple cropping and short-season varieties; (2) Japan's new emphasis on high yield and fertilizer-responsive varieties extended in the 1920s to its colonies, Korea and Taiwan; and (3) tropical Asia's continued dependency on the opening up of new rice lands. For Japan alone, the turn of the century marked a major turning point in terms of change in rice cultivation practices brought about through rice research. In discussing these differing patterns of growth, we begin with a brief history of Chinese varietal improvement.

Chinese Antecedents

The history of rice improvement in China extends back over several centuries. Since the beginning of rice cultivation probably less than 9,000 years ago, water control, cultural improvement, and varietal selection have been key elements in the process of development and improvement.[5] Around A.D. 1000, the introduction of rice varieties with short growth periods complemented irrigation development and permitted more intensive cultivation of the limited land area suitable for rice production.[6]

Available evidence suggests that early-maturing varieties were not commonly grown in China prior to the eleventh century. It was not until this time that land suitable for rice production in the Yangtze Valley became limited. The early-maturing Champa rices, introduced from Vietnam in the eleventh century, facilitated the more intensive use of land through multiple cropping. The expansion of rice production was accompanied by a rapid migration of population into South China.

The "plant breeders" of China were, for the most part, peasant farmers. However, the government, no less aware than today's Chinese leadership of the importance of adequate rice supplies for political stability, took an active role in the selection and dissemination of superior varieties. For example, as a result of drought conditions in the lower Yangtze and lower Huai river valleys during the 1011/12 crop year, Emperor Zheng Zong of the Song Dynasty ordered 30,000 bushels of Champa rice to be brought from Fujian Province and distributed to farmers in the drought-stricken areas, together with instructions for the proper method of cultivation.[7]

The process of selecting and disseminating improved rice seeds appears to have continued to the present century. There were no artificial barriers to the spread of improved seeds. Although hundreds of varieties were grown locally, a few particularly important varieties could be found over a wide area. At the same time, there does not appear to be any rational pattern of distribution, as for example, from highly developed provinces to less developed and more recently settled areas.[8] Although varieties undoubtedly were selected for a wide range of characteristics such as yield, quality, and drought tolerance, early maturity continued to be the most important characteristic not only of rice, but of other crops as well.[9]

The Japanese Initiative

The Meiji era (1868 to 1912) signaled the beginning of the modernization of agriculture in Japan. Japan initially sought to develop agriculture rapidly through the direct import of Anglo-American technologies. But Japanese agriculture was not compatible with American machinery, which was designed for large operating units, and efforts to transplant exotic plants and animals were equally unsuccessful.[10]

A reorientation of agricultural development strategy in 1890 gave primary emphasis in agricultural education and research to German chemistry and soil science in the von Liebig tradition. Following the establishment of the National Agricultural Experiment Station network in 1893, the national government provided subsidies to aid in the development of prefectural experiment stations. However, there was a shortage of both trained personnel and financial support for undertaking substantive research. For more than two decades, the fledgling research network had little, if any, impact on the development of new rice varieties.

One of the first trials undertaken by the National Experiment Station was a test of the response of nitrogen manuring on yield and lodging of native varieties gathered from various districts in Japan.[11] In Japanese rice varieties, as in German small grains, varietal differences had been found in the way yields responded to nitrogen. But the early breeding objectives in the development of pure-line and crossbred varieties emphasized cold tolerance and early maturity. In the 1920s, when chemical fertilizers became more widely used, emphasis was placed on production of strains resistant to lodging and to insects and diseases.[12]

In spite of the lack of research capacity, between

1900 and 1920 national rice production grew at 1.7 percent and rice yields at 1.2 percent per annum—approximately the same rate as population.[13] There are a number of factors that contributed to this yield growth. Meiji Japan inherited a relatively well-developed land infrastructure from the Tokugawa period (1600–1867). Almost all of the rice in Japan was irrigated in 1900, even though irrigation facilities were poor and lack of drainage was a serious problem in many cases. As a result, yields were already 2.8 mt/ha, a ton above the post–World War II level of most countries in South and Southeast Asia. However, a backlog of technical knowledge developed in some parts of Japan during the Tokugawa period. The breakup of the feudal system and the development of transportation during the Meiji restoration facilitated the diffusion of knowledge. Farmers organized into voluntary agricultural societies, and the advanced technology of the veteran rono farmers was widely disseminated. Until the 1920s, rono varieties were predominant. Some of these showed a good response to fertilizer.[14]

The development of modern manufacturing facilities in the second decade of the century led to a shift from Manchurian soybean cake to ammonium sulfate as the main source of nitrogen fertilizer. This shift, coupled with improvements in transportation, sharply lowered the hitherto prohibitively high price of commercial fertilizer. The establishment of cooperatives for handling inputs and credit gave further impetus to the use of fertilizer. In addition to investment in infrastructure, the government encouraged grass-roots agricultural organizations and established an inspection system to ensure that farmers received fertilizer of standard quality.

In summary, the accomplishments of this period (1868–1920) are not so much the result of access to new technology as the development of the capacity to disseminate the technology over a wide area. The difference between the present product of a sector and the product that could be realized if all resources were optimally used is sometimes referred to as "economic slack." Economic slack is increased when a region gains access to new technology. It is reduced when farmers and governmental and other organizations move toward a more efficient use of resources. The period of Japanese agricultural development extending from the Meiji restoration to the end of World War I can be seen as a period in which slack was reduced.[15]

The achievements of the Meiji period show that Japan had developed the capacity to rapidly disseminate and effectively use modern agricultural technology, but these achievements owed much to the feudal heritage of the Tokugawa period. Much of the technology that was diffused in the Meiji period had already taken root in some localities during this earlier period. Under feudal rule, the organizational capacity of the rural people was highly developed. Farmers learned to cooperate in the construction and maintenance of local irrigation systems. These historical antecedents appear to be highly significant in explaining a society's capacity to rapidly diffuse and effectively use new technology.

The interwar years were marked by stagnation in Japanese agriculture. Between 1920 and 1940, rice production rose by less than 1 percent per year, and rice yields showed no significant increase. The reasons for the lack of growth are not clear. Lack of price incentives and tenant–landlord disputes may have been contributing factors. The shift from "innovative landlords" to "parasitic landlords" progressed through the late Meiji era and into the Taisho era (1912–26).[16] The economic slack created by the backlog of technology from the Tokugawa period was exhausted. Despite the organization of breeding programs based on ecological conditions in 1927 and the gradual transition from pure-line to crossbred varieties, the yield potential of the new varieties did not increase significantly. Part of the explanation can be found in the selection of materials for crossbreeding. As late as 1963, nearly all of the leading varieties in Japan were developed from parent materials that were leading varieties since 1908. Akemine in 1958 suggested that the genetic constitution of rice populations in Japan, subject to natural and artificial selection for 2000 years, was "narrow and shallow."[17] In the interwar period, however, the import of this fact went unnoticed. It appeared that Japanese breeders had reached a yield ceiling at somewhere near 5 mt paddy/ha.

As stagnation set in, Japan turned to its colonies to obtain the additional rice supplies needed to meet domestic demand. In the early 1920s, paddy yields in Korea and Taiwan were 1.7 mt/ha, well below the 2.3 mt/ha yield in Japan in the early Meiji period. Between 1925 and 1940, rice yields in both Korea and Taiwan grew at 2.2 percent per annum. This rapid increase was caused by the combined effect of improvement and expansion of irrigation and drainage facilities and the introduction of fertilizer-responsive varieties.

Because the climate of Korea is similar to that of Japan, Japanese varieties were directly transferred. The South Korean branch of the Japanese National Agricultural Experiment Station was established in 1935 to assist in the selection of fertilizer-responsive, high-yielding varieties.

Taiwan, with its subtropical climate, posed more

of a problem. Japonica varieties were tested in Taiwan in the early 1900s, but gave lower yields than the native varieties except in the hilly areas. As late as 1924, the Taiwan Rice Production Committee recommended that Japanese varieties be grown only in the mountains to supply rice for local Japanese residents.[18] However, research initiated several years earlier showed that the yield of Japanese varieties could be substantially increased in the lowland paddies by lowering the seedling age for transplanting to 30 to 40 days for the first crop and 15 to 20 days for the second crop, or about half the time conventionally practiced for the native varieties. This discovery was accompanied by further technical improvements designed to adapt the Japanese varieties to subtropical conditions.

In 1926, the Japanese varieties adapted to Taiwan's conditions were officially designated as ponlai (heavenly rice), and the colonial administration undertook a campaign to spread these varieties and expand double cropping. Prior to 1926, varietal improvement mainly involved screening Japanese varieties. Varieties from southern China and Southeast Asia were also tested, but had lower yields than the native Taiwan varieties. Between 1931 and 1940, several new fertilizer-responsive varieties were selected from more than 5,000 crosses made mostly in Taiwan. The most important of these was Taichung 65, which was selected in 1927, and remained the most popular variety in Taiwan until 1959. Another important ponlai variety was Chianan 8, which was widely planted in southern Taiwan.

An important feature of the ponlai varieties, in addition to fertilizer-responsiveness and insensitivity to temperature and day length variations, is early maturity. This permitted a significant increase in multiple cropping. From 1911 to 1925, the multiple cropping index (number of crops/ha/year × 100) was fairly stationary at 118. It rose to 130 in 1945 and reached 190 in the mid-1960s.[19]

The rapid expansion of the new technology in Korea and Taiwan was aided by favorable prices, extensive development of irrigation facilities, and by the establishment of farmers' associations to promote the dissemination of new information and inputs. The colonial administration maintained a tight control over local farmer organizations, and the new government programs were energetically implemented, sometimes through the coercion of local farmers.

Tropical Asia in Limbo

In contrast to East Asia, rice yields in tropical South and Southeast Asia remained low and showed no upward trend from the turn of the century until after World War II. This pattern prevailed in all of the major rice growing countries of the region, which, with the exception of Nepal and the northern part of India, Pakistan, Bangladesh, and Burma, lie below the Tropic of Cancer. Nor did the experiment stations in the region develop varieties with high yield potential. In order to understand why this occurred, we need to examine not only the state of rice research during the period, but also the situation with respect to the complementary factors that encourage the development and dissemination of high-yielding varieties.

Population pressures were historically not as severe in tropical Asia as in East Asia, except in parts of India and on Java, because surplus land still existed. Burma, Thailand, and Indochina continued to expand rice exports to Europe and China by bringing new lands into cultivation, and most other areas were able to keep pace with domestic demand. India was an important exception. Between 1910 and 1938, the area in rice and wheat in India increased by only 3.5 and 4.2 percent respectively. But the area in inferior cereal grains grew more rapidly: sorghum by 110 percent, barley by 57 percent, pearl millet by 25 percent, and maize by 5 percent.[20]

Basic knowledge of the relationships between water control, nitrogen application, and high yield taught to the Japanese by the Germans was also understood by the expatriates who manned the various colonial research stations. But the colonial administrations' priorities in research were for export crops—sugar, rubber, tea, and cotton—rather than cereal grains. Until 1930, when the Imperial Council of Agricultural Research (the predecessor of the Indian Council of Agricultural Research) was established, there were only two full-time scientists in India working on rice, one at Dacca in Bengal, and one at Coimbatore in Madras.

The research effort was also very modest in the export crops, but in some instances spectacular results were obtained. For example, the stations in Coimbatore and in Java conducted research on sugarcane as well as rice. By 1930, the tri-hybrid sugarcane varieties developed at these two stations combined climatic adaptability with resistance to disease and were being grown commercially in every cane-producing area of the world. Between 1910 and 1930, sugarcane yields in Java increased by nearly 50 percent and production doubled.[21]

In contrast to sugarcane, the breeding work in rice initially took a rather different turn. In most countries, research started at a single station, and local varieties provided material for the selection work.

No critical tests were performed on the range of adaptability of these varieties.[22] The great diversity of rice varieties was believed to be caused by their narrow adaptability. The tendency was to gradually establish several stations, each in a specific ecological area and to select and recommend varieties for that region. Too many varieties were recommended to permit a practical program of seed multiplication.

Other conventional views of rice production held by scientists seemed to stand in the way of progress in varietal improvement. Through a long process of farmer selection, the japonicas had the inherent capacity to respond to the intensive cultivation practices pursued in East Asia. The indicas, on the other hand, were suited to the extensive cultivation practices of tropical Asia. Indica rice yields, although relatively low, could be sustained year after year without the addition of fertilizer.[23] Furthermore, the best results seemed to be obtained with photoperiod-sensitive varieties with long growth periods.

The use of fertilizer in monsoon Asia was negligible at this time, and, in the case of rice, there was ample evidence to prove that it was unprofitable.[24] Although there is no ready information on early fertilizer/rice price ratios, it is obvious that with a lack of supply and poor transportation facilities, ratios must have been very high. Normally, new varieties were not even screened for fertilizer responsiveness.[25] Crop residues and organic matter, widely used as compost in the Sino-Japanese sphere, were wasted or used for fuel in the rest of Asia.

Finally, irrigation and drainage, so highly complementary to the seed-fertilizer technology, were poorly developed throughout much of the region. Probably less than 25 percent of the total rice growing area was irrigated in India, and the percentage was even lower in most of the other rice growing regions of tropical Asia, with the exception of Indonesia. Irrigation in tropical Asia was also qualitatively inferior to that found in the Sino-Japanese sphere. Large canals and diversion dams were designed to do little more than provide insurance for the main season crop in case the monsoons failed. Drainage remained a critical problem, even in much of the irrigated area, and only a small portion of the irrigated area was double cropped.

A final deterrent to progress in research lay in the general lack of interaction between the research workers and the farmers. In Japan, farmers and farmers' organizations formed an integral part of the research extension network from the very beginning. By contrast, researchers and farmers in tropical Asia were (and to a large degree remain today) a world apart. The literature of the period is replete with references to "indolent," "lazy," "suspicious," and "conservative" peasants unable to comprehend new technology and unresponsive to economic incentives. Progress was made in plantation agriculture, but extending new technology to peasant smallholders was regarded by many as an insurmountable obstacle. Imparting the findings of research was viewed as a top-down exercise of educating the peasants, and there was no thought of consulting outstanding veteran farmers for research ideas. Literally and figuratively, those who conducted research and those who tilled the fields did not speak the same language.

The lack of progress in varietal improvement in tropical Asia can be attributed to several factors: (1) lack of financial support for research; (2) failure of research workers to focus on appropriate breeding strategies and objectives; (3) negligible use of fertilizer and unfavorable economic conditions for the expansion of fertilizer use in cereal grains; (4) poor water control; and (5) lack of high-yielding parents. Taking all of these factors together, there was little economic incentive to create high-yielding, fertilizer-responsive varieties.

It would be wrong to conclude, however, that there was no progress in tropical rice research during the first half of the nineteenth century. As in other parts of the world, knowledge of breeding techniques advanced tremendously, and statistical procedures were developed in the 1930s that improved the accuracy and reliability of varietal screening. There is an extensive Indian literature on increasing tropical rice production, but the Dutch in Indonesia seem to have made the most progress, although the full impact of their work on rice yields was not realized until well after World War II.

Breeding research in Java was started in 1905 with the establishment of the General Agricultural Research Station at Bogor (then named Buitenzorg). The indica variety Cina (meaning China, formerly spelled Tjina) was introduced into Indonesia in 1914 and spread over a substantial area because of its photoperiod insensitivity.[26] This undoubtedly boosted production in the area where rice was double-cropped. Six regional stations were established on Java between 1926 and 1945 to breed photoperiod-insensitive varieties that could be widely adopted. An intensive procedure of breeding and selection, including extensive testing on farmers' fields, produced successful results. Eight strains from the Cina (from China) × Latisail (from Bengal) cross became popularly known as the 40C selections. These were released on the eve of World War II and included several that became widely used as varieties and as parents.[27] Yields showed no progress from 1915 to 1940, but this may have been because rice area was being extended onto marginal lands.

World War II led to a disruption in rice production in many areas, but this was temporary. The more serious long-term consequences lay in the dismantling of the colonial research network as the developing countries gained independence after World War II. The colonial powers left no legacy of trained manpower and no "economic slack" to be exploited by the newly independent developing nations. The population explosion in the aftermath of World War II hastened the disappearance of the arable land frontier and, at midcentury, increased the demand for a high-yielding rice technology.

Achievements Since World War II

Over the past three decades, there has been rapid expansion in research activities designed to increase the level of rice production. A great deal has been learned, not only about the technical aspects of breeding rice varieties for higher yields, but also about developing research networks, transferring technology among and within countries, and extending technology to farmers.

The key objective of research with the indica rices was to increase the yield potential by developing varieties that would respond, as the japonicas did, to increased fertilizer application. Initial efforts met with limited success until scientists working independently in Taiwan and China, and subsequently at the International Rice Research Institute in the Philippines, began to focus on the development of a semidwarf type. The new semidwarf varieties, disseminated in the mid-1960s, had spread to a quarter of the rice growing area of Asia and to many parts of the rest of the world within a decade. These varieties performed best in areas with good water control. In the tropics, greater priority has subsequently been given to developing insect- and disease-resistant varieties, and varieties suitable for the unirrigated areas and less favorable environments that account for the other three-quarters of the tropical rice growing area. In the temperate zone, one of the most important breeding achievements in the 1970s has been the development of the F_1 hybrid rices in China.[28] The Chinese claimed that F_1 hybrid rices raise yields by 20 percent and are grown on about 15 percent of China's rice area. Yield potential has also been increased in Japan with the introduction of foreign genetic material and in Korea through successful crossbreeding of local japonica with semidwarf indica varieties from the International Rice Research Institute. In short, by the 1970s Asian scientists were using a wide range of alternatives to develop higher yielding varieties, and breeding objectives became increasingly dependent on environmental constraints in the particular location.

International Rice Hybridization Program

The first important international undertaking after World War II was the establishment in 1949 of the International Rice Commission (IRC) within the framework of the United Nations Food and Agricultural Organization (FAO).[29] It undertook several projects, such as categorizing and maintaining genetic stocks, japonica-indica hybridization, cooperative variety trials, and wide adaptability tests. Its primary aim was increased yield through selection and breeding.

The Indica-Japonica Hybridization Project was sponsored by the FAO from 1950 to 1957. Because of the lack of technical skills in many locations, crosses were made at the project headquarters, the Central Rice Research Institute (CRRI), Cuttack, India.

Varieties derived from the indica-japonica crosses, such as ADT 27, were distributed for commercial cultivation only in India and Malaysia. ADT 27 became popular in the Tanjore District of Tamil Nadu (Madras) because of its early maturity. In Malaysia, the varieties Malinja and Mahsuri were well suited to the second crop season in the irrigated areas. Mahsuri, a cross between the popular Taiwan ponlai, Taichung 65, and Mayang Ebos 80 and backcrossed to the latter, has subsequently spread into Andhra Pradesh and other parts of eastern India, into Bangladesh (where it is known as Pajam), and into upper Burma, becoming one of the most widely grown varieties in Asia during the 1970s.[30]

Despite the limited initial success in India and Malaysia, the achievements of the Indica-Japonica Hybridization Project were disappointing. The major shortcoming of the project lay in the failure of the scientists to identify the short-statured plant type that should have been the object of their search.[31] In 1958, the IRC member governments called for the establishment of an international rice research institute in the tropics, not only for achieving the identified objectives of rice breeding, but also for training personnel in different disciplines.

National programs were also producing higher yielding varieties of rice independent of the japonica-indica varieties of the FAO. Among those were H-4 from Sri Lanka and BPI-76 from the Philippines, both developed during the mid-1950s. What subsequently proved to be the most important achievement of this decade occurred in Taiwan.

Breeding for Semidwarf Indicas in Taiwan

The development of the ponlai japonica rice varieties under the Japanese colonial administration has been discussed earlier. Despite rapid dissemination of the ponlais, more than 40 percent of the rice growing area in Taiwan was still planted to the indica or *chai-lai* (native) varieties at the end of World War II. A program for pure-line selection of these varieties was initiated in the early 1950s. Hybridization among native and between native and other indica varieties began at the Taichung District Agricultural Experiment Station and at the Taiwan Agricultural Research Institute in the early 1950s. The semidwarf hybrid variety, Taichung Native 1, developed by the Taichung station in 1956 and officially released in 1960, was the first of the crossbred semidwarf indicas in Asia. The variety was selected from a cross between Dee-geo-woo-gen, a semidwarf, and Tsai-yuan-chung, a tall, disease-resistant local variety. Taichung Native 1 outyielded the best local indica varieties in most trials, and its yield compared favorably with that of superior ponlai varieties.[32]

The precise origins of semidwarfs Dee-geo-woo-gen and I-geo-tse are not known. It is generally assumed that the dwarf varieties came from China, probably Fujian Province, sometime before the Japanese occupation. The recent discovery that the semidwarf varieties from China and from Taiwan are allelic, with the same gene-controlling dwarfism in both sets of varieties, lends further credence to this theory. Although varieties such as Woo-gen in Taiwan and Nan-de in China were known before the turn of the century, there was nothing in the names of these varieties (Dee-geo, I-geo, Hsai-geo, or Ai-chiao meaning shortlegged or dwarf) to suggest that dwarf varieties were being grown extensively prior to this century. I-geo-woo-gen, a synonym for Dee-geo-woo-gen, was recorded by the Taiwan Agricultural Station in 1906, the first clear indication of the existence of a semidwarf variety.[33] Probably a spontaneous mutation or a natural crossing caused the dwarfing, and the plant was selected and propagated by some enterprising farmer, ultimately finding its way into the experiment station collection.[34] Later records show that between 1,000 and 5,000 hectares were planted to Dee-geo-woo-gen in 1939. The total area under the variety increased to 10,907 hectares in 1953.

Taichung Native 1, together with several of the leading ponlai varieties, was widely disseminated to other countries in tropical Asia and Africa in the early 1960s. In 1965, 6 tons of Taichung Native 1 (5 from Taiwan and 1 from the Philippines) were shipped to India, and by 1968/69, more than 1 million hectares were planted to this variety.[35] However, the major contribution of the Taiwan semidwarf indicas was the source of the dwarfing gene for the new varieties that were subsequently developed at the International Rice Research Institute and elsewhere in Asia.

China's High-Yield Technology

China's historical focus on cropping intensity through emphasis on water control and on early-maturing varieties continued after 1949. The objective of the new Chinese government was to maximize output per hectare per year under the assumptions: (1) that there was no opportunity for expanding land area; (2) that there was a tight restraint on liquid capital for purchase of inputs; and (3) that there was no constraint on the labor supply.[36] Between 1952 and 1957, the multiple cropping index rose from 167 to 187 in the South China rice growing region.[37] During the same period in Taiwan, it rose from 174 to 179, eventually reaching a peak of 189 during the mid-1960s.

During the 1950s, most new varieties were developed through pure-line selection, with only a handful being developed by hybridization (table 5.1). Pure-line selection was apparently preferred because it required a minimum time to achieve the quick results demanded by the political system. Of the ninety-five improved rice varieties distributed in South China before 1959, only 20 percent were the product of post-1949 breeding by professionals.[38] In a country the size of China, the capacity to effectively implement a hybridization program and to develop and disseminate new crosses or foreign materials requires an extensive network of cooperating experiment stations. This network was not established until the late 1950s.[39]

Guangdong Province led in research on yield improvement. Chinese scientists, like their professional colleagues in the International Hybridization Project, initially failed to identify the appropriate plant type for fertilizer response. Their approach in choosing parent stock was to select strains that had tall and strong stalks, long panicles, and large grains. Observations of the lodging resistance of a glutinous dwarf variety from Guangxi, and the practices that some farmers followed of deliberately stunting plant growth through water control and timing of fertilizer application, seem to have convinced the breeders of the importance of short-statured rice plants.[40]

At the Academy of Agricultural Sciences in Guangdong Province, Ai-zai-zhan-4 was crossed with Guang-chang-13 in 1956, and a new semidwarf Chinese variety, Guang-chang-ai, was released in 1959.[41] Thus, Guang-chang-ai was the first short-statured, high-

Table 5.1. Number of Improved Varieties Introduced in China, 1949–60

Location	Foreign varieties	Local varieties	Systematic selection	Hybridization	Total
South China	3	11	12	2	28
Central and East China	1	16	18	2	37
North and Northwest China	1	1	3	0	5
The Southwest Plateau	0	11	5	0	16

Source: Zhongguo Nongye Kexue (Chinese Agricultural Science) no. 8 (1961) p. 2; also cited in Kang Chao (1970) p. 173.

yielding indica variety successfully developed by crossbreeding. Ai-zai-zhan, one of the parents of Guang-chang-ai, has become the major dwarfing source of most of the important varieties bred in China.[42]

The large-scale dissemination of the new semi-dwarfs began with the planting of 1 million hectares in 1964. By 1965, a year prior to the release of the first IRRI variety, at least 3.3 million hectares were planted principally in Guangdong, Jiangsu, Hunan, and Fujian provinces. In 1973, 6.7 million hectares or 20 percent of the rice area in those regions with good water control benefited from the new varieties.

One of the most significant technical achievements in rice breeding in the 1970s was the commercial development of F_1 hybrids in China, although there is still not enough information to fully assess the total impact of this breakthrough on rice production. Rice and wheat, unlike maize and sorghum, are self-pollinating plants. As a result, the commercial production of F_1 hybrid seed is very difficult. The process involves: (1) locating a cytoplasmic male-sterile female parent plant; (2) crossing it with a maintainer line to produce offspring with sterility but with desirable genetic traits; and (3) crossing these seeds with a "restorer" line to produce F_1 seeds with normal self-fertilizing power. If successful, the process can offer two important advantages, higher yields as a result of heterosis or "hybrid vigor," and greater facility in combining important dominant genes for resistance to disease and insects.

In the autumn of 1970, a male-sterile wild rice plant (wild aborted) was found on Hainan Island, which led to the breakthrough in the breeding of hybrid rice.[43] In 1971, a hybrid rice breeding program was initiated in Hunan Province, apparently with strong political support. In searching for the best "restorer" lines, the Hunan group found that seeds from Southeast Asia (principally the IRRI varieties IR24 and IR26) were superior.[44] By 1974, the first demonstrations of hybrid rice were grown with good success. Since then, the hectares planted to hybrid rice have increased rapidly, reaching about 2.1 million in 1977 and 4.7 million (13 percent of the total) in 1978. The speed with which the new hybrid rices have been disseminated represents a significant technical and organizational achievement.

The temperate zone environment presents an obstacle to rapid development of varieties. Using conventional breeding methods, more than a decade is required from the first cross to wide dissemination. However, by growing two to three crops a year, including one crop in the winter on subtropical Hainan Island, it is possible to complete in three or four years the six or seven generations needed to stabilize the genetic characteristics of the variety.[45] Since 1966, the Chinese have successfully experimented with haploid breeding through "anther culture," a means of achieving stability in the varietal characteristics or homozygosity in one generation instead of six or seven, saving land, labor, and time.[46]

Organizationally in China, the four-level research network—county, commune, brigade, and production team—provides a mechanism for rapid evaluation of new varieties and seed multiplication. As noted previously, there is often pressure brought to bear through the political system to promote certain technologies, and this is one reason that it is difficult to judge the likely impact of the F_1 hybrid technology on rice production in China.[47] While the yield of the new hybrids is said to be 20 percent or close to 1 ton per hectare higher than the conventional semidwarfs, the Chinese readily admit that there are problems with the technology. Producing the F_1 hybrid seeds requires a considerable amount of land and labor. The seed fields, for example, produced one-tenth the seed of a regular field. Furthermore, the growth time of the existing hybrids is rather long (135 days), and hence hybrids pose a problem in those areas that normally produce two crops of rice each year.

Experiments conducted at the International Rice Research Institute in 1980 and 1981 show that some of the F_1 hybrids can produce yields 20 percent or more above the best standard commercial varieties, supporting Chinese findings.[48] However, most of the other Asian developing countries lack the organizational capacity to rapidly multiply and distribute hybrid seeds annually.

IRRI and the New Plant Type

The idea of an international center for rice research in the tropics germinated in the 1950s from different sources. Exploratory trips to Asia were made by officers of the Rockefeller Foundation in 1952–53. Discussions with government officials revealed a definite interest in the idea, but no formal steps were taken until 1958. On August 18th of that year, the issue was raised rather informally at a joint meeting of several officers of the Ford and Rockefeller foundations. The idea took root and developed into a cooperative agreement between the two foundations to establish the International Rice Research Institute. By 1958 they decided to locate the new center in the Philippines, with its formal foundation in 1960. It began operations in early 1962.

Research on varietal improvement in tropical Asia over the decades of the 1960s and 1970s can be divided into three phases: (1) development of fertilizer-responsive varieties; (2) improvement of insect and disease resistance and food quality of the new varieties; and (3) development of varieties suited to intensifying existing cropping patterns and to less favorable environmental conditions.

The initial breeding objective of IRRI was to create a plant type that would be resistant to lodging and would make efficient use of solar energy and fertilizer to achieve high yields.[49] The ideal modern plant type as opposed to the traditional was believed to have the characteristics shown in the table below.

Most discussions of plant type concentrate on physical features. Short, stiff culms (stalks) and erect leaves are regarded as essential to prevent lodging and achieve high fertilizer response. Although not strictly features of the plant type, nonphotoperiod sensitivity and short growth duration must be considered critical elements of the modern variety. Nonphotoperiod sensitivity increases the flexibility in planting and harvesting dates and makes it possible to extend the variety over wide regions with varying day length. In general, the shorter duration varieties are those with high-yielding ability and nitrogen responsiveness.[50] There is also a relationship between growth duration and plant height, the short-statured varieties being those with a short growth duration.[51] Plant height at flowering is inversely related to yielding ability and nitrogen response. The contribution of shorter growth duration to crop production has been largely ignored in the literature of the Green Revolution, perhaps in part because of the difficulty in obtaining a quantitative estimate of its impact, which has largely been felt through the expansion of multiple cropping.

In late 1966, just four years after research began at IRRI, the first of the IRRI varieties was released. IR8 was a selection made from one of the initial crosses in 1962 between the Taiwan semidwarf, Dee-geo-woo-gen, and the Indonesian variety Peta, one of the 40C selections released on the eve of World War II, but widely grown in the Philippines in the mid-1960s. Neither the founders of IRRI nor the researchers who began work at IRRI in 1962 would have been so optimistic as to suggest that a major breakthrough could be achieved in four years. Also, it would be wrong to assume that this achievement was due to IRRI alone. The new institute served to catalyze and accelerate the process of varietal improvement that was already well under way in tropical Asia in the early 1960s.

The new varieties spread rapidly (table 5.2). The enthusiasm that accompanied their initial release led many to assume that it was just a matter of time before the new seed-fertilizer technology would be disseminated to all parts of Asia. However, it was soon evident that the adoption of the new technology

Characteristic	*Traditional plant type*	*Modern plant type*	*Advantage of modern over traditional*
Stature	Tall: 150–200 cm	Short: 90–110 cm	Reduced lodging
Stem	Tall and weak	Short and sturdy	Reduced lodging
Leaves	Drooping	Erect and narrow	Increased efficiency in light use
Tillering	High	High	
Grain to straw ratio	1.5:1	1:1	Higher partitioning of dry matter as grain
Photoperiod sensitivity	Sensitive	Nonsensitive	Flexibility in planting date and location
Growth duration	Long: 150–200 days	Medium: 125–130 days	Increased output per hectare per day

Table 5.2. Estimated Percentage of Total Rice Area Planted to Modern Varieties in Selected Asian Countries

Region & country	1965/66	1966/67	1967/68	1968/69	1969/70	1970/71	1971/72	1972/73	1973/74	1974/75	1975/76	1976/77	1977/78	1978/79	1979/80
East and Southeast Asia															
Burma	0.0	0.0	0.0	3.5	3.1	4.0	3.9	4.4	5.0	6.3	6.4	6.2	5.7	5.8	—
Indonesia	0.0	0.0	0.0	2.5	10.4	11.1	16.0	24.2	37.3	39.7	44.5	48.4	53.3	55.8	60.6
South Korea	0.0	0.0	0.0	0.0	0.0	0.0	0.2	15.7	11.8	25.5	22.5	43.9	53.7	75.5	60.4
West Malaysia	10.3	15.4	23.1	20.9	26.5	31.4	37.1	38.1	36.7	35.7	37.4	56.0	52.8	—	—
Philippines	0.0	2.7	21.2	40.6	43.5	50.3	56.3	54.0	63.3	61.5	64.3	68.1	70.0	72.4	75.2
Thailand	0.0	0.0	0.0	0.0	0.1	0.4	1.4	4.1	5.0	5.5	7.1	11.3	11.2	11.8	8.8
South Asia															
Bangladesh	0.0	0.0	0.6	1.6	2.6	4.6	6.7	11.1	15.7	14.9	15.0	13.4	15.1	17.5	19.9
India	0.0	2.5	4.9	7.3	11.3	14.5	19.1	22.1	25.4	28.5	32.3	35.7	35.8	43.8	—
Nepal	0.0	0.0	0.0	3.7	4.2	5.7	6.3	16.1	16.7	18.0	17.2	17.5	23.0	24.7	29.4
Pakistan	0.0	0.0	0.3	19.8	30.9	36.6	50.0	43.7	42.1	39.3	38.9	38.8	44.9	50.1	—
Sri Lanka	0.0	0.0	0.0	1.1	4.2	4.3	10.2	36.3	54.8	51.7	50.9	62.5	74.4	68.0	71.8

Source: A. C. Palacpac (1982). 1965/66 to 1975/76 principally from D. G. Dalrymple (1978).

was closely associated with good water control.[52] Unirrigated areas account for over two-thirds of the total rice growing area in Asia, and even today, modern varieties are seldom found in these areas.

However, breeding for semidwarf plant type became increasingly popular throughout the rice growing world in the period from the mid-1960s to the mid-1970s. According to a recent survey of fourteen Asian experiment stations in seven countries (Bangladesh, India, Indonesia, South Korea, Philippines, Sri Lanka, and Thailand), by 1974/75, 84 percent of all crosses made contained a semidwarf parent, and 58 percent of all parents were semidwarf (table 5.3).[53] Among the thirty-six newest varieties released by twenty-seven agricultural experiment stations in ten Asian countries (the seven previously mentioned plus Iran, Nepal, and Pakistan) in 1975, 69 percent were semidwarfs. More than 90 percent of the new releases were developed by hybridization and less than 3 percent by pure-line selection.

Table 5.3. Percentages of Rices of Different Plant Heights Used as Parents in Crosses at Fourteen Agricultural Experiment Stations and Universities in Seven Asian Nations, 1965–75

Plant height	Rices used 1965–67[a]	1970–71[b]	1974–75[c]
	In crosses		
Tall	74	57	45
Intermediate	51	39	35
Semidwarf	61	86	84
Floating or deepwater	2	1	2
	As individual parents		
Tall	40	30	24
Intermediate	31	22	17
Semidwarf	28	48	58
Floating or deepwater	1	—	1

Source: Thomas R. Hargrove (1978).
[a]277 rice varieties and lines used in 119 crosses.
[b]351 rices used in 147 crosses.
[c]191 rices used in 89 crosses.

The principal reason for using intermediate and tall varieties in the crosses was to improve grain quality and pest resistance. Breeders in national programs tended to use local materials as donors of desired grain quality and growth duration. However, IRRI materials were often used as donors of insect and disease resistance.[54]

Increasing productivity per unit area, accompanied by modern production practices, tends to create ideal conditions for the development of diseases and harmful insects. The problem is particularly acute in the monoculture areas, where two or more rice crops are planted in succession. Therefore, if the high yield potential of the modern varieties is to be realized, a high degree of resistance to different plant pests must be incorporated into the new varieties. Breeding for resistance is, of course, not the only method of control, but for a farm population with limited capital resources and little knowledge about the proper method of applying insecticides, this method offers a distinct advantage.

Insect and disease resistance have been major objectives of the IRRI breeding program from the beginning. Over time, changes in the nature and scope of insect and disease problems have allowed no opportunity for a slackening of effort. Bacterial leaf blight and the rice stem borer were considered the major sources of damage in the 1960s. However, an outbreak of tungro virus in the Philippines in 1971 resulted in a loss of one-third of the Central Luzon rice crop and directed attention to this new problem. Subsequently, varieties resistant to both tungro virus and the green leafhopper, which transmits the virus, were developed and released.

Until 1973, the brown planthopper was considered a minor insect throughout most of Asia. Since then, there have been serious outbreaks of brown planthopper damage or "hopperburn" in the Philippines, Indonesia, India, and other parts of Asia. As in the case of tungro virus, plant breeders quickly bred a new variety, IR26, which was resistant to the insect. By 1975, IR26 was an important variety in the Philippines. All seemed well, but soon unbelievable reports began coming in from the Solomon Islands and from Kerala State in India stating that IR26 completely lacked resistance to the brown planthopper.[55] It was subsequently discovered that a different biotype of brown planthopper existed in these areas.[56] The planthoppers were physically identical to those to which IR26 was resistant but physiologically (biochemically) capable of feeding on IR26. The two types were named biotype 1 and biotype 2. In Indonesia, biotype 2 quickly emerged from the natural mix of insects and spread rapidly, causing extensive damage. No varieties were available to farmers in 1976–77 that were resistant to biotype 2. Over 1 million hectares were damaged, with a yield loss estimated at 1 million metric tons. By the following year, 1.6 million hectares were planted to varieties resistant to the new biotype, and, by 1979–80, nearly 3.9 out of Indonesia's 7.9 million hectares of rice were planted to resistant new varieties.

The above account illustrates the complexity of the insect and disease problem. In such a situation, the current solution is to develop and multiply as rapidly as possible varieties with different sources of resistance and have them in reserve, ready to place into the hands of farmers. In some cases, the effective resistance may be short lived in a dynamic insect and disease environment. A more viable, long-term solution would be to develop varieties with more than one source of at least moderate resistance, the strategy being not to eliminate the brown planthopper and the related virus diseases transmitted by the hopper, but rather to keep the insect and disease problem below epidemic levels.[57] Some scientists argue that "rotation" of highly resistant varieties is a better alternative, but this requires a well-developed seed production infrastructure.

As the new varieties of rice disseminated throughout tropical Asia in the 1960s, it became increasingly evident, even to the most enthusiastic advocates, that their adoption was unlikely to spread beyond that one-quarter to one-third of the area that had reasonably good water control.

Breeding for tolerance to unfavorable environmental conditions offers one possible alternative for mitigating the effect of these conditions on yield. While there has been increased emphasis on the development of varieties tolerant to drought or flooding, to temperature extremes, and to certain adverse soil conditions, research in this area remains very much in the formative stage. Environmental variability increases the complexity of the breeding problem and offers an obstacle to the transferability of the technology. In this situation, efficient research systems should be decentralized and stress placed on the development of regionally specific technology.[58] Diffusion of technology *per se* may tend to be low in such situations, although diffusion of scientific knowledge can be quite significant.

One problem, of course, is that the national breeding programs in many Asian countries have as yet very limited capacity for development of regionally specific technology. As our previous discussion has suggested, yield potential and the associated fertilizer responsiveness and lodging resistance have been the dominant breeding objectives at most Asian experiment stations in the 1970s.[59] Breeding for tolerance to adverse environmental conditions, particularly flood and drought, has received little attention, despite their obvious importance in many locations. Furthermore, in most instances, the experiment station environment reflects the most favorable conditions, rather than those of the majority of farmers served by the station. Over the past several years, however, significant research has been undertaken in Bangladesh, eastern India, and Thailand in developing varieties designed to perform better under flooding or deeper than normal water conditions.[60]

The first of the tropical semidwarfs had a growth duration of 120 to 130 days from transplanting to harvest compared with 150 days or more for traditional indicas. The development of rice varieties that mature in 100 days without a significant loss in crop yield undoubtedly has been one of the most significant breeding achievements in the 1970s. In the more favorable environments with intensive crop production, reducing the maturity date by three weeks or more can permit further intensification of the cropping pattern. In the less favorable environments, earlier ripening makes it possible to escape the adverse effects of drought, flood, typhoons, or cold weather. Photoperiod-insensitive varieties with early maturity can be readily disseminated across a wide range of environments, and it seems that their impact has not yet been fully exploited in tropical Asia.[61] Research on varietal improvement at IRRI is summarized in table 5.4, which lists several of the important IRRI varieties and their characteristics. Peta, a tall indica parent of IR8, is also included in the list.

Table 5.4. Trends in Plant Characteristics for Selected IRRI Varieties and for Peta, a Tall Parent of IR8 (1966–77)

Variety	Year released	Plant height (cm)	Maturity (days)	Yield (mt/ha)[a] Wet	Dry	Remarks
Peta	1940	155	150	3.7	4.3	Parent of IR8
IR8	1966	100	130	4.3	7.0	First IRRI semidwarf bred for fertilizer response
IR20	1969	110	125	4.9	6.1	Bred for insect and disease resistance
IR26	1973	105	125	4.8	7.0	Bred for improved insect and disease resistance
IR36	1976	85	110	5.0	6.1	Early maturing, broad spectrum of pest resistance
IR42	1976	105	130	5.3	6.7	High yield with low inputs in the wet season

Source: IRRI, *Annual Reports,* various years.
[a]From experiments on nitrogen response at Maligaya Rice Research and Training Center and Bicol Rice Corn Experiment Station, Philippines, 1976.

The Spread of Modern Varieties

As indicated in table 5.2, modern semidwarf varieties spread rapidly to many of the developing countries of Asia. Documentation of that spread is somewhat obscured by differences in the definitions used for various classes of varieties.[62] The initial IRRI varieties, such as IR8 and IR20, were semidwarf, but some more recently developed varieties are intermediate in height. One of the objectives of plant breeders in the early 1960s was to design varieties responsive to fertilizer and not sensitive to day length. Most of the first generation new varieties had these characteristics, but fertilizer response is a much more variable character than plant height and hence is not an easy characteristic by which to classify varieties. In some countries, the success of research, extension, or other programs is judged by the area of new varieties, leading to tendencies to overstate the area. This section reviews the available information on the spread of modern varieties.

Classification of varieties raises questions in several countries that had already developed improved "modern" varieties before IRRI introduced semidwarfs. In Sri Lanka, the "H" series of varieties was developed by the Sri Lanka Department of Agriculture and released during the late 1950s, well before the semidwarf, nonphotoperiod varieties. Some reports do not include the H series as modern varieties. Malaysia released similar improved varieties in the 1950s that are also reported as modern varieties. The general term modern variety is used here to include semidwarf and intermediate stature, fertilizer-responsive, mainly nonphotoperiod-sensitive rice varieties developed since the 1960s.

The Spread of Modern Varieties

Table 5.5 shows the area planted to modern varieties in South Korea and in five Southeast Asian countries. Rather steady growth in the area planted is shown in most countries. Korea and Thailand show sharp downturns in 1979/80. In Thailand, this can be traced to the extreme drought that year, which reduced the area planted to the dry-season crop, the season in which modern varieties (MVs) are widely grown.

South Korea released a series of highly fertilizer-responsive varieties in the early 1970s. By 1977, over half of that country's rice area was planted to such types. By the mid-1960s, Malaysia had a significant fraction of its area devoted to modern varieties developed from research conducted within Malaysia. By 1976, those and other modern varieties covered over half of Malaysia's rice area. The Philippines and Indonesia adopted IRRI varieties at a rapid pace—surpassing 50 percent of the Philippines' rice area by 1970 and reaching 50 percent of Indonesia's rice area by 1976. Adoption in Thailand and Burma has lagged behind other countries of the region, with less than 10 percent of the total rice area in MVs in 1979. However, Burma's special production program sharply accelerated the adoption of MVs in subsequent years.

Table 5.6 shows the area planted to modern varieties in South Asia. These varieties have spread at a rather steady rate through India, increasing to nearly 50 percent of the total rice area by 1980. They were more rapidly adopted in Pakistan, covering 50 percent of the total area within five years of their introduction. However, since then, their use has fluctuated at about that level or slightly less. In 1965, Sri Lanka had over half its area planted to H varieties. The

Table 5.5. Area Planted to Modern Rice Varieties, South Korea and Southeast Asian Countries
(thousand ha)

Year	South Korea	West Malaysia	Thailand	Philippines	Indonesia	Burma
1965/66	0	43	0	0	0	0
1966/67	0	63	0	83	0	0
1967/68	0	91	0	702	0	3
1968/69	0	96	0	1,012	198	167
1969/70	0	132	3	1,360	831	143
1970/71	0	165	30	1,565	903	191
1971/72	3	197	100	1,827	1,323	185
1972/73	186	212	300	1,680	1,914	199
1973/74	121	217	400	2,177	3,135	246
1974/75	307	213	450	2,175	3,387	328
1975/76	274	222	600	2,300	n.a.	407
1976/77	533	318	960	2,417	4,049	450
1977/78	660	316	960	2,457	4,454	496
1978/79	929	—	1,100	2,512	4,982	651
1979/80	744	—	800	2,708	5,366	948
1980/81	604	—	—	2,678	5,416	1,502

Sources: All countries except Burma—R.W. Herdt and C. Capule (1983). Burma—A. Palacpac (1982).

introduced MVs added to this area during the early 1970s, and the new "BG" varieties developed within Sri Lanka were rapidly adopted in the late 1970s, so that by 1978, nearly the entire rice area was planted to MVs. Adoption was slower in Bangladesh, with a substantial concentration during the *boro* (summer) season. The MVs reached about 15 percent by 1973 and stabilized at about that level since that time. In Nepal, MVs were adopted later but have continued to spread and by 1980 covered nearly 30 percent of the total rice area.

The Dynamics of Varietal Change

The data reviewed above give some indication of the speed with which various countries adopted modern varieties. The data, however, fail to convey an adequate appreciation of how dynamic the process really is. Contrary to the stereotypic image of peasant farmers resisting change, many Asian farmers have been extremely anxious and willing to experiment with new varieties as they are made available. They seem to be motivated both by a desire to gain the

Table 5.6. Area of Rice Planted to Modern Varieties, South Asia
(thousand ha)

Year	India	Pakistan	Sri Lanka (1)	Sri Lanka (2)	Bangladesh	Nepal
1965/66	0	0	0	292	0	—
1966/67	888	0	0	366	0	—
1967/68	1,785	4	0	389	63	—
1968/69	2,681	308	10	448	152	43
1969/70	4,253	501	31	478	264	50
1970/71	5,454	550	74	498	406	68
1971/72	7,199	729	119	430	624	82
1972/73	8,607	647	251	437	1,065	177
1973/74	9,718	637	396	559	1,549	205
1974/75	10,780	631	293	450	1,444	223
1975/76	12,742	665	300	487	1,552	216
1976/77	13,731	678	437	605	1,280	220
1977/78	15,516	852	496	643	1,204	291
1978/79	17,619	1,015	491	611	1,373	313
1979/80	—	—	562	704	1,998	315
1980/81	—	—	612	748	2,194	326

Sources: R.W. Herdt and C. Capule (1983). Time series (1) for Sri Lanka is from Dalrymple (1978) up to 1975/76, thereafter from government sources. Time series (2) comes from government of Sri Lanka data and includes the H series of varieties as well as the IR and BG varieties included in (1).

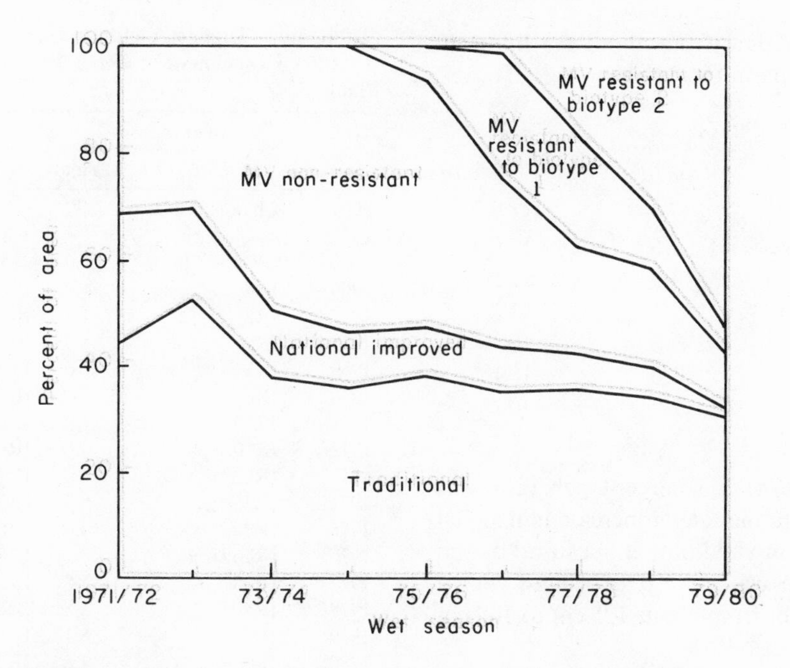

Figure 5.1. Percent rice area by varietal type, Indonesia, wet season, 1971/72–1979/80

possible advantages of newer varieties, and by a desire to avoid the deficiencies of the previous "new" varieties.

For example, a study of rice production in Central Luzon, Philippines, showed that no farmers had yet used the new varieties in the wet season of 1966, a year after their release. By 1970, 64 percent of the area owned by the same farmers was planted to MVs, with IR5 the most common variety. After the tungro virus epidemic of 1971–72, IR20 became the most popular variety, and by 1979, IR36 was the most popular variety. Furthermore, only one of the four most frequently planted varieties in 1979 had been developed and released for farm cultivation in 1974.[63]

Another example is illustrated in data from Indonesia shown in figure 5.1. MVs were grown on a significant portion of Indonesia's area in 1973/74. Government efforts to increase production by encouraging improvement of irrigation systems to expand the area planted to two or more crops of rice per year created conditions ideal for the development of serious infestations of brown planthopper (BPH) by 1973 and 1974. "All the varieties grown in Indonesia at this time (traditional, national improved, and modern) were susceptible to this pest."[64] Rices resistant to BPH were obtained from IRRI and released to farmers after evaluations by national scientists. Wide-scale planting of these varieties showed that while they were resistant to certain biotypes of the BPH, they were susceptible to other biotypes, so new varieties were developed and released.[65] As shown in figure 5.1, each new generation of varieties was rapidly adopted by farmers, replacing the previous generation, and, in the process, belying the idea that farmers were unwilling to adopt new technologies.

A similar dynamic picture exists for Sri Lanka. In 1964, about 30 percent of the country's total area was planted to H series varieties, with the rest in traditional varieties. IR varieties were introduced in 1968/69 and by 1970/71 were planted on 30,000 ha. However in 1970/71, BG varieties were released. Developed within Sri Lanka, these varieties proved to be very popular, replacing both the recently introduced IR varieties, and the older H varieties, to cover 55 percent of the total rice area in 1973/74—only four years after their introduction.

The Search for Higher Yield

Rice yields in the tropics remained stagnant until the conclusion of World War II. Then, almost overnight, the introduction of the semidwarfs established, under ideal conditions, a yield potential for tropical rice that was comparable to yields achieved as a result of more than a half century of steady progress in the temperate zone.

An unusual experiment conducted in Japan in 1967 compared the nitrogen response of several Japanese varieties released over a period of more than 60 years.[66] Rice yield potential, as reflected in maximum yields from these fertilizer experiments, rose steadily in the temperate zone of Japan, from about 3.5 mt/ha in 1900 to over 7.5 mt/ha in the 1960s.

Year of release	*Variety*	*Maximum yield (mt/ha)*	*Plant height (cm)*
1900	Bozu 6	5.6	123
1905	Akage	3.9	122
1935	Fukoku	7.1	110
1940	Eiku	6.8	107
1959	Mimisari	7.6	100
1962	Yukara	7.9	103

This increase rate of 1.3 percent per year was approximately equal to the rate of increase in national yields. The increase in yield potential caused by the response to nitrogen was closely associated with a decline in plant height from about 125 cm to 100 cm over the sixty year period.

There has been further progress in temperate zone rice potential in the 1970s, most notably the development of the previously mentioned Chinese hybrids and the development of indica-japonica crosses in South Korea.[67] The new Korean rices, shorter in stature than the local japonicas, with more grains per panicle, sturdier stems, and heavier tillering capacity, have 20 to 40 percent higher yields. In 1976, the national yield average of South Korea reached 6 mt/ha, surpassing average Japanese yields for the first time. However, subsequent problems have been encountered because of the lack of cold tolerance and blast resistance in these new varieties.

The speed with which yields can improve is clearly seen by progress made at the Central Rice Research Institute in Cuttack, India. Until 1964, the best results of the maximum potential yield trials conducted on about 2 ha of land remained fairly constant and never reached 4 mt/ha. In 1968, the maximum yield using IR8 in these trials was 5–6 mt/ha. The yields for the entire research station at Cuttack between 1966 and 1969 were as follows:[68]

Year	*Wet*	*Dry*
1966	1.4	1.4
1967	2.4	3.1
1968	1.9	3.8
1969	2.5	4.3

The figures also reveal the sharp contrast between yield potential in the wet season (with low solar energy) and the dry season, even under experiment station conditions.

Table 5.7. Highest Yielding Variety in a Long-term Fertility Experiment at IRRI, 1964–83
(mt/ha)

	Wet season		Dry Season	
Year	Variety	Yield	Variety	Yield
1964	Chainung 242	4.2	—	—
1965	Taichung Native I	7.0	Taichung Native I	7.6
1966	IR8	4.9	IR8	9.4
1967	IR8	4.8	IR8	8.4
1968	IR8	4.9	IR8	9.9
1969	IR8	5.6	IR8	9.0
1970	IR8	5.5	IR8	8.1
1971	IR20	4.9	IR8	8.4
1972	IR20	6.0	IR8	8.0
1973	IR22	4.0	IR20	7.6
1974	IR26	5.1	IR26	9.3
1975	IR26	6.0	IR8	8.0
1976	IR26/IR36	5.1	IR8	8.8
1977	IR36	5.8	IR36	7.4
1978	IR8/IR42	3.8	IR42	6.7
1979	IR36	5.1	IR8	7.3
1980	IR42	4.8	IR42	7.4
1981	IR42	5.5	IR42	7.0
1982	IR36	4.6	IR42	7.1
1983	IR36	4.4	IR8	7.8

Source: IRRI, *Annual Reports.*

At IRRI, maximum yield was determined on the basis of 154 fertilizer trials conducted by the agronomy department between 1966 and 1972. A mean maximum yield of 4.8 mt/ha was attained in the wet season with 75 kilograms of nitrogen per hectare (N/ha), and a mean maximum yield of 6.6 mt/ha was obtained in the dry season with 12 kilograms of N/ha.

Under the favorable dry season environment, there appears to have been no gain in yield potential since the release of the first semidwarfs in 1966, as shown by the long-term fertility experiments at IRRI (table 5.7).[69] IR8 has been included in all experiments, along with two other leading varieties selected each year. In 1983, IR8 obtained the highest yield in the dry season. IR8 and a closely related Indian variety, Jaya, continue to be grown in many parts of India in the dry season because of their high yield ability.

The national average yields of tropical Asian countries remain in the 2–3 mt/ha range. The physical, biological, and socioeconomic factors that limit yield differ among rice growing environments. The problem of identifying the gap between potential and actual yield is discussed in chapter 15. However, potential yields are much lower under the unfavorable environmental conditions that predominate in tropical Asia.

Notes

1. Hybridization is the act of producing a plant from unrelated parents. Rice is a self-pollinating plant, and considerable skill is needed to effect cross-pollination. This process should not be confused with the more complex procedure of developing F_1 hybrid rice recently accomplished in China.

2. See Takane Matsuo, *Rice Culture in Japan* (Tokyo, Ministry of Agriculture and Forestry, 1954) p. 43. Matsuo states that the first successful cross was made by Kyushiro Takahashi, Chief of the Shiga Prefecture Agricultural Experiment Station in 1898. The first work on hybridization in the United States was done in California. See J. W. Jones, "Hybrid Vigor in Rice," *Journal of the American Society of Agronomy* vol. 18 (May 1926) pp. 423–428; and U.S. Department of Agriculture, "Improvement in Rice," *Yearbook of Agriculture, 1936* (Washington, D.C., Government Printing Office, 1936) pp. 433, 444–445.

3. See Shiro Okabe, "Breeding for High-Yielding Varieties in Japan," in *Rice Breeding* (Los Banos, Philippines, International Rice Research Institute, 1972) pp. 53–54.

4. See, for example, Lawrence Klein and Kazushi Ohkawa, eds., *Economic Growth: The Japanese Experience Since the Meiji Era* (Homewood, Ill., Richard D. Irwin, 1968); Kazushi Ohkawa and Henry Rosovsky, "A Century of Japanese Economic Growth," in William W. Lockwood, ed., *The State of Economic Enterprise in Japan* (Princeton, N.J., Princeton University Press, 1965); and William W. Lockwood, ed., *The Economic Development of Japan: Growth and Structural Change, 1868–1938* (Princeton, N.J., Princeton University Press, 1954).

5. Te-tzu Chang, "The Origin, Evolution, Cultivation, Dissemination, and Diversification of Asian and African Rices," *Euphytica* vol. 25 (1976) pp. 425–440.

6. Ping-ti Ho, "Early Ripening Rice in Chinese History," *The Economic History Review* vol. 9 (1965) pp. 200–218, provides a detailed account of the introduction of new rice varieties.

7. Te-tzu Chang and H. I. Oka, "Genetic Variousness in the Climatic Adoption of Rice Cultivars," in IRRI, *Proceedings of the Symposium on Climate and Rice* (Los Banos, Philippines, International Rice Research Institute, 1976) p. 207.

8. Dwight Perkins, *Agricultural Development in China 1368–1968* (Chicago, Ill., Aldine, 1969) pp. 38–41.

9. Tsung-han Shen, *Agricultural Resources of China* (Ithaca, N.Y., Cornell University Press, 1951) pp. 40–41. Shen comments that in comparing Chinese varieties of wheat, cotton, rice, barley, maize, gaoliang (sorghum), soybeans, millet, sweet potatoes, and vegetables with varieties introduced from foreign countries, the native varieties usually show their superiority in early maturity. Further evidence is found in the *National Academy of Sciences Plant Science Delegation Trip Report* (Washington, D.C., 1975) p. 53.

10. Yujiro Hayami, *A Century of Agricultural Growth in Japan* (Tokyo, University of Tokyo Press, 1975) p. 50.

11. Isamu Baba, "Breeding of Rice Varieties Suitable for Heavy Manuring," in *Report of the Fifth Meeting of the International Rice Commission's Working Party on Rice Breeding* (Tokyo, Ministry of Agriculture and Forestry, 1954) pp. 167–185.

12. Shigeru Ishikawa, *Economic Development in Asian Perspective* (Tokyo, Kinokuniya Co., 1967) p. 100.

13. There has been much debate about the growth rate of agriculture during the Meiji period. Based on the earlier Ohkawa estimate, agriculture was thought to have expanded at 2.5 percent per annum from 1878 to 1922. See Kazushi Ohkawa and Henry Rosovsky, eds., *A Century of Japanese Economic Growth*. Nakamura challenged these results, asserting that the official Japanese agricultural statistics were understated in the earlier years. In Nakamura's view, the agricultural growth rate could not have been more than 1 percent per annum in this period. James I. Nakamura, *Agricultural Production and the Economic Development of Japan 1873–1922* (Princeton, N.J., Princeton University Press, 1966). Using the revised figures of Kazushi Ohkawa et al., *Estimates of Long-Term Economic Statistics of Japan Since 1868*, Vol. 9 of "Agriculture and Forestry" (Tokyo, Toyo Keizai Shinposha, 1965), Hayami and Yamada provide a revised estimate of agricultural growth of 2 percent. (The estimates for rice used here are based on these revised figures. See Yujiro Hayami and Saburo Yamada, "Agricultural Productivity at the Beginning of Industrialization," in Kazushi Ohkawa, Bruce Johnston, and Hiromitsu Kaneda, eds., *Agricultural and Economic Growth: Japan's Experience* (Princeton, N.J., Princeton University Press, 1970) pp. 105–136.

14. Isama Baba, "Breeding of Rice Varieties for Heavy Manuring."

15. The interpretation of the Japanese experience in this period is largely based on the work of Yujiro Hayami and his colleagues. See Hayami, *A Century of Agricultural Growth in Japan*, pp. 49–50 and 206–215, and Hayami and Yamada, "Agricultural Productivity at the Beginning of Industrialization." There is still considerable controversy concerning the "backlog of technology" concept emphasized by Hayami and by Ohkawa and Henry Rosovsky, *A Century of Japanese Economic Growth*. See, for example, Kee Il Choi, "Technological Diffusion in Agriculture under the Bakuhan System," *Journal of Asian Studies* vol. 19 (August 1971) pp. 749–759. In contrast to Hayami-Yamada and Ohkawa-Rosovsky, Choi argues that, during the Tokugawa era, agricultural technology diffused between regions.

16. Hayami, *A Century of Agricultural Growth in Japan*, p. 60.

17. Okabe, "Breeding for High-Yielding Varieties in Japan."

18. C. H. Huang, W. L. Chang, and Te-tzu Chang, "Ponlai Varieties and Taichung Native 1," in IRRI, *Rice Breeding* (Los Banos, Philippines, International Rice Research Institute, 1972) p. 32.

19. Ibid.

20. Amar N. Agrawal, *Indian Agriculture and its Problems* (Delhi, Ramjas College, Economics Society, 1951) p. 111.

21. Robert E. Evenson, James P. Houck, Jr., and Vernon W. Ruttan, "Technical Change and Agricultural Trade: Three Examples—Sugarcane, Bananas, and Rice," in Robert Vernon, ed., *The Technology Factor in International Trade* (New York, Columbia University Press for the National Bureau of Economic Research, 1970) p. 420.

22. N. Parthasarathy, "Rice Breeding in Tropical Asia up to 1960," in *Rice Breeding* (Los Banos, Philippines, International Rice Research Institute, 1972) p. 7.

23. Albert Howard, *Crop Production in India, a Critical Survey of its Problems* (London, Oxford University Press, 1924) p. 35.

24. Vernon P. Wickizer and M. K. Bennett, *The Rice Economy of Monsoon Asia* (Stanford, Calif., Stanford University Press, 1941) p. 234.

25. S. Harahap, H. Siregar, and B. H. Siwi, "Breeding Rice Varieties for Indonesia," in *Rice Breeding* (Los Banos, Philippines, International Rice Research Institute, 1972) pp. 141–146.

26. Parthasarathy, "Rice Breeding in Tropical Asia," pp. 21–22.

27. Four of the varieties from this cross spread to the Philippines and Sri Lanka, one of them being Peta, a parent of the initial IRRI release, IR8, another being Mas, a parent of the popular Sri Lankan variety, H-4.

28. F_1 hybrid refers to the seed that results from a cross between two different pure-line selections. When the F_1 is planted, all individuals appear to be uniform, but the resulting seed (F_2) is heterogeneous and has many different combinations of traits from the two parents. In conventional varieties, desirable individuals from subsequent generations are selected. After about seven or eight generations, the genes are stable. With F_1 hybrids, the product of the cross is the seed planted by the farmers.

29. An excellent account of rice breeding in the tropics in the period prior to the establishment of IRRI in 1962 is found in Parthasarathy, "Rice Breeding in Tropical Asia up to 1960." Further details are contained in the annual reports of the International Rice Commission's "Working Party on Rice Breeding," beginning in 1950.

30. Mahsuri was one of the more widely grown varieties of rice in Asia in the 1970s. It is medium in stature and does possess many of the plant type characteristics that breeders prefer. In addition to high grain quality, it has the capacity to give yields slightly higher than traditional varieties with low levels of fertilizer application and at water depths of up to 30 or more centimeters, which inhibit the growth of the semidwarf varieties.

31. Parthasarathy, "Rice Breeding in Tropical Asia," p. 10.

32. C. H. Chang, W. L. Chang, and Te-tzu Chang, "Ponlai Varieties and Taichung Native 1," p. 40.

33. C. S. Huang, "Evolution of Rice Culture in Taiwan," JCRR/PID-SC-37 (Taipei, Chinese-American Joint Commission on Rural Reconstruction, 1970 [mimeo]). Huang reviews various documents dealing with rice culture in China and Taiwan and the possible origin of the semidwarfs.

34. Dilbagh S. Athwal, "Semi-dwarf Rice and Wheat in Global Food Needs," *Quarterly Review of Biology* vol. 46 (1971) pp. 1–34. This article contains an excellent account of the origin and development of the semidwarf rices and wheats.

35. Dana G. Dalrymple, *Development and Spread of High-Yielding Varieties of Wheat and Rice in the Less Developed Nations*, Foreign Agricultural Economic Report No. 95 (Washington, D.C., U.S. Department of Agriculture, 1978) p. 74.

36. Thomas B. Wiens, "The Evolution of Policy and Capabilities in China's Agricultural Technology," in *Chinese Economy Post-Mao Policy Performance*, vol. 1 Committee Print, Joint Economic Committee, 95 Cong. 2 sess. (1978) pp. 671–703.

37. Kang Chao, *Agricultural Production in Communist China 1949–1965* (Madison, University of Wisconsin Press, 1970) p. 166.

38. Wiens, "The Evolution of Policy and Capabilities," p. 675.

39. Many visitors to China today comment on the capacity of the research-extension network to rapidly disseminate new technology. This, however, is a very recent development. The failure of efforts to promote untested technology in the 1950s is described in Kang Chao, *Agricultural Production*, pp. 172–179. Chao describes the loss in production that resulted when attempts were made to substitute japonica for indica varieties in Central China and Sichuan Province in 1965. The recent shift from double to single cropping of rice in Sichuan and other areas suggests that, even under the new research-extension network, there have been significant inefficiencies in the use of resources and dissemination of new technology.

40. Guangdong Provincial Academy of Agricultural Sciences, "Principal Experience in the Breeding of Dwarfed Paddy Rice," *Renmin Ribao* (Beijing, December 1964) translated in Joint Publications Research Service 28139 (January 1965) pp. 26–35.

41. Jin-lua Shan, "Rice Breeding in China," in *Rice Improvement in China and Other Asian Countries* (Los Banos, Philippines, International Rice Research Institute, 1980) pp. 9–30.

42. Shan, "Rice Breeding in China;" and Dalrymple, *Development and Spread of High-Yielding Varieties*, p. 87.

43. A discussion of the development of the hybrid rice program is found in Shih-cheng Lin and Loung-ping Yuan, "Hybrid Rice Breeding in China," in *Innovative Approaches to Rice Breeding* (Los Banos, Philippines, International Rice Research Institute, 1980). Another recent summary on hybrid rice is found in Te-tzu Chang, "Hybrid Rice," in J. Sneep and A. J. T. Hendriksen, eds., *Plant Breeding Perspectives* (Wageningen, PUDOC, 1979) pp. 173–174.

44. Wiens, "The Evolution of Policy and Capabilities in Chinese Agricultural Technology," pp. 679–680. The author speculates that "hybrid vigor" may in fact be the result of crossing with the tropical indica semidwarfs. If this were so, conventional hybridization could replace the more difficult and costly process of producing F_1 hybrids. Shih-cheng Lin and Loung-ping Yuan in "Hybrid Rice Breeding in China" identify heterosis in a number of agronomic, physiological, and biochemical characters.

45. The Chinese were not alone in this achievement. In Korea, it is reported that during the period from the 1920s to the 1960s, conventional breeding of japonica varieties required fifteen years from hybridization to dissemination. However, by breeding two crops a year and making maximum utilization of greenhouses, the South Koreans were able to develop new indica and japonica hybrids in six or seven years. See Korea, Office of Rural Development, *Success in the Green Revolution 1964–1977* (Suweon Korea, Office of Rural Development, Ministry of Agriculture and Fisheries, Republic of Korea, n.d.).

46. International Rice Research Institute, *Rice Research and Production in China: An IRRI Team's View* (Los Banos, Philippines, International Rice Research Institute, 1978) p. 59. According to the Chinese, another advantage for rice is that the diploid lines obtained through the anther culture of indica-japonica F_1 hybrids rarely show sterility. When crosses between indica and japonica are handled in the conventional manner, the progeny show varying degrees of sterility.

47. Following the ouster of the Gang of Four in 1976, much of the political pressure seems to have been relaxed. In Sichuan Province, there are reports of uneconomically double-cropped land reverting to single cropping of rice. However, because of international significance, there must be a strong interest in seeing the hybrid program succeed.

48. See R. C. Aquino, S. S. Virmani, and G. S. Khush, "Heterosis in Rice" (Los Banos, Philippines, Saturday Seminar, International Rice Research Institute, August 8, 1981 [mimeo]).

49. Peter R. Jennings, "Plant Type as a Rice Breeding Objective," *Crop Science* vol. 4 (1964) pp. 13–15; Akira Tanaka, K. Kawano, and J. Yamaguchi, *Photosynthesis, Respiration, and Plant Type of the Tropical Rice Plant*, IRRI Technical Bulletin, No. 7 (Los Banos, Philippines, 1966); and S. K. DeDatta, A. C. Tauro, and S. M. Balaoing, "Effect of Plant Type and Nitrogen Level in the Growth Characteristics and Grain Yield of Indian Rice in the Tropics" *Agronomy Journal* vol. 60 (November–December 1968) pp. 643–647.

50. Tanaka, Kawano, and Yamaguchi, *Photosynthesis, Respiration and Plant Type*.

51. Benito S. Vergara, Roberto Lilis, and Akira Tanaka, "Studies of Internode Elongation of the Rice Plant: In Relationship Between Growth Duration and Internode Elongation," *Soil Science and Plant Nutrition* vol. 11 (1965) pp. 26–30.

52. The interaction between the modern seed-fertilizer technology and water control is discussed in Shigeru Ishikawa, *Economic Development in Asian Perspective* and Sam-chung Hsieh and Vernon W. Ruttan, "Environmental, Technological, and Institutional Factors in the Growth of Rice Production: Philippines, Thailand, and Taiwan," *Food Research Institute Studies* vol. 7 (1967) pp. 307–341. The authors clearly foresaw that the spread of the new technology in the tropics would be restricted to areas with good water control. Aquino and Jennings also pointed out that for areas in which land preparation, weed control, and water control are less than adequate, varieties approximately 120–130 cm tall may be considered superior to the semidwarfs (100 cm tall). See Rodolfo C. Aquino and Peter R. Jennings, "Inheritance and Significance of Dwarfism in the Indica Rice Variety," *Crop Science* vol. 6 (1966) pp. 551–554.

53. See Thomas R. Hargrove, W. Ronnie Coffman, and Victoria L. Cabanilla, "Genetic Interrelationships of Improved Varieties in Asia," IRRI Research Paper Series No. 23 (Los Banos, Philippines, 1979). The authors draw attention to the narrow genetic base of the new semidwarfs. The Taiwan semidwarfs remain the predominant if not the exclusive source of the dwarfing gene (and the semidwarf genes of the Chinese varieties are identical to those from Taiwan). Furthermore, of the first 20,000 crosses made at IRRI from 1962 to 1977, 44 percent of the female parents were derived maternally from Cina, the female parent of Peta, which was in turn the female parent of IR8. This implies that components of their cytoplasm might be similar. The authors recommend that steps be taken to immediately broaden the genetic base.

54. Thomas R. Hargrove, "Diffusion and Adoption of Genetic Materials Among Rice Breeding Programs in Asia," IRRI Research Paper Series No. 18 (Los Banos, Philippines, 1978).

55. Arlando S. Varca and Reeshon Feurer, "The Brown Planthopper and Its Biotypes in the Philippines," Paper read before the National Conference of Farmer's Associations, Bacolod City, Philippines, April 21, 1976.

56. A biotype is an insect variety or type of a given genus and species that is physically indistinguishable from another of the same species but that reacts differently to certain varieties of rice. IRRI scientists knew as early as 1971 of the existence of at least two biotypes of brown planthopper (a third was subsequently discovered). See Dilbagh S. Athwal and Mano D. Pathak, "Genetics of Resistance to Rice Insects," in *Rice Breeding* (Los Banos, Philippines, IRRI, 1972) pp. 375–386.

57. Distinction was drawn between "vertical" or major gene resistance and "horizontal" or minor gene resistance. Major gene resistance is short lived in some cereals, but until recently little was known about the stability of minor and major gene resistance in rice. See Gurdev S. Khush and Henry M. Beachell, "Breeding for Disease and Insect Resistance at IRRI," in *Rice Breeding* (Los Banos, Philippines, IRRI, 1972) pp. 318–319. Major gene resistance has proved to be stable in the case of tungro virus, but unstable in the case of the brown planthopper. One alternative strategy is to breed so-called multiline varieties, where the plant(s) of a given variety are identical in every aspect except the source of resistance. The mixture of plants in a field carrying two or three separate sources of resistance to a given disease reduces the likelihood that the disease can spread rapidly to cause severe crop damage. In the current situation, where emphasis has been placed for several years on breeding for major gene resistance, identifying minor gene resistance to the brown planthopper is extremely difficult.

58. A discussion of this issue is in Robert E. Evenson and Hans P. A. Binswanger, "Technology Transfer and Research Resource Allocation," in Hans P. A. Binswanger, Vernon W. Ruttan, et al., eds., *Induced Innovation, Technology, Institutions, and Development* (Baltimore, Johns Hopkins University Press, 1978) pp. 164–211.

59. Initially, research was directed toward the shallow flooded areas rather than the deepwater areas, where water depth exceeds 1m. Since 1974, a series of international deepwater rice workshops have been held in various countries to exchange information on research in this area.

60. See Bangladesh Rice Research Institute, *Proceedings of the International Seminar on Deepwater Rice* (Dacca, Bangladesh Rice Research Institute, 1975); and International Rice Research Institute, *Proceedings of the Workshop on Deepwater Rice* (Los Banos, Philippines, IRRI, 1977). The term "deepwater" is somewhat misleading, since much of the discussion at these conferences dealt with the development of varieties suited to shallow-to-medium water areas. This is in contrast to areas where floating rices are prevalent and water depths are in excess of 1m. With respect to unirrigated rice, the potential for increased production lies in the shallow water areas. See Randolph Barker and Robert W. Herdt, "Rainfed Lowland Rice as a Research Priority—An Economist's View," in International Rice Research Institute, *Rainfed Lowland Rice: Selected Papers from 1978 International Rice Research Conference* (Los Banos, Philippines, IRRI, 1979) pp. 3–51.

61. There is also research under way to develop improved photoperiod-sensitive varieties for areas with deeper water. There has been some debate as to whether photoperiod-insensitive or photoperiod-sensitive varieties are best suited to certain environments. Photoperiod-sensitive varieties are highly location specific.

62. For this reason, and because we have used national statistics, some of the data in this section may differ from the standard source, which is Dana Dalrymple, *Development and Spread of High-Yielding Wheat and Rice Varieties in the Less Developed Nations*. A full set of the collected data on modern rice varieties is available in Robert W. Herdt and Celia Capule, *Adoption, Spread, and Production Impact of Modern Rice Varieties in Asia*, (Los Banos, Philippines, International Rice Research Institute, 1983).

63. Violeta Cordova, Aida Papag, Sylvia Sardido, and Leonida Yambao, "Changes in Practices of Rice Farmers in Central Luzon: 1966–79," a paper presented to the Crop Science Society of the Philippines, Bacnotan, La Union, April 1981.

64. Richard H. Bernsten, Bernard H. Siwi, and Henry M. Beachell, "The Development and Diffusion of Rice Varieties in Indonesia," IRRI Research Paper Series No. 71 (Los Banos, Philippines, International Rice Research Institute, January 1982).

65. Z. Harahap, "Breeding for Resistance to Brown Planthopper and Grassy Stunt Virus in Indonesia," in International Rice Research Institute, *Brown Planthopper: Threat to Rice Production in Asia* (Los Banos, Philippines, IRRI, 1979) pp. 201–208.

66. See A. Tanaka, J. Yamaguchi, Y. Shimazaki, and K. Shibaty, "Historical Changes in Plant Type of Rice Varieties in Hokkaido," in *Soil Science and Manure* vol. 39 (1968) p. 11 (in Japanese).

67. The first of these new Korean varieties, Tong-il, is a triple-cross variety with the F_1 hybrid of the cross between the Japanese variety Yukura and Taichung Native 1 crossed with IR8. The original cross was made in 1965, and Tong-il was released in 1971.

68. Central Rice Research Institute, ed., *Annual Report for 1969* (Cuttack, India, 1970). See also W. David Hopper, "Mainsprings of Agricultural Growth in India," *Indian Journal of Agricultural Science* vol. 35 (June 1965) pp. 3–28. Hopper reports on the static nature of yield levels over time for the period 1948–63.

69. Caution must be exercised in interpreting these figures. There is some indication that yield potentials at the IRRI experiment station may have declined over the years because of an increase in insect and disease problems, rise in soil pH and salinity, and the depletion of trace elements in the soils, all a consequence of intensive year-round cropping.

6

Fertilizers and Agricultural Chemicals

Three significant factors have had a critical influence on productivity growth in Asian agriculture. They are an increase in fertilizer use, development and adoption of improved seeds (chapter 5), and improved irrigation and drainage (chapter 7). In this chapter, we look at the use of fertilizer and associated agricultural chemicals in Asia. Beginning in the pre–World War II period, we trace the growth of chemical fertilizer usage from almost nil to the present relatively high rates of application. We briefly discuss the use of other agricultural chemicals, chiefly pesticides, but also herbicides and fungicides, although these chemicals are not in wide usage except in the developed countries of East Asia. Considerable attention is also devoted to a number of important questions about fertilizer use. What is the outlook for future consumption? What is the productivity effect of fertilizer on rice and how is this affected by varietal types? What will the future demand for fertilizer be and how will this be affected by rising oil prices? How is fertilizer marketed and priced in Asia? Discussion of these problems will help to shed light on the overarching problem of prospective productivity growth in Asian agriculture.

Sources of Plant Nutrients

The word "fertilizer" brings to mind chemical fertilizer, but organic sources of plant nutrients have been indispensable in the past, are important in some countries today, and for nitrogen could well replace chemical sources when fossil-fuel-based supplies are depleted. Organic fertilizers have been used in China since the beginning of organized agriculture several thousand years ago and are still widely used throughout Asia. Animal manure, green manure crops, and compost are important traditional sources of local fertilizer.

The commercialization of the fertilizer industry in Japan in the late nineteenth century made new sources of nutrients, principally nitrogen, available. Nitrogen and its price in real terms in commercial trade declined continuously in Japan from 1880 to 1970 (excluding World War II). From 1883 to 1887, the nitrogen to milled rice price ratio was 10.7, but by 1958–62 it had fallen to 1.2.[1] On world markets, the price of imported urea fell to an all time low of about $US 60/ton in 1970 to 1971.[2]

The decline of fertilizer prices was made possible by a steady stream of innovations in the production and marketing of fertilizer.[3] Dried sardines and cotton and rapeseed meals were the traditional commercial fertilizers in Japan. With the growth of the fishing industry and improvements in transportation facilities on Hokkaido in the middle of the nineteenth century, herring meal was gradually adopted as a fertilizer source, and Manchurian soybean cakes supplanted fishmeal in the early part of the twentieth century. After World War I, chemical fertilizer (initially ammonium sulfate) began to replace organic sources as the price of chemical fertilizer continued to decline. However, even in the late 1950s, approximately one-third of Japan's fertilizer nutrients came from organic sources—amounting to 100 kg/ha. By the 1970s, organic sources were contributing less than 50 kg/ha. The development of the Japanese fertilizer industry had little impact on the rest of Asia, except Taiwan and Korea, which were Japanese colonies. In South and Southeast Asia, there was no demand, and in China, the limited transportation system made it difficult to support a commercial fertilizer industry.

Steady improvements in fertilizer production technology resulting in declines in fertilizer prices, the closing of the land frontier in tropical Asia, and the gradual improvement of transportation facilities along with the development of fertilizer-responsive rice varieties in the 1960s led to a rapidly growing demand for chemical fertilizer. China, with an underdeveloped transport system, called on surplus rural labor to increase the supply of organic fertilizers to augment inadequate supplies from local, small-scale chemical fertilizer plants.[4] However, by 1972, China emerged as the world's largest importer of nitrogen fertilizer (1.5 million metric tons). In the 1970s, thirteen modern large-scale ammonia-urea plants were constructed to supplement domestic production of chemical fertilizers from small-scale plants. At the end of the decade, organic sources still accounted for about half of the Chinese supply of plant nutrients.[5]

With the sharp upward movement of oil prices in 1973–74, the historic downtrend in chemical fertilizer prices ended abruptly. Most Asian countries responded to higher prices and the pending shortage of supply by increasing imports beyond normal levels, helping to drive prices to unprecedented heights.[6] When high fertilizer prices were passed on to farmers, consumption was at least temporarily dampened, but government policies in many countries tended to buffer farmers from the full impact of the rapid price increases. By 1976, world fertilizer prices returned to slightly above 1973 levels, but the fertilizer scare forced policymakers and scientists to reexamine the growing dependence of agriculture on fossil-fuel-based chemical sources of nitrogen. Historically, parts of India, China, and East Asia have been very dependent on organic sources of plant nutrients. This difference from the rest of Asia can be explained by relative population densities. Tropical Asia typically had lower man-to-land ratios, land was farmed more extensively, and supplemental fertilization was less necessary. Also, organic fertilizers require much labor, making them expensive in real terms. Even at the significantly higher prices that have prevailed since the early 1970s, chemical fertilizer remains an attractive economic alternative to the more labor-intensive production of organic fertilizers.

Because increases in grain production depend so heavily on additional supplies of plant nutrients, a significant rise in chemical fertilizer prices means that fertilizer-dependent gains will be expensive. The search for alternatives to chemical fertilizers and for more efficient ways to use chemical fertilizer has been intensified.[7] Experiments on proper timing and placement of fertilizer and on alternative forms of fertilizer suggest that fertilizer inputs can be reduced as much as one-third without lowering yields.[8] Considerable research is under way on nitrogen-fixing crops such as azolla and blue-green algae, which can be grown in the paddy fields. Geneticists and microbiologists are working to create a new breed of cereal crops that will flourish without artificial nitrogenous fertilizers by fixing atmospheric nitrogen much as legumes do.[9]

For the foreseeable future, however, chemical fertilizers will continue to be the major source of commercial plant nutrients for rice. The remainder of the chapter deals almost exclusively with the use of chemical fertilizers and focuses principally on nitrogen.

Consumption Trends

Until after World War II, most Asian rice farmers produced relatively little surplus rice and used few purchased inputs. Since the 1960s, however, government development programs have encouraged farmers to grow more rice by using inputs, especially chemical fertilizer produced off the farm. In the early 1950s, Asia's total consumption of fertilizer nutrients was about 1 million metric tons per year; by the late 1960s it had reached 20 million metric tons per year.

The recent history of chemical fertilizer use in Asia can be viewed from two perspectives: one of astonishment at the speed with which fertilizer use has doubled and redoubled in country after country in a relatively short period of time, or one of disappointment at the low levels of per hectare application. Even with the rapid growth since the early 1960s, only about 20 kg of fertilizer nutrients were applied per hectare of arable land in South and Southeast Asia by the mid 1970s, or less than one-tenth the level in East Asia. International agencies and Asian governments, often abetted by the fertilizer industry, have viewed any slowing of the growth rate of fertilizer consumption with alarm. Nevertheless, the growth in fertilizer use has been remarkably steady despite alternating periods of surplus and shortage in the world market and wide fluctuations in world prices.

Between the early 1960s and the mid-1970s, fertilizer consumption increased sevenfold in South Asia,

Table 6.1. Consumption of Fertilizer Nutrients ($N + P_2O_5 + K_2O$) in Asia

Country	1950/51–1954/55 (thousand mt)	1960/61–1964/65 (thousand mt)	1970/71–1974/75 (thousand mt)	1975/76–1979/80 (thousand mt)	Annual growth rate 1960/61–1964/65 to 1975/76–1979/80 (percent)
South Asia	125.1	631.4	3,421.7	5,170.8	15.0
India	85.1	469.5	2,707.0	3,974.7	15.3
Pakistan	6.1	60.8	440.0	753.7	18.3
Sri Lanka	31.8	67.4	93.3	113.3	3.5
Bangladesh	2.1	33.2	170.3	312.3	16.1
Nepal	—	0.5	11.1	16.8	26.4
Southeast Asia	66.1	405.8	1,359.7	1,894.0	10.8
Malaysia	5.3[a]	44.4	210.3	334.0	14.4
Thailand	3.2	24.2	163.9	260.5	17.2
Philippines	36.7	89.7	234.5	275.7	7.8
Indonesia	20.4	110.6	426.3	643.2	12.5
Burma	0.5	6.2	46.6	69.3	17.5
Vietnam	—	130.6	277.9	311.2	6.0
Laos	—	0.1	0.2	0.1	0
East Asia	1,308.1	3,334.4	8,543.1	13,063.7	9.5
Japan	1,013.8	1,788.1	2,045.6	2,124.5	1.2
Taiwan	106.8	186.2	300.8	385.2	5.0
North Korea	—	149.2	359.3	622.1	10.0
South Korea	100.5	323.9	743.3	793.4	6.2
China,	87.0[b]	887.0	5,094.1	9,138.5	16.8

Sources: A. Palacpac (1982); Food and Agriculture Organization, *Fertilizer Yearbook* (1978); and Food and Agriculture Organization, *Monthly Bulletin of Statistics* vol. 5, no. 3 (March 1982).

[a] 1952/53 to 1954/55.

[b] 1953/54 to 1954/55.

fourfold in Southeast Asia, and eightfold in China (table 6.1). Of the seventeen countries for which data are available, ten sustained rates of increase in fertilizer use of over 10 percent per year during the period. Only Japan, Taiwan, and South Korea can be said to have reached a high enough level of nutrient use per hectare so that an aggregate growth rate of 10 percent could not reasonably be sustained.

Early post–World War II efforts to increase fertilizer use in programs such as India's Community Development program met modest success. Between 1955 and 1960, India's fertilizer use went from 130,000 mt to 240,000 mt. This compares with over a million tons per year being used in Japan on a fraction of the area in the late 1950s. Thus, despite the growth in use, the level of application was low. There were many reasons for this—lack of fertilizer, inadequate distribution systems for moving the fertilizer from the ports to the farming areas, limited research on response to fertilizer, poor understanding on the part of farmers of the use and potential value of fertilizers, and in some cases, pricing policies that discouraged the use of fertilizer. An even greater problem was the lack of fertilizer-responsive rice and other cereals in the tropics. Figure 6.1 illustrates the contrast between the response of indigenous Indian rices and U.S. rices to fertilizer in the 1940s and 1950s. The indigenous varieties grown in the tropics simply did not have the genetic potential to respond to high levels of fertilizer. Thus, in the 1950s, farmers had little incentive to apply fertilizers. The development of fertilizer-responsive tropical rice varieties, together with installation of fertilizer production capacity and improved distribution systems, led to a virtual explosion of fertilizer use in Asia.

There are few regular national data series that show how much fertilizer is used on rice and how much is used on other crops. In some countries, like Bangladesh, most of the cultivated land is used for rice, so most of the fertilizer is probably applied to rice. In other countries, like Pakistan and India, a relatively small fraction of the land is used for rice, so it is likely that only a small amount of fertilizer is used for rice. Some countries have developed estimates of fertilizer use on rice based on single- or multiple-year surveys. These have been extrapolated to 1976–79 assuming that fertilizer use on rice increased at the same rate as total use. The data are summarized in table 6.2.

Although these aggregate figures leave much to

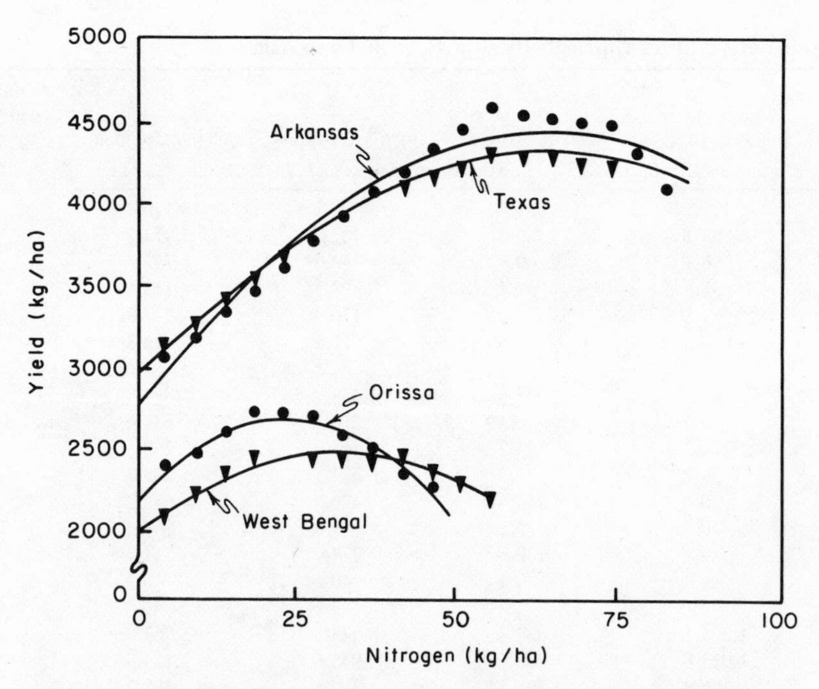

Figure 6.1. Yield response of rice to nitrogen fertilizer in two states of India and two states in the United States (Source: R. Herdt and J. Mellor, "The Contrasting Response of Rice to Nitrogen: India and the United States," *American Journal of Agricultural Economics* vol. 46, no. 1 [1964] p. 152, reprinted by permission of the publisher)

be desired, they do provide some indication of the trends in fertilizer use during the post–World War II decades. In the early 1950s, only Japan, Taiwan, and Korea used substantial amounts of chemical fertilizer on rice (relatively large quantities of organic fertilizers were also being used). In Malaysia and Sri Lanka, some measurable levels of chemical fertilizers were used in the late 1950s, but in most developing Asian countries the rates were below 5 kg/ha—contributing at most 100 kg of grain/ha to yields. By the late 1970s, Malaysia was using almost 100 kg/ha, Indonesia was using more than 50 kg/ha, the Philippines were using about 30 kg/ha, and Thailand, Bangladesh, and Burma were only applying about 10 kg/ha of rice.

The Productivity of Fertilizer on Rice

Nitrogen fertilizer has made a substantial contribution to the rice output increases achieved by many countries. The general relationship between nitrogen fertilizer input and yield is indicated in figure 6.2. At low levels, yields increase rapidly with additional fertilizer, eventually reach a maximum, and if fertilizer is applied beyond that level, decline. This response pattern can be represented by a quadratic equation

$$\text{Yield} = a + b\,(\text{fertilizer}) + c\,(\text{fertilizer})^2$$

where a and b are positive coefficients, c is a negative coefficient, and fertilizer and yield are measured per acre or per hectare.

Given the numerical values for the coefficients a, b, and c, the following measures of fertilizer productivity are illustrated in figure 6.2.

Yield (Y) = the output per hectare for a given level of fertilizer

Yield increase (ΔY) = the change in yield from zero fertilizer to a given level of fertilizer application

Maximum yield increase (ΔY_m) = the highest increased yield obtainable: the difference between yield with zero fertilizer and maximum point on the response function

Maximum yield fertilizer (F_m) = the amount of fertilizer needed to get the maximum yield increase

Average product ($\Delta Y/F$) = the ratio of the yield increase to the amount of fertilizer applied

Marginal product (MP_f) = the change in yield that would be obtained from a very small change (say

Table 6.2. Estimated Application of Fertilizer Nutrients per Hectare of Rice
(kg/ha)

	$N + P_2O_5 + K_2O$				
Country	1956–60	1961–65	1966–70	1971–75	1976–79[a]
South Asia					
India	2	5	15	23	32
Pakistan	4	5	15	29	46
Sri Lanka	13	20	32	74	65
Bangladesh	0	1	4	5	11
Nepal	[b] —[b]	—	2	5	8
Southeast Asia					
West Malaysia	—	47	53	67	97
Thailand	0	2	6	6	11
Philippines	4	8	13	25	29
Indonesia	0	8	13	37	57
Burma	0	1	3	6	9
East Asia					
Japan	217	229	286	288	340
Taiwan	168	193	201	189	205
South Korea	—	93	156	189	311

Note: Fertilizer statistics are not widely or uniformly collected throughout Asia. Thus, the numbers provided here must be seen as estimates of total use.

Sources:

India: Fertilizer Association of India, *Fertilizer Statistics*. Rice is assumed to use 32 percent of all fertilizer.

Pakistan: Esso Pakistan Fertilizer Co., *Pakistan Nitrogen Demand Forecast Study* (1974).

Sri Lanka—1958/59 to 1964/65: Food and Agriculture Organization (1968); 1965/66 to 1973/74: Ministry of Plantation Industries.

Bangladesh—1950/71: Bangladesh Agricultural Development Corporation; 1972 to 1979: *FAO Fertilizer Yearbook*. Rice is assumed to use 32 percent of all fertilizer applied.

Nepal: R. Panta, "Nepal" (1979).

West Malaysia: K. Y. Ming (1977).

Thailand: J. Intachairi and S. Pradithavanij (1975) gives total consumption. Percentage used on rice is from Department of Economic Research, Bank of Thailand.

Philippines: The Fertilizer and Pesticide Authority has estimated that the following percentages of total fertilizer were used on rice in the years mentioned: 27 percent up to 1969; 34 percent in 1970; 38 percent in 1971; 37 percent in 1972 and 1973; 35 percent thereafter.

Indonesia—1956 to 1960: Agrar-Und Hydrotechnik GmbH (1972); 1961 to 1963: World Bank, *Agricultural Sector Survey*; 1964 to 1975: Secretariat Fertilizer Committee, Department of Agriculture (1976). Rice is assumed to use 81 percent of total food crop fertilizer.

Burma: Ministry of Planning and Finance, *Report to the Pyithu Hluttaw* (1976).

Japan: Ministry of Agriculture, Forestry and Fisheries, *Survey on Production Cost of Rice*. Only the nutrients supplied from chemicals are shown.

Taiwan: Taiwan Provincial Food Bureau, *Taiwan Statistical Data Book*.

South Korea: P. Y. Moon and B. S. Yoo (1974).

[a]Except for Japan and Taiwan, this column is obtained by extrapolating the per hectare use data from earlier periods at the same rate as total fertilizer use grew between 1971–75 and 1976–79.

[b]— indicates no data are available for this period.

one kg) in fertilizer; in mathematical terms $\frac{\partial Y}{\partial F}$, which in the case of a quadratic function is $b + 2cF$. Note that MP_f varies depending on the level of F, and that in particular it equals 0 at F_m in the case of a quadratic response function

Profit maximizing fertilizer (F^*) = the level of fertilizer that would be applied in order to maximize the increased *value* of yield above fertilizer cost

Profit maximizing (Y^*) = the yield corresponding to F^*; note that F^* and Y^* depend not only on the yield response but also on the ratio of the prices of fertilizer and rice.

$F^* = (\frac{1}{2}c)\ [(P_f/P_r) - b]$, where P_f = price of fertilizer and P_r = price of rice; i.e., F^* is the level of F that equates marginal cost (P_f) and marginal returns $[P_r\ (b + 2\ cF)]$ from fertilizer

Of the above, the average product comes closest to the agronomic idea of "fertilizer efficiency." It may be measured at the slope of a straight line from the point with value a to the curve. However, the average product depends on the level of fertilizer at which it is measured (see figure 6.2) and therefore

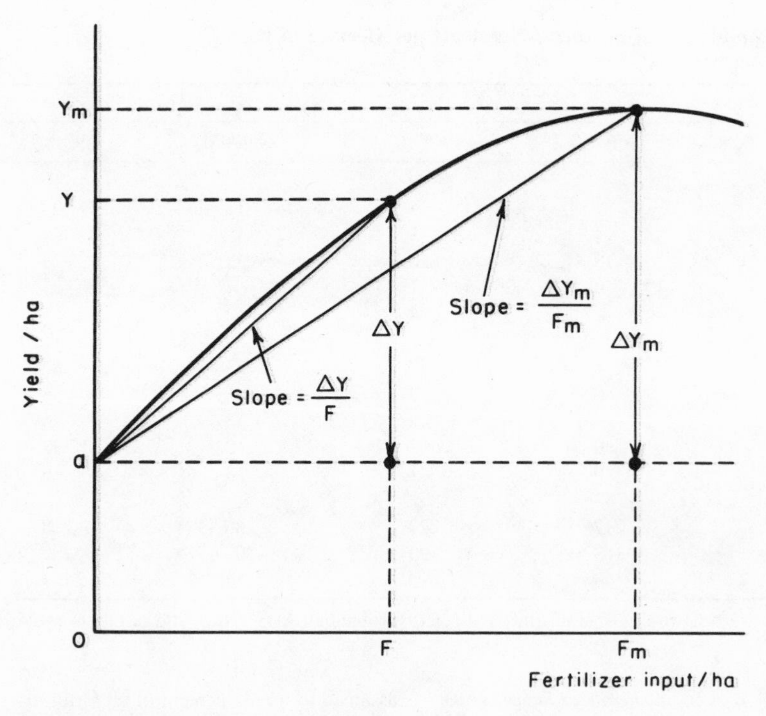

Figure 6.2. Measures of fertilizer productivity

may be inappropriate for certain kinds of comparisons. For example, examination of figure 6.1 shows that it would be misleading to compare the average product of the Orissa curve and the Texas curve at 40 pounds of N per acre, and it would also be misleading to compare their average products at 100 pounds of N per acre. Neither comparison would fairly reflect the differences in the physical yield response to fertilizer.

Comparison of yields, marginal productivity, and average productivity at profit-maximizing fertilizer levels has a certain appeal. It incorporates both the effect of physical response differences, as well as economic considerations that have varying effects with different types of response curves. However, using profit-maximizing concepts can be somewhat misleading when comparing physical productivity across countries because price ratios vary widely across countries.

For these reasons, in the comparisons intended to reflect physical productivity differences, the maximum yield increase (ΔY_m) and maximum yield fertilizer level (F_m) are used. These measures reflect differences in the physical yield response and provide a direct reflection of the extent to which yield can be increased by applying fertilizer. They do not, however, reflect the economically attainable productivity differences.

There is a large body of literature on the productivity of nitrogen fertilizer in the Asian tropics, much of it resulting from agronomic research attempting to characterize the nature of the response function. Economic planners use national average productivity ratios across various production locations and conditions to summarize the average productivity of fertilizer for planning and projection purposes. But these average ratios ignore the many factors that affect the productivity of fertilizer.

The response to fertilizer can be measured from data reflecting fertilizer–output relationships across nations, regions, farms, experiment plots in farmers' fields, and experiment plots at research stations. Researchers prefer the last approach because it allows measurement of the fertilizer response alone. Although yields achieved under fertilization experiments in farmers' fields may be more representative of actual responses to fertilizer applications, year-to-year and field-to-field variations cannot be controlled. Factor response functions may also be based on farm management survey data. Problems can arise when farmers are asked to record or recall the use of inputs rather than employing quantitative measurement. Aggregate analyses at the regional or international level, such as those reported in chapter 4, can help to provide overall impressions of the response to fertilizer, but they generally cannot adequately measure responses in particular countries or the effect of separate factors. Those are reflected in the studies reviewed here, most of which are derived from experimental data.

The response of rice to nitrogen fertilizer depends on many factors, including the amount of available water and soil resources. The presence or absence of insect pests, plant diseases, and unfavorable weather also affect fertilizer response, and some rice varieties have a greater capacity to respond to high fertilizer applications than others. The type of fertilizer, method of application, time of application, and nutrient balance also influence the response. Factors such as these make the response functions highly variable from one experiment to another.

The effect of the many factors besides fertilizer that affect rice yields can be accounted for by various means. One way is holding those factors constant while varying fertilizer levels. Another way is by classing fertilizer-yield data into groups with constant levels of other factors. A third way is to measure factor levels and include them as variables in a comprehensive multifactor response equation. The first approach is sometimes possible with experiment station data; the second is possible if only one or two independent factors other than fertilizer are important. The third is most attractive when there are many factors that cannot be controlled, as happens when experiments are spread over many farmers' fields and several years. If such an approach is used with a quadratic fertilizer term, one has a function

$$\text{Yield} = f\,(\text{fertilizer, fertilizer}^2\text{, factor 2, factor 3, } \ldots \text{ factor } N)$$

The inclusion of many independent variables sometimes permits the researcher to measure the interactive effects of two or more variables. For example, if the rice plant suffers from lack of water, yields will be reduced. Theoretically, the yield reduction may be intensified or reduced through the use of fertilizer. Functions that include interactions can reflect the effect of these relationships, but there are relatively few examples of such detailed analyses for the tropical rice-producing countries.

Impact of Varietal Type

We have stressed the differences between traditional rice varieties and modern fertilizer-responsive varieties, but is this disparity consistent across countries and time? David and Barker have compiled a large number of experimentally based fertilizer response equations that compare the two types for sites in India and the Philippines.[10] Additional data from many similar trials in Nepal, Bangladesh, Burma, and Thailand were assembled from the literature.

Unfortunately, most agronomic trials do not systematically compare traditional and modern varieties—most include only one type. The available information from a large number of experiment station trials and thousands of simple trials in farmers' fields is reported in table 6.3. As one might expect, there is considerable variability in the response functions even though each represents the average of a large number of trials. However, the comparison of modern and traditional varieties within each country shows that the modern varieties consistently have higher yields, higher yield increases (ΔY_m), higher maximum yield fertilizer levels (F_m), and higher average productivity of fertilizer ($\Delta Y/F_m$). The average maximum yield increase for modern varieties (MVs) was about 1,200 kg/ha in the wet season compared to 300 for traditional varieties (TVs), while the average maximum yield increase was 3,100 kg/ha for MVs compared to 700 kg/ha for TVs in the dry season. Thus, modern tropical rice varieties are clearly more responsive to fertilizer than the previously available, traditional varieties.

Has this greater responsiveness been achieved by developing a variety that requires higher levels of fertilizer? This can be answered by comparing the values of the *a* coefficients in the response functions, which measure the yield with no added fertilizer. If the modern varieties required higher fertilizer, their *a* coefficients would be lower than those of the traditional varieties. The wet season data for Thailand show this pattern, perhaps helping to explain why adoption of MVs has lagged in Thailand. In all the other cases the MVs have higher yields without fertilizer. The original, unaveraged data confirm the same thing—of the eighty comparisons possible for India and the Philippines, the modern varieties had higher *a* values in fifty-three cases. Thus, one can find some situations where the modern varieties had lower yields than traditional varieties in the absence of applied fertilizer, but one can say, in general, that the new varieties do not require more fertilizer.

How do modern and traditional varieties compare in their efficiency of fertilizer use? In each pair of comparisons possible in table 6.3 (except for the Burma data), the average product ($\Delta Y_m/F_m$) of the MVs was higher—by an average of 6.2 kg rice/kg of fertilizer in the wet season and by 8.9 kg rice/kg fertilizer in the dry season. These data indicate that the MVs do not require more fertilizer but rather use the fertilizer they receive more efficiently.

Has the greater fertilizer responsiveness of the new

Table 6.3. Average Value of Fertilizer–Yield Response Functions for Modern and Traditional Varieties[a]

Country	Variety	No. of trials	Coefficients of response function			ΔY_m	F_m	$\Delta Y_m/F_m$
			a	*b*	*c*			
			Wet season					
Philippines[b]	Modern	31	3,337	31.1	−0.178	1,347	87	15.5
	Traditional	30	2,873	−2.5	−0.001	0	0	0
India[c]	Modern	13	3,444	18.1	−0.058	1,411	156	9.0
	Traditional	18	3,169	12.2	−0.084	447	73	6.1
Nepal	Modern	30	2,987	17.2	−0.062	1,180	138	8.6
	Traditional	15	2,081	7.6	−0.150	93	25	3.8
Bangladesh	Modern	large[d]	2,036	29.7	−0.172	1,272	86	14.8
	Traditional	large[d]	1,437	22.4	−0.172	726	65	11.1
Thailand	Modern	large[e]	2,385	17.7	−0.071	1,124	125	9.0
	Traditional	large[e]	2,615	9.8	−0.056	428	88	4.8
	Modern	24	2,495	20.7	−0.076	1,406	136	10.3
Burma	Modern	large[f]	2,267	13.7	−0.070	670	98	6.8
	Local	large[f]	1,563	14.9	−0.190	292	39	7.5
			Dry season					
Philippines	Modern	29	4,053	46.1	−0.130	4,086	177	23.0
	Traditional	29	4,018	5.9	−0.090	96	33	5.9
India[c]	Modern	8	3,115	30.9	−0.060	3,978	257	15.4
	Traditional	7	2,283	19.4	−0.070	1,344	138	9.7
Bangladesh	Modern	large[d]	2,275	29.1	−0.172	2,146	85	14.6
	Traditional	large[d]	1,677	21.5	−0.172	681	62	10.8
Thailand	Modern	16	3,206	22.4	−0.104	1,213	108	11.2

Sources:

Philippines and India: C. C. David and R. Barker (1978).

Nepal: S. P. Pandey (1977); G. P. Deo and R. N. Shah, "Review of Fertilizer Investigation Work in Rice at Parwanipur (1958–1976)" (1978); N. K. Rajbhandany (1978); G. P. Deo and R. N. Shah, "Performance of Different Promising Lines at Different Levels of N under Rainfed Conditions" (1978).

Bangladesh: K. M. Badruddoza (1976).

Thailand: Rice Division and Planning Division, Department of Agriculture (1974).

Burma: Agricultural Corporation, Agricultural Research Institute (1975).

[a]Data are fitted to agronomic trial data. Response function was $Y = a + bF + cF^2$ where Y is yield/ha and F is fertilizer applied/ha. For measures of fertilizer productivity, Y_m is the maximum yield, F_m is the fertilizer level at which yield is maximized, and Y/F is the average productivity of fertilizer—the yield increase divided by F_m.

[b]Trials at four experiment stations over an eight-year period.

[c]Trials at twenty-two experiment stations for three years and eight stations for one year. Thus, the actual number of trials is greater than indicated. This is the number of separate functions reported in the source.

[d]Derived from data on over 8,000 farmers' field trials.

[e]Derived from trials run over a three-year period in at least ten locations throughout the country.

[f]Source states 2,363 observations for local, 1,539 for modern.

varieties been achieved at the cost of higher risk? If one's concept of risk is the amount of money that could be invested in the form of applied fertilizer, then perhaps MVs involve a greater risk, but the more conventional definition of risk requires a consideration of the probability of obtaining low yields with high levels of fertilizer and, as a consequence, losing money. To compare the modern and traditional varieties on this basis, one may examine the variability of their response coefficients. The coefficient of variation (*CV*) is a measure of the variability of an estimated number (the standard deviation of a coefficient divided by its mean). The CVs for the average *a*, *b*, and *c* coefficients for the Philippine, Indian, and Nepalese data in table 6.3 are shown in the table below.

		CV of coefficients		
		a	*b*	*c*
Wet season				
Philippines	MV	0.29	0.45	0.50
	TV	0.41	4.82	43.80
India	MV	0.21	0.49	0.81
	TV	0.24	1.25	0.94
Nepal	MV	0.19	0.66	1.00
	TV	0.23	1.47	2.54
Dry season				
Philippines	MV	0.23	2.03	1.15
	TV	0.29	2.89	0.89
India	MV	0.17	0.40	0.67
	TV	0.25	0.37	0.71

In every comparison, the *a* and *b* coefficients of

the MVs have lower variability than the TVs, and in all but one comparison the *c* coefficients of the modern variety are less variable than those for the traditional varieties. This indicates that the response to fertilizer of the MVs is generally less variable than the response of TVs. In the wet season Philippine situation, the data strongly support the hypothesis that MV yields are less variable than TV yields. Unfortunately, large sets of data are not available from other countries to test the same issue, but these data indicate there is no support for the contention that the modern varieties entail greater risk. Indeed, for any given level of applied fertilizer, there is a higher probability of receiving the expected yield (and thereby a higher net return) with the modern varieties than with the traditional varieties. Of course, because the optimal level of fertilizer is higher with the MVs, it is likely that farmers will apply higher levels of fertilizer on MVs than on TVs, and thereby they have some probability of losing larger amounts of cash, but an equal expenditure on fertilizer will result in a lower risk with MVs than with TVs.

Trials on Farms Compared with Trials on Experiment Stations

The results of experiments conducted under the controlled environmental conditions of experiment stations are relatively accurate, but their value for agricultural extension or for judging response under field conditions may be limited because many of the factors that are uncontrollable on farms are controlled on experiment stations. Unfortunately, fertilizer-response trials that can be used to compare results on experiment stations with those on farmers' fields are available for only a relatively limited number of situations. Available data suggest that there is a higher response to fertilizer on farmers' fields in Bangladesh than at the experiment stations, but data on relatively few trials at stations were available.[11] In the Philippines and Thailand, the response to fertilizer as reflected in average productivity was somewhat higher at experiment stations than in farmers' fields.[12] Still, the data do not give strong support to one view as opposed to the other, so no general conclusion can be drawn.

Nutrients Other Than Nitrogen

The above discussion deals exclusively with nitrogen fertilizer. But rice farmers often apply a combination of fertilizer nutrients. For example, in the Philippines, data from the Fertilizer and Pesticide Authority show that in 1973 rice and corn farmers applied 60 thousand metric tons of nitrogen, 18 thousand metric tons of phosphate (P_2O_5), and 8 thousand metric tons of potassium (K_2O),[13] with all of the phosphate and potassium contained in compound fertilizers.

The yield response of rice to the other nutrients is much less predictable than its response to nitrogen. Unless a soil lacks phosphorus or potassium, there will be no response to those elements, and many analyses show no response or a much lower response to phosphorus or potassium compared with nitrogen. For example, Rosegrant's analysis showed an average product of P_2O_5 of 3.8 compared to an average product of about 10 for N.[14] An analysis of Bangladesh fertilizer-response trials gave an average product of NPK of 5 at the yield maximizing level of 71 kg N.[15] Long-term trials in the Philippines show that after more than twenty-five crops grown on the same plots, there was no response to P_2O_5 and K_2O at IRRI and a response only half the time at other research stations in the country.[16] In all of the literature surveyed above, there is not enough systematic research on P_2O_5 and K_2O response to permit a rigorous estimate of a continuous response function, even in soils where a response is evident. However, the evidence shows that in nearly all instances, N response is greater than P_2O_5 response, which in turn is greater than K_2O response.

Impact of Climatic Factors

The yield response to fertilizer is affected by the climatic conditions under which the crop grows. In Asia, most rice is grown under wet season monsoon conditions where rainfall is relatively heavy throughout the growing season (200 to 1000 mm/month), cloud cover is thick, and the sunlight or solar radiation level is relatively low. In some insular and coastal areas, typhoons occur frequently throughout the growing season. In contrast, some rice is grown during the dry season when irrigation provides most of the water, skies are clear, solar radiation levels are high, and storms seldom appear. There is considerable evidence that shows more consistent responses to fertilizer under the high levels of solar radiation that characterize dry season conditions (figure 6.3).

The data reported in table 6.3 also support this view. The average productivity of fertilizer on MVs in the dry season is 50 percent higher in the Philippines and 70 percent higher in India than in the wet season, and on TVs in the dry season is 100 percent higher in the Philippines and 50 percent higher in India than in the wet season. The basic difference in fertilizer productivity arises from the higher solar

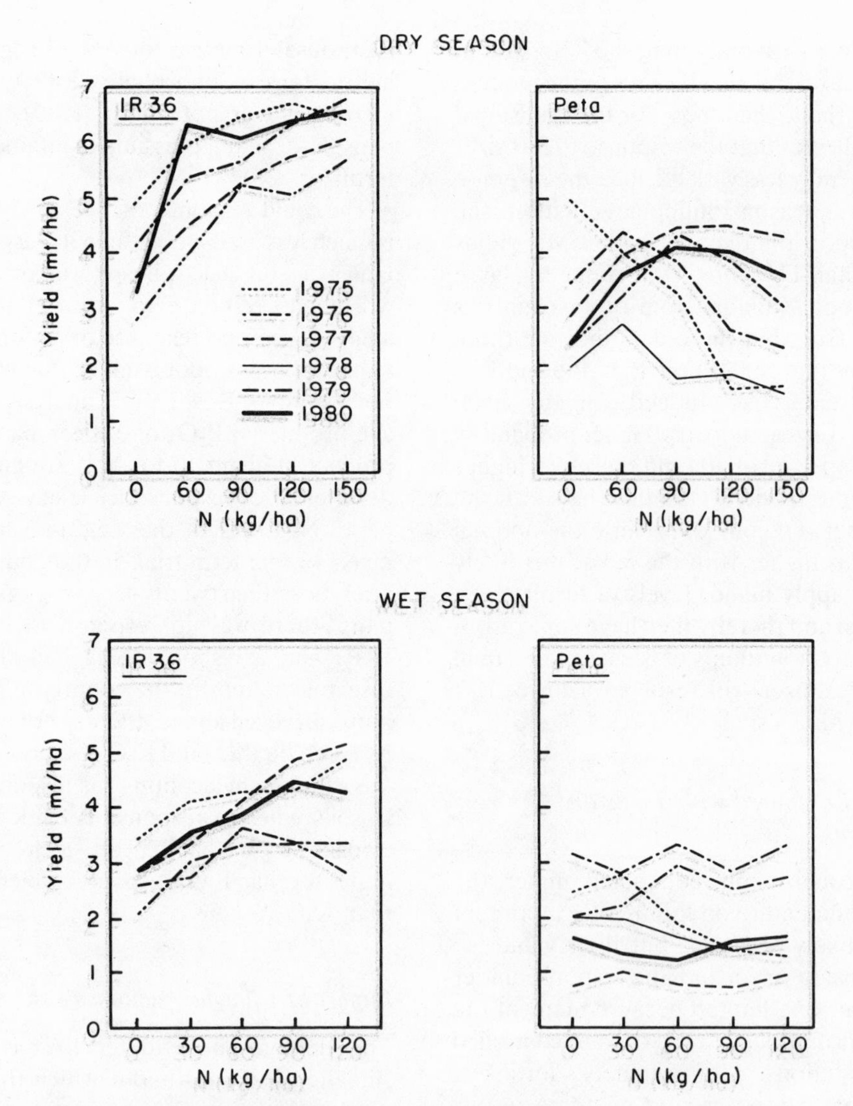

Figure 6.3. Nitrogen response of one modern (IR36) and one traditional rice variety (Peta) in trials at IRRI, 1975–80

radiation available for plant growth in the dry season. Where the plant has adequate water, this higher level of solar radiation leads to better nutrient uptake and higher yields—as reflected in the Philippine and Indian data cited above. However, if water stress occurs, that potential is not reached.

Comprehensive analyses of a number of data sets from experiments both in farmers' fields and at research stations in the Philippines have resulted in response functions that can be used to represent a wide range of rice production conditions. Figure 6.4 shows yield responses for modern varieties in the Philippines under wet and dry season conditions with high (20 days) and low (5 days) levels of moisture stress.[17] These functions represent relatively favorable (low stress) and unfavorable production conditions. The analyses from which they were synthesized encompass some functions that explicitly include the effects of insects, diseases, weeds, soil factors, and other variables that affect yields, but the major climatic effects are reflected in the impacts of season and stress.

The wet season typically has lower levels of solar radiation, hence maximum yields are obtained with smaller fertilizer applications than in the dry season. In either season, moisture stress dramatically reduces yields. High levels of stress are probably more frequently experienced in the dry season so the extra sunlight is only an advantage when excellent irrigation systems are available. Fertilizer response is greater in the dry than in the wet season if water is adequate. However, most Asian irrigation systems provide water less reliably in the dry season.

Another related factor contributing to stress is the

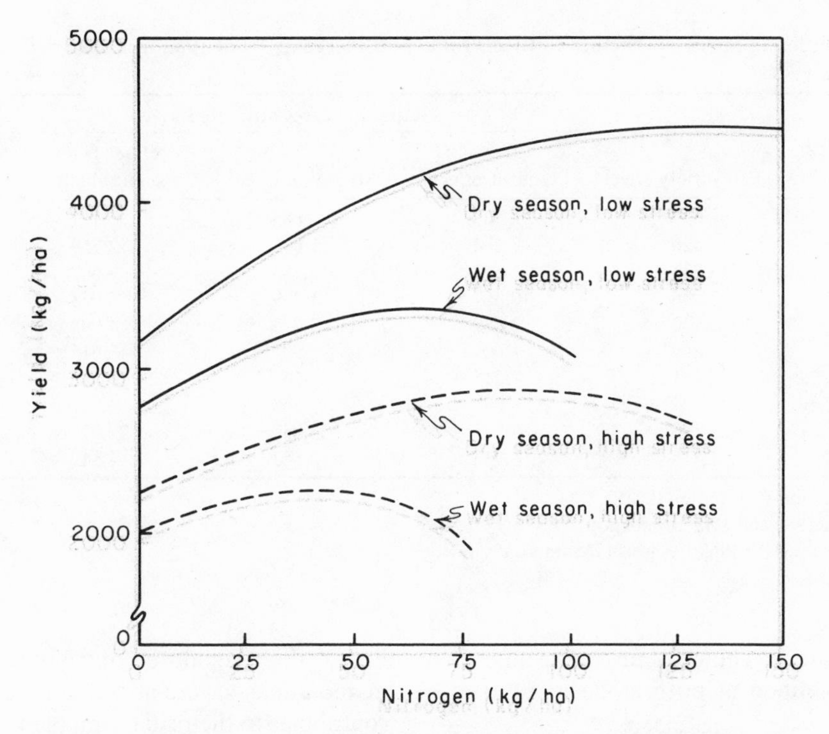

Figure 6.4. Synthesized response curves of modern varieties reflecting the impact of fertilizer and climatic factors on rice yields

soil texture, which determines seepage and percolation (S & P) losses of water. In their simulation of the effect of fertilizer, Wickham, Barker, and Rosegrant identify "ideal," "good," "average," and "poor" irrigation performance in the dry season and "ideal," "irrigated," and "rainfed" performance in the wet season for soils typical of Central Luzon, Philippines.[18] (They omitted poor irrigation in the wet season as a separate category because its performance is virtually the same as rainfed.) The mean number of stress days was zero with ideal irrigation, about five days with minimum S & P and average quality irrigation, and about fifteen days with high S & P and average quality irrigation in the dry season. Yield disparities between the two situations result from differences in both the *a* coefficient and the yield response (ΔY) and amount to about 1 mt/ha of rough rice. Wet season yield differences between minimum and high S & P conditions are about half as large.

There are few data sets that document the level of stress experienced on farmers' fields. In one comparison of rainfed and irrigated rice farms in two provinces of the Philippines, it was found that the rainfed farms experienced 13.5 days of drought and 0.8 days of flood while the irrigated farms experienced 8.5 days of drought and 0.6 days of flood. The extent to which this is true for other locations is unknown, but without doubt it contributes to a national average response of rice to fertilizer much below the agronomic potential of available varieties.

Aggregate Contribution of Fertilizer to Rice Production

The foregoing sets of fertilizer-response functions provide some indication of the factors that affect the yield response of rice to fertilizer. To derive the contribution of fertilizer to increases in rice output, one must either have an overall average fertilizer-response function or a limited number of well-specified disaggregated response functions. We have argued that the response of rice varies with variety type, quality of irrigation, and season. The average response may therefore vary for different countries, but all of the response functions examined above give little basis for any categorical statement about the differences between countries.

Countries are distinctly different in the level of rice production technology they use, especially in the proportion of area irrigated and in the extent to which modern fertilizer-responsive varieties are planted. The analysis of the contribution of fertilizer to rice output made in this section was linked to the irrigation and variety factors (table 6.4). We used a more complex approach than the growth rate technique used in

Table 6.4. Contribution of Specified Factors to Rice Production Increases from 1965 to 1980
(output in thousand mt paddy)

	Output increases attributed to				
Country	Modern variety effect	Fertilizer effect	Irrigation effect	Other factors (residual)	Total growth in output[a]
Burma	647	353	685	167	1,852
Bangladesh	420	1,284	1,091	2,759	5,554
China	13,231	11,507	16,153	9,609	50,500
India	7,998	10,867	11,209	5,078	35,152
Indonesia	3,162	2,680	2,773	4,998	13,613
Philippines	849	1,009	801	615	3,274
Sri Lanka	241	215	262	316	1,034[b]
Thailand	822	682	865	4,031	6,400
Total of above	27,370	28,597	33,839	27,573	117,379
Value ($US million)[c]	4,516	4,718	5,583	4,549	19,367

Source: R. W. Herdt and C. Capule (1983).
[a]Difference between 1980 and 1965 production.
[b]Three-year average used for 1965 because 1965 unusually low.
[c]Valued at $US165/mt.

chapter 4 and has the advantage of providing an estimate of the contribution of both modern varieties and fertilizers.

A key concept in the exercise is the response of rice yield to fertilizer under the specified conditions. We have shown above that yield response is greater with modern varieties than with traditional varieties and that moisture stress reduces the response. To calculate the effect of all changes in technology, the 1965 levels of irrigated land, fertilizer, and modern variety adoption were substituted in the model with actual levels of all other variables. The difference between actual production and that estimated assuming the 1965 levels of the three factors was taken as the measure of the total effect of changes in those three factors. Actual change in output reflects the impact of all changes and includes the effects of increases in the three specified factors plus other, unmeasured factors such as changes in land area, labor, and complementarity among factors.

The separate effects of irrigation, fertilizer, and modern varieties were calculated as follows. Substituting the 1965 level of irrigated area and fertilizer into the model with the 1980 level of all other factors to estimate production, and subtracting this from estimated production with the 1980 level of all factors gave a measure of the contribution of MVs. Substituting the 1965 level of MVs and irrigated area and subtracting estimated production from the 1980 level gives a measure of the contribution of fertilizer. Substituting the 1965 level of MVs and fertilizer and following a similar procedure leads to a measure of the irrigation effect. However, the sum of the three "effects" exceeded their total measured contribution because of their complementarity. To derive a measure of their separate contributions, the three estimated impacts were added, and the proportion each contributes to their sum computed. Clearly, the results depended on the land data in each category and the fertilizer applied to rice as well as the area in MVs.

Table 6.4 shows the results of the exercise. The total values are quite large—rice output in eight countries was nearly 120 million metric tons higher in 1980 than it was in 1965. Roughly equal amounts of the increase are attributed to varieties, fertilizer, irrigation, and residual unmeasured factors. The value of the increased production from each factor is $US 4.5 to 5.0 billion. The proportions attributed to the four factors are different in each country, reflecting differences in the levels of MVs, fertilizer, and irrigation, as well as in their productivity.

Given the estimated allocation of land between modern and traditional varieties, and irrigated and rainfed, if no fertilizer had been applied to rice, output growth would have been 89 million metric tons instead of the reported 117 million metric tons. Thus, it is clear that fertilizer was a major contributing input to rice production in 1980. Its contribution to output growth has been smallest in Thailand (11 percent) and Burma (19 percent), where increases have been rather small, but fertilizer contributed an average of 24 percent of output growth since 1960 for the group as a whole.

The Demand for Fertilizer

One of the most important policy questions is the degree to which the level of fertilizer application depends on price. Timmer and Falcon drew attention to the strong correlation between the relative price of rice to fertilizer and per hectare fertilizer applications across Asian countries. They suggested that "price may be more important in the development process than any of us have realized."[19] An alternative explanation is that only the more developed countries, such as Japan and South Korea, can afford high prices of rice relative to fertilizer and that prior investments in water control, human capital, and research have raised the fertilizer-response curve to a high level while in less developed countries the response curve is much lower.

A study of fertilizer-response functions permits an estimation of the effect of modern varieties on fertilizer consumption, and from such response functions, one may compute the effect of price changes on fertilizer consumption. One may also estimate a fertilizer demand function directly from data on fertilizer price and consumption. The hypothetical relationship between varying levels of fertilizer response function and the corresponding demand functions are shown in figure 6.5. The advantage of the demand functions (based on survey data) over response functions (derived from experimental data) is that the former reflect farmers' decisions and incorporate their response to factors such as risk. Furthermore, the demand formulation permits a direct estimation of farmers' response to price and the use of other inputs complementary to fertilizers. Thus, by taking into account factors responsible for shifts in fertilizer-response functions as would be reflected in curves d_1, d_2, and d_3 in figure 6.5, estimates can be made of short-run price elasticity of demand for fertilizer use on rice (percentage change in fertilizer input due to a given percentage change in the relative price of fertilizer to rice). The long-term response, for example, d_1 may represent the situation with no irrigation and no MVs, d_2 the situation with irrigation and no MVs, and d_3 the situation with both irrigation and MVs. The corresponding yield response curves are P_1, P_2, and P_3.

The relationship between price and fertilizer input that Falcon and Timmer observed reflects the long-run relationship D in figure 6.5. It cannot be interpreted as the response of farmers to a unit change in price in any particular country because the correlation is based on farmers' behavior in situations of varying fertilizer productivity.

David estimated fertilizer-demand functions for the Asian rice economy using one aggregate and two farm-level sets of data.[20] Variation in fertilizer consumption among countries, across villages, and over time was estimated as a function of the fertilizer–rice price ratio, proportion of area in modern varieties, and other factors, such as weather and irrigation. The price elasticities derived from the simple relationship between fertilizer use per hectare and the fertilizer–rice price ratio were remarkably stable at around −0.8 across three sets of data (meaning that one percent change in the price ratio results in

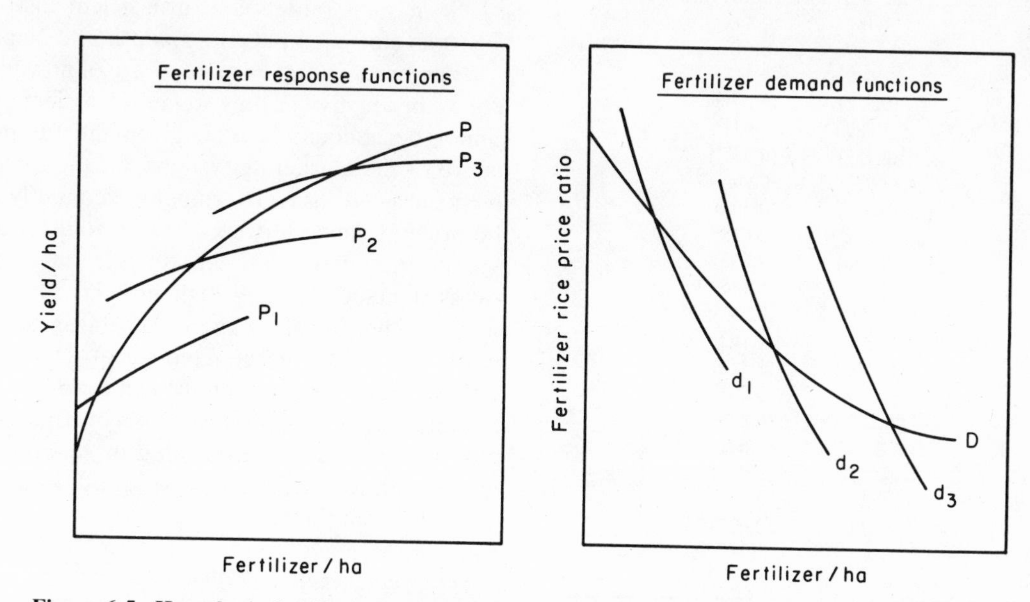

Figure 6.5. Hypothetical shifts in fertilizer-response functions and their corresponding demand functions (Source: C. C. David, "Factors Affecting Fertilizer Consumption," in International Rice Research Institute, *Interpretative Analysis of Selected Papers from Changes in Rice Farming in Selected Areas of Asia* [Los Banos, Philippines, IRRI, 1978] p. 69, reprinted by permission of the publisher)

a 0.8 percent change in fertilizer input in the opposite direction). This price elasticity can be regarded as a long-run response, D in figure 6.5, although one may be unwilling to accept the causal relationship implied. The estimated short-run elasticity of demand, which takes into account shifts in fertilizer-response functions due to factors such as modern varieties (d_1, d_2, d_3 in figure 6.5) had the expected lower values of from -0.4 to -0.7.

The estimated price elasticities of demand for fertilizer varied considerably among countries as shown in table 6.5. For seven countries the price elasticities of demand have the expected negative sign; that is to say, in the aggregate, farmers respond as anticipated by decreasing fertilizer inputs with a rise in the price of fertilizer relative to the price of rice. Where fertilizer levels are high and relatively more important in the farm budget—Japan, Taiwan, and South Korea—there is a greater sensitivity to price changes in contrast to countries where fertilizer applications are much lower—the Philippines and Indonesia. The estimated price elasticities of demand are positive in three cases—Burma, Pakistan-Bangladesh, and Thailand. Extremely low levels of fertilizer are used, and a small portion of rice land is irrigated in these areas, so the true value of the elasticity of demand is probably close to zero.

The study by David gives a much clearer understanding of the sources of change in fertilizer demand. Estimates were made of the relative contribution of each of the explanatory factors to the gap in fertilizer consumption between the average and heaviest fertilizer users. While significant differences exist in the estimated contribution of each factor across the three sets of data, the results generally indicate that differences in fertilizer-response functions provide the major explanation for the wide gap in fertilizer application. The differences in the response functions reflect the spread of modern varieties and other factors correlated with modern varieties such as irrigation. Differences in the fertilizer–rice price ratio explain up to about one-third of the difference in fertilizer inputs. Hence, we conclude that farmers in many Asian countries still apply low levels of fertilizer not only because of unfavorable prices, but also because of the smaller yield response of rice to fertilizer under their environmental conditions.

Table 6.5. Fertilizer Demand Function Estimated Using Aggregate Asian Data, 1950–72

Country	Intercept	Fertilizer–rice price	Modern varieties
Philippines	1.482	−0.492	1.191
		(0.416)	(3.927)
Japan	1.660	−0.723	
	(0.312)	(0.191)	
South Korea	1.389	−0.931	
	(0.157)	(0.345)	
Taiwan	1.727	−0.968	
	(0.397)	(0.382)	
Sri Lanka	2.332	−0.818	
	(1.230)	(0.262)	
Indonesia	1.198	−0.186	
	(0.402)	(0.243)	
Thailand	−0.277	1.192	
	(2.563)	(1.412)	
Burma	−0.200	0.503	
	(2.394)	(0.875)	
India	2.045	−1.671	
	(0.704)	(0.845)	
Pakistan-Bangladesh	0.217	2.309	
	(−1.781)	(2.078)	
For reference:			
$R^2 = 0.928$			

Note: Numbers in parentheses for the Philippines show the t value of the variables. For other countries, they show the t value of the dummy variable for each country and thus test the significance of difference between that country and the Philippines.

Source: C. C. David and R. Barker (1978).

Marketing, Distribution, and Pricing of Fertilizer

The governments of most developing Asian countries would like fertilizer use to expand more rapidly than it has in the past. For several countries still at an early stage of use, this expansion is hindered by a lack of sales outlets and insufficient distribution networks (for example, Bangladesh and Nepal).

When farmers first become aware of how fertilizer affects productivity, they generate a demand that cannot be adequately met, given the rudimentary nature of the marketing systems. At the same time, to encourage its use, governments frequently control the price at which fertilizer can be sold to farmers. A fixed price below the market price may result in excess demand, causing high margins for distribution; another result may be shortages and black markets (sales above the fixed price). One response to these problems has been the institution of a public marketing system or strict controls over pricing; this may, in turn, lead to a backlog of supply at ports and warehouses (because of inadequate legal marketing margins).

Government Fertilizer Policies

The degree of public control over fertilizer varies. In China, fertilizer distribution, like all enterprises of any size, is managed by publicly owned firms. Bangladesh, Nepal, Sri Lanka, and Burma also have public sector firms that import and distribute fertilizer and agricultural chemicals. Generally, these have been set up and operated as public sector monopolies. After some experience with public monopoly, some nations have changed their rules to permit private firms to enter the distribution process.

Indonesia, Pakistan, and the Philippines have a mixture of private and government distribution, with the level of imports determined by government and the wholesale distribution carried out by government firms, but with retail sales made by village cooperatives and private dealers. In South Korea, fertilizer was distributed by government firms through the nation's village cooperatives during the late 1970s. Thailand and Malaysia had the least government intervention in fertilizer marketing during the 1970s. The Malaysian government had no control over fertilizer marketing, but did encourage distribution through cooperatives during the 1970s. In 1980, in an attempt to increase fertilizer use, the government provided free fertilizer to any rice farmer to cover 6 acres at the recommended rate. In Thailand, fertilizer is distributed through the private sector. The government imposed a tariff on imported fertilizers, but that was removed in the late 1970s.

Government intervention in fertilizer merchandising in Asia is often complex. Indonesia, a case in point, controls fertilizer pricing and maintains a mixture of public and private distribution. On the demand side, the government finances farmers' purchases of fertilizer through the BIMAS program. On the supply side, the government awards quotas to licensed fertilizer importers.[21] In addition, large agricultural estates are licensed to import directly from overseas. Indonesia produces a substantial amount of fertilizer through five government enterprises. P.T. Pusri, the primary company, is responsible for distributing all fertilizer at the national level. Prices, distribution margins, and retail prices are fixed by the government, with substantial subsidies in some years. The government appoints distributors and retailers, and establishes village cooperatives to handle retail distribution.

Such a high degree of control forces government accountants to calculate costs and set profit margins at every stage in the marketing process. As a result, numerous problems arise. For example, the distribution margins fixed by the government are attractive for sellers near ports since transport costs are lower than the official allowances. Because the distribution system is based on consignments, it does not encourage sellers to take responsibility for losses arising from poor handling. Wholesalers and retailers receive no allowances for overhead costs, and thus have no incentive to become actively involved in fertilizer distribution.[22]

In the Philippines, the fertilizer distribution system is somewhat less rigid. The Fertilizer and Pesticide Authority sets regional exwarehouse prices for fertilizer and authorizes imports to meet estimated needs. Local distribution costs are determined for each province, with maximum retail prices controlled by official provincial price stabilization councils.[23] The difference between costs and sales prices is periodically adjusted to maintain a desirable ratio between fertilizer and rice prices while keeping subsidies within reasonable levels. In late 1981, the government announced a plan to phase out fertilizer subsidies because of the fiscal burden.

One result of the desire to keep fertilizer prices at a controlled low level and to expand the use of fertilizer has been rather large subsidies. During most of the 1970s, Sri Lanka was selling fertilizer to the small-scale agricultural sector (which includes rice farmers) at a high subsidy rate, which amounted to over $US 200/mt of urea.[24] Bangladesh subsidized 47 percent of the cost of fertilizer in 1978/79, with a total subsidy that amounted to three-quarters of the total development budget of the government.[25] About one-third of Bangladesh's fertilizer needs are obtained from foreign donors at a low cost to the government. Nepal also receives much of its fertilizer as foreign aid and has government policies that mandate a uniform price through the country. This requires a large transportation subsidy to keep prices lower in the hills. Fertilizer subsidies cost the Fertilizer and Pesticide Authority of the Philippines over $US 50 million in 1975 when world fertilizer prices peaked.[26] In 1977, it was estimated that the Indonesian government paid a subsidy equal to 12 percent of the actual cost of urea, or $US 24/mt, which totaled over $US 30 million.[27]

One result of extensive government intervention in fertilizer marketing is that prices of fertilizer vary widely across countries. This, plus the sharp fluctuations in the world price, means that it is not very meaningful to talk about a single fertilizer price. Table 6.6 shows some of these effects. In 1960, nitrogen was priced at $US 0.12/kg in Pakistan and was four times as costly in South Korea. In 1970, most countries had lower prices for nitrogen fertilizer than they had in 1960, but Thailand had a higher price. The lowest price was $US 0.25/kg in Indonesia, while the Philippines, Thailand, and South Korea had a price twice as high, and in Japan it was higher still. Thus, the differences across countries seem to have declined, perhaps because as fertilizer use grows, the financial burden of large subsidies becomes more and more difficult for governments to bear.

Table 6.6. Prices Paid by Farmers for Nitrogen
($US/kg)

Country	1960	1970	1980
Indonesia	0.29	0.16	0.24
Sri Lanka	0.25	0.15	0.28
Pakistan	0.12	0.26	0.41
Bangladesh	0.13	0.15	0.35
Taiwan	0.37	0.25	0.37[a]
India	0.34	0.28[b]	0.55
Philippines	0.31	0.19	0.59
Thailand	0.27	0.32	0.56[a]
South Korea	0.53	0.19	0.56[c]
Japan	0.25	0.22	0.86

Source: A. Palacpac, (1982) pp. 73–75.
[a]Refers to 1978.
[b]Refers to 1971.
[c]Refers to 1977.

To counteract the growing dependence on imported fertilizer, many Asian countries have developed their own fertilizer production capacity. This development was given impetus by the sharp rise in fuel oil prices in 1974 and the discovery during the 1970s of substantial deposits of natural gas in Bangladesh, Indonesia, and India. However, local production is frequently more costly than imported fertilizer, especially when one takes into account the subsidies given for fertilizer plant construction.

Production and Imports of Fertilizer

Among the developing countries of Asia that became independent after World War II, India and China led in the development of fertilizer production capacity (table 6.7). China pioneered in the development of small-scale chemical fertilizer plants operated at the commune level.[28] By 1970, these plants had a production capacity of almost 2 million metric tons. In the 1970s, the Chinese began construction of conventional large-scale urea plants, tripling production to over 6 million metric tons by 1980.[29] During the late 1950s, surplus electrical power from one of India's first postindependence dams was used to produce nitrogeneous fertilizers. From that modest start, India has built the second largest production capacity in the region, surpassing Japan in 1976–77. Both the private and public sectors participate in production, distribution, and sales. The Fertilizer Association of India, a trade organization, has an active program promoting fertilizer use, and Indian fertilizer experts provide consulting services to other developing countries. Despite controls on prices and a complex system of nonmarket distribution rules, fertilizer production in India increased very rapidly during the 1970s, apparently because the opportunities for profitable production and sales continue. In 1979, India was reported to have installed production capacity of 3.3 million metric tons of nitrogen and 1.1 million mt of phosphate. The country planned to raise capacity to 5.3 million mt of nitrogen and 1.3 million mt of phosphate by 1981/82.[30]

Other countries of the region have also increased fertilizer production rapidly, especially in the 1970s. Bangladesh, Malaysia, Burma, and Vietnam all nearly doubled their production from the first half to the second half of the 1970s. Indonesia increased its production by a factor of four over the period (table 6.6). Despite this rapid growth in production, however, the area remains dependent on imports for a large fraction of its fertilizer needs.

Only Japan and South Korea did not import fertilizer during the 1970s. Sri Lanka and Nepal imported all their needs. Pakistan, Malaysia, Thailand, the Philippines, and Vietnam all imported more than they produced domestically. Even China and India, the two biggest producers, imported over a million tons each in the late 1970s. Considering the rapid rate of increase in use and the large capital investment needed for modern fertilizer plants, it is likely that the region will continue to import a substantial amount of fertilizer in the foreseeable future.

Agricultural Chemicals

The developing Asian countries use relatively small amounts of agricultural chemicals to control insects, weeds, and plant diseases, especially compared with Japan and Korea. The data in table 6.8 show, for example, that India, with 168 million hectares of agricultural land, used less than 55,000 metric tons of pesticides or approximately 0.33 kg/ha of active ingredients of all types of pesticides during the 1970s. Levels in Indonesia, Burma, Sri Lanka, and Bangladesh were similar. Japan, with 5.5 million hectares, used about 79,000 metric tons, or 14.3 kg/ha. Korea used about the same level per hectare as Japan.

Agricultural chemicals can be classified into six types. Insecticides, used to control insects, make up the bulk of the chemicals used, with fungicides not far behind. Most of the fungicides are used in Japan and Korea where, for example, in the 1970–74 period insecticide consumption was 18,016 mt/yr while fungicide consumption was 24,738 mt/yr. Fungicide use was relatively minor in the developing Asian countries. Other agricultural chemicals to control seedborne diseases, weeds, rats, and other pests are

Table 6.7. Annual Average Asian Fertilizer Production (1961–79) and Imports (1975–79)

	Thousand mt of N + P_2O_5 + K_2O produced				Imports
Country	1961–65	1966–69	1970–74	1975–79	1975–79
South Asia	376	723	1,642	3,071	2,048
India	302	606	1,316	2,558	1,335
Pakistan	41	74	252	357	404
Sri Lanka	0	0	0	1	115
Bangladesh	33	43	74	155	176
Nepal	0	0	0	0	18
Southeast Asia	115	150	323	818	1,195
Malaysia	0	13	38	64	283
Thailand	0	5	9	5	255
Philippines	15	59	95	79	211
Indonesia	21	42	81	491	201
Burma	0	0	33	55	14
Vietnam	79	31	67	124	231
East Asia	3,007	4,863	7,704	9,630	—
Japan	1,849	2,620	2,856	2,089	(805)
North Korea	149	195	325	556	69
South Korea	50	263	599	970	(284)
China	959	1,785	3,924	6,015[a]	1,352[a]

Note: Figures in parentheses indicate exports.
Sources: FAO, *Fertilizer Yearbook*; FAO, *Monthly Bulletin of Statistics* vol. 5, no. 3 (March 1982).
[a]1975–78 average.

Table 6.8. Pesticide Consumption in Agriculture in Selected Asian Countries, 1960–78

	Metric tons of active ingredients				kg/ha agricultural land
Country	1960–64	1965–69	1970–74	1975–78	1970s
India	20,316	34,742	22,833	54,305	0.33
Sri Lanka	1,934	869	224	232	0.11
Bangladesh	n.a.	n.a.	166	222	0.02
Malaysia	6,164	5,766	n.a.	5,599	1.92
Thailand	1,665	5,991	n.a.	16,847	0.97
Philippines	698	2,968	8,729	9,501	1.36
Indonesia	846	n.a.	937	6,255	0.38
Burma	370	332	128	3,138	0.16
Japan	57,784	69,758	64,450	78,673	14.30
South Korea	13,776	12,556	37,508	10,415	10.70
Total	103,553	132,982	134,975	185,187	

Sources: United Nations, *Statistical Yearbook for Asia and the Pacific*, 1970; FAO, *Production Yearbooks*. Philippines: National Census and Statistics Office, *Foreign Trade Statistics*. South Korea—1961–72: P.Y. Moon and B.S. Yoo (1974).

used in very small quantities. There is, however, a rising interest in agricultural chemicals. Much of this is in Japan and Korea, but usage is also increasing in the developing countries, although at a much lower level. In the developing nations listed in table 6.8, total use increased from about 32,000 metric tons in the early 1960s to 96,000 in the late 1970s, an annual compound rate of growth of 7.6. This is below the 10 to 15 percent rate of growth in fertilizer use, but is still appreciable and a cause for ecological concern.

Pesticide Use on Rice

There are no national aggregate data that show pesticide use on rice. Tables 6.9 and 6.10 illustrate the kind of information that is available about pesticide application rates on rice. These data, derived from farm surveys, show a significant increase in the use of pesticides on rice in Taiwan between the early 1960s and the early 1970s. The same was true for a small sample of farms in the Philippines. In both

Table 6.9. Selected Costs and Returns on Rice Farms in Central Taiwan

	1961		1967		1972	
	$NT/ha	Percent	$NT/ha	Percent	$NT/ha	Percent
Total paid-out costs	5,164	30	9,879	52	12,869	61
Fertilizer	2,544	15	3,296	17	2,605	12
Insectides	242	1	1,190	6	1,290	6
Herbicides	0	0	0	0	285	1
Gross crop return	17,489	100	18,965	100	21,081	100

Source: L.S. Tsai (1976).

Table 6.10. Selected Costs and Returns on Rice Farms in Central Luzon, Philippines

	1966		1970		1974		1979	
	Pesos/ha	Percent	Pesos/ha	Percent	Pesos/ha	Percent	Pesos/ha	Percent
Total costs	233	24	364	31	844	40	1,383	38
Fertilizer	28	3	57	5	245	12	338	9
Insecticides	2	0	9	1	50	2	87	2
Herbicides	0	0	1	0	18	1	30	1
Gross crop return	984	100	1,177	100	2,092	100	3,593	100

Source: V.G. Cordova, A.M. Mandac, and F. Gascon (1980).

locations, expenditure on insecticides was much lower than on fertilizer, and expenditure on herbicides was even smaller. The situation is similar in other rice-producing areas. Use of pesticides on rice is at a relatively low level.

The Effect of Pesticides on Yields

Pesticides have a somewhat different effect on rice yields than do fertilizers. The initial application of fertilizer almost always adds to crop yield while application of pesticides prevents losses in situations where pests occur. That is, the level of pest incidence is highly variable from crop to crop and so the benefits from applying insecticides are also highly variable. Also, some rice varieties are resistant to some insects. As a result, scientists concerned with pest protection have followed several strategies. From the farmers' viewpoint, the lowest-cost and simplest is the development of pest-resistant varieties. However, resistance is usually specific to particular insects. Rices with resistance to several pests are difficult to develop and may not incorporate other desirable characteristics. Aside from resistant varieties, insect protection can be obtained from (1) prophylactic treatments designed to prevent damage or prevent pest buildup, (2) application in response to a preidentified "threshold" level of pests.

Generalizations about yield response to pesticide application levels are difficult to make because of the extremely large number of materials and pests, each of which interacts somewhat differently, because of the problems in equating pesticides with different types of effectiveness, and because of the absence of a universal method for measuring pest incidence or pressure. Despite these problems, it is useful to examine the economics of pesticide use.

A Case Study of Insect Control on Rice

In the following analysis, insect control alternatives are examined to determine whether some strategy among those tested can be identified that effectively prevents a substantial portion of the yield losses and provides a reasonable economic return to the farmer with a reasonable level of risk. Then a more general formulation of the economics of insecticide application is derived.

IRRI entomologists cooperated with scientists at three Philippine Bureau of Plant Industry research stations between 1972 and 1974 in testing various insecticide treatments on a number of rice varieties. Both resistant and nonresistant varieties were included. The entomologists concluded after three years of trials that insect problems change from one year and location to another, that significant yield increases are usually obtained with insecticides, that in some cases insecticides pay for themselves while in other cases they do not, and that insecticide treatments must be tailored for each season, year, location, and variety.

Because of the uncertain pattern of insect attacks, entomologists have developed the concept of thresholds—levels of insect activity used to indicate when protective measures should be taken. In the studies being considered, the threshold concept was used in some years but not in others. Treatments tested always included a maximum protection plot and an unprotected plot. Variability among locations, years, and varieties, as reflected in the standard deviation of yields, was substantial. For our analysis, we used the number of applications and costs of insecticide as the measure of insecticide input. Table 6.11 shows the results. As would be expected, as input increased, yields tended to increase, but the relationship was not smooth and regular as with fertilizer, especially on the resistant rices. This raises the question of whether the use of high levels of pesticides raises or reduces risk.

The risk associated with using various insect control treatments can be reflected in the probability that a loss will occur, where a loss is defined as a treatment that gives a lower net benefit than the untreated control. The amount of the loss is another part of the risk. One may take the product of these and calculate the expected value of the loss as shown in table 6.12.

On nonresistant varieties, four applications of insecticide always gave an increase in net benefits, so the probability of loss and the expected loss are zero for that treatment. Other treatments had probability of losses between .20 and .36. The expected loss was highest for three applications, but only four observations are available for that treatment so it was omitted from the risk evaluation in the table. On resistant varieties, risk was low for one, two, and four applications and much higher for five or more. The contrast in results between resistant and nonresistant varieties is as expected—a low level of protection is less risky on resistant varieties.

The pattern of risk among treatments for one variety type also seems consistent with our expectations. On resistant varieties, low levels of application entail low risk because both the probability of loss and the average loss are low. High levels of application are much more risky on resistant varieties because these varieties do not give as high a yield response to insecticide, so both the probability of loss and the average loss are high. The risk pattern on nonresistant varieties shows that when the high-cost treatments result in losses, these losses are fairly large because of the costs. Low-cost treatments have smaller losses. The probabilities of loss are low but similar for high- and

Table 6.11. Insecticide Treatments and Resulting Yields with Five Levels of Insect Protection, 1972–74 Dry Seasons, Philippines

Treatment		Nonresistant rices			Resistant rices		
Cost ($US/ha)	Applications (no.)	Yield (mt/ha) Average	Yield (mt/ha) S.D.	No. of observations	Yield (mt/ha) Average	Yield (mt/ha) S.D.	No. of observations
0	0	3.6	0.79	14	4.2	0.76	21
9	1	4.1	0.64	10	4.8	0.79	17
19	2	4.2	0.35	16	4.9	0.58	30
31	3	4.2	0.54	4	—	—	0
47	4	4.6	0.95	18	5.0	0.80	33
109	5	5.2	1.50	12	4.8	0.81	10
187	7	4.9	0.87	4	5.5	0.83	4
251	10	5.7	0.48	6	6.1	0.63	11

Source: Calculated from data obtained from the IRRI Entomology Department.

Table 6.12. Risk Associated with Insecticide Applications on Resistant and Nonresistant Rices, Dry Seasons 1972–74, Philippines

	Nonresistant rices			Resistant rices		
Insecticide applications (no.)	Frequency of loss[a] (percent)	Loss[b] when present $US/ha	Expected value of loss $US/ha	Frequency of loss[a] (percent)	Loss[b] when present $US/ha	Expected value of loss $US/ha
Maximum[c]	.36	60	22	.67	91	61
5	.25	71	18	.75	79	59
4	0	—	0	.21	28	6
2	.37	32	12	.17	23	4
1	.20	23	5	.18	40	7

[a]Defined as net return lower than zero insecticide application.
[b]Average loss in those cases where losses occurred.
[c]In some cases this was seven, in others it was ten applications.

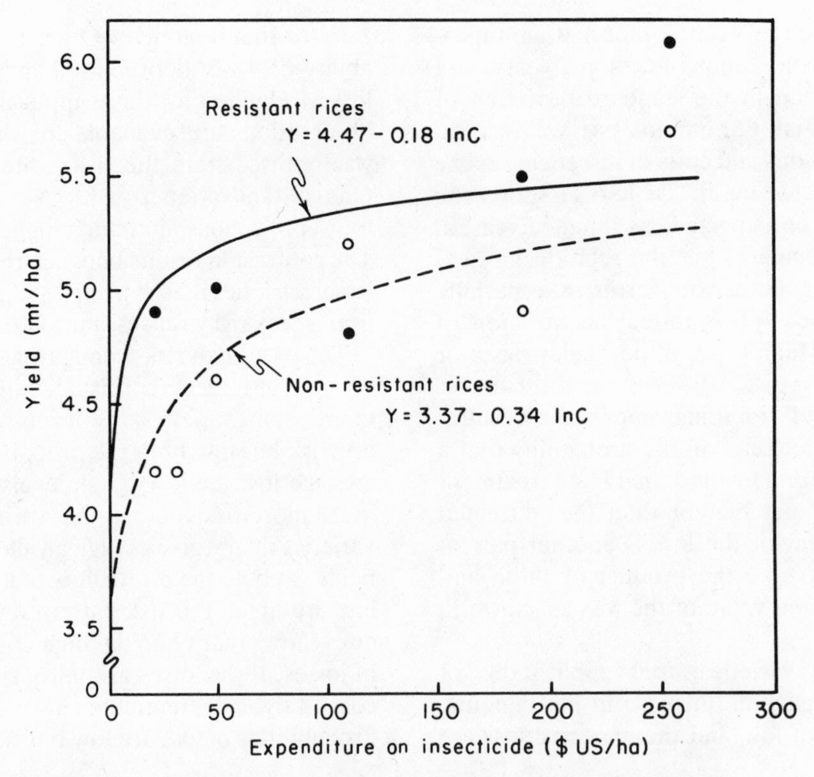

Figure 6.6. Yield response of insect-resistant and nonresistant rices to applications of insecticide in experiments at four Philippine research stations, 1971–74

low-cost treatments, so expected losses are somewhat higher for high-cost treatments.

The data provide fairly convincing evidence for the conclusion that in the dry season, one or two applications of insecticide will provide a high payoff with a low risk on moderately insect-resistant varieties, and that up to four applications of insecticide will give a high payoff with a low risk on nonresistant varieties.

The data were fitted to a logarithmic equation with results shown in figure 6.6. The level of insecticide protection that would maximize the return over insecticide cost, given the two equations, is $US 51 for the nonresistant rices and $US 27 for the resistant rices (assuming a rice price of $US 150/mt). These correspond to four treatments of the nonresistant rices and two to three of the resistant rices—the same levels indicated by the risk analysis. These show the value of resistant rices: without insecticide the resistant rices yielded 0.6 mt/ha more than the nonresistant ones, and with modest levels of insecticide maintained that advantage.

Thus, although there is considerable variability in the incidence of insect pests and therefore in the response of rice yields to insecticide, it appears that with adequate data, the contribution of insecticides can be measured using response functions something like those in figure 6.4. A mass of data like those analyzed for fertilizer in the early section of the chapter are not available for insecticide,[31] however, so we do not attempt to calculate the contribution of insecticides to rice production.

Notes

1. Yujiro Hayami and Vernon W. Ruttan, *Agricultural Development: An International Perspective* (Baltimore, Md., Johns Hopkins University Press, 1971) p. 195.

2. Mohinder S. Mudahar and Travis P. Hignett, *Energy and Fertilizer: Policy Implications and Options for Developing Countries* (Muscle Shoals, Ala., International Fertilizer Center, 1982) p. 64.

3. Anthony M. Tang and Bruce Stone, *Food Production in the People's Republic of China*, IFPRI Research Report No. 15 (Washington, D.C., International Food Policy Research Institute, 1980) p. 47.

4. Ibid.

5. This assumption is based on trends in organic fertilizer use between 1970 and 1977 documented in Tang and Stone, *Food Production in the People's Republic of China*.

6. The world price of urea went from $US 46/mt FOB to western

European ports in 1972 to $US 390/mt at its peak in 1974, before falling back to $US 100/mt in 1976. Food and Agriculture Organization, *Food Outlook: 1981 Statistical Supplement* (Rome, January 1982).

7. This is reflected in two recent research symposia, "Nitrogen and Rice" and "Organic Matter and Rice," held at the International Rice Research Institute in 1978 and 1982, respectively.

8. International Rice Research Institute, *Annual Report for 1980* (Los Banos, Philippines, IRRI, 1981) p. 270.

9. R. C. Valentine, "Genetic Engineering in Agriculture with Emphasis on Biological Nitrogen Fixation," in National Academy of Sciences, *Research with Recombinant DNA* (Washington, D.C., NAS, 1977).

10. C. C. David and R. Barker, "Modern Rice Varieties and Fertilizer Consumption," in International Rice Research Institute, *Economic Consequences of New Rice Technology* (Los Banos, Philippines, IRRI, 1978).

11. Data showing yield response at Bangladesh experiment stations are found in A. C. Roy, "Fertilizer Response of Rice at BRRI Farms in Different Seasons," in Bangladesh Rice Research Institute, *Workshop on Ten Years of Modern Rice and Wheat Cultivation in Bangladesh*, (Dacca, BRRI, March 7–10, 1977); on-farm data analyzed in C. Capule and R. W. Herdt, "Response of Non-irrigated Rice to Fertilizer in Farmers' Fields in Bangladesh, 1970–75," IRRI Agricultural Economics Paper 81-01 (Los Banos, Philippines, International Rice Research Institute, 1981).

12. Data showing yield response at experiment stations in the Philippines and Thailand are shown in table 6.3. On-farm yield response functions for the Philippines are reported in M. W. Rosegrant and R. W. Herdt, "Simulating the Impacts of Credit Policy and Fertilizer Subsidy on Central Luzon Rice Farms: The Philippines," *American Journal of Agricultural Economics* vol. 63 no. 4 (November 1981); and in Robert W. Herdt and Abraham M. Mandac, "Modern Technology and Economic Efficiency of Philippine Rice Farmers," *Economic Development and Cultural Change* vol. 29 no. 2 (January 1981). Experiment station data for Thailand reported in table 6.3. Farm-level data reported in Thailand, Rice Division and Planning Division, Department of Agriculture, *Annual Research Report* (Bangkok, 1971, 1973, and 1974).

13. Philippine Council of Agricultural Resources Research, *Data Series on Rice Statistics in the Philippines* (Los Banos, Philippines, IRRI, 1981) p. 226.

14. Rosegrant and Herdt, "Simulating the Impacts of Credit Policy and Fertilizer Subsidy on Central Luzon Rice Farms: The Philippines."

15. Capule and Herdt, "Response of Non-Irrigated Rice to Fertilizer in Farmers' Fields in Bangladesh, 1970–75."

16. International Rice Research Institute, *Annual Report for 1980* (Los Banos, Philippines, IRRI) p. 282.

17. The wet season curves were derived from the following equation: $Y = 3000 + 22.5N - .175N^2 - .50SD - .45SD^* N$; the dry season curves from $Y = 3500 + 22.0N - .085N^2 - 63SD - .38SD * N$; low stress was $SD = 5$, high stress was $SD = 20$. The analyses on which these were based are reported in Cal Montano and Randolph Barker, "Economic Returns from Fertilizer Application in Tropical Rice in Relation to Solar Energy Level," *The Philippine Economic Journal* vol. 13, no. 1 (1974) pp. 24–40; Mark Rosegrant, "The Impact of Irrigation on the Yield of Modern Varieties," IRRI Agricultural Economics Paper 76-28 (Los Banos, Philippines, International Rice Research Institute, 1976); Abraham Mandac and R. W. Herdt, "Environmental and Management Constraints to High Rice Yields in Nueva Ecija, Philippines," IRRI Saturday Seminar Paper (Los Banos, Philippines, International Rice Research Institute, September 8, 1974); and Thomas H. Wickham, "Predicting Yield Benefits in Lowland Rice through a Water Balance Model," in *Water Management in Philippine Irrigation Systems: Research and Operation* (Los Banos, Philippines, International Rice Research Institute, 1973).

18. T. H. Wickham, R. Barker, and M. W. Rosegrant, "Complementarities Among Irrigation, Fertilizers and Modern Rice Varieties" in *Economic Consequences of New Rice Technology* (Los Banos, Philippines, International Rice Research Institute, 1978).

19. Walter P. Falcon and C. Peter Timmer, "The Political Economy of Rice Production and Trade in Asia," in Lloyd Reynolds, ed., *Agriculture in Development Theory* (New Haven, Conn., Yale University Press, 1975) pp. 373–408.

20. C. C. David, "A Model of Fertilizer Demand of the Asian Rice Economy (a Macro-Micro Analysis)" (Ph.D. dissertation, Stanford University, 1975).

21. Much of the information in this section comes from the report: Agricultural Requisites Scheme for Asia and the Pacific," Fertilizer Marketing, Distribution and Use in Indonesia" (Bangkok, January 1978).

22. Ibid.

23. Economic and Social Commission for Asia and the Pacific, *Agro-Chemical News in Brief* vol. 2, no. 2 (April 1979).

24. Economic and Social Commission for Asia and the Pacific, *Agro-Chemicals News in Brief* vol. 1, no. 3 (November 1978).

25. Fertilizer Advisory, Development and Information Network for Asia and the Pacific, Economic and Social Commission for Asia and the Pacific, "Marketing, Distribution and Use of Fertilizers in Bangladesh" (Bangkok, 1980).

26. Economic and Social Commission for Asia and the Pacific, *Agro-Chemicals News in Brief* vol. 2, no. 2 (April 1979).

27. Agricultural Requisites Scheme for Asia and the Pacific, "Fertilizer Marketing, Distribution and Use in Indonesia."

28. International Rice Research Institute, *Rice Research and Production in China: An IRRI Team's View* (Los Banos, Philippines, IRRI, 1978).

29. Paul J. Stangel, "World Fertilizer Sector—at a Crossroads," paper presented at the Symposium on Food Situation in Asia and the Pacific Region April 24–29, 1980 (Taipei, Asian and Pacific Council, Food and Fertilizer Technology).

30. Economic and Social Commission for Asia and the Pacific, *Agro-Chemicals News in Brief* vol. 2, no. 3 (July 1979).

31. One substantial effort to analyze yield response to insecticide application is available in Ralph Young, "An Economic Analysis of Uncertainty in Rice Production" (Ph.D. dissertation, Cornell University, 1981).

7

Development of Irrigation

Historical Perspective

The irrigation of rice has a long and continuous history in Asia. Many highly sophisticated water control schemes developed independently in the pre-Christian era. Thousands of years ago the rice area in the upper reaches of the Yellow River, a cradle of Chinese civilization, began to spread, partly as a result of significant improvements in flood control. The beginnings of the institution of large cooperative water projects, a hallmark of Chinese agriculture on the Yellow River flood plains, is frequently attributed to Yu, founder of the first Chinese dynasty (Xia).[1] Centralized irrigation in Henan Province and in the Central Yangtze Valley began as early as 563 B.C. and 548 B.C., respectively.[2]

At about the same time, the ancient Sri Lankan Sinhala society began the development of a system of large tanks, or reservoirs, to irrigate portions of the fertile Dry Zone. The construction of this irrigation system required an amazingly high level of engineering skill and knowledge. The astounding complexity of the Sri Lankan tank system is described by R. L. Brohier:

> Most of the irrigation schemes are confined to tracts of land which when estimated by the eye appear to all purposes quite flat. Yet we know from such evidence which remains that channels were traced mile upon mile on gradients that would call into use the most precise instruments of the modern age to establish; that baffling ingenuity which cannot be surpassed by any conceivable means available at the present day traced out the bunds and the contours of the large tanks.[3]

Although the irrigation works, along with the Sinhalese civilization, collapsed in the twelfth century A.D., many new irrigation projects in Sri Lanka today are successfully rehabilitating those systems.

A small communal form of irrigation organization that has provoked the interest of scholars is the Balinese *subaks*, locally controlled irrigation works that usually service about 200 hectares. Each subak includes the rice terraces irrigated from a single dam and major canal. Reference to subak can be found in Balinese texts as far back as the first millenium A.D.[4] The term subak is commonly translated as "irrigation society" because of the central role this institution plays in the regulation of water supply.[5] However, the role of the subaks extends beyond the management of water alone to cooperative agricultural planning, autonomous legal decision making, and religious functions.

Many other examples of ancient irrigation systems can be found throughout Asia, including the rice terraces in the Philippines, the Gio-linh irrigation works in Vietnam, the South Indian tanks similar to those in Sri Lanka, and irrigation works in the Indus Valley.

In the modern era, the expansion of irrigated area first became an important source of increased food production in the latter part of the nineteenth century in India under the British and in Java under the

Dutch.[6] Major expansion of irrigated area began about 1700 in Japan and reached a much higher level of capital and management intensity than in the rest of Asia. The Japanese irrigation experience provides insights into the future direction of irrigation development in other parts of Asia.

Almost all the irrigated area in Japan is planted to rice and is in the form of river-canal networks (82 percent) or pond-canal networks (15 percent).[7] Rice was first introduced to Japan in about the second or third century A.D., when waves of immigrants from the Korean peninsula brought new technology and new varieties of rice. Gradually over the next ten centuries, irrigation expanded from the small basins and river plains in central Japan. The major growth in irrigated area occurred in the seventeenth and eighteenth centuries during the Tokugawa period (1600–1867). This expansion depended on major improvements in river control and irrigation technology that permitted the development of irrigation in the northeast region, which has high-discharge rivers and extensive downstream plains. Most of the large river-canal networks originated in this period. Although there has been little subsequent expansion of irrigated area, there have been successive improvements in water control and management, which have permitted a relatively steady increase in land productivity.

The more modern rice technology of the Meiji period required major improvements in water control for its potential to be realized. The most common type of basic investment in land infrastructure during this period was the conversion of "wet rice fields" (rice lands that are left in a wet condition throughout the year because of the lack of drainage facilities) into "dry rice fields" (rice lands which are kept in a dry condition during the nonrice growing season).[8] This conversion required draft animals rather than human labor for plowing and was first practiced in northern Kyushu, spreading to 60 percent of the total rice area in Japan by the end of World War I. It facilitated not only the use of modern inputs such as fertilizer, but also a second crop of wheat or barley following rice. Introduction of small-scale electric irrigation pumps, beginning in 1922, replaced treadle irrigation wheels and greatly reduced the labor required for irrigation and tillage.

The Arable Land Replotment, enacted in 1899, attacked the problem of fragmentation by compulsory consolidation of holdings into uniform plots of one-tenth of a hectare in size. The Rules of Subsidization of Irrigation and Drainage Projects were promulgated in 1923 to permit government financing of large-scale projects beyond the means of the small, village-level irrigation associations. The government provided most of the funds required for physical improvements, particularly after World War II. However, in contrast to much of the rest of Asia, management and control of water resources remained in local hands. That is to say, most Japanese systems can be viewed as "community irrigation systems," as opposed to "agency-operated systems."[9]

Many western analysts have been critical of the inefficiencies of the government-managed irrigation systems in contemporary developing Asia, contrasting them with the high efficiency of Japanese systems. It is surprising that most Japanese analysts up to the early postwar period, regardless of academic discipline or ideological persuasion, reached a strong negative evaluation of the locally managed Japanese systems.[10] It was argued that the rigid local irrigation customs prevented rational operation and maintenance and efficient allocation of water. Marxist writers in particular viewed the irrigation organizations as carryovers from the "feudal" period.

Kagato Shinzawa, an agricultural economist and author of *Treatise on Irrigation*, provides a strong rebuttal to this pessimistic view.[11] He argues that irrigation problems can be traced, not to a particular form of political economy, but to a fundamental conflict between upstream and downstream users. Such conflicts can be resolved by investment in physical improvements. In fact, subsequent government investment in Japan in storage dams and more efficient delivery systems helped establish a new and more equitable distribution procedure.

Ironically, however, local-level irrigation organizations in Japan currently face a new problem. The complete separation of delivery and drainage ditching, finer tuning of field-water levels, and other changes have increased the demand per unit area, raising the need for intranetwork water recirculation and reuse. At the same time, the increasing mechanization of agriculture and growth of part-time farming has undermined the village unit and its ability to sustain operation, maintenance, and water allocation activities.

The Japanese experience suggests that the modernization and development of irrigation systems is a continuous process. By contrast with Japan and other East Asian countries, irrigation in the rest of Asia is at the beginning stages of this process. The Japanese experience also suggests that irrigation development can be greatly enhanced by a suitable combination of local resource mobilization and government subsidy.

Evolution of Irrigation Development

Irrigation systems have typically been divided into large-scale, centrally managed systems and small-scale, local systems. According to Wittfogel in his well-known work *Oriental Despotism,* large-scale systems involving vast tracts of land, many people, and the presence of a centralized bureaucracy have been instrumental in the development of the state in Asia.[12] Although the large systems, such as those mentioned in the examples above, were undoubtedly important in the development of irrigated agriculture, it is now clear that smaller, community-managed systems have historically been equally important to improvements in water control. The more recent development of the Japanese irrigation systems is a graphic example of this point. Although the central government invested considerable sums in the physical development of irrigation facilities, management of the water resources remained in the hands of small, village-level irrigation associations.

Several broad changes are typically associated with the instigation of improvement in irrigation facilities.[13] The first is the stabilization of harvest fluctuations, with attendant improvements in average yields. This is brought about by providing a dependable water source throughout the growing season. However, the simple expansion of irrigated area does not automatically lead to greater overall stability in production. Second, in some circumstances, improved control of available water resources may make a second or even third crop possible. Finally, the availability of reliable water supplies makes it possible to use improved seeds, to introduce new farming techniques, and to increase use of chemical fertilizer—all of which require adequate water but supply large relative increases in productivity.

The evolutionary process of land intensification through the introduction of irrigation and new technology can be described in economic terms as follows:[14] As population pressure pushes the cultivation frontier into marginal areas, the marginal cost of increasing agricultural production through expansion of cultivated area rises. Eventually, a point is reached where investment in irrigation becomes the most economically feasible way to increase agricultural production. This is shown in figure 7.1 at point *P* where curve *A*, representing the marginal cost of increasing agricultural output by opening up new lands, crosses curve *I*, representing the marginal cost of increasing production through irrigation. This occurred at a much later date in South and Southeast Asia than in East Asia. Although the cost of installing irrigation systems rises as irrigation expands into

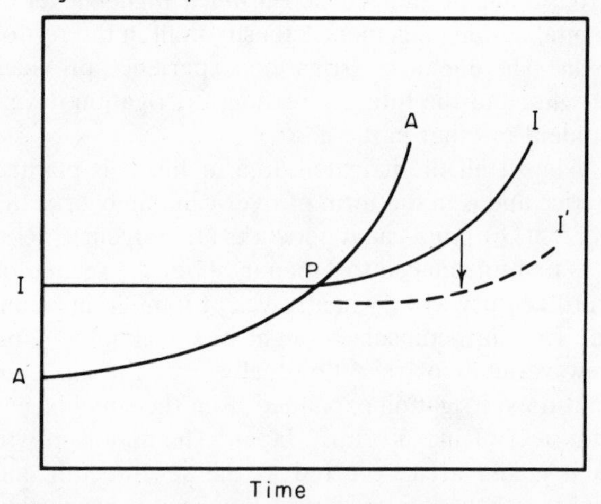

Figure 7.1. Hypothetical relations between the marginal cost of agricultural production by opening new land (A) and by building irrigation systems (I). When A climbs above I (at P), irrigation becomes a more profitable means of agricultural growth than opening new land. I′ represents the reduced marginal cost of irrigation due to the impact of new seed-fertilizer technology. (Source: Y. Hayami, et al., "Agricultural Growth Against a Land Resource Constraint: The Philippine Experience," *Australian Journal of Agricultural Economics* vol. 20 (1976) p. 150, reprinted by permission of the publisher)

increasingly marginal areas, the development of new technologies specifically adapted to irrigated agriculture shifts the marginal cost curve downward from *I* to I^1. As the marginal cost of agricultural production under irrigation decreases, irrigated area is further developed.

Prior to World War II, irrigation capabilities in East, South, and Southeast Asia were at very different levels. Nearly all of Japan's rice area was well irrigated, and national rice yields were at an average of 2.5 mt/ha. Crop intensification had begun, but modern inputs, such as fertilizer, were not yet widely used. China was in a similar position, with advanced irrigation facilities and high cropping intensities in the rice-producing region of South China. In contrast, most of South and Southeast Asia had barely begun to develop irrigation facilities in the lowland flood plains where rice is commonly grown. By the end of World War II, the bulk of the rice area was still unirrigated, and national yields were only about 1.5 mt/ha, one metric ton below average Japanese yields in 1880.

The orderly development of irrigation in successive stages that occurred in East Asia has not happened in South and Southeast Asia. Rapid expansion of irrigated area, crop intensification, and the introduction of modern inputs all occurred almost simultaneously in the period following World War II. This has had serious equity implications for those who do not have access to irrigated area and the attendant improvements in land productivity. Areas that were already well irrigated were the first to benefit from the spread of the new improved rice varieties. Risks associated with rainfed rice discouraged cash expenditures for chemical fertilizer. Thus, in most cases, access to irrigation has become a precondition for the use of modern inputs. Even today, rainfed areas of Asia have largely been bypassed by the benefits of modern technology. The situation for rainfed farms is worsened by the fact that technological advances in the irrigated areas have lowered the cost of protection and, hence, the relative farm price of rice.

Evolution of Irrigation Systems Management

Early Asian irrigation systems developed in proximity to flowing streams and other easily developed resources.[15] These systems required little more than simple diversion structures, usually of a temporary character. Management and distribution of water were also simple, with the main focus on the timely repair of temporary structures and diversion channels. As the area served by these simple systems increased, the normal variation in river flow and natural rainfall prevented adequate supplies of irrigation water from reaching all lands. Furthermore, since these systems irrigated only a small fraction of the dry season crop, the potential benefits from increased solar radiation and reduced incidence of insects and disease that typically characterize the dry season were lost.

Given the potential economic advantage of the dry season crop, in recent years major investments have been made in storage systems and tubewells that permit irrigation throughout most of the year. The new systems, often superimposed on the old, permit a considerable increase in cropping intensity; however, they require not only a higher level of capital investment, but also a higher level of management.

In assessing the relative efficiency of a given irrigation system, one must consider its environmental setting and its historical context.[16] Irrigation development generally takes place in an atmosphere of opposing pressures. On the one hand, there is a need for increased agricultural production to offset production growth and the scarcity of available land. The availability of modern inputs encourages this development. This is offset by the rapid escalation of costs for irrigation development and the complications caused by the necessity for more sophisticated management. In the past, change has come about slowly, permitting gradual accommodation to technologies and institutional change. More recently, however, high population growth, the rapid dissemination of new technologies, and the concerns of the international lending community have created a more dynamic situation in which the demand for irrigation is accelerated.

Particularly in the centralized and technically more complex government systems, as opposed to the local communal systems, management seems to be an obstacle to progress. Irrigation systems management is characterized by Levine as typically evolving through three stages: (1) hydrologic-hydraulic, (2) agricultural-based, and (3) farmer-oriented.[17] In the first stage, the emphasis is on capture, transport, and allocation of water. General equity in access to water is a primary design and management objective, but water is not managed as the scarce resource. In fact, there is typically little understanding of the agricultural use of water, and the design, construction, and operation of the systems are the responsibility of an engineering-based government organization. This management stage prevails in most of South and Southeast Asia today, but is becoming increasingly inappropriate.

When the agricultural utility of water is realized, management enters the second, agricultural-based stage. Information about local soils, crops, and other agronomic elements is incorporated into the design and operation of the system. In the third stage, the farmer is an active participant in the system's design, construction, and operation, and water is designated as the scarcest resource. Planners recognize that farmer participation in management can result in both greater equity and increased efficiency in water use. Levine sees Sri Lanka as just starting to move from the first to the second stage and the Philippines and Indonesia as just starting to move from the second to the third stage.[18] Taiwan, China, and Japan are in the third stage and are farmer-oriented in that they have highly developed farmer-run irrigation associations.

Growth in Irrigated Area in Contemporary Asia

As already noted, most of the growth in irrigated area prior to World War II occurred in the East Asian rice-producing regions. Essentially all of the rice area in Japan was irrigated by the end of the Tokugawa period (1868), and Buck (1937) notes that almost all of the rice land in China was irrigated.[19] By 1940, approximately two-thirds of the rice area in Taiwan and Korea was irrigated (table 7.1). With the devel-

Table 7.1. Growth in Irrigated Area in Korea and Taiwan, 1920–40

	Irrigated area 1940 (thousand ha)	Additional area irrigated, 1920–40 (thousand ha)	Annual growth in irrigated area (1920–40) (percent)	Area irrigated as percent of total paddy area	
				1920	1940
				(percent)	
Korea	1,237	817	5.0	27	70
Taiwan	530	225	2.8	41	62

Source: M. Kikuchi (1975).

opment of storage irrigation and the expansion of multiple cropping, total irrigated area continued to grow slowly in East Asia after World War II.

Prior to World War II in Southeast Asia, the most significant developments occurred in Java. In 1880, the irrigated rice land in Java and Madura exceeded one million hectares, accounting for approximately 50 percent of the total rice area.[20] The Dutch colonial administration made substantial investments in irrigation throughout the early 1900s, although the decision was controversial. One important issue was the failure of irrigation to raise rice yields. In contrast to East Asia, complementary yield-increasing technology, such as fertilizer, was not available there. In South Asia, developments in irrigation were relatively minor in the prewar period. In the latter part of the nineteenth century, the Punjab, the great granary of the Indian subcontinent, was opened up through the development of irrigation facilities, although rice was only a minor crop there.[21]

The growth in irrigated area from 1960 to 1980 is shown for three groups of Asian countries in table

Table 7.2. Growth in Irrigated Area in Asia

Country	Irrigated area 1960 (thousand ha)	Irrigated area 1980 (thousand ha)	Annual growth in irrigated area 1960–80 (percent)	Net crop area irrigated[a] 1980 (percent)	Gross rice area irrigated[b] late 1970s (percent)
Island and peninsula countries					
South Korea	663	1,150	2.8	56	91
Malaysia	214	370	2.8	37	66
Philippines	808	1,300	2.4	18	43
Sri Lanka	255[c]	525	3.7	51	63
Indonesia	4,100[d]	5,418	1.4	38	63
Major river delta countries					
Bangladesh	316	1,620	8.5	18	12
Burma	545	999	3.1	10	17
Thailand	1,636	2,650	2.4	16	14
Vietnam/ Kampuchea/ Laos	n.a.	1,700	—	30	40
Continental diversified-grain countries					
China	32,900	48,000	1.9	49	93
Nepal	n.a.	230	—	10	21
Pakistan	10,234	14,300	1.7	71	100
India	23,393	36,665	2.3	24	35
South and Southeast Asia	41,501	65,777	2.2	12	33
Total	75,064[e]	112,997	2.1[e]	32	51

Note: n.a. = not available.
Sources: Cols. 1–4: FAO, *Production Yearbook*, except China. China: A. Tang and B. Stone (1980). Col. 5: R. E. Huke (1982).
[a]Net area is the physical area.
[b]Gross area includes first and second crops.
[c]1959.
[d]1961.
[e]Excluding Vietnam, Kampuchea, Laos, and Nepal.

7.2: (1) the island and peninsula economies, (2) the major river delta economies, and (3) the continental diversified-grain countries. Irrigated area grew at just over 2 percent per annum, or more than 50 percent over the twenty year period. By 1980, a third of the rice area in South and Southeast Asia was irrigated. The lowest percentage of irrigated area is in the major river deltas, with the exception of the Mekong (Vietnam, Kampuchea, Laos), and in Nepal.

There are few disaggregated data to document the rapid growth in the dry season crop in Asia as a whole. Official statistics for Malaysia show that the proportion of rice area under dry season (off season) paddy grew from 1 percent in 1955 to 90 percent in 1975.[22] In the Philippines, the area with a dry season (second) crop grew from about 20 percent in 1955 to 60 percent by 1975.[23] The expansion elsewhere was in general less dramatic, but by the late 1970s a dry season crop was grown in over one-third of the South and Southeast Asian irrigated rice fields, and accounted for approximately 9 percent of the total rice area (table 7.3) and 16 percent of total rice production (table 7.3, footnote b). This has come about largely through the construction of several medium-sized storage systems with command areas of less than 100,000 ha.

Control of water in the major river deltas of Asia poses a more difficult problem, which explains in part why a smaller portion of cultivated area in these countries is irrigated. Although a large portion of cropped area remains rainfed, the total irrigated area in Bangladesh, Burma, and Thailand increased about 50 percent. With the exception of Vietnam, two crops of irrigated rice are grown on only a small portion of the irrigated area (table 7.3).

In summary, the countries and regions of Asia tend to fall into a bimodal distribution. In most of the island and peninsula economies, North and South India, and Pakistan, well over half of the rice area is irrigated. By contrast, the major river delta countries of Asia, East and Central India, and Nepal have 20 percent or less of the rice area irrigated (with the exception of Vietnam). Given the high degree of complementarity between irrigation and the use of modern inputs, the implications for both productivity and equity are significant.

Table 7.3. Rice Area Irrigated, South and Southeast Asia, Late 1970s

Country	Rice area irrigated		Percent rice area irrigated	
	Wet season	Dry season	Wet season	Dry season
	(thousand ha)		(percent)	
Island and peninsula countries				
Malaysia	266	220	36.2	29.9
Philippines	892	622	25.4	17.7
Sri Lanka	294	182	38.6	24.8
Indonesia	3,274	1,920	39.9	23.4
Major river delta countries				
Bangladesh	170	987	1.7	9.9
Burma	780	115	14.7	2.2
Kampuchea	214	—	10.5	—
Laos	67	9	9.6	1.3
Thailand	866	320	10.0	3.7
Vietnam	1,326	894	23.8	16.0
Continental diversified grain countries				
India[a]	11,134	2,344	28.5	6.0
Central	1,590	0	17.0	0.0
East	984	340	18.1	6.3
North	674	0	100.0	0.0
South	3,808	1,682	47.9	21.2
Nepal	261	0	20.7	0.0
Pakistan	1,710	—	100.0	0.0
South and Southeast Asia[b]	21,254	7,613	24.2	8.7

Source: R. Huke (1982).

[a]Only the major rice growing states of India are represented as follows: Central India: Madhya Pradesh, Uttar Pradesh; East India: Assam, Bihar, Orissa, West Bengal; North India: Haryana, Himachal Pradesh, Jamnu, Kashmir, Punjab; South India: Andhra Pradesh, Karnataka, Kerala, Tamil Nadu.

[b]Estimated percentage of rice production from wet season irrigation in South and Southeast Asia is 35 percent and from dry season irrigation is 16 percent. This assumes yield levels as follows: irrigated wet season 3.0 mt/ha, irrigated dry-season 3.5 mt/ha, shallow rainfed 2.0 mt/ha, all other land (deepwater, floating, and dryland) 1.0 mt/ha.

Investment in Irrigation Development

The investment in land infrastructure and irrigation is shown for Japan, Korea, and Taiwan for the period 1908–12 to 1933–37 in table 7.4. Given the more advanced stage of irrigation development in Japan, the nature of the investment differed markedly from that in Korea and Taiwan. For example, Hayami's estimate of investment in land infrastructure includes expenditures for the compulsory consolidation of land into uniform one-tenth hectare plots, while land consolidation did not occur in Korea and Taiwan until after World War II.[24] In Korea, there was a sizable investment in land reclamation. Between 1933 and 1937, the investment in land reclamation was equal to about one-quarter of the investment shown in table 7.4 for irrigation.

Since World War II, the East Asian, South, and Southeast Asian economies have continued to make substantial investments in improving and expanding irrigated area. Investment in irrigation has been critical to the sustenance of yield increases in Southeast Asia. For example, data for the Philippines show a rapid growth in irrigation investment between 1949–52 and 1973–75 (table 7.5). However, the cost per hectare for irrigation development varies according to the nature of the investment and the stage of the irrigation system. Through time, costs tend to rise as investments in more expensive system components, such as storage facilities, are made. Two recent sources suggest that irrigation of land that is rainfed can cost as little as $US 1,000 or as much as $US 3,000 per hectare, depending on the degree of intensity.[25] In 1975 prices, cost per hectare of irrigation in Taiwan and Korea was approximately $US 450 in 1918–22, rising to $US 1,350 in 1928–33. Cost per hectare in the Philippine government systems was $US 350 in 1958–62 and rose to $US 900 in 1973–75.[26]

When initial government investments are made in water storage systems, a sharp rise occurs, not only in the cost per hectare, but also in the irrigation investment cost as a percentage of the total budget. This problem can be seen clearly in the case of Taiwan, the Philippines, and Malaysia (table 7.6). In each case, at some point, irrigation expenditure as a

Table 7.4. Investment in Irrigation, Japan, Korea, and Taiwan, 1908–12 to 1933–37

	Japan				Korea				Taiwan			
Years	Annual average (thousand yen)	Govt.[a] (percent)	Incremental irrigation (ha)	Cost per ha (yen)	Annual average (thousand yen)	Govt. (percent)	Incremental irrigation (ha)	Cost per ha (yen)	Annual average (thousand yen)	Govt. (percent)	Incremental irrigation (ha)	Cost per ha (yen)
1908–12	51,600	17	89,853	576	—	—	—	—	2,045	94	4,376	478
1913–17	72,000	15	111,000	648	2,803	94	8,154	344	1,794	81	4,039	444
1918–22	87,000	19	127,431	689	1,606	22	6,138	262	1,476	84	4,230	349
1923–27	135,200	20	206,728	654	8,282	—	12,371	669	6,155	30	9,665	637
1928–32	174,800	26	227,000	770	16,497	—	18,613	886	7,797	36	7,942	982
1933–37	158,300	31	248,000	638	2,092	10	2,419	865	1,921	28	5,770	333

Note: Yen are 1934–36 constant yen.
Sources: Japan: Y. Hayami (1975). Korea and Taiwan: M. Kikuchi (1975).
[a] Refers to government vs. private investment.

Table 7.5. Government Investment in Irrigation in the Philippines, 1949–52 to 1973–75

	National systems			Communal systems		
Years	Annual average (1,000 pesos)	Incremental irrigation (ha)	Cost per ha (pesos)	Annual average (1,000 pesos)	Incremental irrigation (ha)	Cost per ha (pesos)
1949–52	7,170	7,450	962	—	—	—
1953–57	18,554	13,331	1,392	1,384	14,046	99
1958–62	29,133	16,038	1,816	1,269	10,774	118
1963–67	6,014	3,463	1,737	1,257	12,730	99
1968–72	37,672	21,700	1,736	1,605	18,030	89
1973–75	296,432	65,977	4,492	5,190	32,497	160

Note: 1970 constant pesos.
Source: M. Kikuchi (1975).

Table 7.6. Percentage of the Agricultural Budget Spent on Irrigation and Drainage, Taiwan, 1911–15 to 1956–60, and Taiwan, Philippines, and Malaysia, 1961–65 to 1976–80

Years	Percent	Years	Percent		
	Taiwan		Taiwan	Philippines	Malaysia
1911–15	23	1961–65	46	10	n.a.
1916–20	11	1966–70	21	10	31
1921–25	23	1971–75	11	34	13
1926–30	33	1976–80	7[a]	34[a]	13
1936–40	5				
1951–55	6				
1956–60	15			million $US (current prices)[a]	
		1961–65	16.3	6.7	8.4
		1966–70	21.4	8.2	22.8
		1971–75	25.4	43.5	17.1
		1976–80	21.9	74.5	41.4

Sources: Taiwan, 1911–15: T. H. Lee (1968); 1961–65 to 1976–77: H. Y. Chang (1980), pp. 133–155. Philippines: M. S. de Leon (1983). Malaysia: D. Taylor (1981).

[a]Exchange rates: Taiwan, $NT 40 = $US 1; Philippines 1961–70 peso 4 = $US 1; Philippines 1971–80 peso 7 = $US 1; Malaysia $M 3 = $US 1.

percentage of the total agricultural budget rose to over 30 percent, and each increase can be identified with a specific large project. The Chainan Ta Tseng Project in Taiwan was completed during the period 1920–31 at a cost of 51,748 yen (1934–36 prices).[27] The Upper Pampanga River Project in the Philippines was completed in 1975 at a capital cost of $US 105.5 million. It covers a command area of 83,000 ha at a cost of $US 1,270 per hectare.[28] The Muda River Project in Malaysia was completed during the period 1966–70 at a cost of $US 82 million. The area irrigated during the main season is 95,950 ha at a cost of $US 850 per hectare. The high per hectare cost of the Muda River system is caused by the large investment in farm-level ditches and land improvement. However, inadequate development of tertiary canals will require an additional investment of $US 2,120 per hectare over a fifteen year period.[29] A second crop is grown on most of the area in all three systems.

The Trilateral Commission Report suggests that there is still opportunity in Asia for substantial gains in yield through modest investments in existing irrigation systems.[30] For situations where storage systems do not exist, this is probably true. However, when storage systems are already in place and grain yields are still low, significant improvements in management capabilities will be necessary to increase yields. Inadequate management capacity may be the most serious bottleneck in many systems.

A substantial amount of irrigation in Asia is under local, as opposed to state or federal government control. In 1975, for example, close to half the irrigated area in the Philippines was defined as communal. However, in contrast to the historical experience of East Asia where the development of irrigation was based on the mobilization of community resources, the bulk of new investment in irrigation is currently being made by national governments, and an increasing proportion of grain is being produced in government-maintained irrigated areas. In Malaysia, for example, the proportion of the rice crop in government-maintained areas increased from less than one-third in 1949 to over two-thirds in 1966.[31] Table 7.7 shows that in the Philippines, communal irrigation systems declined in area relative to other irrigation systems by 70 percent between 1955 and 1975, and this trend is likely to continue.[32]

As investment alternatives for the development of large-scale government systems are fully exploited, national governments and expatriate donor agencies are giving more attention to investment in regions traditionally dominated by small-scale, communal systems. Thus, for example, in the case of the Philippines, although the percentage of area irrigated by communal systems has declined, the percentage of communal systems receiving government assistance has risen (table 7.7).

Complementarity Between Irrigation and Modern Inputs

Growth in irrigated area in Asia has largely been spurred by the development of modern fertilizer-responsive varieties of rice. In this section, we review

Table 7.7. **Percentage of Irrigated Area in the Philippines by Type of System, 1952–75**

Type of system	1952	1955	1960	1965	1970	1975
National gravity system	22.9	24.9	35.3	34.1	36.3	34.9
Communal systems						
government-assisted	—	6.7	11.3	16.4	17.2	20.0
private	69.0	60.2	45.2	40.0	36.2	29.2
Pump irrigation system	2.5	3.2	4.4	6.4	7.7	14.0
Other[a]	5.6	5.0	3.8	3.1	2.6	1.9
Total	100.0	100.0	100.0	100.0	100.0	100.0

Source: M. Kikuchi, G. Dozina, and Y. Hayami (1978).
[a]Includes friar land systems (lands given to the church during the Spanish colonial period) and municipal systems.

the interaction between irrigation and modern inputs from a technical and an economic perspective.

The yield response to additional increments of water is shown for IR8 in figure 7.2. Because rice is a semiaquatic plant, its response to water differs from that of tropical upland crops, and this difference has important implications for irrigation as well as for varietal improvement. Irrespective of the variety planted, rice reaches a maximum yield with the application of approximately 6 mm of water per day.

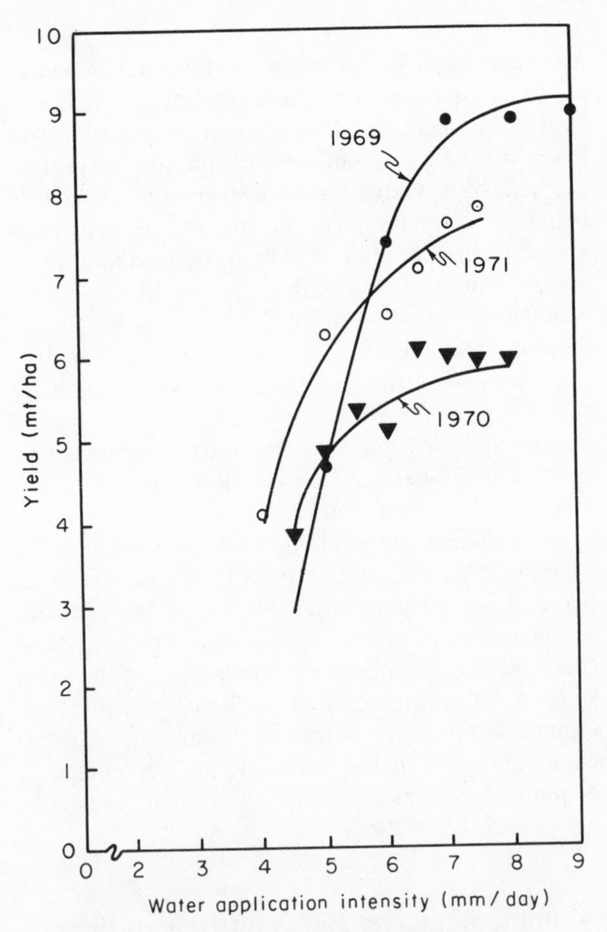

Figure 7.2. Functional relations (logistic functions) between rice yield and water application intensity for IR8 (100 kg/ha N) IRRI, 1969, 1970, and 1971 dry seasons (Source: International Rice Research Institute, *Annual Report for 1972* [Los Banos, Philippines, 1973] p. 58, reprinted by permission of the publisher)

The yield drops very rapidly when available water falls below this level, although the level of maximum yield and the rate of decline will vary by season and variety. The less the cloud cover in the dry season, the higher the rate of evapotranspiration and the more rapid the decline in yield. With the exception of varieties that are drought tolerant, yields will drop sharply at an application of less than 6 mm of water per day. Because of the sensitivity of the rice plant to drought, farmers with inadequate or unreliable supplies of water are in a particularly vulnerable position. Typically, they cannot afford the risk of applying fertilizer if water supplies are not assured.

The yield responses to nitrogen fertilizer under varying conditions of water control in the Philippines are shown in figure 7.3. Traditional rainfed varieties give a maximum yield of about 1.5 mt/ha. Irrigation can raise the yield ceiling to over 2 mt/ha. However, when modern varieties and ideal irrigation conditions obtain, the yield maximum is close to 3 mt/ha in the wet season and over 4 mt/ha in the dry season. Yields under "optimum" levels of fertilizer use (the ratio of the price/kg N to the price/kg of rice is assumed to be 7 to 1) are also shown in figure 7.3. The link between high yields and the use of modern varieties and improved water control is clear. On a per hectare basis (combined seasons), the yield of modern varieties under ideal irrigation is five times greater than the yield of traditional varieties under rainfed conditions.

Using an alternative approach, the economic complementarity between irrigation and the seed-fertilizer technology illustrated hypothetically in figure 7.1 has been measured by Kikuchi for four countries—Japan, Taiwan, Korea, and the Philippines.[33] Figure 7.4 shows the trend in irrigation cost per unit of agricultural income over time at specified levels of nitrogen application. The introduction of modern

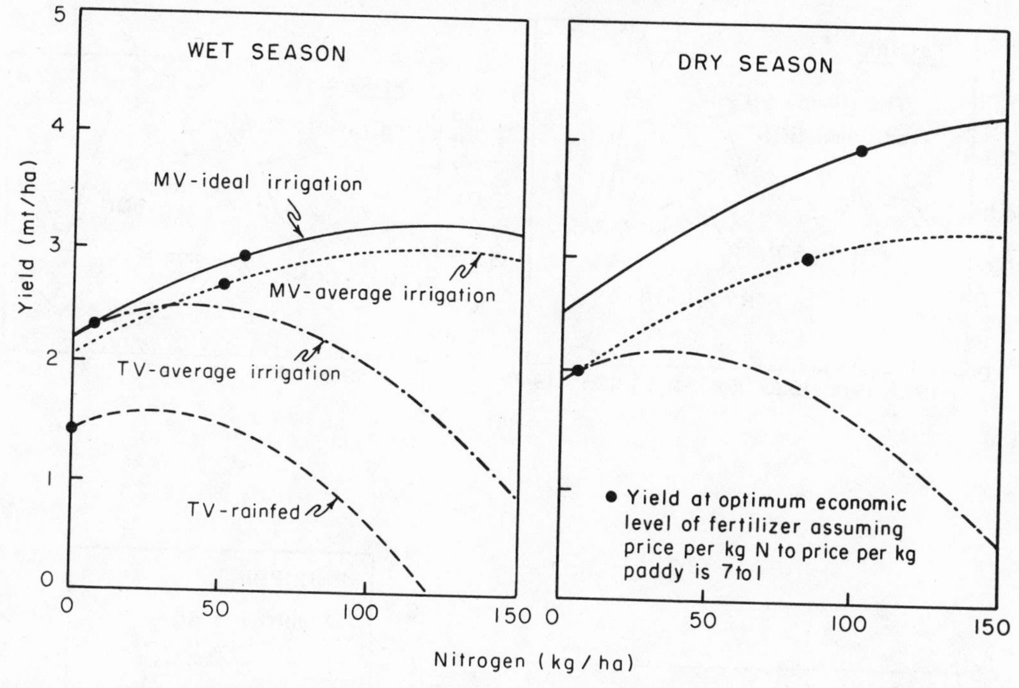

Figure 7.3. Response of modern (MV) and traditional varieties (TV) to nitrogen at different irrigation levels, Philippines, 1976 wet and dry seasons (Source: T. H. Wickham, R. Barker, and M. V. Rosegrant, "Complementarities Among Irrigation, Fertilizer, and Modern Varieties," in International Rice Research Institute, *Economic Consequences of New Rice Technology* [Los Banos, Philippines, 1978] p. 230, reprinted by permission of the publisher)

varieties sharply lowered the irrigation system investment required to produce an additional unit of agricultural income in all four cases. The internal rate of return for investment in irrigation was approximately twice as high with modern as with traditional varieties (table 7.8).

Table 7.8. Internal Rates of Return for Irrigation Investment for Traditional and High-Yielding Variety Rice Technology in Selected Countries and Years

Country/year	Traditional[a]	HYV[b]
Korea/1941	13	38
Taiwan/1938	14	25
Philippines/1970	18	32

Source: M. Kikuchi (1975).
[a]Assuming 5 kg N/ha.
[b]Assuming 50 kg N/ha for Korea and Taiwan, 60 kg N/ha for the Philippines.

Future Investment Requirements

The ability to maintain or accelerate growth in irrigation investment in Asia depends not only on the financial reserves of a country (including foreign loans), but also on the trained manpower available. Delays in construction, often running up to several years, are commonplace, and the recent drain of skilled manpower to the Middle East has exacerbated the problem.

There are a number of recent projections of investment costs based upon various assumptions about "needs" and capacity to expand. The report by the Trilateral Commission suggests that rice production in the developing countries of Asia should double in fifteen years (1978–93) by increasing irrigated area from 32.7 to 86.8 million hectares.[34] The total capital cost is estimated at $US 52.6 billion (1975 dollars) or $US 3.5 billion per year. The proportion of total rice area irrigated would increase from 38 to 79 percent, and there would be substantial improvement in the quality of irrigation, with two-thirds of the irrigated area multiple cropped. Even so, it is extremely doubtful that Asia has the technical manpower and the financial resources to step up the growth rate of irrigated rice area from about 2 percent to 6.7 percent per annum needed to achieve this goal.

In a study conducted at about the same time as the Trilateral Commission's, Herdt, Te, and Barker (1977–78) projected investment costs for rice production from 1974 to 1985 assuming three differ-

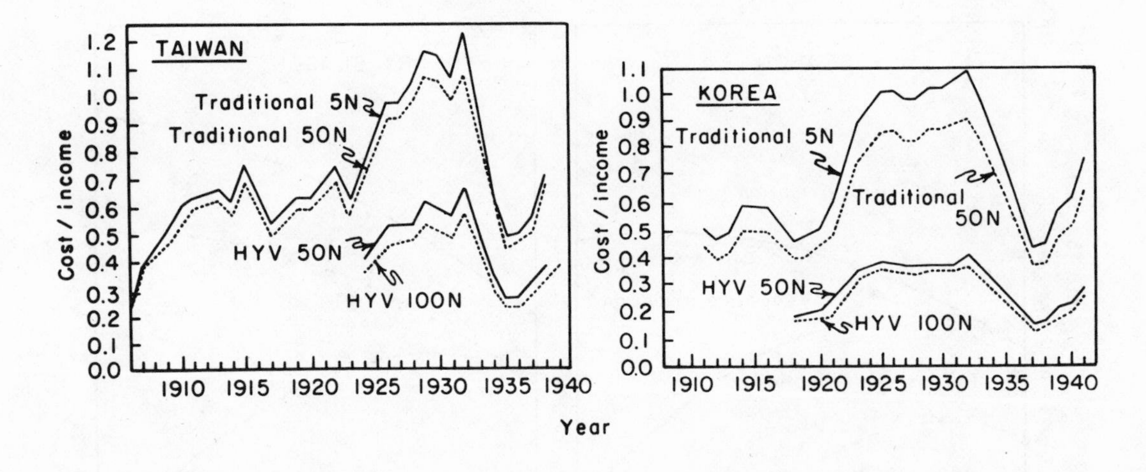

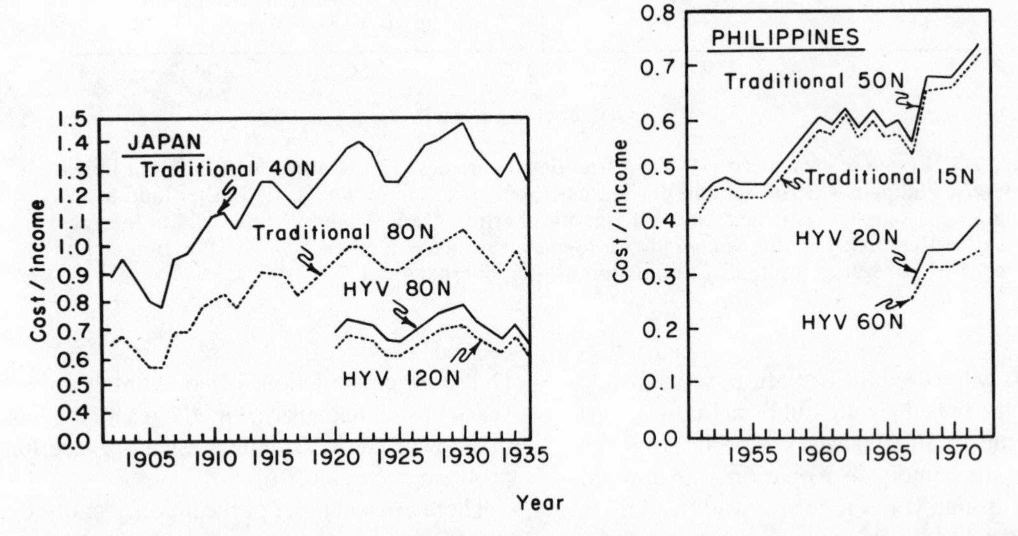

Figure 7.4. Trends in costs of building irrigation systems required to produce an additional unit of agricultural income under alternative levels of rice technology for Taiwan, South Korea, Philippines, and Japan (Source: M. Kikuchi, "Irrigation and Rice Technology in Agricultural Development")

ent rates of growth in irrigated rice area: 1.5 percent, 2 percent, and 3 percent.[35] The annual cost at $US 1,300/hectare would be $US 626 million, $US 863 million, and $US 1,359 million (1975 dollars), respectively.

In a more comprehensive study including eight major countries (Bangladesh, Burma, India, Indonesia, Nepal, Pakistan, Philippines, and Sri Lanka) in South and Southeast Asia, the International Food Policy Research Institute (IFPRI) projected growth in gross irrigated area for all crops at 3 percent and growth in net irrigated area (equipped area) at 2.3 percent.[36] Investment in new irrigation is assumed to be $US 2,200 per hectare on average, ranging from $US 1,500 in Bangladesh to $US 2,900 in Pakistan. The total capital cost for irrigation development is estimated at $US 52 billion or $US 3.5 billion annually (1975 dollars).

The above estimates indicate a broad range of future possible capital needs. Much of this can be attributed to the inclusion of different types of facilities depending on the particular country and level of irrigation. For instance, recent experience in the Muda River Project in Malaysia, the Upper Pampanga River System in the Philippines, and the Mahaweli Project in Sri Lanka suggests that the shift toward investment in storage, coupled with the simultaneous introduction of a range of field-level improvements, can result in a capital cost well above the typically quoted per hectare cost of $US 1,200 and $US 1,500. The trend toward greater emphasis on dry season irrigation can also be expected to add to the per hectare cost. Much of the dry season irrigation is added by superimposing a large storage system on existing run-of-the-river systems.

The International Food Policy Research Institute estimates of costs per hectare for new irrigation ($US 2,200) may appear to be realistic. However, the projected growth in gross irrigated area may be closer to 2 than the 3 percent assumed by the IFPRI study, while the growth rates assumed by the Trilateral Commission appear to be totally unrealistic. The increasing number of projects in the "hard-to-irrigate" areas will add to the capital costs. However, a shortage of manpower and the need for improvement in irrigation organization and management, rather than limited capital, may be the major constraints on future irrigation development. While higher interest rates adversely affect investment in new irrigation facilities, they may have an even more negative effect on investment in the training of personnel for water management.

Implications for Future Irrigation Development

Irrigated area in Asia is growing at an annual rate of 2.1 percent. Over time, the cost per additional irrigated hectare is rising as more emphasis is placed on storage systems and as systems in the easy-to-irrigate areas are completed. Investments must increase in real terms if growth rates are to be maintained, and any significant slackening in the growth of irrigation systems could lead to serious shortfalls in food production. Thus, there is a growing demand for measures to improve system performance, particularly in South and Southeast Asia where the extensive development of irrigated agriculture is a more recent phenomenon. In this regard, there may be important lessons to learn from East Asian experiences.

In South and Southeast Asia, irrigation investment has principally been devoted to the construction of large-scale, state-operated systems. The emphasis on large-scale systems began in the colonial period and continued into the period of national independence. Both national governments and international funding agencies have consistently found it to their advantage to maintain highly centralized administration and control of projects. This mode of administration, although allowing ample central control, was achieved at a sacrifice in overall system management and performance. Most researchers feel that small systems under local control are more efficiently managed and achieve higher standards of productivity and equity, but more field research is needed to substantiate this claim.

If we assume that local financing, management, and control lead to better performance of irrigation systems, then we need to promote institutional arrangements that facilitate interaction between the government and local communities. The government could provide technical support to community irrigation systems without usurping operation and management functions, and large-scale systems could be decentralized to allow local control over segments of the system. East Asian successes seem particularly relevant here. To what degree South and Southeast Asian planners can decentralize control in large systems is not clear. However, national governments in many countries are beginning to provide more technical and financial assistance to small community systems. Examples are Indonesia's Sederhana program, the Philippine National Irrigation Administration's communal projects, and the International Labor Organization's (ILO's) Labor Intensive Public Works Program in Nepal.[37] At this time, researchers need to study the effect of government intervention on the operation and performance of community systems.

Despite what we might conceptually visualize as a close relationship between design and management goals and activities, the realities of the situation are very different. In planning and operating irrigation systems, design and management are treated as separate issues. Those who design and construct systems are usually not responsible for operation and management activities. Furthermore, those who operate and manage systems are usually not familiar with the goals and needs of water users. The farmer is often seen as the culprit who wastes water and needs to be "educated" in proper irrigation practices. However, it is our view that the most serious problems occur well before the water reaches the farm turnout, mainly because the end users are not consulted or involved in the design and management of systems.

In the Asian context, irrigation systems management appears to be gradually evolving, with the more highly developed systems in East Asia becoming "farmer oriented." However, given the rapid development and growing importance of irrigation in South and Southeast Asia, this transformation seems to be well behind schedule. Despite the enormous investment in systems *hardware*, there is an underinvestment in *software*. If the investment costs for irrigation development are to be kept within reasonable limits in Asia, the performance of water systems must be improved. This will require a substantial investment in human resources and a change in institutional structure that will allow farmers a greater opportunity to participate in system design and management.

Notes

1. Te-tzu Chang, "The Rice Cultures," in *The Early History of Agriculture* (London, Oxford University Press, 1977).

2. Ibid.

3. R. L. Brohier, *Ancient Irrigation Works in Ceylon* (Colombo, Sri Lanka, Government Publications Bureau, 1977).

4. Hildred Geertz and Clifford Geertz, *Kinship in Bali* (Chicago, University of Chicago Press, 1975).

5. Clifford Geertz, "Organization of the Balinese Subak," in E. Walter Coward, Jr., ed., *Irrigation and Agricultural Development in Asia* (Ithaca, N.Y., Cornell University Press, 1980).

6. There is considerable debate about the degree to which irrigation in nineteenth-century Java, which was installed principally for sugar cane production, raised rice output. For a discussion of this issue, see Jennifer Alexander and Paul Alexander, "Sugar, Rice and Irrigation in Colonial Java," *Ethnohistory* vol. 25 (1978) pp. 207–223.

7. William A. Kelly, "Japanese Social Science Research on Irrigation Organization: A Review" (New Haven, Conn., Yale University, Department of Anthropology, 1980) unpublished.

8. Shigeru Ishikawa, *Essays on Technology, Employment, and Institutions in Economic Development*, Economic Research Series No. 19 (Tokyo, Kinokuniya Co., 1981).

9. For a discussion of agency and community irrigation systems, see E. Walter Coward, Jr., "Irrigation Development: Institutional and Organizational Issues," in E. Walter Coward, Jr., ed., *Irrigation and Agricultural Development in Asia* (Ithaca, N.Y., Cornell University Press, 1980); Kelly, "Japanese Social Science Research on Irrigation Organization," has applied Coward's theories to the Japanese case.

10. Kelly, "Japanese Social Science Research on Irrigation Organization."

11. Kagato Shinzawa, *Nōgyo Suiri Ron* (Treatise on Irrigation) (Tokyo, Tokyo Daigaku Shuppankai, 1955).

12. Karl Wittfogel, *Oriental Despotism: A Comparative Study of Total Power* (New Haven, Conn., Yale University Press, 1957).

13. Shigeru Ishikawa, *Economic Development in Asian Perspective* (Tokyo, Kato Bunmeisha Printing Co., Ltd., 1967).

14. Yujiro Hayami, Cristina C. David, Piedad Flores, and Masao Kikuchi, "Agricultural Growth Against a Land Resource Constraint: The Philippine Experience," *Australian Journal of Agricultural Economics* vol. 20 (1976) pp. 144–159.

15. This section is based principally on recent writings of Gilbert Levine. See Gilbert Levine, "Perspectives on Integrating Findings from Research on Irrigation Systems in Southeast Asia," no. 26 (New York, Agricultural Development Council, Teaching and Research Forum, 1982) pp. 9–15.

16. Ibid.

17. A more complete description of the three stages in the classic development of irrigation can be found in Ibid.

18. Ibid.

19. John L. Buck, *Land Utilization in China* (New York, Agricultural Development Council, 1956).

20. Anne Booth, "Irrigation in Indonesia Part I," *Bulletin of Indonesian Economic Studies* vol. 13 (1977) pp. 33–74.

21. V. D. Wickizer and M. K. Bennett, *The Rice Economy of Monsoon Asia* (Stanford, Calif., Stanford University Press, 1941).

22. Donald C. Taylor, *The Economics of Malaysian Paddy Production and Irrigation* (Bangkok, Agricultural Development Council, 1981) p. 37.

23. Philippine Council of Agricultural and Resources Research, *Data Series on Rice Statistics in the Philippines* (Los Banos, Philippines, IRRI, 1981) p. 25. Prior to 1967/68, first and second crops were reported as stated by farmers. Subsequently, first crop refers to area harvested during the months of July–December and second crop to the area harvested during January–June.

24. Yujiro Hayami, *A Century of Agricultural Growth in Japan* (Tokyo, University of Tokyo Press, 1975).

25. Umberto Colombo, D. Gale Johnson, and Toshio Shishido, *Reducing Malnutrition in Developing Countries: Increasing Rice Production in South and Southeast Asia,* Trilateral Papers, No. 16 (The Trilateral Commission, 1978); and Kunio Takase and Thomas Wickham, "Irrigation Management and Agricultural Development in Asia," in *Rural Asia Challenge and Opportunity: Second Asian Agricultural Survey,* Supplementary Papers, vol. 1 (Manila, Asian Development Bank, 1978).

26. Assumes that $US 1 = 4 Yen in 1934–36 and that $US 1 = 7 Pesos in 1970. Based on the gross private domestic investment index for the United States, if 1933 = 100, 1970 = 400, and 1975 = 575. A study of four small gravity irrigation systems in Central Luzon, Philippines, showed that they had a capital investment cost in 1980 prices of about $US 700 per hectare. (Piedad Moya, Robert W. Herdt, and Saidique I. Bhuiyan, *Returns to Irrigation Investment in Central Luzon, Philippines,* Agricultural Economics Paper No. 81-23 (Los Banos, Philippines, International Rice Research Institute, 1981).

27. Masao Kikuchi, "Irrigation and Rice Technology in Agricultural Development: A Comparative History of Taiwan, Korea, and the Philippines" (Ph.D. dissertation, University of Hokkaido, 1975).

28. R. W. Tagarino and R. D. Torres, "The Price of Irrigation Water: A Case Study of the Philippines' Upper Pampanga River Project," in *Irrigation Policy and Management in Southeast Asia* (Los Banos, Philippines, International Rice Research Institute, 1978).

29. Colombo, Johnson, and Shishido, *Reducing Malnutrition in Developing Countries.*

30. Ibid.

31. D. E. Short and James C. Jackson, "The Origins of Irrigation Policy in Malaysia," in *Journal of the Malaysian Branch of the Royal Asiatic Society* vol. 44 (1971) pp. 78–103.

32. Masao Kikuchi, Geronimo Dozina, Jr., and Yujiro Hayami,

"Economics of Community Work Programs: A Community Irrigation Project in the Philippines," in *Economic Development and Cultural Change* vol. 26 (Jan 1978) pp. 211–225.

33. Kikuchi, "Irrigation and Rice Technology in Agricultural Development."

34. Colombo, Johnson, and Shishido, *Reducing Malnutrition in Developing Countries.*

35. Robert W. Herdt, Amanda Te, and Randolph Barker, "The Prospects for Asian Rice Production," *Food Research Institute Studies* vol. 16 (1977–78) pp. 184–203.

36. Peter Oram, Juan Zapata, George Alibarubo, and Roy Shyamal, *Investment and Input Requirements for Accelerating Food Production in Low-Income Countries by 1990*, Research Report No. 10 (Washington, D.C., International Food Policy Research Institute, 1979).

37. Desman Djojoadinato, "Indonesia's Simple (Sederhana) Irrigation and Reclamation Program," in *Irrigation Policy and Management in Southeast Asia* (Los Banos, Philippines, International Rice Research Institute, 1978) pp. 25–30; Frances F. Korten, *Building Rational Capacity to Develop Water Users Associations,* World Bank Staff Working Paper No. 58 (Washington, D.C., World Bank, 1982); Louis Rijik, Project Director, Intensive Public Works Program, Nepal, 1982.

8

Mechanization of Rice Production

There is considerable controversy about the desirability of agricultural mechanization in Asia.[1] One extreme view directly equates mechanization with modernization. A more moderate view holds that the functional relationship between power input and agricultural output is analogous to that of fertilizer and yield so that continued development may require additional mechanized power.[2] A third view holds that the major agricultural resource question in the developing countries of Asia is the absorption of agricultural laborers over the next twenty years. Mechanization could be a key to overcoming labor bottlenecks that now prevent increased cropping intensity, which in turn will permit labor to be absorbed at other times during the production cycle.[3] A major benefit of mechanization would be increased agricultural output from the additional acreage harvested and from higher yields that may result from deeper plowing and better cultivation practices. A fourth view, very different from those outlined above, opposes agricultural mechanization in Asia on the grounds that it represents a straightforward substitution of capital for labor, and that under the labor supply circumstances existing in most Asian countries, any such substitution is socially undesirable.[4] In some cases this is supplemented by the idea that distortions in the price ratio of labor to capital have been a primary factor responsible for speeding mechanization and that nonmarket forces have been responsible for a large degree of the distortion.

This chapter provides a perspective on the issues described above by examining the historical experience with mechanization of rice production in East Asia and by considering the evidence on the economic consequences of introducing small-scale machinery into rice production in South and Southeast Asia.

Mechanization means different things in different situations. Our classification is based on power source: human, animal, and mechanical. Machines of some kind can obviously be used with any source of power, and one concept of mechanization is the provision of tools, equipment, and simple machines for use with human and animal power. Our discussion, however, deals with powered equipment, and for the most part with two-wheeled tractors or "power tillers" and power threshers.

The Pattern of Mechanization in Asian Rice Production

Economic, technical, and policy factors are all important in determining the configuration of technology used in production, and mechanical technology seems to be especially sensitive to these factors. Alternative investment opportunities and the prices of land, labor, and capital influence farmers' demand for machinery. The perceived social opportunity costs of these resources influence policies that restrain or encourage mechanization, and the relative abundance of resources influences private and social costs. Technical factors, such as the amount of power required for a given task, the degree of judgment needed to apply the power, and whether the task requires moving through the field, all influence the engineering feasibility and hence the relative cost of mechanizing particular rice production tasks.[5] Climatic or soil conditions may also influence the design of successful machines.

Present Status of Rice Farm Mechanization

Tractors and two-wheeled power tillers are the most important agricultural machines used for rice production, and at least a few are available in all countries. Water pumps, sprayers, dusters, threshers, transplanting machines, and combine harvesters are also used in the more mechanized countries. Table 8.1 gives an overall picture of the level of rice farm mechanization in Asia, showing the number of machines per 100 hectares of arable land.

East Asia is by far the most mechanized, but a number of South and Southeast Asian countries have significant numbers of some machines. India has over four water pumps per 100 ha, and Thailand has two. Malaysia, Thailand, the Philippines, and Sri Lanka all have one or more power tillers per 100 ha. Most countries have less than 0.5 tractors per 100 ha of crop land—a number so small as to probably be insignificant for crop production on a national basis. Manual sprayers in addition to tillers are available in significant numbers outside of East Asia. However, the presence of two to six sprayers per 100 ha in Indonesia, the Philippines, Thailand, and Sri Lanka is far below levels in East Asia. Japan, Taiwan, and South Korea all have more agricultural machines than the other Asian countries. An examination of the historical process of mechanization in East Asia helps to suggest reasons for this difference.

Japan's Experience

Following the Meiji Restoration in 1868, the Japanese government instituted many policies designed to modernize and strengthen Japan's economy. In agriculture, experts from the United States, Britain, and later Germany were employed to teach the technologies being used in those countries. As early as 1873, when the Naito Shinjuku Agricultural Experiment Station was set up, farm machinery imported from the United States and England was tested and evaluated. In 1879, the Mita Farm Machinery Manufacturing Plant was established to produce farm machinery modeled after the imported machines.[6] However, the imported machines were not suited to Japanese conditions, and the Japanese turned their

Table 8.1. Numbers of Agricultural Machines per 100 Hectares of Arable Land

Area	Year	Motors + engines	Water pumps	Power tillers	4-wheeled tractors	Manual sprayers[a]	Power sprayers	Manual threshers	Power threshers
South Asia									
India	1979	[b]	4.1	0.01	0.2	0.5	[c]	[d]	0.4
Pakistan	1978	0.9	0.9	—[e]	0.4	0.1	—	—	0.1
Sri Lanka	1975	—	1.1	1.0	1.8	1.5	0.2	—	0.1
Nepal	1977	—	0.4	—	0.1	0.1	—	0.8	—
Southeast Asia									
Malaysia	1977	—	—	—	0.3	—	—	—	—
Thailand	1978	0.9	2.0	1.4	0.3	5.9	1.2	—	—
Philippines	1978	1.7	1.1	0.9	0.3	3.8	—	—	0.2
Indonesia	1978	0.1	—	—	—	2.1	0.4	—	—
Burma	1976	—	—	—	0.08	—	—	—	—
East Asia									
Japan	1980	52.4	n.a.	50.4	26.9	—	39.2	—	55.5[f]
Taiwan	1978	38.5	19.3	7.3	0.3	34.4	6.1	17.9	6.0
South Korea	1979	9.2	9.0	11.4	0.1	41.3	13.8	13.2	9.7
China	1980	7.3[g]	5.9	1.9	0.8	20.1	0.3	—	2.5

Sources: Except for Bangladesh, Malaysia, Burma, and China: Asian Productivity Organization (1981). Malaysia and Burma: FAO *Production Yearbook*; China: Hua Gouzhu and Yao Jianfu (1982).

[a]Sprayers and dusters.
[b]Those available are used as water pumps.
[c]Very few, included with manual sprayers.
[d]Very few, included with power threshers.
[e]Less than 0.1/100 ha.
[f]Refers to 1967, the latest year for which data are available.
[g]Calculated by assuming 10 hp per motor or engine.

attention to biochemical technologies for the remainder of the nineteenth century.

During those decades, many small electrical motors and engines used for pumping water and threshing were introduced into Japanese agriculture. By 1940, there were nearly 300,000 motors and engines and over 200,000 power threshers easing the work load on Japan's 6 million hectares. The power tiller had been invented but was just being introduced, as were sprayers and dusters. World War II slowed the process of mechanization because of limited capacity to invest, and after the war one would still have described Japan's agriculture as nonmechanized.

Over the next thirty years, under the pull of the industrial sector, Japan's labor force rapidly moved out of agriculture. Farm wage rates rose and the rice sector began to mechanize. The rapid rate of mechanization in the postwar period is shown in table 8.2 and figure 8.1.

By 1950, Japan had about two power tillers for every 1,000 ha of cropland. By 1970, there was one tiller for every 1.2 ha of paddy land, and thereafter, the number of walking tillers declined while there was a rapid increase in the number of riding tillers. The 1970s also saw the rapid introduction of powered rice transplanting machines and combine harvesters and a further increase in numbers of power sprayers and dusters. Power reapers and reaper-binders were also used in significant numbers during the 1960s, although the data are fragmentary.

During the same period, as machines were substituted for human labor, there was a sharp decline in the labor force in Japanese agriculture and a steady reduction in the hours of labor used per hectare of rice.[7] Some observers clearly saw these events as a drive to achieve economic efficiency pushed by rising labor costs, rather than as a continuation of the effort to increase yields.[8]

Differences Within East Asia

By 1980, Korea and Taiwan had also achieved a significant level of mechanization. The pattern of power tiller introduction there and in the Philippines and Thailand is shown in figure 8.2. In 1960, Taiwan had as many tillers per 1,000 ha as Japan had in 1950, but the number increased more slowly. After ten years of modestly increasing numbers, Taiwan still had about two tillers for every 100 ha. However, by 1977, twenty years after the initial introduction, there was nearly one tiller for every 10 ha, and by the 1980s, nearly all of the rice land in Taiwan was prepared by machines. Power tillers were introduced into Korea's rice sector about a decade after they were first used in Taiwan, and their rapid adoption was similar to the Japanese case, reaching the seven tiller per 100 ha level within ten years, and continuing to increase rapidly thereafter.

In the East Asian cases, it appears that adoption of tillers began to accelerate when the number reached about 2.5 per 1,000 ha (figure 8.2). Before that, only a few farmers were experimenting with tillers, and their use was probably concentrated in small areas near dealers and repair facilities. As facilities expanded and as the economic attractiveness of the machines increased with rising labor costs, the process of adoption progressed. There are strong similarities in the economic conditions that accompanied mechanization in the three East Asian countries, but there are also differences (table 8.3). The data show some differences in the real agricultural wage measured in rice equivalent at the takeoff of mechanization (20

Table 8.2. Agricultural Machinery in Japan, 1921–81
(thousand units)

Year	Motors + engines	Power tillers	Riding tractors	Power sprayers	Power threshers	Combines	Rice planting machines
1921	3	0	0	0	0	0	0
1931	92	0	0	0	56	0	0
1939	293	3	0	5	211	0	0
1951	1,003	16	0	20	972	0	0
1961	2,825	1,020	0	305[a]	2,702	0	0
1971	n.a.	3,210	267	2,400	n.a.	84	46
1981	n.a.	2,812	1,413	3,364	n.a.	916	1,601[b]

Note: n.a. = not available.

Sources: All data but number of rice planting machines: Farm Machinery Industrial Research Corp. (1982). Rice planting machines from Japan, Ministry of Agriculture, Forestry and Fisheries, *Monthly Statistics of Agriculture*.

[a]Refers to 1960.

[b]Refers to 1979.

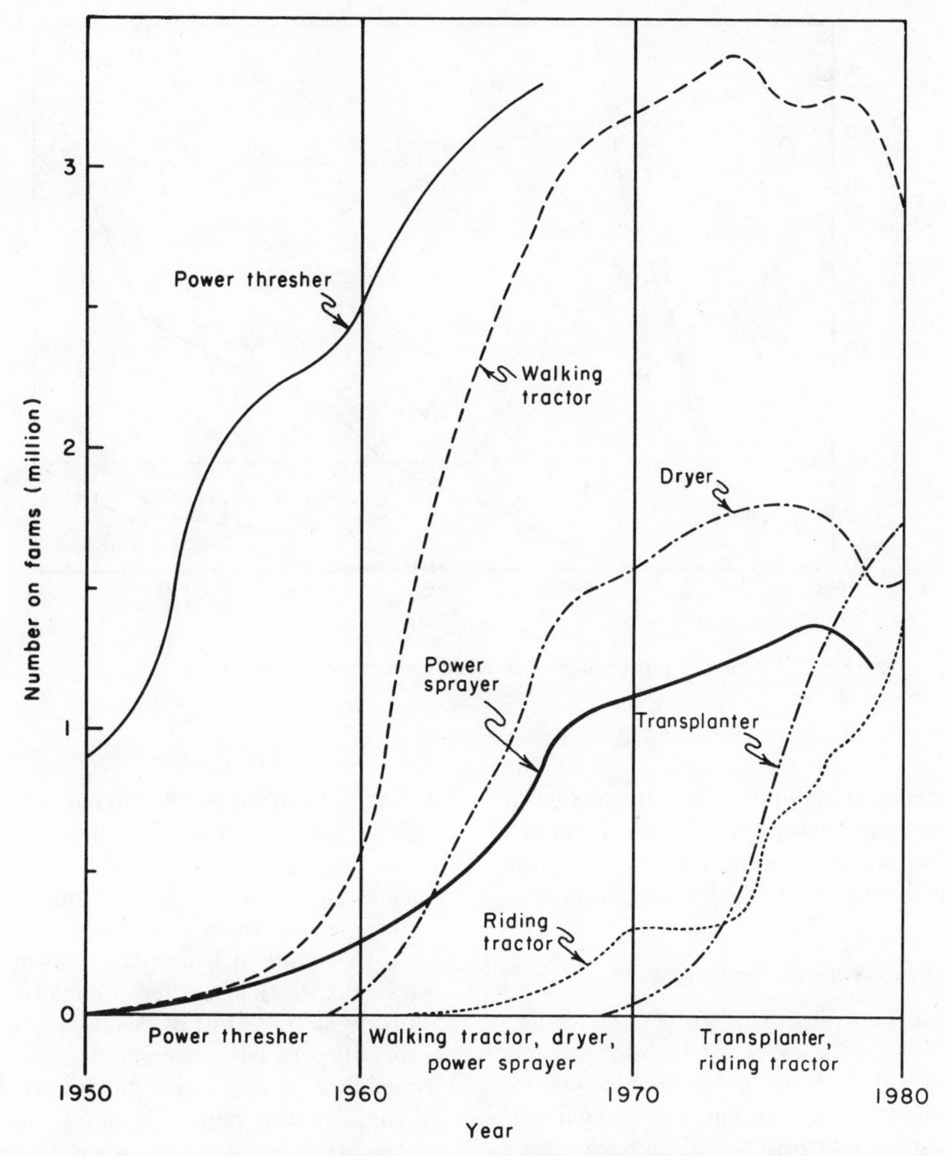

Figure 8.1. Development and diffusion of major farm machinery in Japan, 1950–80
(Source: M. Kisu, "Mechanization of Rice Farming in Japan")

tillers/1,000 hectares). This suggests that there are other factors in addition to the cost of labor that affected the rate of machinery adoption in these areas. The wage rates, when converted into paddy, reflect the domestic price of rice as well as the wage paid to labor.

All three countries insulate their rice prices from the world market, thereby distorting them. The industrial sectors of Japan, Korea, and Taiwan, by contrast, have been well attuned to opportunities in the international market. Thus, it is argued that the costs of mechanization reflected world market conditions and the price of rice reflected policy views on how to achieve the desired pace of development. Japan, with a high rice price relative to the world market, mechanized rapidly even though farm wages were only 3 to 6 kg of paddy per day in the 1950s. In 1956, when Japan reached 20 tillers/1,000 ha, farm wage rates were $US 1.00/day. Both labor and rice were valuable as potential foreign exchange earners, and there was a strong drive to mechanize. In addition, institutional factors in the form of restraints on land sales and farm consolidation encouraged the development of part-time farming that could only conveniently be carried on with machinery. Korea followed a similar path, but with an even higher rice price, encouraging rapid mechanization in the 1970s. Taiwan, by contrast, maintained a low rice price during the early 1970s, reducing the incentive for farm mechanization. By the late 1970s, when Taiwan was experiencing very rapid industrialization, wage rates, both in terms of kilograms of rice and foreign

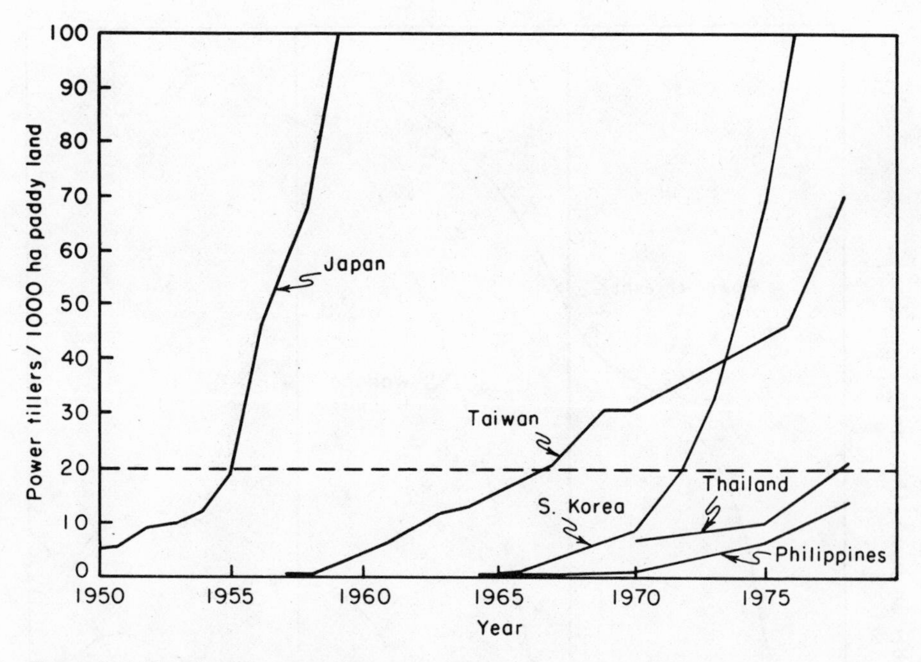

Figure 8.2. Number of power tillers per 1,000 hectares of paddy land

exchange, became so high that it was impossible to stop mechanization. Taiwan's policies of rural industrialization also made part-time farming relatively attractive, which also encouraged mechanization.

A Generalized Sequence of Mechanization

Broad similarities in relative factor abundance and in the tasks required for wet rice cultivation encouraged a general pattern of rice production techniques in Asia, although to date only Japan has fully mechanized. Australian and North American rice production is fully mechanized following a capital-intensive pattern that uses large tractors, airplanes, and combines. It is clear, however, that most Asian countries will follow a small-scale Japanese rather than a Western model. Taiwan and Korea are well started toward mechanization following the Japanese pattern, and a number of other Asian countries are beginning to move in the same direction.

In East Asia, investment in land improvement and water control preceded any move to mechanized production operations. This was partly an accident of history: water control was improved by using human and draft animal power and was one of the few ways to improve the productivity of agricultural land in the high man-to-land economies of East Asia. Most of Japan's rice fields were supplied with irrigation facilities by the nineteenth century, and the major subsequent improvement was investment in drainage, which made it possible to control water to a very high degree.[9] This was supplemented by small pumps in the first half of the twentieth century when small engines and electric motors became available.

Similar investments in water control occurred somewhat later in Korea and Taiwan, where rural infrastructures had existed for many years prior to World War II.[10] Most of this infrastructure took the form of gravity irrigation and drainage. But with the availability of electricity and internal combustion engines, power pumps became one of the first machinery investments for many rice producers. In areas of South and Southeast Asia where gravity systems do not permit efficient water control and groundwater resources are available, there has been substantial investment in private pumps.[11] Electric irrigation pumps replaced foot-operated pumps in Japan during the 1920s, long before power tillers were used.[12] A 1966 study in an intensive double rice cropping area in Taiwan noted that one water pump was available for every three farms, while there was only one power tiller for every eighteen farms.[13]

In addition to pumps, other investments in land were important prerequisites for successful development of mechanized systems in East Asia. Drainage helped to provide a firm soil structure to support rubber-tired machines. Plots were enlarged and consolidated, and new roads improved access to farmers' fields.[14]

After the development of a high-quality land base,

Table 8.3. Farm-level Rice Prices and Wage Rates during Comparable Periods of Agricultural Mechanization, Japan, Korea, and Taiwan

		Farm prices in domestic currency			Prices		
State of mechanization	Period	Paddy (per/mt)	Wages (per/day)	Real wage (kg rice/ day)	Rice[a] ($US/mt)	Wages ($US/day)	World rice ($US/mt)
Japan							
Introduction	pre-1950	n.a.	n.a.	n.a.	n.a.	n.a.	n.a.
2.5 tillers/1,000 ha	1950–51	73,000	250	3.4	311	0.70	n.a.
20 tillers/1,000 ha	1956	77,000	363	4.7	328	1.00	134
100 tillers/1,000 ha	1961	77,000	530	6.9	327	1.47	137
Korea							
Introduction	1961–62	14,640	106	7.2	173	0.85	154
2.5 tillers/1,000 ha	1968	39,510	381	9.6	216	1.36	201
20 tillers/1,000 ha	1972	87,680	803	9.2	338	2.02	148
100 tillers/1,000 ha	1978	176,000	2,900	16.5	559	5.99	367
Taiwan							
Introduction	1955–56	1,600	n.a.	n.a.	86	n.a.	134
2.5 tillers/1,000 ha	1961[b]	3,595	35	9.7	98	0.88	137
20 tillers/1,000 ha	1970	4,734	71	10.1	176	1.77	143
100 tillers/1,000 ha	1978[c]	9,080	254	27.9	376	—	367

[a]Paddy price shown in first col. converted to rice at 65 percent and converted to $US at official exchange rate.

[b]Taiwan passed 2.5 tillers/1,000 ha in 1958, but wage data are not available until 1961.

[c]Taiwan had 70 tillers/1,000 ha in 1978, the year for which data are shown. By 1979 or 1980, it had undoubtedly surpassed 100 tillers/ 1,000 ha.

small threshing machines were among the first mechanical devices to be widely adopted.[15] At first foot-pedal threshers were used, but these were quickly supplanted by power-operated models. Foot-power threshers were introduced into a number of countries by the Chinese and Japanese, but never became established.[16] When power threshers were developed, however, they were widely adopted in East Asia and in some areas of other countries. Power threshers allow faster postharvest processing and hence may contribute to shorter time between two crops in multiple cropping systems. Threshing seems particularly well suited to mechanization for technical reasons—it requires a fairly large amount of power, is carried out in one location, and requires relatively little skill.

The introduction of land preparation equipment sometimes precedes and sometimes follows threshers. In Japan, two-wheeled power tillers were introduced in the early 1950s, much later than power threshers. Land preparation was the first operation mechanized in parts of the Philippines and Thailand, but in some areas, for example Central Luzon, large mechanical threshers have been in use since before the 1950s.[17] Tillers may initially be very small, as in the case of the 2 to 3 horse power (hp) "iron cow" introduced in Taiwan, but after some years, machines in the 2 to 12 hp range seem to take over.[18]

After mechanization of land preparation and threshing, attention is directed to the other tasks in rice cultivation. It seems to be substantially more difficult to develop appropriate machines for planting, fertilizing, cultivating, and grain drying. An economical, efficient combine has also proven elusive. These operations present formidable technical problems. Weeding, for example, requires considerable judgment and relatively little power. Some mechanical weeders have been developed, but herbicides have proved to be cheaper and more effective at distinguishing weeds from rice. Transplanting has been mechanized in Japan, but it requires special techniques for raising seedlings and is still quite labor intensive. There is a continuing discussion over whether direct seeding methods can be developed that are competitive with transplanting, but yields usually suffer. Despite various problems discussed above, by the late 1970s the Japanese had developed commercial machines for each major operation in rice cultivation.

Thus, one can see a general sequence of mechanization beginning with irrigation, then threshing, then land preparation, then other cultural operations. There are substantial divergences from this path. Four-wheeled tractors of the 35–60 hp size have been the first mechanically powered devices introduced into Thailand, Malaysia, the Philippines, and Pakistan. These units are used on a rental basis by small farmers in some areas for initial land preparation, with secondary land preparation carried out using draft animals. In such cases, old and new technologies coexist. The introduction of large tractors has sometimes preceded the development of a high degree of water control, as in Central Luzon or Central Thailand, where their presence in sugarcane farming may have stimulated their adaptation for rice. In other cases, government authorities (such as the Muda River Development Authority of Malaysia) may own the tractors and provide their services to farmers on a rental basis.

A Model of Agricultural Mechanization

The East Asian countries began to mechanize when labor used in rice production found alternative employment in the industrial sector. This suggests that the pace of mechanization can be explained within a supply and demand framework. Rice production machinery can be supplied either by imports or domestically. Imports are generally controlled by government quotas or licensing. Domestic production may occur through private initiative, but experience in Japan and Taiwan shows that concessional government credit, subsidies, tax exemptions, and government development efforts have been major forces in speeding the development of machinery appropriate to rice production conditions in those countries.[19]

The level of private sector machinery production is a function of the development stage of the industrial sector and opportunities for alternative earnings in industrial plants and equipment. The latter are related to the scale of investment needed to begin production of farm machinery compared with other industrial products. The potential earnings of export industries may make governments willing to set policies that encourage or discourage the mechanization of agriculture. Government investment in research and development of farm machinery is an obvious encouragement, while taxation and import restriction are barriers.

Thus, there are many ways in which government policies can affect the supply of agricultural machinery, and indeed, it may seem that policymaking is the main determinate. In addition, there are limitations to the extent to which governments will subsidize imported machinery. Clearly, the supply of agricultural machinery is determined not only by market forces in the industrial sector, but also by government decisions to tax or subsidize imported machinery and to invest in research and development of new machinery.

The demand for rice production machinery is determined by technical productivity (the degree to which the machine substitutes for labor or other inputs), the price of labor relative to substitutes, and the price of rice. Machinery well adapted to the technical requirements of a particular agricultural setting will, by definition, be more productive and substitute for a greater value of alternative inputs. A high price for rice will also increase the demand for machinery.

The data in table 8.4 are broadly consistent with the above static equilibrium concepts, although there are many other forces within each country that determine the level of mechanization. The data show that Japan is far and away the most highly mechanized country of the region. In 1970, Japan had more than one tiller for every hectare of rice land and almost fifty tractors per 1,000 ha of agricultural land. Farm wage rates were approaching $US 5 per day, and the price of rice was the world's highest. Taiwan and Korea had wage rates about four times as high as any other country except Japan and were well started toward the adoption of power tillers. Korea had the second highest rice price, and Taiwan's was not far behind.

Several countries had a small number of tractors, but these were either used for road transportation, plantation, or other nonrice crops (Sri Lanka and Malaysia), or on a contract basis for the initial preparation of land for rice (Malaysia, Philippines, Thailand). Power tillers were introduced in small numbers in many countries, but aside from Thailand, Malaysia, and the Philippines, they were still mainly used by exceptional farmers. Wage rates and rice prices in most South and Southeast Asian countries, with the possible exception of Malaysia, were far below those existing in the East Asian countries, and hence it appeared that rice production in most of South and Southeast Asia was not ready for rapid mechanization. China is somewhat of an anomaly in having a much larger number of tractors and power tillers than other developing countries in the region. Even though precise data are not available, wage rates and rice prices are both low and cannot be used to explain the situation.[20]

Burma has a high wage rate measured in paddy terms, but that is because of policies that ensure an unusually low paddy price. Thailand has a similar policy. Other countries have farm-gate prices around $US 100/mt. Low prices and the relatively slow growth of the nonagricultural sector resulted in small increases in real wages over time and little incentive to mechanize. This is in sharp contrast to the situation in East Asia. For example, wage rates increased by about 700 percent in Japan between 1961 and 1976

Table 8.4. Land Preparation Machines and Economic Factors Associated with Mechanization in Asia

Area	Year	Tillers/ 1,000 ha rice area	Tractors/ 1,000 ha agric. land	Wage rate/day		Farm-gate price rough rice ($US/mt)
				$US	kg rough rice	
South Asia						
India	1970	n.a.	0.61	0.38	4.67	82
	1979–80[a]	0.57	2.28	0.47	4.15	112
Pakistan	1972	n.a.	1.67	n.a.	n.a.	158
	1978–79	n.a.	3.80	1.00	12.53	80
Sri Lanka	1970	0.35	6.82	0.57	4.79	119
	1980[a]	12.41	4.43	1.00	4.73	155
Bangladesh	1970–71	0.26	0.23	0.65	5.96	109
	1979–80[a]	0.42	0.38	0.67	7.55	89
Southeast Asia						
Malaysia	1969	3.85	1.37	0.84	12.44	68
	1978		2.76	6.02	30.93	195
Thailand	1970	5.48	1.31	n.a.	n.a.	30
	1980	26.24	5.92	2.45	23.26	105
Philippines	1970	2.97	1.63	0.50	8.89	56
	1978	13.84	4.11	1.33	8.84	150
Indonesia	1970	0.08	0.03	n.a.	n.a.	73
	1978	0.26	0.11	0.56	3.52	158
Burma	1970	0.05	0.53	1.51	41.26	37
	1978[b]	n.a.	0.82	1.15	17.59	65
East Asia						
Japan	1970	1,095.11	46.07	4.22	11.26	375
	1980	1,157.62	619.03	22.48	15.11	1,488
South Korea	1970	9.88	0.03	1.83	10.54	174
	1980	198.28	0.92	11.71	12.77	917
Taiwan	1970	27.16	0.73	1.77	15.00	118
	1978	73.00	3.00	8.65	34.25	252
China	1970	—	2.38	n.a.	n.a.	n.a.
	1980	56.47	7.52	n.a.	n.a.	n.a.

Sources: Asian Productivity Organization (1983) various years, appendix tables.
[a]Prices refer to 1977.
[b]Prices refer to 1975.

and 1,700 percent in Korea over the same period, while in Burma and the Philippines they increased by only 40 and 220 percent, respectively. In most of the South and Southeast Asian countries, real agricultural wages have been constant, so the incentive to mechanize is generally lacking.

During the 1970s, economic progress in East Asia pushed the pace of mechanization—not only in land preparation, as shown in table 8.4, but also in other implements, as indicated in table 8.1. In South and Southeast Asia, economic conditions were less dynamic. Wage rates remained below $US 1.00 per day in most of South Asia and Indonesia. Malaysia experienced a rapid wage increase, but Thailand and the Philippines had less dramatic increases. Rice prices increased by 50 percent, but far less than in East Asia. Thus, a review of the economic indicators important for mechanization suggests that most countries in South and Southeast Asia will not undergo rapid mechanization during the 1980s.

Evidence on the Impact of Mechanization

Tractors and other farm machines have become the symbol of modern agriculture because they are the most visible difference between peasant farms in developing countries and farms in Japan, the United States, and other developed countries. Tractors are also often identified as important components of development assistance programs because they generate business in the donor country and they are one means of giving a large amount of "assistance" in a relatively efficient package, thereby minimizing the overhead of the assistance agency. Tractors are also promoted by commercial firms from developed countries seeking new markets.

Of course, these reasons are not often mentioned in the discussion and promotion of farm machinery for developing countries. The main point stressed is that machinery will save farmers money, improving profits. Increased production is hypothesized to come from higher cropping intensities, which result in more harvests from a plot of land each year, or from higher yields generated either by better land preparation or because crops are planted earlier, and therefore are harvested during a more favorable time of year. In this section, we use empirical studies to examine and summarize the evidence on these two issues.

Impact on Rice Output

There are many types of machinery, each with specific functions, so any evaluation should focus on particular machines. It is clear that mechanical irrigation devices directly improve yields and may also expand the area harvested without displacing labor. However, questions of relative investment efficiency invariably arise.[21] Similarly, mechanical sprayers are used to carry out tasks that cannot be done by hand. Such mechanical devices may be considered as *yield increasing* rather than labor displacing, even though in some cases where water is extremely scarce and labor extremely cheap, as in China, water may be moved with human labor or insects may be manually removed. There are important questions about the most appropriate scale and design of yield-increasing devices, but because of the clear output-increasing purpose, they are not reviewed here.

We focus on the impact of mechanized land preparation and threshing in rice or rice-based systems. There are a large number of studies examining wheat-based systems in South Asia, although we do not consider them here.[22] Machinery for preparing land and threshing has recently been introduced in a number of countries. The machines have obvious labor-saving properties and are promoted on the basis that they improve output by increasing intensity, or raise yields by ensuring that the crop is planted or harvested at a more optimal time.

There are two ways to empirically determine the impact of a change in technology: (1) compare users with nonusers, holding all other factors constant, or (2) trace the impact of a change on a sample of users that have experienced only that change. Because the world is not a social laboratory in which other factors are constant, all empirical studies have some level of imperfection, but it is hoped that they approximate the *ceteris paribus* conditions.

Cropping Intensity

Studies measuring cropping intensity impacts are summarized in table 8.5. The first line for each study shows the cropping intensity under the unmechanized system, the second shows the change in cropping intensity associated with mechanization.

The King study clearly indicates the dominance of irrigation quality as contrasted with power tillers.[23] The subsample of farmers with good irrigation who used *carabao* (water buffalo) for land preparation harvested 1.67 crops per year. The farms with poor irrigation had just about one crop per year regardless of the use of carabao or power tillers. The Deomampo and Torres study shows similar results. Evidence from a more recent study in Central Luzon shows a larger difference between irrigated and rainfed farms and indicates that irrigated farms using tractors or power tillers for land preparation had higher cropping intensities than those using carabao. The average cropping intensity increase due to mechanization for the five Philippines cases was 12 percent.

Pudasaini found somewhat similar results in the Bara District of Nepal—farms with mechanized land preparation had about 12 or 14 percent higher cropping intensity than those using animals, whether irrigated with pump sets or not. Two studies in India and one from Bangladesh showed modestly higher cropping intensity on mechanized farms, but the samples were not separated by water control. Farms using tractors had as much as 10 percent higher cropping intensities than those using bullocks for land preparation.

Studies in South Sulawesi and West Java, Indonesia, reported the highest overall cropping intensities. In South Sulawesi, one of the sparsely populated outer islands, the researchers determined that almost no change occurred in cropping intensity on either rainfed or irrigated farms after the introduction of tractors for land preparation. It is clear that an extremely high level of intensity was achieved without tractors. The same result was reported in West Java.

The historical pattern of cropping intensification and introduction of power tillers in East Asia also fails to show any relationship between the two. Japan reached a peak cropping intensity of 1.3 in 1957 and declined thereafter, even as mechanization rapidly proceeded (figure 8.3). Taiwan reached its peak cropping intensity of 1.9 in 1962 before mechanization really got under way. A similar situation occurred in Korea. Few studies evaluating the impact of mechanized threshing are available and those few show highly variable results. Until more research is available, it will be difficult to make generalizations.

Yields

There are suggestions in the literature that mechanization of land preparation can lead to increased crop yields, but empirical data supporting these views are difficult to find. One field experiment comparing

Table 8.5. Summary of Studies on the Impact of Mechanized Land Preparation on Cropping Intensity in Rice-based Systems, Asia

Study	Area	Comparison	Cropping intensity premechanization	Cropping intensity postmechanization
King	Central Luzon, Philippines	Carabao vs. power tillers	0.98 (poor irrigated subsample)	1.03
King	Central Luzon, Philippines	Carabao vs. power tillers	1.67 (good irrigated subsample)	1.70
Deomampo and Torres	Central Luzon, Philippines	Before vs. after tractors	1.56	1.68
IRRI	Central Luzon, Philippines	Carabao vs. tractors	0.77	1.37
IRRI	Central Luzon, Philippines	Carabao vs. tractors	1.00	1.00
Pudasaini	Bara District, Nepal	Animal vs. tractors[a]	1.25	1.45
Pudasaini	Bara District, Nepal	Pumpset vs. tractors and pumpsets	1.36	1.55
NCAER	Andhra Pradesh, India	Bullocks vs. tractors	1.15	1.22
NCAER	Tamil Nadu, India	Bullocks vs. tractors	1.22	1.34
Gill	Five areas of Bangladesh	Bullocks vs. power tillers & tractors	2.01	2.12
Jabbar, et al.	Bangladesh	Bullocks vs. power tillers	1.48	1.63
Alam	Bangladesh	Bullocks vs. power tillers	1.17	1.19
Consequences team	South Sulawesi, Indonesia	Before vs. after tractor	1.78 (rainfed subsample)	1.83 (rainfed subsample)
Consequences team	South Sulawesi, Indonesia	Before vs. after tractor	1.96 (irrigated subsample)	1.92 (irrigated subsample)
Bagyo	West Java, Indonesia	Manual vs. tractor	2.16	1.95
Bagyo	West Java, Indonesia	Animal vs. tractor	2.12	1.93

Sources: A. H. M. Mahbudul Alam (1981); Ali Sri Bagyo (1981); Consequences Team, "Consequences of Land Preparation Mechanization in Indonesia: South Sulawesi and West Java" (1981); N. R. Deomampo and R. D. Torres (1975); IRRI, Loop Survey (unpublished) (1980); G. J. Gill (1981); M. A. Jabbar, S. R. Bhuiyan and A. K. Maksudul Bari (1981); F. King (1974); India, National Council of Applied Economic Research (NCAER), "Implications of Tractorization on Farm Employment, Productivity and Income: Summary and Highlights" (n.d.); S. P. Pudasaini (1979).

[a]Combined tractor-owning and tractor-hiring farms.

alternative land preparation techniques failed to show any difference in wetland rice yields (table 8.6). There was a clear substitution of fuel (and capital) for labor hours, and there seemed to be some indication of fewer weeds with certain techniques, but yields were essentially identical for each of the five tested techniques.

A number of the studies of cropping intensity also evaluated yield differences between farms with mechanized and nonmechanized land preparation. Some of the authors who found that mechanized farms had higher yields, noted that those farms also had higher rates of fertilizer application, apparently because of the greater financial capabilities of the owners. Given the undisputed yield-raising effect of fertilizer, it seems appropriate to compare the yields of mechanized and nonmechanized farms after adjusting the differences in fertilizer applications. For this purpose, it is assumed that each kg of fertilizer (nutrient) produces 10 kg of paddy. Table 8.7 shows data for ten studies for which comparisons were possible. In 90 percent of the cases, after accounting for differences in fertilizer applied, yields were no higher on mechanized than on nonmechanized farms.

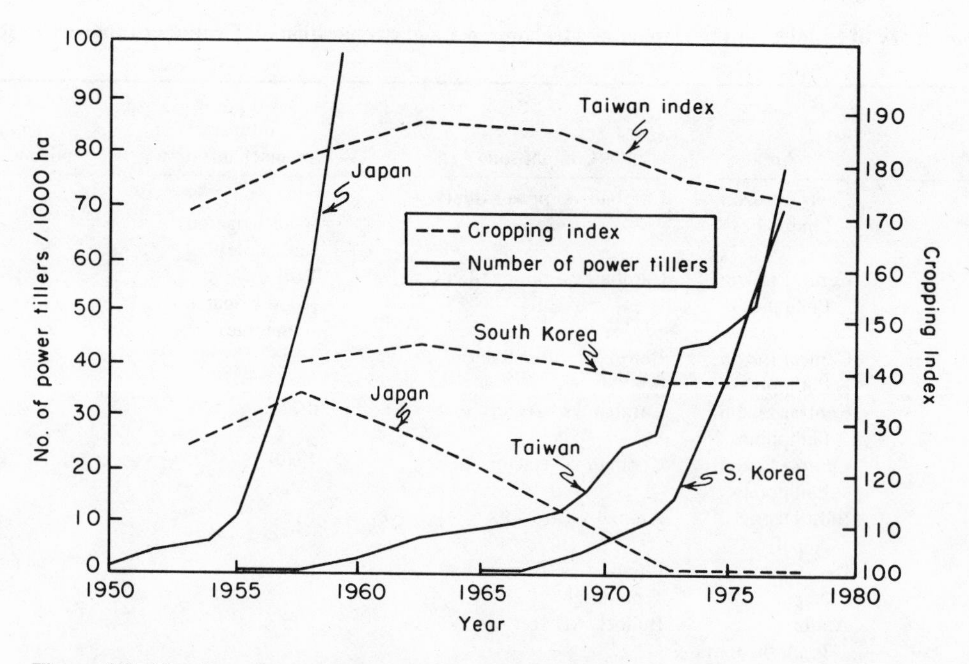

Figure 8.3. Number of power tillers and cropping index (Source: Based on data from S. K. Jayasuriya, A. Te, and R. W. Herdt, "Mechanization and Cropping Intensification: Economic Viability of Power Tillers in the Philippines")

Timeliness

It is hypothesized that timeliness may also be improved on mechanized units, thereby permitting farmers to take advantage of more optimal crop conditions. Pudasaini found almost no difference in the paddy transplanting schedule of farmers who hired tractors. This is in contrast to earlier wheat planting practiced after farmers adopted tractors.[24] The Consequences Team (Indonesia) study referred to earlier (table 8.5) compared the percentage of the area in which land preparation was completed each week on farms using various techniques. These data fail to show any consistent support for the hypothesis that timeliness is significantly improved by mechanization.[25]

A recent study of farm mechanization in Bangladesh resulted in similar findings. In areas where tractors were used, the turnaround period between successive crops was not reduced. When sowing dates of a given crop were compared, no significant difference was found between tractor-cultivated and animal-cultivated plots in the same area.[26]

Table 8.8 summarizes data collected in Iloilo and Pangasinan provinces in the Philippines concerning turnaround time between rice crops. Relatively few farmers used tractor power, but there is no evidence that tractor users enjoyed any timeliness advantage over those using carabao power. In fact, the first crop was generally planted earlier on the plots cultivated by carabao than on the tractor farms. The turnaround time between the harvest of the first crop and planting of the second crop averaged 1.7 weeks on carabao-cultivated plots and 3.1 weeks on the tractor-cultivated plots.

Table 8.6. Labor and Fuel Used in Experiments with Five Alternative Land Preparation Methods for Rice Production (Averages of Four Soil Types) Philippines, 1973 Wet Season

Tillage treatment					
Primary tillage	Secondary tillage	Labor hrs/ha	Fuel lt/ha	Weeds g/.2m²	Yield mt/ha
64 hp tractor	Carabao	45	20	16.7	3.61
14 hp tiller	Carabao	49	15	12.3	3.56
7 hp tiller	7 hp tiller	29	36	11.0	3.81
Carabao	7 hp tiller	56	26	14.4	3.60
Carabao	Carabao	81	0	17.9	3.60

Source: B. Duff (1978).

Table 8.7. Summary of Studies Comparing Rice Yields on Farms with Animal or Hand Land Preparation with Farms Using Mechanical Methods

Study	Area	Comparison	Reported yield (mt/ha)	Fertilizer (urea) (kg/ha)	Adjusted yield[a]/ (mt/ha)
Alam	Bangladesh	Bullock vs.	1.5	n.a.	1.5
		power tiller	1.5	n.a.	1.5
Antiporta and Deomampo	Philippines	Animals vs.	2.6	86	2.6
	8 provinces	tractors and tillers	2.8	117	2.5
Anuwat	Central Thailand	Bullock vs.	2.6	32	2.6
	(irrig., transplanted)	tractor	2.8	48	2.6
Anuwat	Central Thailand	Bullock vs.	0.2	3	0.2
	(rainfed, broadcast)	tractor	0.2	2	0.2
Deomampo and Torres	Central Luzon, Philippines	Before vs.	2.2	57	2.2
		after tractors and tillers	2.6	79	2.1
Gill	Three areas of Bangladesh	Bullock vs.	2.4	n.a.	2.4
		tractors and tillers	2.4	n.a.	2.4
Pudasaini	Nepal	Bullocks vs.	1.7	16	1.7
	(without pumps)	tractors	2.1	164	1.4
Pudasaini	Nepal	Bullocks vs.	2.1	214	2.1
	(with pumps)	tractors	2.3	264	2.1
Sinaga	West Java, Indonesia	Animal vs.	4.9	323	4.9
	(wet, 1979/80)	tractors	4.9	323	4.9
Sinaga	West Java, Indonesia	Manual vs.	3.8	285	3.8
	(3 seasons, 1979–80)	tractors	3.9	308	3.8
Sinaga	South Sulawesi, Indonesia	Animal vs.	2.7	138	2.7
	(3 seasons, 1979–80)	tractors	2.9	227	2.5
Tan and Wicks	Nueva Ecija, Philippines	Carabao vs.	2.6	89	2.6
	(wet, 1979)	tractors	4.0	129	3.8

Sources: A. H. M. Mahbudul Alam (1981); D. B. Antiporta and N. R. Deomampo (n.d.); N. R. Deomampo and R. D. Torres (1975); G. J. Gill (1981); S. P. Pudasaini (1979); R. Sinaga, "Effects of Mechanization on Productivity: West Java, Indonesia" (1981); R. Sinaga, "Effects of Mechanization on Productivity: South Sulawesi, Indonesia" (1981); Anuwat Wongsangaroonsri (1981); Yolanda Tan and John A. Wicks (1981).

[a]The yield of the mechanized group was adjusted by subtracting the estimated yield contribution of the difference in fertilizer applied to the two groups.

Available evidence suggests that power tillers and threshers may contribute to an increase in cropping of about 10 percent. Neither seem to have any clear effect on yields through timeliness or turnaround time.

Who Gains from Mechanization

When mechanization results in increased output, rice prices will tend to fall and benefits will be widely shared by rice consumers whether they are landless agricultural workers, farmers, or urban people.[27] The absolute benefits to individuals are positively related to the proportion of their incomes spent on rice.

If output does not increase, machinery may shift earnings from one group to another. That is, the owner of a machine that replaces labor will receive the wage formerly paid to workers. There is an inherent difference in the ownership pattern of capital and labor. In the absence of slavery, labor can be owned at a rate of only one unit per person, or at most five

Table 8.8. Average "Turnaround" Time Between Harvest of First Rice Crop and Planting of Second Rice Crop by Land Preparation Power Source, Philippines

Location	Water control	Power source	Number of weeks				
			1976–77	1977–78	1978–79	1979–80	Average
Iloilo	Rainfed	Carabao	1.9	1.2	1.2	2.1	1.6
		Tractor	3.3	3.0	0.3	4.2	2.7
Iloilo	Irrigated	Carabao	2.6	3.4	1.9	2.8	2.7
		Tractor	n.a.	4.1	2.7	5.3	4.0
Pangasinan	Irrigated	Carabao	0.2	1.4	0.5	n.a.	0.7
		Tractor	3.0	1.8	3.3	n.a.	2.7

Source: International Rice Research Institute, Cropping Systems Economics Farm Record Keeping Project.

to ten units per household. Ownership of capital, on the other hand, is usually concentrated in the hands of a few, usually through inheritance, political power, or business acumen. Concentrated capital ownership means that the income earned by capital is also centralized. Thus, the introduction of machinery will tend to cause a redistribution of benefits, although the magnitude of the change will depend on whether absolute output is increased.

When a task can be accomplished more cheaply using machinery rather than labor, farmers will be motivated to adopt machines. However, if machines are cheaper because of purchase price or low-cost credit, then farmers are responding to artificially low (policy-induced) market prices. If machines result in a faster rate of output growth than would be achieved in their absence, then the policymaker must evaluate the tradeoff between more output and redistribution of income from labor to owners of capital. But if there is no output effect, the decision to promote mechanization is a decision to support a transfer of income from labor to machinery owners, without offsetting benefits.

The redistributive effects of labor-saving (private), cost-reducing machines are substantially different in the market-directed economies of South and Southeast Asia from what they are in communally organized agriculture in China. In the latter, cost reductions are shared among members of the commune in a manner decided upon by the commune. Thus, if introduction of a machine cuts down on costs, all benefit in proportion to work points earned. In market-directed economies, the earnings of factors of production go to machine owners.

Under many conditions, landless agricultural workers are highly dependent on agriculture, and perhaps even more dependent on the earnings they obtain during the harvest season. Data from a small study of sixteen families in Iloilo, Philippines, show that two groups of landless worker families obtained about 60 percent of their season's income from the share payment they received for harvesting. Clearly, the actual effect on laborers of any reduction in the demand for labor depends on the proportion of their incomes that derives from operations that are mechanized and the opportunities for alternative earnings. This varies from place to place, making the need for a careful assessment of the likely impact of mechanization that much more important.

Impact on Employment and Other Sectors

The foregoing discussion deals only with the direct impact of mechanization on yield and cropping intensity. Those effects are modest, but the incentive to mechanize is generated by the possibility of the farmer's saving on costs of production. From the viewpoint of the individual farmer, mechanization may make sense because of the savings generated by not hiring labor. The farmer may even prefer to hire a machine rather than supervise a team of laborers. Thus, it seems that employment will definitely be reduced when farm machines are introduced.

However, there are offsetting effects. Machines must be produced and serviced, and these operations generate income and employment in other sectors. As discussed in chapter 9, a rate of growth adequate to absorb the growing labor force from agriculture depends on the rates of growth in the agricultural and nonagricultural sectors and the proportion of labor in each sector. Hence, the contribution of the agricultural machinery production and sales sector may be an important factor.

The indirect effects of farm mechanization arise as demand is generated for agricultural machinery and in turn for engines, steel, bearings, and manufacturing labor. The magnitude of the demand for labor depends on the intensity with which labor is employed in the production of rice (direct) and farm machinery (indirect, first round), and the production process used to produce machines needed in manufacturing farm machinery (indirect, second round). There are also income distribution effects as the extra income from mechanization (or redistribution of income) changes the demands of those receiving the income. Depending on their patterns of demand, consumption will change.

These effects can be quantitatively measured within an input–output model of the type developed and popularized by Leontief.[28] The technical details of such a model are not discussed here, but one interesting example of such a model adapted to evaluate the impacts of rice farm mechanization is discussed.[29]

A model was developed in which the entire rice output of the Philippines was assumed to be produced in thirteen distinct production subsectors identified by four mechanization intensity levels and four water control intensity levels. (The least intensive water control level was paired only with the least intensive mechanization level, but all other combinations of three water control and four mechanization levels were included.) The model first solves for employment and income distribution based on production among the thirteen sectors. The consumption demand for rice is increased by 1 percent from subsectors 1, then 2, 3, and so forth in turn. The model solves for direct employment in rice production and indirect employment in the rest of the economy. The conclusions, while not entirely surprising, are interesting.

The calculated employment effect from meeting

the 1 percent increase in demand by producing rice in a gravity irrigation system with a power tiller rather than a carabao is an increase in total employment in the economy of about 1,000 workers. The increase was about the same if the system was pump irrigated. The increase was the net effect of a direct reduction of 1,000 workers in rice production and an indirect increase of about 2,000 workers in the machinery sectors and in the sectors that experienced activity because of other indirect effects. The calculated employment effect from meeting the 1 percent increase in rice demand by producing rice in a gravity-irrigated system rather than a rainfed system is an increase in total employment of 25,900 workers. This increase was the total of 8,300 direct effect and 17,700 indirect effect.

The analysis provides especially dramatic evidence of the different employment and income distribution effects of increasing rice production in more intensively irrigated systems in contrast to increasing production in more intensively mechanized systems. The results suggest that most of the developing South and Southeast Asian rice-producing countries that need to increase rice production, employment, and income equity would be more likely to achieve those goals through increasing the proportion of output produced with intensified irrigation rather than through intensified mechanization of rice production.

Future Directions

The examination of research findings on mechanization in rice indicates that mechanization has saved labor, but has not normally led to an increase in production, higher yields, or improved timeliness. Thus, increased productivity or efficiency through mechanization can be achieved only in the degree to which mechanization reduces the cost of production. Adoption of machinery has been independent of the adoption of modern rice varieties and of changes in cropping intensity.

These findings appear to be consistent with evidence that the demand for mechanization is closely related to the economy-wide wage rate. The real wage (expressed as the kilograms of rice that can be purchased with one day of labor) and the current wage both probably influence the demand for mechanization, but the latter shows a more consistent relationship across countries. Thus, machinery is neither a constraint nor a factor facilitating output growth in most South and Southeast Asian countries.

Among the countries of South and Southeast Asia, only Malaysia appears ready to follow the path of East Asia toward rapid mechanization in the decade ahead. The farm wage rate in Malaysia already exceeded $US 6.00 in 1978. Although there has been growth in mechanization in both Thailand and the Philippines, the continued rapid expansion of the agricultural labor force will likely depress farm wages there. In most of the rest of Asia, the farm wage rate remains below $US 1.00 per day, and there is little prospect for rapid growth in mechanization in the near future. For other countries, government attempts to encourage mechanization will further reduce the already low incomes of those dependent on wage earnings from agricultural labor for their livelihood.

This prognosis does not imply that the growth rate of agricultural output will be slowed by the lack of mechanization. To the contrary, there is little evidence to show that the mechanization process contributed to increased cropping intensity or increased yields, either historically in East Asia where the process is virtually complete, or on those farms in South and Southeast Asia where it has already occurred. The growth rate of output is controlled by inputs that release the binding constraints on nutrients, water, and biological factors. Until labor becomes a binding constraint in Asia, mechanization is unlikely to contribute to increased output.

Notes

1. A part of this chapter appears in R. Herdt, "Perspectives, Issues, and Evidence on Rice Farm Mechanization in Developing Asian Countries," in Asian Productivity Organization, *Farm Mechanization in Asia* (Tokyo, APO, 1983) pp. 111–148.

2. Javed Hamimd, "Agricultural Mechanization: A Case for Fractional Technology," in Tan Bock Thiam and Shao-er Ong, eds., *Readings in Asian Farm Management* (Singapore, University of Singapore Press, 1979). Hans Binswanger calls this the "net contribution" view in *The Economics of Tractors in South Asia* (New York, Agricultural Development Council and International Crops Research Institute for the Semi-Arid Tropics, 1978).

3. H. M. Southworth states: "Substitution of machines for labor in peak periods could make possible more productive use of labor over the whole year," in "Some Dilemmas of Agricultural Mechanization," in H. M. Southworth and M. Barnett, eds., *Experience in Farm Mechanization in Southeast Asia* (New York, Agricultural Development Council, 1974).

4. One of the more moderate statements of this position is K. C. Abercrombie, "Agricultural Mechanization and Employment in Developing Countries," in *Effects of Farm Mechanization on Production and Employment* (Rome, Food and Agriculture Organization, 1975).

5. Thanks to Dr. Gilbert Levine (Ford Foundation) for his succinct identification of power, judgment, and mobility as critical factors in determining the technical feasibility of machinery design.

6. Yujiro Hayami, *A Century of Agricultural Growth in Japan* (Tokyo, University of Tokyo Press, 1975).

7. Labor used for rice production in Japan declined from 229 workdays to 141 workdays per hectare between 1956 and 1971. Shigeru Ishikawa, *Essays on Technology, Employment and Institutions in Economic Development*, Economic Research Series No. 19 (Tokyo, Kinokuniya Company Ltd., 1981).

8. Keizo Tsuchiya, "Mechanization and Relations Between Farm, Non-Farm and Government Sectors," in Herman Southworth, ed., *Farm Mechanization in East Asia* (New York, Agricultural Development Council, 1972).

9. Ishikawa identifies the conversion of "wet rice fields" (*shit suden*: rice land left in a wet condition throughout the year because of a lack of drainage facilities) into "dry rice fields" (*kanden:* rice lands kept in a dry condition during the non-rice-growing season) by installation of drainage facilities as the most common type of basic investment in land during the early phase of improvements in Japanese rice production. Ishikawa, *Essays on Technology, Employment and Institutions*, p. 157.

10. John C. H. Fei and Gustav Ranis, "Agriculture in Two Types of Open Economies," in Lloyd Reynolds, ed., *Agriculture in Development Theory* (New Haven, Yale University Press, 1971).

11. S. M. Patel and K. U. Patel, *Economics of Tubewell Irrigation* (Ahmedabad, Indian Institute of Management, 1971).

12. Ishikawa, *Essays on Technology, Employment and Institutions*, p. 44.

13. Weng-chieh Lai, "Current Problems of Farm Management on Mechanized Farms," in Herman Southworth, ed., *Farm Mechanization in East Asia* (New York, Agricultural Development Council, 1972).

14. Tsuchiya, "Mechanization and Relations Between Farm, Non-Farm and Government Sectors."

15. The pattern also holds in Korea. C. C. Lee states that the manual thresher worked by two persons (one keeps the drum turning and the other feeds the grain) is the most usual type of thresher in Korea. "Economic and Engineering Aspects of Mechanization of Rice Harvesting in Korea," in Southworth, ed., *Farm Mechanization in East Asia.*

16. The pedal thresher, although widely used in East Asia, never became popular in South and Southeast Asia. The japonica rices grown in East Asia are much more difficult to thresh by hand than indica rices. Thus, in South and Southeast Asia the pedal thresher offered no significant time and energy savings over hand threshing.

17. Mechanical threshers in Central Luzon were introduced to increase landlord control over harvest shares rather than to save labor. After the change from share tenancy to cash rent under the land reform, there was a rapid shift from mechanical back to hand threshing. See Yujiro Hayami and Masao Kikuchi, *Asian Village Economy at the Crossroads* (Tokyo, University of Tokyo Press, 1981).

18. Carson Kung-Hsien Wu, "Analysis of Machinery—Labor Relationship in Farm Mechanization," in Herman Southworth, ed., *Farm Mechanization in East Asia.*

19. Zyuro Kudo, "Implications of Farm Management Research for Government Mechanization Programs," and Trong-chuang Wu, "Government Policies Promoting Farm Mechanization," in Herman Southworth, ed., *Farm Mechanization in East Asia.*

20. The demand for tractors has dropped sharply in the last few years in China following the introduction of the household responsibility system. Thus, policy under "communization" created a demand for mechanization, which, at least in part, explains this anomaly.

21. Piedad F. Moya, Robert W. Herdt, and Sadique I. Bhuiyan, "Returns to Irrigation Investments in Central Luzon," Philippines Agricultural Economics Paper No. 81-23 (Los Banos, Philippines, International Rice Research Institute, 1981).

22. These are covered by Hans P. Binswanger, *The Economics of Tractors in South Asia.*

23. Cited in J. Bart Duff, "Mechanization and Use of Modern Rice Varieties," in *Economic Consequences of the New Rice Technology* (Los Banos, Philippines, International Rice Research Institute, 1978).

24. Som P. Pudasaini, "Farm Mechanization, Employment and Income in Nepal," *IRRI Research Paper Series 38* (Los Banos, Philippines, International Rice Research Institute, 1979).

25. Consequences Team, "Consequences of Land Preparation Mechanization in Indonesia: South Sulawesi and West Java" (Jakarta, Regional Seminar on Appropriate Mechanization for Rural Development, Jan. 26–30, 1980).

26. G. J. Gill, *Farm Power in Bangladesh*, volume 1 (Reading, England, Department of Agricultural Economics and Management, University of Reading, 1981) p. 188.

27. Yujiro Hayami and Robert W. Herdt, "Market Price Effects of Technological Change on Income Distribution in Semisubsistence Agriculture," *American Journal of Agricultural Economics* vol. 59, no. 2 (May 1977).

28. Wassily W. Leontief, "The World Economy of the Year 2000," *Scientific American*, vol. 243, no. 3 (September 1980) pp. 206–231.

29. Chowdhury S. Ahammed and Robert W. Herdt, "Farm Mechanization in a Semi-Closed Input–Output Model: The Philippines," *American Journal of Agricultural Economics* vol. 65, no. 3 (August 1983).

9

Trends in Labor Use and Productivity

A central issue for the South and Southeast Asian countries is the necessity to increase employment among the poorer segments of the population. Despite a substantial decline over the past twenty years in the share of the labor force employed in agriculture, approximately two-thirds of the labor force still remains in this sector (table 9.1). Population growth rates have declined only slightly, and industrialization has not absorbed the major share of the growing labor force. Policymakers are now convinced that a significant portion of new workers will have to continue to find employment in agriculture.

Rice is a relatively labor-intensive crop. Often about twice as much labor is employed per hectare compared with other grain crops and legumes.[1] Thus, the utilization and productivity of labor in rice is a matter of much concern.[2] Across regions and over time in Asia there have been enormous differences in the way in which rice is grown and in the amount of labor employed per hectare and per ton of rice produced. There are differences not only in the total labor input, but also in the amount of family versus hired labor and in the sex composition of the labor force. There are also variations in the regulations and laws that govern the share of profit received by labor.

There are, of course, many reasons for these differences. However, relative endowments of land and labor play an overriding role. Growing population pressure on the land is associated with changes in agrarian institutions as well as technology. There is much debate about cause and effect. Boserup argues that it is more sensible to regard change as an adaption to increasing population density caused by natural population growth or immigration.[3] Boserup treats population growth as an exogenous variable. There is no determination of the origin of population pressure, and no complete theory of institutional determination.[4] Our discussion of labor use in rice production suffers in this respect.

This chapter reviews the evidence relating to

Table 9.1. Percentage of Asian Labor Force in Agriculture, 1960–80

Country	1960	1980
East Asia		
China	—	64
Japan	33	12
South Korea	66	34
North Korea	62	49
Southeast Asia		
Burma	—	67
Indonesia	75	58
Kampuchea	82	—
Laos	83	75
Malaysia	63	50
Philippines	61	46
Thailand	84	76
Vietnam	—	71
South Asia		
Bangladesh	87	74
India	74	69
Nepal	95	93
Pakistan	61	57
Sri Lanka	56	54

Source: World Bank, *World Development Report, 1982.*

potential employment capacity in the rice sector. It assesses the effect of a growing labor supply and technical changes on the demand for labor in rice production, on labor productivity, on real wages, and on labor contracts. We attempt to relate the differences in the amount of labor supplied by the rice farm family (including landless workers) to differences in the socioeconomic environment.

Historical Trends in Labor Use and Productivity

Historical data on labor use in Asia are most complete for the East Asian countries, where records extend back to the nineteenth century. In this section, we summarize historical trends in labor use with East Asian and Javanese data. This is a useful exercise not only because of the importance of the historical experience to the individual economies, but also because of the similarities between East Asia and South and Southeast Asian economies, which are currently facing similar problems of land scarcity and rapid population growth.[5]

A simple measure of the productivity of labor is the kilograms of rice produced per day of labor. For four areas where long-term data were available, labor productivity measured in kilograms per man-day was plotted against man-days per hectare (figure 9.1). A rise in labor productivity can be accompanied by an increase in labor input per hectare only if yield per hectare (kg/man-day × man-day/ha) is increasing rapidly. Otherwise, an increase in labor productivity will be associated with a decline in labor input per hectare.

Data from Japan show that labor productivity in rice has been increasing and labor input declining for more than a century. Labor input fell by about 50 percent over the period, from 278 man-days per hectare in 1874 to 146 man-days per hectare in 1970. South Korea and Taiwan show a similar pattern in the post–World War II period, although in both cases labor inputs have been lower than in Japan. In the early part of the century, however, labor input per hectare rose in Taiwan. At this time, the new fertilizer-responsive varieties of rice (ponlai varieties) were introduced. Ishikawa hypothesizes that labor input per hectare must also have risen in Japan prior to 1874 when irrigation was being expanded.[6]

Why was labor input per hectare so much higher in Japan than in Taiwan, particularly in the 1960s when both regions had highly developed irrigation and when both had access to much the same technology? The answer seems due in part to the protection provided the rice sector by the Japanese government, which allowed artificially high returns to primary factors and to the retention and attraction of labor from less favored sectors.[7]

Java, like Japan, has shown a steady growth in labor productivity for a century, although the data are less reliable. What is distressing about the Javanese case, however, is that labor input per hectare has been declining nearly as rapidly as in Japan without a corresponding rise in the real wage rate. Even though rice still requires substantial amounts of labor, and labor input is high compared with most other crops, the labor absorption per hectare has been declining.

For much of the rest of South and Southeast Asia, the labor input per hectare is lower than in Java (sometimes less than 100 man-days per crop), and the yields are also lower (about 2 mt/ha). The major question is whether there is opportunity for these countries to increase labor absorption per hectare in the process of increasing yield per hectare. This issue will be examined in the next section by studying the impact of the introduction of modern rice technology on labor productivity and employment per hectare. But first, it is necessary to understand more clearly the historical reasons for the decline in labor input per hectare.

There are three levels of technical conditions that combine to determine the level of per hectare labor input in rice production in any area at any time.[8] These are:

1. Labor-using technology factors such as irrigation, new varieties, and improved crop management practices
2. Labor-saving technology factors, including mechanical devices and herbicides
3. Natural and institutional conditions existing at a particular time, including physical factors such as soil and water conditions, and institutional factors such as traditional labor practices and labor contract arrangements

The influence of these factors can be seen by examining the difference in labor input for specific tasks over time and space.

Table 9.2 shows the change in labor input per hectare by task over time for the major operations in rice production in Japan, Central Taiwan, and Java. The data must be viewed with some caution, as the labor categories are not always comparable, and some values are obviously inconsistent. Qualifications notwithstanding, there are some marked trends in these data.

A decline in labor used for land preparation occurs over time in all three cases. In Taiwan between 1961 and 1972, this signals the introduction of the power

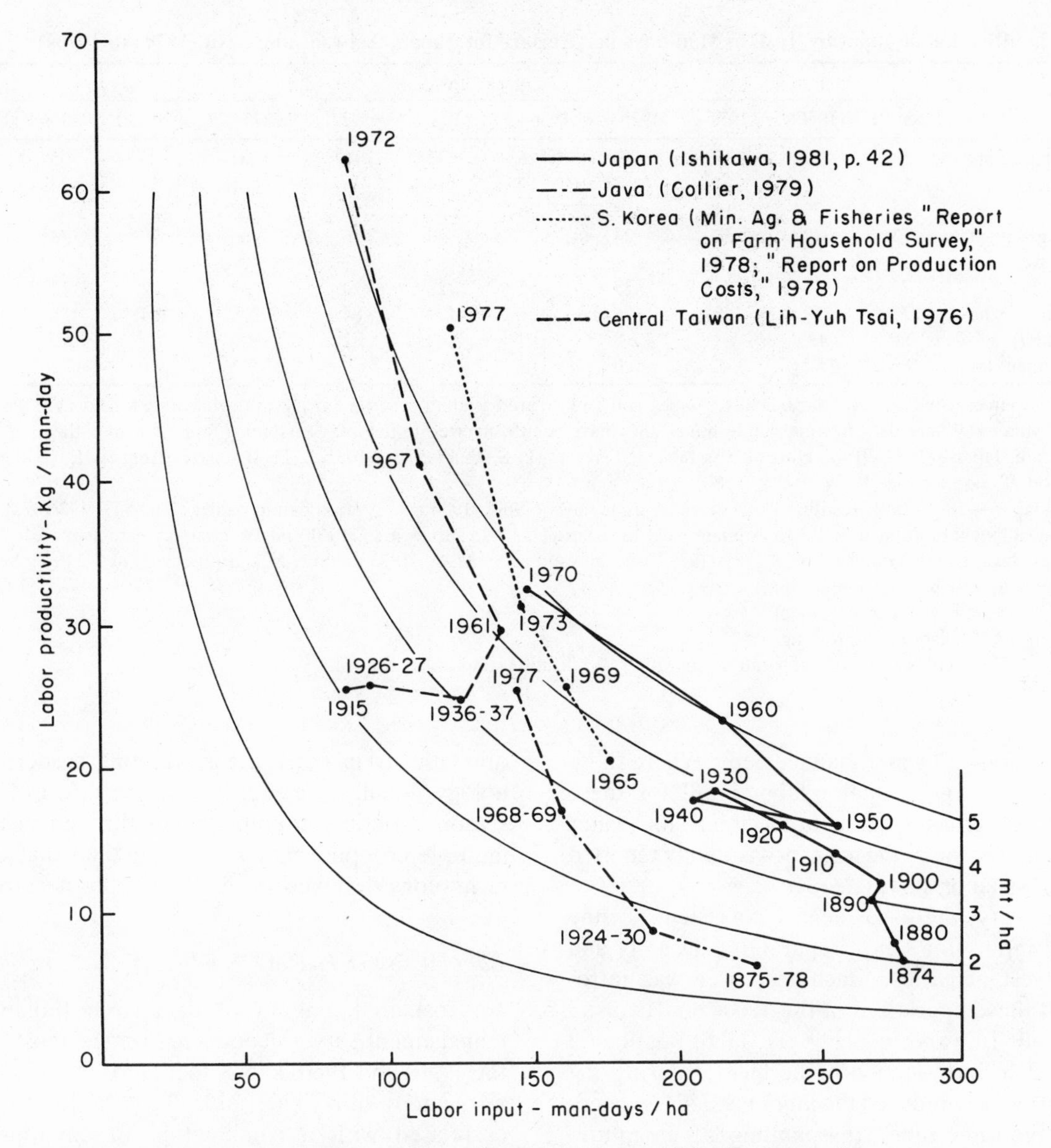

Figure 9.1. Trends in per hectare labor input and labor productivity for Japan, Java, South Korea, and Central Taiwan, 1870s to 1970s

tiller. In Java, on the other hand, mechanization in land preparation reflects the shift from human to animal power. There is essentially no trend in labor used for transplanting. It is not clear why labor use is so much lower in Taiwan than in Japan and Java. The pattern of labor use in weeding is mixed. The widespread use of herbicides marked a sharp decline in weeding labor in Japan and Taiwan in the 1970s. However, labor for weeding increased in Java, where herbicides are not extensively used.

Harvest and post-harvest activities are difficult to compare because the technologies used and the operations performed are substantially different over time and across regions. In Japan and Central Taiwan, the major operations in this category are harvesting and threshing. The decline in labor input is largely a result of the mechanization of threshing. In Java, rice was traditionally harvested one stalk at a time with an ani-ani (hand knife). The rice was not threshed, but stalks containing the panicles were bundled in preparation for milling, which was completed by hand pounding. The decline in labor input for harvest and post-harvest operation in the 1970s reflects a shift from the ani-ani to the sickle associated with the introduction of modern varieties. There has also been a rapid diffusion of small rice mills, displacing most of the labor used for hand pounding.[9]

The category "other" includes a wide range of activities that are not always consistent among studies, but refers largely to other cultural practices such

Table 9.2. Rice Labor Input by Task in Man-days per Hectare for Japan, Taiwan, and Java, 1870s to 1970s

	Japan			Central Taiwan				Java			
	1888–91	1956	1971	1926–27	1936–37	1961	1972	1875–78	1920–30	1968–71	1977–80
Land preparation	53	29	15	17	6*	37	10	61	36	35	23
Transplanting	19*	45	37	7*	12	14	15	31	51*	33	26
Weeding	51	39	14	20	28	29	19	28[a]	27[a]	42[a]	42[a]
Harvesting—post-harvest	73	79	50	24	27	30	16	66[b]	78[b]	34[b]	43[b]
Other[c]	75	37	25	27	53	28	24	39	12	14	10
Total man-days/ha	271	229	141	96	126	138	84	225	204	158	144
Yield (mt/ha)	2.6	4.2	5.1	2.5	3.1	4.1	5.2	1.6	1.8	2.7	3.7
Yield (kg/man-day)	9.6	18.3	36.2	26.0	24.6	29.7	61.9	7.1	9.0	17.0	25.6

Sources: Sources listed are secondary. Data for Japan and Taiwan are based on government surveys, while data for Java are the average of several studies. Where data are reported in hours, they have been converted to man-days assuming an 8-hour man-day.

Labor data: Japan—1888–91; Y. Hara (1980), table 4. Taiwan—L. S. Tsai (1976) table 5.2. Java: adapted from W. L. Collier (1979). For method of adoption, see R. W. Herdt (1980).

Yields: Japan—1888–1971: Institute of Developing Economies (1969), 1956 and 1971: appendix tables. Taiwan—1936/37: appendix tables. Taiwan yield multiplied by 1.2 to estimate yield for Central Taiwan survey area. 1961 and 1972: survey data as reported in L. S. Tsai (1976). Java: appendix tables except 1885–1978 based on yield estimates for 1873 for Java-Madura reported by K. F. Holle (1882).

*Not consistent with other observations.

[a]Includes labor for water management.

[b]Does not include labor for threshing.

[c]Includes items such as seed bed preparation, irrigation, fertilization, and insect control.

as fertilization and water management. Where these numbers are large, a major labor input for these activities is reflected. The labor input for water management is one of the most inconsistently reported tasks across labor surveys.

In summary, there has been a consistent decline in labor inputs for land preparation, harvest, and post-harvest activities. Machine power was introduced in these operations in the 1960s and 1970s in Japan and Taiwan, reducing the labor input still further. The introduction of herbicides during the same period also reduced the labor needed for weeding. At the same time, transplanting labor requirements have remained relatively constant. This is one of the last tasks to be mechanized. By 1980, Java appeared to be employing roughly the same technology and level of labor input used in Central Taiwan in 1960 and achieving comparable yields. The question we now turn to is what impact recent technological changes have had on the employment of labor in rice production.

The Impact of Modern Technology on Labor Use and Productivity

In this section, we first show that with the introduction of modern rice technology, labor use per hectare has tended to increase while labor per kilogram of rice produced has tended to decrease. We then show that employment of hired labor has risen dramatically in those areas adopting modern technology. Finally, we suggest that the effect of short-season varieties on employment through increased multiple cropping may be greater than that due to technology designed for high yield per hectare.

Labor Use per Hectare

A substantial amount of data from Indian farm management surveys conducted for the most part in the 1950s and 1960s shows that labor input is closely associated with yield (table 9.3). Yield in turn is correlated with a combination of environmental

Table 9.3. Labor Input in Man-days per Hectare for Ten Districts of India Classified by Yield

	Yield mt/ha		
	Below 1.5	1.6 to 2.4	Above 2.4
Number of observations	7	8	3
Land preparation	17	26	30
Cultural practices	52	54	81
Harvest and post-harvest	39	34	50
Total man-days	108	114	161
Yield (mt/ha)	1.2	2.1	3.4
Yield (mt/man-day)	11.1	18.4	21.1

Note: In several instances data are taken for selected seasons, irrigation conditions, or varieties. Thus, there is more than one observation per district for six of the districts.

Source: Adapted from tables 11 and 14 in A. Viadyanathan (1978).

factors, including water control, high-yielding varieties, and other modern inputs. At the time of the surveys, only the three observation points with yields of 2.4 mt/ha or greater were extensively using modern varieties. Labor input and productivity are consistent with Central Taiwan in 1961 and with Java in 1977–80 (table 9.2).

Prior to the introduction of modern varieties, most Asian farmers had yields of 2 mt/ha or less and probably required about 100 man-days of labor to grow a crop of rice. Those farmers with good water control have subsequently been able to take advantage of the new technology and presumably have increased labor inputs. Table 9.4 summarizes a number of studies in which labor inputs were compared for modern and local varieties. Most of these consist of single surveys. The Philippine study, however, compares changes in the same farms over time. Analysis of changes in labor input by task in the Philippines studies shows that the principal increase was for weed control, and that this was partially offset by a decrease in labor input for land preparation.[10]

In summary, with the introduction of modern technology, there is an opportunity for increasing labor productivity and input per hectare. However, with the technological alternatives available today, labor use in rice will not reach the level of 200 or more man-days per hectare prevalent in Japan and Java for a decade or more after World War II. On the contrary, there is a danger that, in many parts of Asia, premature introduction of labor-saving technology such as tractors, threshers, and herbicides could keep labor input per hectare from rising significantly above the current level of around 100 man-days per hectare (chapter 8).

Hired Labor

The proportion of hired labor in rice production tends to vary widely throughout Asia. Within a given location, however, large farms typically have a higher proportion of hired labor than small farms.[11] A number of factors explain the ratio of hired to family labor. These include: (1) alternative employment opportunities for family labor, (2) inadequacy of family labor at peak periods, (3) the desire on the part of the operators to avoid arduous tasks involved in rice production (that is, preference of leisure), (4) a land tenure structure that encourages the family to work off the farm, and (5) the social pressure not to engage

Table 9.4. Labor Used per Hectare by Farmers Growing Modern Rice Varieties Compared with Labor Used by Farmers Growing Local Rice Varieties

Location	Man-days/ha		Ratio	Kg/man-day		Ratio		
	Modern varieties	Local varieties	MV/LV	Modern varieties	Local varieties	MV/LV	Study years	References
Indonesia[a]								
West Java	290	138	2.1	21.5	39.8	0.5	1968/69	Palmer, 1977
West Java	340	218	1.6	15.3	13.3	1.2	1969/70	Sajogyo and Collier, 1973
Central Java	236	235	1.0	—	25.5	—	1968/69	Palmer, 1977
Central Java	202	214	0.9	—	—	—	1968/69	Soejono, 1976
Central Java	197	234	0.8	24.4	14.9	1.6	1969/70	Sajogyo and Collier, 1973
Central Java	244	187	1.3	—	—	—	1973/74	Soejono, 1976
East Java	224	209	1.1	20.1	16.3	1.2	1968/69	Sajogyo and Collier, 1973
East Java	247	253	0.9	31.2	12.2	2.6	1969/70	Palmer, 1977
Other								
Suphan Buri, Thailand	117	81	1.4	29.3	25.6	0.9	1971/72	Sriswasdilek, 1973
Central Luzon, Philippines[b]	82	60	1.4	35.6	27.3	1.1	1966–74	Barker and Cordova, 1978
Laguna, Philippines[c]	106	88	1.2	33.1	28.4	1.2	1966–75	Barker and Cordova, 1978
Hwasunggan, South Korea	139	126	1.1	51.8	44.4	1.2	1974	Suh, 1976
Mymensingh, Bangladesh	194	137	1.4	17.5	16.1	1.1	1969/70	Muqtada, 1975
West Godavari, India	90	98	0.9	—	—	—	1969/70	George and Choukidar, 1972
Ferozepur, India	92	92	1.0	—	—	—	1969/70	Mehra, 1976
Punjab, Pakistan	43	45	1.0	66.6	39.4	1.7	1972	Khan, 1975
Sind, Pakistan	35	20	1.8	57.8	68.0	0.8	1972	Khan, 1975
Kanpur, India	105	91	1.2	—	—	—	1966–71	Mehra, 1976
Palamau, India	279	143	1.7	—	—	—	1970/71	Mehra, 1976
Dry Zone, Sri Lanka	169	127	1.3	21.6	17.8	1.2	1970/71	Amerasinghe, 1972

Note: MV = modern varieties, LV = local varieties.

[a]Preharvest labor only.

[b]Compares sixty-three farms with no MVs in 1966 with the same farms in 1974 when 64 percent were planting MVs.

[c]Compares sixty-two farms growing MVs in 1966 with 94 percent growing MVs in 1975.

in farm labor and to provide employment for relatives, neighbors, and landless laborers. All of these factors may be important in varying degrees at different locations and at different times in the same locations. Some illustrations will help to shed light on the complexity of these relationships.

The amount of hired labor employed on rice farms in East Asia has historically been very low. Prior to the 1960s, the proportion of hired to total labor on Taiwan farms was less than 20 percent.[12] However, as the opportunities for nonfarm employment have increased over the last two decades, the proportion of hired labor rose from 32 percent in 1967 to 59 percent in 1972.

In the densely populated areas of South and Southeast Asia, off-farm employment opportunities are limited. The number of landless laborers is rising, and these workers seem to be relatively mobile, searching out job opportunities in the more productive and progressive farming areas. Thus, in the case of rice, we might expect to find a higher proportion of landless or near landless people working principally as hired laborers in the irrigated areas.

Table 9.5 illustrates the variability across three different rice growing environments in the annual labor requirement and in the proportion of labor that is hired. The data are taken from farm surveys in six locations, four in the Philippines and two in Indonesia. Villages have been paired according to the method of rice cultivation: (1) *upland* or unbunded rice fields, (2) *rainfed* rice paddies that have been bunded but depend on natural rainfall for moisture, and (3) *irrigated* rice paddies. The labor input is about twice as high on the irrigated as on the rainfed areas because two crops of rice are grown on the former. In the upland areas, crops other than rice are grown following the rice harvest, which accounts for the fact that labor is higher than in the rainfed areas. The proportion of labor hired varies from less than 20 percent in one of the upland villages to over 80 percent in one of the irrigated villages. The total labor requirement and proportion hired are similar among each of the pairs.

Comparing villages in Bantul (Central Java), Maurer observed that where the overall economic and resource situation was worst, conditions were more uniform across households.[13] The proportion of landless households was highest in the village with the best resource situation.

As the productivity of land increases with the spread of irrigation and the introduction of new varieties and fertilizer, the demand for total labor rises, but the demand for hired labor rises even more rapidly. Table 9.6 compares the input of hired labor per hectare for modern and local varieties. In fifteen of the twenty-one cases, the amount of hired labor was higher under modern varieties. In all but two of the fifteen cases, yields were also higher.

Multiple Cropping

Multiple cropping, or the production of more than one crop on a single piece of land in a year, has been practiced in Asia for centuries. The advantage of multiple cropping is that it makes use of normally underemployed workers, leading to a greater number of work days per worker per year.

Multiple cropping is closely linked to irrigation throughout most of Asia because there is normally inadequate water in the dry season to grow a second crop. Table 9.5 shows that as a result of multiple cropping, the average number of man-days per farm was approximately twice as high in the irrigated compared with the rainfed and upland locations.

The effect of multiple cropping on employment is clearly indicated in the Taiwan experience.[14] Table

Table 9.5. Characteristics of Cropping Systems with Respect to Labor Use in Three Different Rice Growing Environments in the Philippines and Indonesia, 1973/74 to 1976/77

Site	Year	Av. no. man-days per farm	Farmland operated (ha)	Av. no. man-days per ha	Percentage labor by source		
					Family	Hired	Exchange
Upland							
Bantangas, Philippines	1973/76	231	1.4	165	77	18	4
Lampung, Indonesia	1976/77	208	1.2	173	76	20	4
Rainfed							
Pangasinan, Philippines	1975/76	232	2.2	105	47	34	16
Iloilo, Philippines	1975/76	240	2.1	114	44	54	3
Irrigated							
Laguna, Philippines	1975	443	1.9	233	25	70	5
Indramayu, Indonesia	1975/76	560	2.4	233	14	86	0

Source: E. C. Price and R. Barker (1976).

Table 9.6. Hired Labor Used per Hectare by Farmers Growing Modern Rice Varieties Compared with Hired Labor Used by Farmers Growing Local Rice Varieties in Asia

Location	Hired labor (man-days/ha)			Yields (mt/ha)				
	Modern varieties	Local varieties	Ratio MV/LV	Modern varieties	Local varieties	Ratio MV/LV	Study years	References
India								
Cuttack, Orissa	260	110	2.4	4.3	2.1	2.0	1966/67	Desai, 1977
Varanasi, U. P.	230	225	1.0	1.5	1.6	0.9	1966/67	Desai, 1971
Saharanpur, U. P.	94	50	1.9	3.5	2.0	1.7	1966/67	Desai, 1971
Raipur, M. P.	99	115	0.9	0.8	1.9	0.4	1966/67	Desai, 1971
Kolaba, Maharashtra	94	77	1.2	1.2	1.4	0.9	1966/67	Desai, 1971
Amritsar, Punjab	199	178	1.1	2.6	2.9	0.9	1967/68	Desai, 1971
Krishna, Andhra	178	133	1.3	4.2	4.0	1.1	1966/67	Desai, 1971
East Godavari, Andhra	390	316	1.2	5.8	2.2	2.6	1968/69	Dasgupta, 1977
West Godavari, Andhra	373	328	1.1	5.7	3.2	1.8	1967/68	Desai, 1971
West Godavari, Andhra	659	588	1.1	5.5	2.8	2.0	1968/69	Dasgupta, 1977
Ernakulam, Kerala	354	294	1.2	3.3	2.2	1.5	1966/67	Desai, 1971
Thanjavur, Tamil Nadu	98	116	0.8	3.0	1.9	1.6	1966/67	Desai, 1971
Thanjavur, Tamil Nadu	195	186	1.0	3.5	2.7	1.3	1967/68	Desai, 1971
Birbhun, West Bengal	221	144	1.5	3.5	2.5	1.4	1968/69	Dasgupta, 1977
Indonesia[a]								
West Java	274	107	2.3	6.3	5.5	1.1	1968/69	Palmer, 1977
West Java	241	172	1.4	5.2	2.9	1.8	1969/70	Sajogyo and Collier, 1973
Central Java	221[b]	239[b]	0.9	3.7	4.8	0.8	1968/73	Sinaga and Sinaga, 1978
Central Java	171	168	1.0	4.5	3.4	1.3	1968/69	Sajogyo and Collier, 1973
Other								
Laguna, Philippines	85[c]	51[c]	1.7	3.5	2.5	1.4	1966/75	Barker and Cordova, 1978
Mymensingh, Bangladesh	114	71	1.6	3.4	2.2	1.5	—	Muqtada, 1975

Note: Labor units are days.
[a]Preharvest labor only.
[b]Compares thirty farms, of which two grew MVs in 1968/69 and twenty-eight grew MVs in 1973/74.
[c]Compares sixty-two farms, of which none grew MVs in 1966 and 95 percent grew MVs in 1975.

9.7 shows the close association between increase in work days per hectare and the multiple cropping index. Work days per hectare and the multiple cropping index (harvested area of crops grown per year divided by geographic area times 100) rose steadily, reaching a peak in the 1960s and then began to decline as rural labor was attracted to nonfarm employment opportunities and farm mechanization became more prevalent.

Modern rice varieties have made an important contribution to the growth in the multiple cropping index in other parts of Asia, although the evidence is largely undocumented. The first of the modern varieties (IR8) had a crop growth duration (transplanting to harvest) of about 130 days compared with 150 to 200 for most traditional varieties. By the mid-1970s, varieties with a growth duration of about 100 days, such as IR36, were released. There is a little question that these short-season varieties have had a major effect on cropping intensity and labor use per hectare in the irrigated areas. It is perhaps reasonable to suggest that, with the introduction of modern rice technology, short-growth-duration varieties have had a greater effect than high-yielding varieties on employment in the rice growing regions of tropical Asia.

Table 9.7. Labor Intensity, Labor Utilization, and Multiple Cropping Index in Taiwan

Period	Labor intensity (workdays/ha)	Labor utilization (workdays/ farmworker)	Multiple cropping index
1911–15	200.8	120.5	120.9
1916–20	210.4	136.9	121.7
1921–25	212.4	143.0	123.0
1926–30	212.6	143.6	123.6
1931–35	202.0	141.7	132.7
1936–40	229.0	142.7	133.6
1941–45	223.1	160.0	131.4
1946–50	223.0	139.1	150.7
1951–55	276.1	165.1	171.9
1956–60	306.0	183.5	180.0
1961–65	318.2	186.7	186.6
1966–70	334.2	195.3	186.5
1971–75	323.4	197.0	177.7

Source: T. Lee, H. Chan, and Y. Chen (1980) p. 207. Reprinted by permission of the Asian Employment Programme.

Table 9.8. Percentage of Rice Cultivation Labor Performed by Women

Country	Seed selection	Seedbed	Land preparation	Trans-planting
Nepal (hill systems)	X		0	64
India				
Andhra Pradesh			7	78
Tamil Nadu			0	69
Sri Lanka (Kandy)				
Irrigated			0	100
Rainfed			0	91
Thailand	X		16	29
Malaysia	X	19	6	55
Indonesia				
(East Java)				X
Philippines				
(Central Luzon)			0	45
China				
(southern region)		X		X
South Korea	33	27	2	48
Japan				

Note: X = an unknown amount of labor by women.
Sources:
Nepal: Pradhan (1983)
India: Agrawal (1983)
Sri Lanka: de Alwis (1983)
Thailand: Chandrara (1983)

Women in Asian Rice Cultivation[15]

Women in Asia play a major role in all facets of rice production. There is evidence from China and India to suggest that female labor inputs are much greater in rice-based cropping systems than in dryland farming.[16] Within most Asian rice farming systems, women contribute from one-third to one-half of the total labor, and they frequently contribute most of the transplanting, harvesting, and weeding labor (table 9.8). An exception to this pattern is in Bangladesh where women do not participate in the field work but are responsible for almost all of the post harvest processing.[17]

Much of the labor contributed by women to rice cultivation is hired by the farmer, frequently from landless labor households. Thus, modern technologies that increase the demand for hired labor, including new varieties and irrigation, should increase the demand for female labor.

Empirical evidence is, however, rather scarce on an Asia-wide basis. Most labor use data are not disaggregated by sex. Agrawal compared labor use for modern variety and non–modern variety farms in three states of India. She found that the use of hired female labor was greater for farms cultivating modern varieties, primarily for weeding and harvesting.[18] Sen did not examine labor use in modern varieties directly but did find that women's participation was greatest in the more intensive irrigated paddy farming systems.[19] However, Res found a decline in female labor use in a rainfed area in the Philippines because direct seeding and mechanical threshing were also adopted to reduce seasonal demand for labor.[20] Sajogyo and White both noted a decline in female labor use in Indonesia, not as a consequence of the adoption of modern varieties, but as a result of the adoption of new technology in harvesting and milling brought about in part by institutional changes.[21]

Despite the fact that women play a major role not only in supplying labor but also in decision making in rice production, they do not seem to have direct access to information on new rice technologies and to other resources such as credit. The effect of this on their productivity and ability to gain from technological change is largely unknown. Furthermore, we know very little about the intrahousehold distribution of resources and sharing of income. These factors notwithstanding, a strong case can be made for involving women more directly in the research and extension process. As we shall see in the following section, the critical problem facing Asian agriculture is the slow growth in demand for labor irrespective of sex. Gains in labor demand from the

Table Table 9.8. *(continued)*

Crop Care					Other		
Weed-ing	Ferti-lizing	Insect & others	Harvest-ing	Thresh-ing	post-harvest	Market-ing	Total
72			52				54
73	8	3	63	25			48
85	39	8	64	41			55
80	0	0	—49—		46		n.a.
80	0	0	—37—		67		n.a.
	33		—73—			61	n.a.
27	19		46	24	76	15	37
X			X				35
—7—			—15—				19
	X		X				33
23	12	24	34	34	43		26
							40

Malaysia: Yap (1981)
Indonesia: Collier, Hidayat, Soentoro and Yuliati (1983)
Philippines: Sison, et al. (1983)
China: Croll (1979); Xue-bin (1983)
Korea: Lee (1983)
Japan: Yoshida (1983)

new rice technology have not offset the effects of population growth and slow growth in the industrial sector.

Trends in Real Wages and Factor Shares of Labor

Thus far, partial productivity of labor has been used to indicate changes in labor productivity. However, the partial productivity index may not be an accurate reflection of changes in labor value of marginal product because it includes the contribution of other inputs that can be important in periods of rapid technological change. In addition, the change in labor productivity may not be closely associated with a change in real wages if labor is in surplus and labor's bargaining power is weak. This section examines the trends in real wages and in factor shares of labor in Asian agriculture.

Table 9.9 records kilograms of milled rice that can be purchased with one day's rice farming wage during 1976–79 for selected locations in Asia. The grain purchasing power of a day's wage in Japan and Taiwan is almost six times that in the majority of Asian countries, while in South Korea and Malaysia it is three times. In the exporting countries, Burma and Thailand, the purchasing level is also high in part because the rice price is low. Locations in Eastern India, Java, and Nepal share the lowest earning power for labor because population pressure has forced wages down, and rice production has lagged behind demand.

Table 9.10 shows recent trends in real wages (estimated as the nominal wage deflated by consumer price index) for seven Asian countries. The index of real wages has been rising sharply in Japan and South Korea, where labor is moving rapidly out of agriculture and into industry. As a result, there is a strong demand for labor-saving technology in agriculture. In South and Southeast Asia, the trend in growth of real wages has been mixed. No country seems to have shown the steady rise in real wages that occurred in East Asia. Rather, the general picture seems to be one of slow growth or stagnation.

Studies in India and the Philippines have compared the trends in real wages in different regions of the country.[22] The decline in real agricultural wages in the Philippines has been pervasive throughout all regions of the country. The general conclusion of the Indian studies covering the period from the 1950s to the early 1970s is that real wages are higher in the states where technological progress has occurred (for example, Punjab) and in the states where rural labor has bargaining power (for example, Kerala). However,

Table 9.9. The Rice Purchasing Power of Hired Labor in Asia, 1976–79

Area	Milled rice ($US/kg)	Farm wage ($US/day)	Kg rice per day of work
East Asia			
Japan	1.22	19.02	15.6
South Korea (Suweon)	0.67	4.94	7.4
Taiwan (Taichung)	0.38	7.60	20.0
Southeast Asia			
Burma (Rangoon)	0.13	0.92	7.1
Indonesia (Central Java)	0.31	0.65	2.1
Malaysia (Selangor)	0.41	3.84	9.3
Philippines (Central Luzon)	0.27	0.87	3.2
Thailand (Suphanburi)	0.20	1.58	7.9
South Asia			
Bangladesh (Joydebpur)	0.21	0.69	3.3
India: (Coimbatore)	0.22	0.65	3.0
(Cuttack)	0.22	0.33	1.5
(Waltair)	0.22	0.46	2.1
Nepal (Kathmandu)	0.26	0.62	2.4
Pakistan (Punjab)	0.32	1.07	3.3
Sri Lanka (Kurunegala)	0.27	0.71	2.6

Source: A. Palacpac (1982) table 41.

even in Punjab, where technological advances have been most pronounced, real wages seem to have declined in the 1970s (table 9.10).

The basic difficulty in measuring labor's share in production growth lies in how to evaluate the family labor input. Most studies evaluate family labor at market wages, but if the hired component of total farm labor use rises with intensive cropping and production growth (as indicated previously), and if the implicit supply price of family labor is lower than that of hired labor, then labor's share before the introduction of modern technology would tend to be overestimated.[23] The general conclusion of most Indian studies is that labor input has increased with the introduction of modern varieties, but that labor's share of total income has decreased because yields have risen more rapidly than labor input.

In the Philippines, there does not seem to be any indication of a major difference in real wage trends by region. The sharp decline in real wages in all regions in the early 1970s was caused by the steep rise in the consumer price index as a result of rapid inflation led by oil price hikes and grain shortages.[24]

The trends in real wages depicted by the aggregate data were also observed in the panel surveys of rice farms conducted in Central Luzon and in Laguna province.[25] The analysis of factor shares from these panel surveys showed that despite the decline in real wages from the mid-1960s to 1970s, income share of labor remained fairly constant because labor use per hectare increased.

Family Labor Use

Allocation of family time among various tasks and between production work and leisure is also an important consideration in an overall evaluation of labor in Asia. One question is how much of the avail-

Table 9.10. Index of Real Wages of Agricultural Laborers in Asia
1965 = 100[a]

Year	Japan	S. Korea	Malaysia	Philippines	Sri Lanka	Bangladesh	Selected states of India: Punjab	Bihar	Tamil Nadu	Kerala
1965	100	100	—	100	100	100	100	100	100	100
1966	104	108	100	102	99	94	130	138	96	113
1967	111	117	95	104	99	91	126	164	131	106
1968	128	129	97	102	107	104	134	135	133	115
1969	133	141	100	92	100	98	134	100	130	101
1970	139	152	98	82	95	101	142	134	121	98
1971	147	159	93	81	103	97	143	110	159	103
1972	161	165	91	76	103	105	146	127	159	115
1973	168	177	89	79	97	148	148	—	144	125
1974	176	189	100	85	102	130	139	—	153	128
1975	—	189	90	92	117	102	118	77	68	80
1976	190	212	112	103	130	114	129	—	—	96
1977	192	238	102	106	120	110	131	—	—	108
1978	197	300	107	95	154	112	119	138	150	111
1979	197	385	119	100	—	113	118	113	124	109
1980	—	377	—	84	—	112	107	—	—	147

Sources: A. Palacpac (1983); International Labor Office, *Yearbook of Agricultural Statistics*; Indian Ministry of Agriculture, Directorate of Economics, *Agricultural Wages in India*, and *Agricultural Situation in India*, various issues.

[a]Nominal wages of agricultural workers are deflated by the consumer price index.

able manpower is actually productively employed. It is not uncommon for peasants in a country such as the Philippines to be described as "lazy." This is simply another way of saying that they place a higher value on leisure than on work, thus restricting the potential supply of labor.

In this section, family labor input is analyzed for typical irrigated rice growing areas in three island regions: Java, Luzon, and Taiwan. These areas differ widely in resource use, potential production, and nonfarm employment opportunities. Taiwan, despite a comparatively small average farm size (less than 1 ha) has superior irrigation systems and excellent opportunities for nonfarm employment. Luzon has larger farms (over 2 ha average), a comparatively poor irrigation system, and limited nonfarm employment opportunities. Java has extremely small farms (less than 0.5 ha average) and poor nonfarm employment opportunities.

We hypothesize that farm families in Java and Taiwan will have higher family employment than in the Philippines, but for different reasons. The Javanese farm family will work long hours at a relatively low productivity to earn a simple subsistence living, while the Taiwanese farm family will also work long hours, but its labor will be highly productive. The Philippine farm family will probably work fewer hours on a family basis because appropriate incentives do not exist.

Two studies have recently been completed that show a detailed breakdown of family labor allocation: one by Hayami for a village in Laguna province, 80 km southeast of Manila, and the other by Hart for a village in the northern lowland plain of Central Java. It is difficult to make generalizations based on these case studies because neither is typical of the area as a whole. Nevertheless, the data from these villages, combined with less thorough observations made by other researchers, provide some evidence to support our hypothesis regarding the intensity of labor use.

In their collection and analysis of data, both Hayami and Hart divided their samples according to large farms, small farms, and landless workers (table 9.11). The small farm in the Hayami study is approximately the same size as the large farm in the Hart study. Farms in Luzon are, on average, four to five times the size of those in Java.

The Hart investigation confirms the findings of other studies regarding labor use in Java: (1) villagers work hard and are fully employed, often working more than forty hours per week;[26] (2) poorer members of the village tend to work longer hours;[27] and (3) the amount of family labor devoted to *sawah* (paddy rice) cultivation is relatively small.[28] Figures are not available from the Hart study to show the total amount of labor (own and hired) in the village devoted to rice production.

In the Philippine village studies by Hayami, the economically active worked on average less than 100 days per year. For the large farm family, over 50 percent of the labor was employed in family paddy production.

In a survey of three separate locations in Central Luzon, Guino found that at two of the three locations, average working days per farm family of about seven members were approximately 500, and per worker close to 120, with 50 percent of the labor employed in their own rice cultivation.[29] The number of working days is about 50 percent higher than found by Hayami, but the proportion of family labor devoted to rice production is fairly comparable. In a third

Table 9.11. Working Days per Family and per Economically Active Member for Rice Farms in the Philippines and in Indonesia

Study & farm type	Farm size (ha)	Family members: Total	Family members: Economically active[a]	Working days: Total household	Working days: Economically active[a]	Own rice (—percent—)[b]	Hired rice (—percent—)[b]
Hayami—Laguna Province, Luzon, Philippines							
Large	3.2	7.5	4.4	284	64	53	23
Small	1.3	5.3	2.3	378	147	33	14
Landless	0	4.8	3.2	231	72	5	78
Average	2.1	5.9	3.5	269	77	32	42
Hart—North Central Java, Indonesia							
Large	1.4	4.9	3.6	634	175	23	—
Small	0.3	5.6	3.6	717	197	24	—
Landless	0.04	4.9	3.2	713	221	3	—
Average	0.5	5.1	3.4	695	202	15	—

Sources: Y. Hayami (1978); G. Hart (1978).

[a]Persons age 13–65 in Hayami and persons over 10 years of age in Hart.

[b]Total percentage of time, own and hired, devoted to rice production.

atypical location, over 90 percent of family labor was devoted to rice production, the family also worked 590 man-days per year, and each wage earner worked an average of 150 man-days. Daily earnings from farming were higher than for hired farm labor or for nonfarm work.

One needs to be cautious in comparing the data from Java and the Philippines because different procedures and definitions were used. Distinctions in procedure notwithstanding, the difference in total labor input for the two villages is striking. Even if we were to accept the higher values given by Guino for Central Luzon as being more representative, the working days per economically active person in Java are almost twice the level in Luzon.

Table 9.12 illustrates the seasonal distribution of labor in Hayami's and Hart's studies. Not surprisingly, labor input is more evenly distributed over the year in Java, while in the Luzon study the peaks and troughs for labor demand in the rice crop are reflected. For the farm family, the peak periods are land preparation for the first crop in June, harvesting the first crop and planting the second in October-November, and harvesting in March. For the landless, the pattern is slightly different; January and July are key periods for weeding the first and second crop, and April and October the months of peak demand for harvesting and threshing.

Data from Taiwan comparable to these studies are not available. However, a study of agricultural employment in postwar Taiwan[30] shows that the number of work days (converted from 8 to 10 hours) a year per worker rose from 91 in 1946–47 to 185 in 1969–70. Labor input per worker in 1946–47 was close to the current levels of the Philippines, and in the latter period to current levels in Java. The increased labor input is a reflection of the opportunity for productive employment and a widespread desire to increase earnings for purchase of a wider range of consumer goods.

Falling Wages, Landless Laborers, and Labor Contracts

Virtually all models of tenancy suggest that contractual terms move against tenants and agricultural laborers as wages fall.[31] Conversely, contractual terms improve when wages rates rise. Contrasting examples are seen in the Hayami and Kikuchi case study of two Philippine villages (falling wages), and in a study of an Indian Punjab village by Leaf (rising wages).[32]

As a consequence of falling wages throughout most of South and Southeast Asia, there appears to be a growing gap between the middle and upper peasantry (which in many instances includes small land owners and tenants of 1 to 3 ha of land) and the landless or near landless. The former have been able to share in the increased profits and rent as a result of the productivity gains of the new technology. But there have been insufficient increases in labor demand to absorb the rapidly growing population of landless laborers without a fall in wages. The lack of technological advancement in the rainfed areas has aggravated the problem, and there is evidence that rural workers are migrating to the irrigated rice areas to become landless laborers.[33]

The rapid increase in landless laborer households is documented in recent studies by Cornell University; The Institute of Developing Economies, Tokyo; and the International Labour Office.[34] It is difficult to estimate the number of landless labor households in rural areas because the census definitions are not always consistent across time and space. Rosenberg and Rosenberg report the following estimate of the

Table 9.12. Seasonal Distribution of Family Labor for Large Farms, Small Farms, and Landless Households, 1975/76[a]
(percent of family work days per household)

Farm size	Jan.	Feb.	March	April	May	June	July	Aug.	Sept.	Oct.	Nov.	Dec.	Total
Luzon, Philippines													
Large	7.1	4.6	5.8	4.9	4.8	12.2	11.9	11.7	9.4	9.7	8.5	9.4	100.0
Small	6.4	7.0	8.7	8.3	6.4	10.3	9.7	7.4	7.0	11.4	8.9	8.5	100.0
Landless	10.3	7.6	7.4	10.6	4.0	6.4	11.2	7.3	7.6	13.3	6.7	7.6	100.0
North Central Java, Indonesia													
Large	8.0	7.1	7.7	8.5	7.4	9.1	9.9	9.5	9.6	9.0	6.6	7.6	100.0
Small	9.6	7.5	7.7	6.8	7.3	9.1	9.2	9.0	9.2	8.0	7.6	9.0	100.0
Landless	8.5	7.1	8.0	6.6	8.2	8.9	9.0	8.6	10.0	9.6	7.1	8.4	100.0

Sources: Philippines: Y. Hayami (1978). Indonesia: G. Hart (1978).
[a]Does not include in-household work.

percentage of landless rural households in five Asian countries.[35]

		Percent of landless rural households	
Country	*Year*	*Agricultural workers*	*Nonagricultural workers*
Bangladesh	1977	5.3	19.9
India	1971	27.0	16.0
Java (Indonesia)	1971	41	
Philippines	1971	11.3	17.1
Sri Lanka	1973	12.6	39.8

Landless agricultural workers are defined as "rural workers in agriculture with no ownership or usufruct rights to land, who earn a livelihood from the proceeds of their own labor."[36] *Landless nonagricultural workers* are "self-employed and hired workers outside agriculture (but in the rural area) with no ownership rights to land, who earn their livelihood from the proceeds of their own labor."[37] Unfortunately, not very much is known about the composition and growth of the landless population. However, it seems fair to say that only a generation ago landless agricultural workers did not exist as a definable class in many parts of Asia.[38]

A decline in wages can occur in a number of ways, such as lowering the share or cash payment for work, increasing input required for the same pay, or raising risk by shifting from a share to a cash payment. A brief examination of three situations where labor contracts for harvesting have been undergoing change will be illustrative.

Harvesting Contracts in Java

Under the traditional Javanese system of rice harvesting, *bawon*, the harvest is a community activity in which all can participate and receive a share in kind.[39] In a study conducted in Central Java, Utami and Ihalauw report that it is common practice for farmers to give one-fourth of the rice harvested to relatives, one-fourth to one-sixth to close neighbors, and one-tenth to fellow villagers.[40] When the rice is ripe, a horde of harvesters enters the field and cuts the stalks about 6 inches below the panicles with a hand knife (ani-ani). The harvested stalk paddy is bundled and brought to the house, where the harvester receives his share.

During the 1970s, the bawon system was replaced in many parts of Java by a system called *tebasen*. Under tebasen, the standing crop is sold to a middleman about 10 days before harvest. Because the middleman is from outside the village, he is not bound by village customs. He hires a few workers to harvest the crop, and this is normally done with a sickle to reduce the work requirement. The workers are paid in cash after the harvest. Collier et al., reported that the tebasen system was used in 25 percent of the Agro-economic Survey sample villages in 1973.[41]

Two major factors, population pressure and technological change, underlie the shift from bawon to tebasen. With large numbers of harvesters entering the fields, there are significant losses from trampling, stealing, and handling. Utami and Ihalauw estimated that as a result of these losses and the high share paid to relatives and neighbors, the farmer is left with less than 60 percent of the crop after harvest.[42] New technology in the form of modern semidwarf varieties seems to have facilitated the introduction of the sickle and a reduction in the number of rice harvesters because the new varieties shatter very easily and are difficult to harvest with the ani-ani.

In their study of a hamlet in Subang District in West Java, Hayami and Kikuchi found that bawon was replaced not by tebasen, but by *ceblokan*, a system that is widely practiced in West Java.[43] Under this system, workers who are employed for harvest must also perform extra services without pay, such as transplanting and weeding. Real wages are effectively reduced by requiring more labor for the same harvest share.

In the hamlet under investigation, ceblokan was first introduced in 1964, but change has been very gradual, beginning with the restriction of bawon first to villagers and finally only to invitees (table 9.13). The shift to ceblokan, requiring more than a decade to complete, was accompanied by a reduction of the harvester's share from one-sixth to one-seventh on more than a quarter of the farms. Ceblokan initially required only the provision of free transplanting in exchange for the right to harvest, but this has been extended on several farms to include weeding, and in some instances, even harrowing. The results of this investigation clearly show that institutional transition was well under way before the introduction of modern varieties.

Collier et al. conclude that the tebasen system represents an institutional innovation designed to relieve the well-to-do members of the village community from their traditional obligation to give a share to the poor.[44] Both farmers and middlemen gain at the expense of the laborers, many of whom can no longer find employment during the harvest. Hayami and Kikuchi take issue with the conclusion that the tebasen (or the ceblokan) system promotes a polarization.[45] While not denying that the system reduces the share of returns going to labor, they argue that,

Table 9.13. Changes in a Rice Harvesting System in a South Subang Village
(percent of farmer adopters)

Year	Bawon[a]				Ceblokan[b]					Total
	PO	OV	OM	LI	1/6(T)	1/7(T)	1/7(T+W)	1/7(H+T)	1/7(H+T+W)	
1950s	35	29	18	18						100
1960–61	29	31	21	19						100
1962–63	16	34	33	17						100
1964–65	9	16	16	32	27					100
1966–67	3	10	8	27	52					100
1968–69	1	4	6	19	44	24	2			100
1970–71			2	10	33	51	4			100
1972–73				8	17	67	8			100
1974–75				7	15	67	10	1		100
1976–77				4	7	67	18	2	2	100
1978				4		72	19	1	4	100

Source: Y. Hayami and M. Kikuchi (1978) table 8-10.
[a]Bawon system: PO = purely open, OV = open for villagers only, OM = open with maximum limit, LI = limited to invitees.
[b]Ceblokan system: 1/6, 1/7 = harvester's share; T,W,H = obligatory work to establish the harvesting right (T = transplanting, W = weeding, H = harrowing).

given the basic economic factors that result from continuing population pressure on the land, it would be difficult to stop this Ricardian process in the absence of effective policies to raise the relative productivity of labor. Greater investment in land infrastructure and in labor-using technology, such as modern varieties, is needed to raise labor productivity.

Harvesting Contracts in the Philippines

Although population pressure on the land is far less severe in the Philippines, a number of recent studies show that there have been significant changes in labor contracts over the past decade. These are the result of growth in the number of landless laborers, introduction of new technology, and implementation of land reform legislation.

Under the traditional system, the rice harvest, as in Java, was open to all who wished to participate. The individual worker harvested with a sickle, threshed the crop by hand under an arrangement known as *hunusan*, and received one-sixth of the paddy as payment. A new arrangement, known as *gama* and analogous to the ceblokan system in Java, has now come into prominence in Laguna Province. Weeding labor is provided, often without compensation, to establish the worker's right to participate in harvesting and the threshing, and to receive one-sixth of the crop. The farm operator may have several gama contracts, and the hired laborer, who usually has no land of his own, may harvest a field or two on several farms. In a sample survey conducted in part of Laguna Province, the number of farmers with gama contracts rose from 33 to 85 percent between 1970 and 1975.[46]

The economic rationale for the gama system is illustrated in Hayami and Kikuchi's village study.[47] As the supply of labor increased and crop yields improved, the one-sixth share of the output became substantially larger than the marginal product of the labor for harvesting work. Farmers could have reduced the share or changed to a cash payment. However, the shift to gama, where workers contribute more labor for the same share, is socially more acceptable.

Systems similar to gama have appeared elsewhere in the Philippines. Ladesma, in his study of a village in Iloilo, finds the same practice being introduced for the first time in 1973 under the name *sagod*.[48] In some parts of Bicol, the system is known as *hilani*.[49] There are also cases where either the harvester's share is reduced or there is a shift from in-kind to cash payment.

Harvesting Contracts in Bangladesh

Although the situation in Bangladesh is not as completely documented as in parts of Indonesia and the Philippines, there is also evidence of a transition in contractual arrangements governing harvest shares. As in the previous cases, the traditional practice is the payment of a fixed share in kind for harvesting and threshing. However, in a study undertaken in Joydepur area north of Dacca, Clay encountered a bewildering range of modes of payment, including the traditional share payments, daily wages with prepared food, and fixed contract payments in cash.[50] The labor market seems to be relatively imperfect, with payments received by harvesters varying as much as 25 percent within a short distance.

A common way of reducing real wages is to shift from share to cash payment. In the Clay study, 90 percent of the village labor was paid a share while close to half of the labor from outside the village was paid in cash through a daily wage or contract at a significantly lower rate. In short, migrant workers from outside the village bid down the wage rate. There is evidence from other parts of Bangladesh that the wage rate is also being lowered by not using local labor.[51]

The social and cultural relationships between hired and family farm labor are vastly different in South compared with Southeast Asia. In Southeast Asia, there is a strong patron–client relationship between the landed and the landless. The latter are often relatives of the farmer. In South Asia by contrast, vestiges of the caste system are still dominant in labor relationships. Thus, one is more likely in the latter case (as, for example, in the Bangladesh study) to find falling wages with changes in the labor contracts that reduce the laborers' share of the harvest.

The Elasticity of Demand for Labor

This section reports on empirical econometric estimates of labor demand in rice production. Variables affecting the demand for labor can be grouped into several categories:

Prices: for example, wage rate, output price

Labor-using technology: for example, irrigation, seed-fertilizer technology

Labor-saving technology: for example, farm power, herbicides, mechanical threshers

Institutional factors: for example, farm size, tenure, labor contracts

Locational and time factors: for example, area, season, year.

These variables have been combined in various ways by researchers to estimate the demand for total and for hired labor. Two of the most comprehensive recent studies are by Evenson and Binswanger for Indian agriculture and by David and Barker for Philippine rice.[52]

The former study is based on surveys of selected districts conducted from 1954–55 to 1971–72 throughout India. Summary reports have been published for twenty-six or more studies conducted in twenty-two different districts in fourteen states. Labor demand equations were estimated by grouping data into three regions: (1) northern wheat, (2) eastern rainfed rice, and (3) coastal irrigated rice. The Philippine analyses are based on three separate data sets from surveys conducted by the International Rice Research Institute from 1966 to 1978: (1) Central Luzon/Laguna, 1966, 1970, 1974; (2) Laguna survey, 1966, 1970, 1978; and (3) Laguna, Nueva Ecija, Camarines Sur survey , 1975/76. Demand equations were estimated separately for each set of data for both total and hired labor.

In general, the results of both studies reinforced our present understanding of the way in which specific factors influence demand. For example, irrigation and inputs such as fertilizer increased labor demand, while labor input per hectare declined with farm size. In neither India nor the Philippines were high-yielding varieties significantly related to labor input. This leads to the conclusion that other variables associated with modern varieties, such as fertilizer and irrigation, probably capture the effect of modern varieties in the increase of labor use in the regression models.

The most significant finding is the relatively low elasticity of demand for labor with respect to wage and output price. The findings were extremely consistent across studies, with elasticities of demand with respect to wage falling in the range of −0.2 to −0.4 and with respect to output price in the range of 0.2 to 0.4.[53]

What are the implications of an inelastic demand for labor? First, we can expect that a 10 percent increase in wages will reduce labor input by 2 to 4 percent. Where labor is in short supply either seasonally or locally, that is to say, where the labor supply function is inelastic, there will be a substantial sensitivity of rural wage rates to shifts in labor demand or supply or to policies that induce them. This is in sharp contrast to the earlier theoretical literature on rural labor markets, which stressed institutional wage fixities and income-sharing mechanisms. These would jointly imply a capacity of agriculture to absorb additional labor without substantial deterioration of rural wages or standard of living.[54]

Progress and Poverty

With respect to labor use and productivity, the picture that emerges is mixed. Labor productivity has been steadily increasing throughout the Asian rice growing world. However, only in East Asia (excluding China) has growth in productivity been matched by sustained growth in real wages.

In South and Southeast Asia, growth in labor demand has generally not kept pace with supply. There are exceptions, such as Malaysia and the Indian Punjab, where real wages are rising, but the key to rapidly rising labor demand in these areas, as in East Asia, has been broad-based agricultural and economic development.

Throughout most of the region, the rapidly growing population of landless or near landless people, some of them migrating from upland to irrigated rice areas, has experienced only small increases in labor demand from the technological advances in rice production. The situation in South Asia appears to be worse than in Southeast Asia, where there is less social class distinction between the landed and the landless in the village, and the traditional patron–client relationships offer the poor and disadvantaged greater protection against falling wages.

There appears to be little likelihood of an accelerated pace of labor absorption in the rice sector. There is opportunity for a modest increase in labor input per hectare, but this may be largely offset by government policies that favor low interest rates and mechanization. The greatest potential for labor absorption lies in increases in irrigation and multiple cropping. There have been substantial gains in expansion of irrigation in the past decade, but an accelerated pace of investment seems unlikely (chapter 7). In fact, in the face of the current financial crises, many countries will be hard pressed to maintain the present level of investments. Furthermore, the slow growth of production in the rainfed areas is putting increased population pressure on the irrigated rice areas through migration.

In short, if real wages are to rise, it appears that continued technological development must be accompanied by policies and social reforms that help on the one hand to increase labor demand, demand for wage goods, and access to resources among the rural poor, and, on the other hand, to slow population growth. Given the urgency of the problem in many countries, a wide range of alternatives must be implemented simultaneously. A substantial rise in agricultural production is a necessary but not a sufficient condition for improvement in the standard of living of agricultural workers.

Notes

1. Comparisons of labor requirement by crop are found for prewar China in John Lossing Buck, *Land Utilization in China* (New York, Agricultural Development Council, 1956) p. 302, and for contemporary India in A. Viadyanathan, "Labor Use in Indian Agriculture: An Analysis Based on Farm Management Survey Data," in P. K. Bardhan, A. Viadyanathan, Y. Alagh, G. S. Bhalla, and A. L. Bhadem, eds., *Labour Absorption in Indian Agriculture, Some Exploratory Investigations* (Bangkok, International Labour Office, 1978) pp. 33–118.

2. A considerable amount of data on the utilization and productivity of labor in agriculture and rice production have become available in the past few years as a result of the research sponsored by the Asian Programme for Employment Production, International Labour Organization, Bangkok. Research on labor use and employment under this program was initiated in 1977 under the direction of Dr. K. N. Raj. An initial publication by Shigeru Ishikawa, *Labour Absorption in Asian Agriculture* (Bangkok, International Labour Office, 1978), has been followed by a number of other important documents that have included a wealth of historical and contemporary data on labor utilization in rice and other crops.

3. Esther Boserup, *The Conditions of Agricultural Growth: The Economics of Agrarian Change Under Population Pressure* (London, George Allen and Unwin, Ltd., 1965).

4. James Roumasset, "Fundamental Explanation of Farmer's Behavior and Agricultural Contracts," (Banff, Canada, Conference of the International Association of Agricultural Economics, 1979).

5. Shigeru Ishikawa, *Essays on Technology, Employment and Institutions in Economic Development: Comparative Asian Experience,* Economic Research Series No. 19 (Tokyo, Kinokuniya Company Ltd., 1981).

6. Ibid., p. 37.

7. For further discussion, see Hiromitsu Kanela, "Measurement of Labor Inputs: Data and Methods," in Yujiro Hayami, Vernon W. Ruttan, and Herman M. Southworth, eds., *Agricultural Growth in Japan, Taiwan, Korea and the Philippines* (Honolulu, University of Hawaii Press, 1979) p. 175–176.

8. Ishikawa, *Essays on Technology, Employment and Institutions in Economic Development,* p. 36.

9. The bulk of labor replaced through the introduction of new technology in harvest and post-harvest activities was female labor. The issue of labor displacement has been the subject of much concern. See for example, C. Peter Timmer, "Choice of Technique in Rice Milling in Java," *Bulletin of Indonesian Economic Studies* vol. 9 (1973) pp. 57–76; W. A. Collier, J. Colter, Sinarhadi, and R. d'A. Shaw, "Choice of Technique in Rice Milling in Java: A Comment," *Bulletin of Indonesian Economic Studies* vol. 10 (1974) pp. 106–120; and C. Peter Timmer, "Choice of Technique in Rice Milling in Java, A Reply," *Bulletin of Indonesian Economic Studies* vol. 10 (1974) pp. 121–126.

10. Randolph Barker and Violeta G. Cordova, "Labor Utilization in Rice Production," in *Economic Consequences of the New Rice Technology* (Los Banos, Philippines, International Rice Research Institute, 1978) pp. 113–126.

11. Barker and Cordova, "Labor Utilization in Rice Production," p. 125.

12. Lih-Yuh S. Tsai, "Production Costs and Returns for Rice Farmers in Taiwan" (M.S. thesis, University of the Philippines at Los Banos, 1978).

13. Jean-Lue Maurer, "Some Consequences of Land Shortage in Four Kelurahan of the Kabupaten Bantul" (Yogyakarta, Indonesia, Gadja Mada University, 1978) mimeo.

14. Teng-hui Lee, Hsi-huang Chan, and Yueh-eh Chen, "Labour Absorption in Taiwan Agriculture," *Labour Absorption in Agriculture, the East Asian Experience* (Bangkok, International Labour Office, 1978) pp. 167–236.

15. This section draws heavily from L. J. Unnevehr and M. L. Stanford, "Technology and the Demand for Women's Labor in Asian Rice Farming," paper presented at the Conference on Women in Rice Farming Systems, International Rice Research Institute, September 26–30, 1983.

16. For China in the 1930s, see Buck, *Land Utilization in China*. For China under collectivization, see Elizabeth Croll, *Women in Rural Development—the People's Republic of China* (Geneva, International Labour Office, 1979). For India see, Esther Boserup, *Women's Role in Economic Development* (New York, St. Martin's Press, 1970); Mark T. Rosenzweig, and T. Paul Schultz, "Market Opportunities, Genetic Endowments, and Intrafamily Resource Distribution: Child Survival in Rural India," *The American Economic Review* vol. 72 (1982) pp. 803–815; and Gita Sen, "Paddy Production, Processing, and Women Workers in India—the South Versus the Northeast," paper presented at the Conference on Women in Rice Farming Systems, International Rice Research Institute, September 26–30, 1983.

17. Tahrunnesa A. Abdullah and Sondra Ziedenstein, *Village Women in Bangladesh: Prospects for Change* (Oxford, Pergamon Press, 1982).

18. Bina Agrawal, "Rural Women and High-yielding Rice Technology in India," paper presented at the Conference on Women in Rice Farming Systems, International Rice Research Institute, September 26–30, 1983.

19. Gita Sen, "Paddy Production, Processing, and Women Workers in India."

20. Alida Res, "Changing Labor Allocation Patterns of Women in Iloilo Rice Farm Households," paper presented at the Conference on Women in Rice Farming Systems, International Rice Research Institute, September 26–30, 1983.

21. Pujiwati Sajogyo, "Impact of New Farming Technology in Women's Employment" and Benjamin White, "Women and the Modernization of Rice Agriculture: Some General Issues and a Javanese Case Study," papers presented at the Conference on Women in Rice Farming Systems, International Rice Research Institute, September 26–30, 1983.

22. Kalpana Bardhan, "Rural Employment, Wages and Labour Markets in India: A Survey of Research," in Wilbert Gooneratne, ed., *Labor Absorbation in Rice-Based Agriculture* (Bangkok, International Labor Organization, 1982) pp. 119–157. Richard Hooley, "An Assessment of the Macroeconomic Policy Framework for Employment Generation in the Philippines," a report submitted to the U.S. Agency for International Development/Philippines, April 1981; and Cristina C. David and Randolph Barker, "Labor Demand in the Philippines Rice Sector" (Los Banos, Philippines, University of the Philippines, College of Development Economics and Management, 1981) mimeo.

23. Bardhan, "Rural Employment, Wages, and Labour Markets in India," p. 1072.

24. David and Barker, "Labor Demand in the Philippines Rice Sector," p. 11.

25. These panel surveys are discussed in Robert W. Herdt, "Costs and Returns for Rice Production" in *Economic Consequences of New Rice Technology* (Los Banos, Philippines, International Rice Research Institute, 1978) pp. 63–80; and the factor shares are analyzed in Chandra G. Ranade and Robert W. Herdt, "Shares of Farm Earnings from Rice Productions" in *Economic Consequences of the New Rice Technology* (Los Banos, Philippines, International Rice Research Institute, 1978) pp. 87–104.

26. Wade Edmundson, "Land, Food, and Work in Three Javanese Villages" (Ph.D. dissertation, University of Hawaii, 1972); Benjamin White, "Population, Involution, and Employment in Rural Java," *Development and Change* vol. 7 (1976) pp. 267–290.

27. C. Alexander and C. Saleh, "The Distribution of Production Factor Inputs by Representative Farm Crops in Jati, West Java," *Agro-Economic Survey Research Notes* (Bogor, Indonesia, May 1974); Wade Edmundson, "Land, Food, and Work."

28. David H. Penny and Masri Sigharimbun, "A Case Study of Rural Poverty," *Bulletin of Indonesian Economic Studies* vol. 8 (1972) pp. 79–88; Benjamin White, "Population, Involution, and Employment in Rural Java."

29. Ricardo A. Guino, "Time Allocation Among Rice Farm Households in Central Luzon, Philippines" (M.S. thesis, University of the Philippines, 1978).

30. Wen-hui Lai, "Trends in Agricultural Employment in Post-War Taiwan," China Council of Sino-American Cooperation in Humanities and Social Sciences, Sino-American Conference in Manpower in Taiwan (Taipei, Academica Sinica, 1972) pp. 127–134.

31. Hans P. Binswanger and Mark. R. Rosenzweig, "Contractual Arrangements, Employment and Wages in Rural Labor Markets: A Critical Review" (New York, Agricultural Development Council, 1981) pp. 33–34.

32. Yujiro Hayami and Masao Kikuchi, *Asian Village Economy at the Crossroads, An Economic Approach to Institutional Change* (Tokyo, Tokyo University Press, and Baltimore, Md., Johns Hopkins University Press, 1981); and Murray J. Leaf, "The Green Revolution and Cultural Change in a Punjab Village, 1965–1978," *Economic Development and Cultural Change* vol. 31 (1983) pp. 227–270. Hayami and Kikuchi also compare the harvest contract changes in two Indonesian villages, one with falling wage rates and one with rising wage rates.

33. See for example, Hayami and Kikuchi, *Asian Village Economy at the Crossroads,* p. 105.

34. Cornell University, Rural Development Committee Special Series on Landlessness and Near-Landlessness; S. Hirashina, *Hired Labor in Rural Asia* (Tokyo, Institute of Developing Economies, 1977); and International Labour Office, *Poverty and Landlessness in Rural Asia* (Geneva, ILO, 1977).

35. David A. Rosenberg and Jean G. Rosenberg, *Landless Peasants and Rural Poverty in Selected Asian Countries,* Rural Development Committee Special Series on Landless and Near-Landlessness LNL3 (Ithaca, N.Y., Cornell University, 1978) p. 3 and table 1. They also estimate that for the five countries, about 75 percent of the households are landless and near landless. However, their definition of near landless is too broad to be very meaningful in identifying households that are impoverished.

36. Ibid.

37. Ibid.

38. In a survey of rural households in three municipalities of the Central Luzon, Philippines, Guino found that 75 percent of the fathers of the landless were farmers, 20 percent had nonagricultural employment, and only 4 percent were themselves agricultural laborers. Ricardo A. Guino, "Time Allocation Among Rice Farm Households in Central Luzon, Philippines."

39. Hayami and Kikuchi, *Asian Village Economy at the Crossroads,* pp. 155–157.

40. Widya Utami and John Ihalauw, "Some Consequences of

Small Farm Size," *Bulletin of Indonesian Economic Studies* vol. 9 (July 1973) pp. 46–56.

41. William L. Collier, Soentoro, Gunawan Wiradi, and Makali, "Agricultural Technology and Institutional Change in Java," *Food Research Institute Studies* vol. 13 (1974) pp. 169–194.

42. Utami and Ihalauw, "Some Consequences of Small Farm Size."

43. Hayami and Kikuchi, *Asian Village Economy at the Crossroads*. For a discussion of the geographic distribution of tebasen and ceblokan in Java, see pp. 158–161. While tebasen is more common in Central and East Java, it is normally practiced in conjunction with bawon.

44. Collier, Soentoro, Wiradi, and Makali, "Agricultural Technology and Institutional Change in Java," pp. 191–193.

45. Hayami and Kikuchi, *Asian Village Economy at the Crossroads,* pp. 161–170.

46. Barker and Cordova, "Labor Utilization in Rice Production," p. 124.

47. Hayami and Kikuchi, *Asian Village Economy at the Crossroads,* chapter 5.

48. Antonio J. Ladesma, *Landless Workers and Rice Farmers: Peasant Subclasses Under Agrarian Reform in Two Philippine Villages* (Los Banos, Philippines, International Rice Research Institute, 1982).

49. Jose Barramela, Jr., "A Case Study of a Coconut Tenant-Farmer in the Bicol River Basin," paper presented at the National Workshop on Small Farm Credit (Albay, Philippines, Food and Agriculture Organization, 1977).

50. Edward J. Clay, "Institutional Change and Agricultural Wages in Bangladesh," *Bangladesh Development Studies* vol. 4 (1976) pp. 423–440.

51. Edward J. Clay, "Institutional Change and Agricultural Wages in Bangladesh"; and G. Wood, "Class Differentiation and Power in Bandakgram: The Minifundist Case," in M. N. Hoq, ed., *Exploitation and the Rural Poor* (Camilla, Bangladesh, Bangladesh Academy of Rural Development, 1976).

52. Robert F. Evenson and Hans P. Binswanger, "Estimating Labor Demand Functions for Indian Agriculture," Economic Growth Center Discussion Paper No. 356 (New Haven, Conn., Yale University, August 1980); and David and Barker, "Labor Demand in the Philippine Rice Sector."

53. Evenson and Binswanger, "Estimating Labor Demand Functions in Indian Agriculture"; and David and Barker, "Labor Demand in the Philippine Rice Sector." Estimates are presented for the elasticity of wages deflated by rice price for the Philippine data, but equations were also estimated with wage rate and rice price as independent variables. There was a slightly more responsive labor demand to rice price than to wage rates.

54. Evenson and Binswanger, "Estimating Labor Demand Functions for Indian Agriculture," pp. 22–23.

10

Who Benefits from the New Technology?

The widespread adoption of the new rice technology in Asia has generated an extended discussion among development specialists over the distribution of the resulting economic improvements. Some scholars maintain that the new technology has narrowed the income gap between the rich and the poor while others find evidence for the opposite case. There are important arguments on both sides, and this lack of consensus is reflected in the bits of empirical evidence that different scholars cite, on the interpretation of that evidence, and in the divergent models and theoretical frameworks that have been developed. In this chapter, we will simply review the current empirical research rather than make value judgments.

Growth, Equity, and Technological Change

It seems fair to say that those who developed the initial Green Revolution technology, principally biological scientists, gave little consideration to the socioeconomic implications of their work, although their broad goal was that of increasing food production so as to reduce human misery. The new technology was enthusiastically embraced by policymakers and other scientists alike because it offered a quick solution to Asia's critical physical land problems caused by a rapidly shrinking land frontier and an accompanying rise in the man–land ratio. The American and Japanese experience suggested that continuous technological innovation in seeds, inputs, and implements could be the cutting edge of the agricultural transformation,[1] although the distinct differences in the historical patterns of development in countries with resource endowments as different as Japan's and the United States's were not well understood. Still, the emphasis on technological innovations proved to be suitable to the high labor–land ratios in most Asian countries.

The new seed-fertilizer technology was developed in experiment stations in Asia that were favored with fertile soils, well-controlled water sources, and other factors suitable for high production. There was little perception of the complexity and diversity of farmers' physical environments, let alone the diversity of the economic and social environment in rural South and Southeast Asia. In retrospect, this proved to be a mixed blessing. If the modest resources available to the international agricultural research system in the 1960s had been concentrated in the less favorable environments, it is likely that no major breakthrough would have been made. Unfortunately, scientists frequently saw their responsibility as ending at the experiment station gate. That the modern rice technology did not gain acceptance by farmers in many areas was initially attributed to peasant conservatism and backwardness and to the failure of extension to do its job in disseminating the technology.[2]

As the new rice technology spread, largely in more favorable environments, social scientists developed their own interpretation of events. Two contrasting points of view arose, which we can broadly classify as the *dependency theory* model and the *induced innovation* model. Dependency theorists argued that the new technology widened the gap between the rich and the poor and that this was no accident.[3] The developed countries gained at the expense of the developing, large farmers at the expense of small, and landlords at the expense of tenants. The widely acclaimed theory that benefits would "trickle down" to the poor did not work and perhaps was never intended to work. Without major institutional reforms, efforts to introduce new technology were misspent. Furthermore, the dependency theorists argued, there was an inherent bias in the exotic technology of the international centers in favor of the rich. They argued that technologies to help resource-poor peasants had to be developed by the indigenous scientific community working hand-in-hand with local farmers.

Advocates of the induced innovation model saw the course of events in a rather different light.[4] They argued that the situation in developing countries is similar to that described by classical economists such as Ricardo. When population growth presses on limited land resources and technology does not change, the frontiers of cultivation must be expanded to more marginal areas and greater amounts of labor must be expended per unit of cultivated land; the cost of food production increases and food prices rise. In the long run, laborers' incomes will be driven down to a subsistence minimum barely sufficient to maintain a stationary population, and all the surplus will be confiscated by landlords in the form of increased land rent. The Green Revolution was seen as being induced by incentives to develop technologies that effectively utilize large surpluses of labor and conserve increasingly scarce land while responding to the deteriorating food-population balance brought on by the rapid rising population growth of the mid-twentieth century.

The induced-innovation advocates generally also argued that if free market prices reflecting the true social value of resources were allowed to prevail, development problems would be solved by allowing response to market prices to dictate appropriate investments in scientific technology (price purists). Alternatively, others, not trusting the "trickle down" of economic benefits, argued that wealth should be redistributed through welfare programs.[5] Another group maintained that whether the new technology promoted equity or reinforced inequity was determined by the nature of the technology, by the pattern of resource ownership, and by the institutional setting.

Three aspects of a new agricultural production technology influence its effects on equity:[6] (1) the crop, (2) the environment in which the new technology is effective, and (3) the factor bias of the technology. Thus, research work on staple foods that are consumed in large quantities by the poor, grown in disadvantaged environments, and produced with labor-using and land-saving technologies tends to promote equity. But the balance between research concentrated on disadvantaged environments and that concentrated on areas with greater potential has been debated in developed as well as developing countries. The appropriate balance will depend on careful analysis of the social benefits and costs in specific situations.

Opinions in the early Green Revolution literature were divided between ardent support and severe reservation. Scholars studying the same events, and in some cases the same data, drew opposite conclusions about equity effects of technology. This dialectical debate has, however, led to a greater understanding of the complexity of the issues by both social and biological scientists and has encouraged a reevaluation and redirection of research emphasis.

One consequence of these debates has been a greater effort to describe the rice growing environments and to develop technology suited for less favorable agroclimatic areas.[7] The need for a more holistic look at the farm family as a decision-making unit, and for a greater interaction between the farmer and the researcher are reflected in the development of farming systems projects throughout the world.[8] Where the farming systems involve many enterprises, the problems are complex, and the productivity and equity effects of research are as yet uncertain.

No clear answer emerges from the theoretical arguments. The view reflected in the semipopular media is that the new technology has had relatively little positive effect. But, is this supported by the evidence? Is there clear evidence that the new rice technology has not increased production? that small farmers or tenant farmers have been unable to adopt new varieties or fertilizer? that new rices need more or less labor than old rices? that people in villages with rapid technical change have been disadvantaged relative to people in villages with less or no technical change?

In this chapter we attempt to assemble the results of as many empirical studies as possible that relate to these questions for rice in Asia. Necessarily this requires a definition of the term "new rice technology." Conceptually, the idea could embrace a wide range of innovations, including seeds, improved water control, farm chemicals, and tractors. However, farmers seldom adopt all innovations as a package, and research studies that classify farmers into users and nonusers of each of the many aspects of production technology that can be identified as "new" result in a very large number of categories. The mass of data generated by this type of approach is very difficult to interpret and few multi-innovation studies have been carried out. An alternative is to consider only those farmers adopting all components of a given combination or "package" of technology as adopters. With a package composed of several components, the resulting number of adopters is usually rather small.

Instead of following either of these approaches, the studies examined here used rice variety as the key component of new technology: adopters used new varieties (variously called "high yielding" or "modern varieties") and non-adopters used traditional or old varieties. The central issue is the impact of new varieties on equity in the agricultural system into which they have been introduced. Because fertilizer is so highly complementary to new varieties, its pattern of use is also examined. And, because new varieties can only generate an effect if they provide a higher yield than old ones, the first issue examined is their aggregate contribution to production.

The Effect of New Technology on Production

The new rice technology has had a marked and permanent effect on Asian rice production. Based on a survey of the literature, Dalrymple concluded that the high-yielding variety and fertilizer package added 7.7 million metric tons of additional rice to Asia's production in 1972–73, or 4.9 percent of total production. This conclusion was based on what was considered to be the best estimate of the ratio of high-yielding varieties (HYV) to traditional yields.[9]

Few countries collect and publish yields of rice by variety, although some data are available (table 10.1). When 2.5 percent of India's rice areas was in modern varieties (MVs) in 1966–67, the yield ratio of MVs to traditional varieties was 2.6, while in 1973–74 when 25 percent of the areas was in MVs, the yield ratio was 1.7. Similar trends were observed for Bangladesh. Thus, these data suggest that as MVs spread to cover a higher proportion of the total rice area, their yield benefit compared with traditional varieties fell somewhat.

A number of farm-level studies report comparable yield data for new and old varieties. Table 10.2 summarizes the results of farm management studies that show that modern varieties yielded 10 to 100 percent more than traditional varieties. Although physiologically the MVs have a greater yield advantage over the traditional varieties (TVs) in the dry season, the survey evidence does not show that advantage because the dry season crop in many areas has inadequate water supplies.[10] Surveys in India and Indonesia indicate that MVs yield about 1.5 times as much as the TVs. Philippine studies give an average yield ratio of 1.2, while in other countries the ratio is about 1.4.

Table 10.1. Ratio of the Yield of Modern Rice Varieties to Local Rice Varieties

Year	India	Bangla-desh	Philippines		Indonesia
			Irrigated	Rainfed	
1968/69	1.9	—	1.1	1.0	—
1969/70	1.8	3.2	1.1	1.0	—
1970/71	1.9	3.3	1.1	1.0	—
1971/72	2.3	3.1	1.2	1.1	1.4
1972/73	2.5	2.7	1.1	1.2	1.4
1973/74	2.3	2.7	1.1	1.2	1.4
1974/75	1.7	2.5	1.2	1.2	1.4
1975/76	2.0	2.3	1.2	1.1	—
1976/77	2.1	2.3	1.2	1.3	—
1977/78	1.8	2.0	1.4	1.3	—
1978/79	—	—	1.3	1.3	—

Source: R. W. Herdt and C. Capule (1983).

Evenson and Flores used regression analysis to look at the effect of major contributing factors on rice production for twelve Asian countries. The area planted to modern varieties was one of the independent variables in the analysis, along with land, fertilizer, and research inputs. This approach gave results that suggested even higher yield benefits from modern varieties than those reported above.[11] Our own analyses are discussed in chapters 4 and 6.

Thus, the evidence consistently shows that the new varieties resulted in an increase in production through higher yields. The effect of this increase can be measured at both the market and the farm level. The market effects can best be examined through a simple supply and demand model that depicts technical change as a shift in the supply function. This indicates the overall effect on price and consumption and suggests how benefits are divided between consumers and producers. The differential effects among classes of producers are best understood by examining the experiences of various classes of producers.

Market Effects of Technical Change

The supply of rice must expand at the same rate as the demand, or market prices will rise. If this happens in a situation without an administered rationing system, the reduced supply is allocated

Table 10.2. Ratio of the Yields of Modern and Traditional Varieties During the Wet and Dry Seasons

Location	Yield MV/Yield TV		Study period	Source
	Wet season	Dry season		
India				
Andhra Pradesh	1.40	1.93	1971/72	Parthasarathy, 1975
Tamil Nadu	1.56	1.61	1971/72	Rajagopalan, 1975
Uttar Pradesh	2.01	—	1971/72	Sharma, 1975
Karnataka	1.89	—	1971/72	Krishna Murthy, 1975
Orissa	—	1.41	1971/72	Pal, 1975
West Bengal	1.47	—	1972/73	Mandal and Ghosh, 1973
Bihar	1.53	—	1972/73	Mandal and Ghosh, 1973
Orissa	1.45	—	1972/73	Mandal and Ghosh, 1973
Indonesia				
West Java	—	1.30	1971/72	Prabowo and Sajogyo, 1975
Central Java	1.08	1.16	1971/72	Ihalauw and Utami, 1975
East Java	—	1.07	1971/72	Prabowo and Sajogyo, 1975
West Java	2.20	1.50	1968/71	Sajogyo and Collier, 1972
Central Java	1.30	1.40	1968/71	Sajogyo and Collier, 1972
East Java	1.39	1.30	1968/71	Sajogyo and Collier, 1972
Philippines				
Nueva Ecija	1.25	—	1971/72	Herrera, 1975
Leyte	1.31	1.36	1971/72	Contado and Jaime, 1975
South Cotabato	1.06	—	1971/72	Tan, 1975
Camarines Sur	1.04	—	1971/72	Mangahas and Librero, 1973
Iloilo	1.08	—	1971/72	Mangahas and Librero, 1973
South Cotabato	1.33	—	1971/72	Mangahas and Librero, 1973
Laguna	1.41	1.24	1966/70	Herdt, 1978
Central Luzon	1.15	—	1966/70	Herdt, 1978
Others				
Punjab, Pakistan	1.60	—	1971/72	Chaudhari et al., 1975
Punjab, Pakistan	1.52	—	1972/73	Khan, 1975
Sind, Pakistan	1.42	—	1972/73	Khan, 1975
Kelantan, Malaysia	1.00	1.21	1971/72	Tamin and Mustapha, 1975
Suphan Buri, Thailand	1.43	—	1971/72	Sriswasdilek et al., 1975
Minipe, Sri Lanka	1.66	1.61	1966/70	Amerasinghe, 1976

according to consumers' desires and abilities to pay. Lower income consumers are forced to modify food consumption patterns to include more low-cost foods. Farmers benefit, in general, by selling their products at higher prices.

Periods of abundance or sharp increases in supply generally have the opposite effect. If technological change is pervasive enough to move supply ahead of demand over a period of time, and if prices are allowed to respond by moving downward, consumers gain relative to producers. There are, of course, subgroups of producers and consumers, and in the rural areas of the developing world there are many households that combine both production and consumption activities.

Technological change shifts the supply function of rice outward, indicating that at any price a greater quantity of output will be supplied than previously. As a consequence, price falls and quantity increases. Consumers gain by having a larger quantity available at a lower price. Producers lose revenue because prices are lower, but they gain because of the reduced cost per unit of output generated by technical change. Society's gain from the indicated change in supply is positive. The division of the gains between producers and consumers depends on the elasticities of demand and supply.

Clearly, the consumers' gain depends on a reduction in the price of rice in the market. This will occur in a "closed economy" in which international trade in rice is restricted so the shift in supply is reflected in prices. In open economies, the price will be dependent on the international market price, so technological change may not result in any market price reduction. Where technological change is so widespread as to have a perceptible effect on the world supply curve, the situation can be interpreted in the same way as above. With the world price of rice falling, benefits are spread throughout the market. In that situation, countries that permit the international price to be reflected in their domestic prices generate consumer gains.

Evenson and Flores used the consumers'-producers' surplus model to estimate total benefits from rice research in Asia from 1950 to 1975.[12] They separately identified the effect of national research and IRRI-type new varieties. Using a range of yield effects, they derived the estimates for producers' and consumers' gains shown in table 10.3. Their analysis assumes a price elasticity of demand of −0.3, and a price elasticity of supply of 0.4. They estimate that rice research increased yields enough to provide benefits valued at from $US 268 million to $US 310 million per year during the 1970s, with consumers gaining at the expense of producers.

A similar analysis of the benefits of new rice varieties was calculated for Colombia, where rice exports were prohibited.[13] The results show an analogous pattern of gains to consumers and losses to producers, with total gross social benefits of about $US 435 million in 1974. Both studies showed high rates of return on the investments made in rice research.

Up to this point in our discussion we have considered producers and consumers to be homogeneous groups. An extension of the above analysis was used to argue that producers, who consume a large portion of the rice that they produce, retain the benefits of technical change internally.[14] By definition, a technological change results in the capacity to produce using fewer inputs per unit of output than before the change. This is true whether the output is sold or consumed. That portion of output that is consumed is therefore available to producers at a lower real expenditure of resources; that is, producers retain some of the consumers' surplus benefits. The output that is sold brings a lower price than it would have without the technological change, and so the net effect on producers depends on how much prices change relative to costs (that is, the relative elasticities of demand and supply). However, producers who sell all their output do not retain any of the increase in consumers' surplus.

When rigorously pursued and given the reasonable assumption of inelastic demand, this line of argument shows that the degree of consumers' surplus internalized by producers is inversely related to the proportion of output sold, if market prices are permitted to react to the technological change. Where the price of the commodity is fixed by the government or determined by an international market unaffected by the technical change, then a price change may not result, and the benefits to producers will be in proportion to their sales.

For the range of parameters specified and with fixed home consumption, small farmers obtained an increased cash income while large farmers suffered a reduced cash income as a consequence of technical change in rice production. An extension of the model allowing for variable home consumption also showed that small farmers gained relative to those who sold a large proportion of their output.

However, these results hold only if the technological change is adopted by small and large farmers and is pervasive enough to shift the aggregate supply function, allowing prices to fall. If the technological change is monopolized by a few large farmers, or if prices are not permitted to respond to the new technology, then large farmers could capture the major part of the gains from technical change, denying them to both small farmers and consumers. If small farmers do not adopt the new technology, then they will not receive benefits. Thus, an important issue is whether certain groups of farmers are systematically excluded from participation in the new technology. This issue is considered in the next part of this chapter.

Farm-Level Experience with New Technology

We have argued that the gains from technological changes affecting basic foods that have price inelastic demands are ultimately transmitted to consumers

Table 10.3. Estimated Annual Benefits from Rice Research
(million US $)

Research institutions	Years	Producers' gain		Consumers' gain		Social gains	
		High	Low	High	Low	High	Low
National varieties	1950–60	−25.9	—	−52.0	—	26.0	—
	1961–65	−53.1	—	107.0	—	53.9	—
	1966–71	−211.4	−186.6	431.7	374.1	220.3	190.4
	1972–75	−403.2	−190.1	414.3	387.0	211.0	196.9
IRRI HYV	1966–71	−133.4	−70.4	270.2	141.7	136.8	71.3
	1972–75	87.3	−60.7	176.0	141.6	88.7	71.9

Source: R. Evenson and P. Flores (1978).

through the lower prices that occur when supply moves ahead of demand. This analysis suggests that the distribution of gains and losses among producers depends on how fully they can adopt the technology and the extent to which they participate in the market. However, in nearly every Asian country, rice prices are controlled or limited by government policies, and, in addition, many would argue that most developing countries have marketing systems that do not follow the perfect competition model. Other challengers to the above analysis argue that the model is static and aggregated—it does not reflect the fact that some individuals adopt a new technology sooner than others nor does it reflect the fact that certain technologies are never adopted in some environments because of their inherent unsuitability.

One way to gain insights into the likely effect of a new technology under such conditions is to construct an analytical model that would be dynamic and disaggregated, with policies and institutions endogenous. Although many analytical attempts exist, and in fact, the bulk of the literature on the income distribution effects of the Green Revolution is speculative and theoretical in nature, few of the theoretical treatments are satisfactory because most focus on only one or two of the determinants of the effect of economic change.

One of the more useful discussions on the effect of technological change following the development of new varieties is Carl Gotsch's paper on technical change and income distribution.[15] Gotsch reiterates the importance of four basic environmental and social components that determine the impact of change on income distribution:

1. the characteristics of the technology
2. the absolute magnitude and distribution of productive resources
3. the type and distribution of institutional services
4. local social custom and traditions

Item 1 on Gotsch's list refers to the effect the technology has on the demand for factors of production; item 2 refers to the supply of productive factors; and items 3 and 4 refer to the arrangements governing the ownership and allocation of the earnings of the productive factors.

Much of the disagreement on the effect of technical change has occurred because some analysts focus on the adoption of technology or the elasticities of factor substitution existing in the old and new technology (no. 1), others focus on rates of population growth and man–land ratios (no. 2), others on land tenure (no. 3), and still others on institutional rules for employment of wage labor (no. 4). All are important, and only a partial picture can emerge from scrutinizing them one at a time. However, given intellectual limitations to examining a situation where everything changes simultaneously, much can be understood from even a partial picture. Thus, subsequent sections of the chapter review the evidence about the use of new rice technology at the farm level, considering the questions of who has adopted the technology, the effect on incomes, and the apparent effect on those rural people who depend on being hired by farmers—the agricultural laborers. In a final section, we attempt an integrative analysis that shows the total effect of income distribution.

Adoption and Farm Size

Frankel's book on the economic gains and political costs of India's experience appeared soon after the initial release of new rice varieties and was widely quoted.[16] Published in 1971, it was based on observations of India's Intensive Agricultural District Program (IADP) implemented in 15 of India's 450 districts. Frankel's observations were made when modern varieties were first being introduced and when there was a general shortage of fertilizer in the country. Thus, they do not provide a comprehensive view of the impact of new rice varieties on Indian agriculture, but they do give a general picture. And that picture is gloomy indeed.

Discussing the agricultural system of Burdwan, West Bengal, Frankel reports that:

> the overwhelming majority of farm families in Burdan district are either completely landless or operate uneconomic holdings. Many families with land operate it under sharecropping arrangements. The sharecropping system as it is practiced in Burdan district is particularly unfavorable for investment in improved practices. . . . Many landlords are actually absentees, living in other villages or in Calcutta. . . . If the landlord lacks the ability or interest to invest in modern inputs, *the sharecropper plainly finds it impossible to do so. . . . Sharecroppers rarely used any chemical fertilizers*, or only small doses. They reported no increase in crop yields with local varieties over the last five years. *None had tried the high yielding varieties*, even though some parts of their holdings were topographically suitable, because of the *high production costs involved. As subsistence farmers, they had no surplus to sell*, and could not benefit from rising prices for food grains. . . . The circumstances of farmers with ownership holdings of less than 3 acres were only moderately better than those of sharecroppers. Generally, they achieved some increase in yields, about 25 to 30 percent over the last five or six years from the use of *small doses of chemical fertilizer*. But at most, this permitted them to maintain their existing standard of living

in the face of rising prices for farm inputs and essential commodities. They reported no improvements and *virtually no adoption of the high-yielding varieties*. The vast majority of agriculturists in Burdwan district, i.e., sharecroppers and small farmers with holdings of less than 3 acres, have received very little help from the cooperatives.[17] (emphasis added)

A picture of two distinct sectors emerges—the big farmers who monopolize new technology, fertilizer, and credit, and the small farmers who ignore innovations or who cannot employ them because of restrictions imposed by their tenure. However, that is a picture from a very limited study area. More comprehensive studies show quite a different scene.

One of the earliest comprehensive studies of adoption of new technology as related to farm size in India was reported by Lockwood, Mukherjee, and Shand and was based on large surveys conducted by a research group in India's Planning Commission.[18] They found a strong positive linear association between proportion of farmers adopting HYV seed and the size of the farm (figure 10.1). That is, a high proportion of large farmers planted the modern varieties. However, the data also show that there was a clear upward shift in adoption from year to year and that there was an inverse association between farm size groups and proportion of rice area each group planted to modern varieties.

The authors conclude that smaller farmers were slower to adopt the new varieties, but that once they made the decision to use the varieties, they were likely to sow much the same proportion of the crop to HYVs, to use as much fertilizer, to spend almost as much on cash inputs and hired labor, and to achieve yields similar to those of larger farmers. Further, they found no relationship between farm size and the proportion of total farm input expenditures made using credit.

A similar conclusion emerges from a series of studies conducted by various agroeconomic research centers in India. M. Schluter analyzed these reports and found a statistically significant positive relationship between farm size and proportion of farms in each size group that adopted new varieties in forty-five of the seventy areas studied. However, he also found that the proportion of adopters among farms in all size groups increased over time. Furthermore, he found that small rice farmers who adopted new varieties did so on a greater portion of their acreage than large farmers.[19]

A six-country research project coordinated by IRRI surveyed Asian rice producers in thirty-six villages

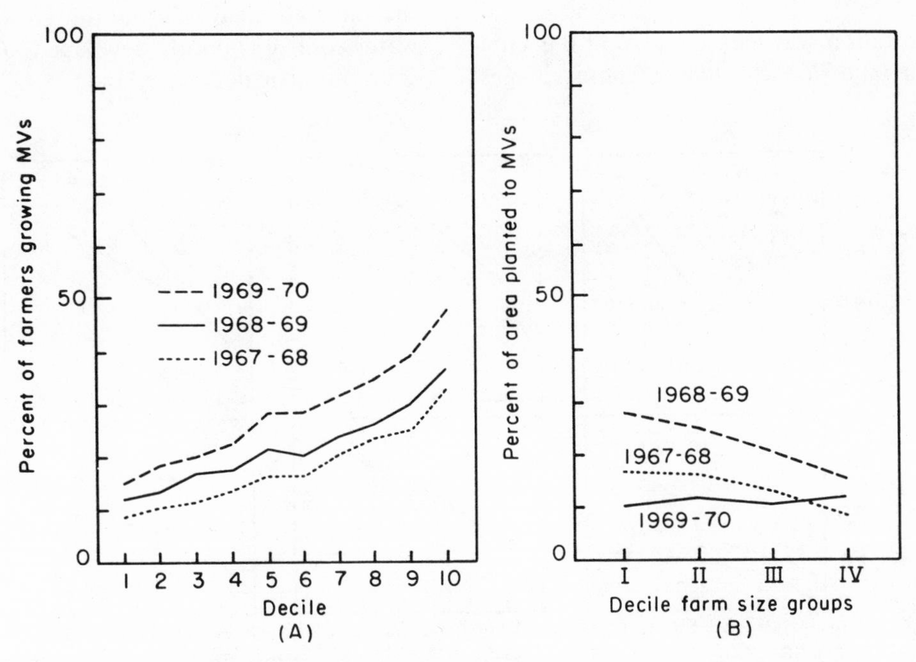

Figure 10.1. Proportion of farmers in various size groups growing MV (A) and proportion of adopters' holdings under MV seed (B) for various size groups of rice farmers in India (Source: B. Lockwood, P. K. Mukherjee, and R. Shand, *The High-Yielding Varieties Programme in India, Part I* [Canberra, Planning Commission, Government of India and the Australian National Government, 1971] reprinted by permission of the publisher)

where the new varieties had been widely adopted. The data show that the smallest farms adopted new varieties first and the largest farms last. The pattern for adoption of fertilizer and insecticide was similar. Large farms led in the adoption of tractors and mechanical threshers (figure 10.2), partly because the larger farms and mechanical innovations are concentrated in certain countries, the smaller farms in others. For example, tractors were widely used in two villages in Thailand where 85 percent of all farms were over 3 ha, while in four study villages in Indonesia, 85 percent of all farms were less than 1 ha, and no tractors were used.

A second analysis of the same data compared the adoption pattern of modern varieties on the smaller farms with that for larger farms in each village, a method that more adequately reflects intravillage differences (figure 10.3). This analysis shows little or no difference in rates of adoption of modern varieties on small and large farms, in contrast to the data from the Indian study reported in figure 10.1. However, in a few villages the small farmers adopted later, and it was hypothesized that this might have been caused by the difficulties small farmers experienced in obtaining seed or inputs complementary to MVs in villages where a few large farmers dominate the distribution channels. To test this hypothesis, villages were grouped by the degree of concentration of land holdings.

The concentration was measured using the Gini coefficient for land in each village. Those villages with land more highly concentrated in a few large farms (group I in figure 10.3) were contrasted with villages with less highly concentrated holdings (groups II and III). There is no substantial pattern of earlier adoption by large farmers where land is more highly concentrated, although such a pattern is observed in Pedapulleru, Andhra Pradesh, one out of the thirty-six villages in which small farmers clearly lagged behind large.

Even where one group of farms lags behind in adoption, the logical consequence of a continuation of the process is for the lagging group to eventually catch up to the leaders. Recent studies in Bangladesh and India indicate that adoption is no longer affected by farm size. India's National Council of Applied Economic Research (NCAER) conducted a comprehensive study of the use of modern varieties and fertilizers, interviewing over 25,000 farmers in every state in India.[20] A summary of their findings on the relationship between farm size and adoption of modern varieties is shown in table 10.4. In most states, there was no relationship between farm size and percentage of rice area in MVs. The striking variability in rate of adoption is among states rather than across size within a state, and this appears to be associated with irrigation. Where irrigation is well developed, adoption is high for all farm size groups (Punjab and Haryana). By contrast, in the eastern states where the rains are heavy during the growing season and water control is poorly developed, adoption is low, even on "irrigated" rice land.

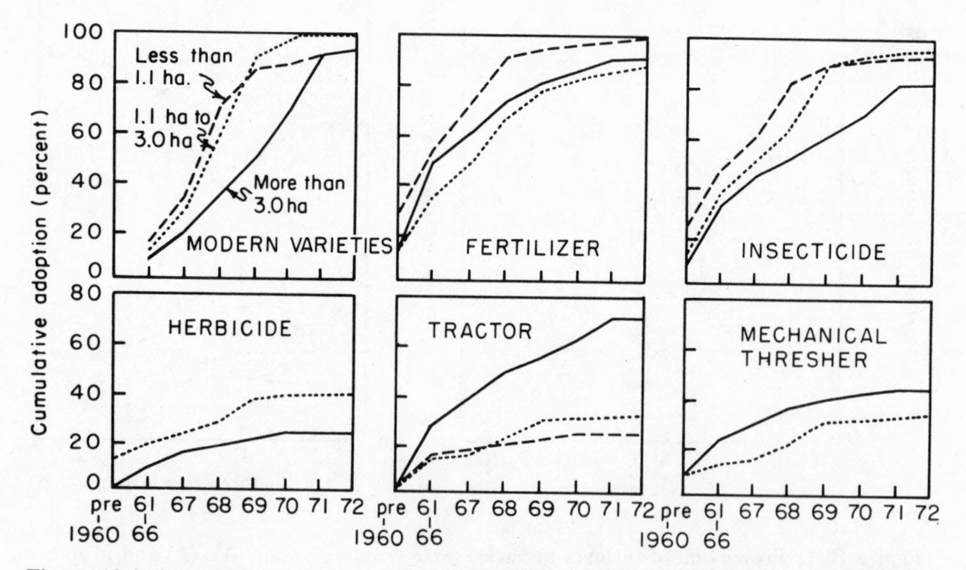

Figure 10.2 Cumulative percentage of farms in three size classes adopting specific innovations (Source: R. Barker and R. Herdt, "Equity Implications of Technology Changes," in International Rice Research Institute, *Interpretative Analysis of Selected Papers from Changes in Rice Farming in Selected Areas of Asia* [Los Banos, Philippines, IRRI, 1978] p. 91, reprinted by permission of the publisher)

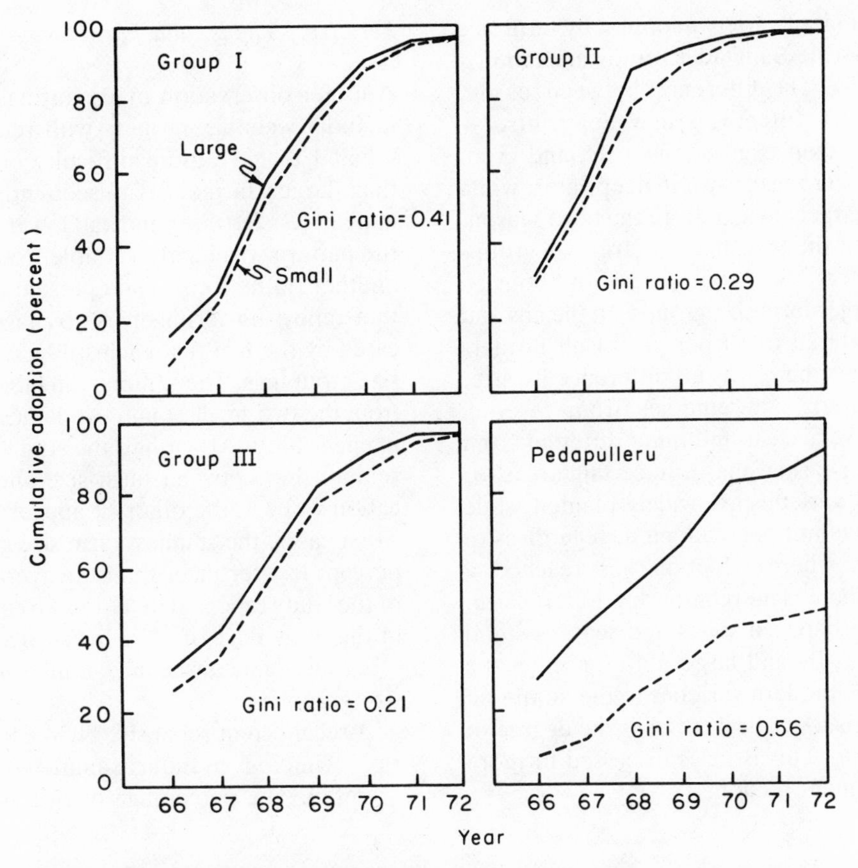

Figure 10.3. Gini coefficients, farm size, and year of adoption of modern varieties on large and small farms (classed within each village) (Source: R. Barker and R. Herdt, "Equity Implications of Technology Changes," in International Rice Research Institute, *Interpretative Analysis of Selected Papers from Changes in Rice Farming in Selected Areas of Asia* [Los Banos, Philippines, IRRI, 1978] p. 94, reprinted by permission of the publisher)

Table 10.4. State Rice Area and Proportion Planted to Modern Varieties by Farm Size, India, 1975–76

State	Total rice area (thousand ha)	Percent of area in modern varieties on farms below 1 ha	1–2	2–4	4–10	over 10
West Bengal	5,426	18	13	17	15	50
Bihar	5,257	34	33	38	24	16
Orissa	4,684	28	30	33	35	38
Uttar Pradesh	4,622	34	34	24	31	21
Madhya Pradesh	4,588	0	0	0	0	1
Andhra Pradesh	3,894	34	42	54	54	49
Tamil Nadu	2,564	70	60	50	67	19
Assam	2,241	1	1	2	0	0
Maharashtra	1,417	4	6	7	12	0
Karnataka	1,194	39	46	37	52	23
Kerala	885	48	39	51	100	100
Punjab	567	99	100	100	100	100
Gujarat	459	5	2	3	4	2
Haryana	304	71	91	92	86	93
Jammu and Kashmir	252	88	75	74	80	0
Rajasthan	155	0	0	2	12	56

Sources: Percent of area in MVs by size of farm: National Council of Applied Economic Research, *Fertilizer Demand Study: Interim Report*. Total rice area: India, Directorate of Economics and Statistics, "All India Estimates of Rice" (1977).

The data on modern variety adoption by farm size in a recent large-scale Bangladesh study are summarized in table 10.5.[21] The differences between season, as in India, reflect differences in water control—much of the aus crop is grown as a dryland crop, much of the *aman* crop is grown in deepwater, while the boro crop is largely irrigated. In the boro season, over 70 percent of the two smallest farm size groups planted modern varieties, compared with about 50 percent of the larger farm size groups. In the aus and aman season, only 20 to 30 percent of all farmers grew the MVs, but there was no difference by size.

Thus, the pattern that emerges from recently collected data is very clear and quite different from studies done at an earlier phase. In certain areas, or seasons, the new varieties are widely planted while in others they have not been adopted, regardless of farm size. Even where adoption has reached a moderate level, there is no relationship to farm size. Large differences between states and seasons stand out. These reflect, by and large, differences in the adaptability of the modern varieties to the dominant growing conditions. A critical factor is the degree of control over water. This issue is discussed in more depth in a subsequent section.

Fertilizer, Yields, and Farm Size

Another observation made during the early studies in India was that farmers with relatively little land seemed to have more difficulty obtaining fertilizer than larger farmers. Consequently, small farmers applied less fertilizer and had lower yields. However, the pattern was highly variable from one location to another (table 10.6). Data presentation from studies that report on fertilizer use by farm size is complicated by the fact that each study used different farm size groupings. The summary in table 10.7 shows data from the two smallest and two largest groups reported in each study. About half the studies show a positive relationship between farm size and fertilizer application while in the others it appears negative. In the worst cases, the smallest farm size groups applied 20 percent less fertilizer than the average. The authors of the study reported in table 10.6 comment that even in the early days of MVs in the rice cropping areas, "there was no consistency at all" in fertilizer use by size group.[22]

A recent comprehensive NCAER study of modern rice technology in India examined the use of fertilizer in detail. The percentage of rice area fertilized and

Table 10.5. Proportion of Rice Area Planted to Modern Varieties on Farms of Various Sizes, Bangladesh, 1979–80

Season	Percent of rice area in MVs on farms					Number of sample farms
	below 0.4 ha	0.4–1.0 ha	1.0–2.0 ha	2.0–3.0 ha	over 3.0 ha	
Boro 1979–80	86	71	47	47	52	1,812
Aus 1980	24	21	19	20	31	1,897
Aman 1980	31	27	27	27	23	1,872

Source: Unpublished data from the Bangladesh Agricultural Research Council and International Fertilizer Development Council study on fertilizer use in Bangladesh (provided by IFDC).

Table 10.6. Application of Chemical Fertilizers on MV Rice by Farm Size, India, 1967–69
(kg/ha)

Location	No. of districts	Deciles 1–3 (smallest)	Deciles 4–6	Deciles 7–9	Decile 10 (largest)
Tamil Nadu	3	299	320	286	313
Karnataka	1	466	318	257	333
Orissa	2	196	359	479	399
West Bengal	3	246	386	340	516
Kerala	2	104	332	273	465
Andhra Pradesh	4	506	388	388	597
Uttar Pradesh	2	212	503	643	353
Punjab	1	0	689	701	280
Bihar	2	322	345	365	419
Maharashtra	2	103	198	195	259

Source: B. Lockwood, P. K. Mukherjee, and R. T. Shand (1971).

Table 10.7. Relative per Hectare Fertilizer Application on Rice by the Smallest and Largest Farms in Farm Surveys in Asia

Location	Relative level applied by[a] Smallest	Second smallest	Second largest	Largest	Reference
West Bengal, India	123	93	102	78	Ghosh
Orissa, India	104	104	109	80	AERC, 1967
Punjab, India	76	—	—	123	Kahlon/Singh, 1973
West Bengal, India	85	78	119	116	Mandal/Ghosh, 1976
Orissa, India	79	86	119	113	Mandal/Ghosh, 1976
India	94	89	100	125	India, NCAER, 1978
3 districts, Bangladesh	72	—	—	127	Quasem, 1978
12 villages, Bangladesh	110	—	—	89	Quasem, 1978
Bangladesh	123	109	91	85	BARC/IFDC, 1982
Punjab, Pakistan	90	103	96	109	Khan, 1975
Sind, Pakistan	92	105	116	86	Khan, 1975
Permatang Bogak, Malaysia	114	85	100	100	Bhati, 1976, p. 105
Central Java, Indonesia	113	108	71	106	Palmer, 1977
Laguna, Philippines	111	—	—	88	Kikuchi et al., 1982

[a] Shows the rate applied by each size group compared with the average for all groups in each study. Where data are shown only for smallest and largest, only three size groups were given in the original.

the rate of fertilization are reported for farms of five different sizes (table 10.8). There seems to be a slight positive association between farm size and proportion of land receiving fertilizer. The data on fertilizer applied per hectare show a somewhat different trend. In Bihar, Karnataka, Madhya Pradesh, Maharashtra, and Tamil Nadu, there was a distinct inverse relationship between farm size and rate of application, with small farms applying more than large. In Andhra Pradesh, Uttar Pradesh, and West Bengal, the opposite was true. On the average, the smallest farms that applied fertilizer used 92 kg/ha, compared with 75 kg/ha for the largest and 78 kg/ha overall.

A Bangladesh study of fertilizer use in 1979–80

Table 10.8. Percent of Rice Area Fertilized and Fertilizer Nutrient Consumption per Hectare of Rice (N, P_2O_5, K_2O) in India, 1975–76

State	Percent of area fertilized: below 1 ha	1–2	2–4	4–10	over 10 ha	kg/ha (nutrients) per fertilized area: below 1 ha	1–2	2–4	4–10	over 10 ha
Andhra Pradesh[a]	70.4	73.9	87.8	79.6	68.8	112.0	117.9	109.4	101.4	119.5
Assam[a]	1.6	8.6	5.6	3.6	67.0	108.6	43.9	58.7	48.4	8.4
Bihar[b]	27.9	47.0	61.1	46.0	48.2	54.6	51.6	37.0	36.4	31.0
Gujarat	49.1	56.0	64.9	55.8	95.3	72.1	49.1	62.4	63.4	43.9
Haryana	83.2	93.2	96.1	95.0	99.0	91.1	91.7	77.9	96.7	116.8
Himachal Pradesh	41.9	31.2	9.0	40.2	—	38.1	20.1	25.7	28.0	—
Jammu and Kashmir	77.7	72.0	82.2	61.3	—	46.1	46.7	47.8	34.6	—
Karnataka[a]	79.8	85.6	89.3	96.4	100.0	194.2	165.8	133.4	142.3	72.1
Kerala[a]	84.0	86.3	85.4	100.0	—	100.4	103.0	117.4	173.6	—
Madhya Pradesh	0.9	6.6	13.2	15.5	42.2	52.6	50.3	35.8	23.1	17.4
Maharashtra	43.1	53.2	55.9	52.8	63.1	83.5	76.5	53.5	64.3	62.6
Orissa[a]	33.2	38.4	42.6	46.0	60.9	82.2	85.1	105.9	106.5	—
Punjab	56.9	71.8	88.4	97.1	100.0	87.3	92.3	96.0	114.7	—
Rajasthan	2.1	7.6	32.4	48.2	100.0	143.3	63.1	32.0	44.4	50.5
Tamil Nadu[a]	82.2	89.0	90.3	92.2	100.0	134.7	137.4	121.1	122.6	108.2
Uttar Pradesh	21.6	30.8	44.0	55.2	28.5	47.9	46.1	39.3	38.8	71.4
West Bengal[a]	40.4	38.5	46.5	44.7	100.0	99.9	95.7	77.2	63.4	133.2
Average	46.8	52.3	58.0	60.6	76.6	91.9	78.1	70.9	75.4	75.4

Source: India, National Council of Applied Economic Research, *Fertilizer Demand Study: Interim Report.*

[a] Average of summer, winter, and autumn paddy crops.

[b] Average of autumn and winter paddy crops.

gave similar results. Table 10.9 shows the results for the boro season in which the use of fertilizer is widespread. In this study, covering farmers in sixteen districts, fertilizer use was reported by farm size and tenure category. Contrary to some expectations, fertilizer was applied to a higher percentage of cash-rented land than owner-operated land, and cash-rented land received a higher rate of fertilizer. The largest farm size group applied a slightly lower level of fertilizer per hectare than the smallest farm size group. A higher proportion of the larger owner operators and cash-rent farmers applied fertilizer, but among share tenants, the smaller farmers used more fertilizer.

The fertilizer data from various studies support the hypothesis that there is little generalizable difference in fertilizer use between large and small farmers. Despite this, there may be unequal access to other inputs that result in higher yields on the large farms. We examined the available data to see what could be learned about this issue.

The yield data from the 1967–69 study of India's Planning Commission are shown in table 10.10 for four decile size groups (ranked from largest to smallest farm area). In seven cases, the small farm group had the lowest yield, but when averaged across cases, the yields differed only by 0.2 mt/ha.

In an analysis of data from the study of thirty-six villages in six different rice growing countries referred to previously, we compared the yields of farmers with smaller than average farms with those with larger than average farms.[23] In eight villages, all in India

Table 10.9. Chemical Fertilizer Applied by Farmers in Bangladesh
(1,800 sample farmers, boro season, 1979–80)

Farm size group (ha)	Owner-operated		Share tenant		Cash-rented	
	Percent users[a]	kg/ha[b]	Percent users	kg/ha	Percent users	kg/ha
Below 0.2	56	114	68	130	84	276
0.2–0.4	61	114	64	130	84	226
0.4–1.0	64	111	56	101	77	203
1.0–2.0	66	86	55	96	75	235
Over 2.0	78	103	60	88	100	69

Source: Bangladesh Agricultural Research Council and International Fertilizer Development Center, "Agricultural Production, Fertilizer Use, and Equity Considerations." (1982).

[a] Percent of the number of sampled farmers with crops on the respective parcels applying chemical fertilizers.

[b] Average level of use computed as fertilizer used divided by land planted to crops.

Table 10.10. Paddy Yields Reported by Farmers in Four Decile Groupings, India 1967–69
(mt/ha)

Location	Crop season	Deciles 1–3 (smallest)	Deciles 4–6	Deciles 7–9	Deciles 10 (largest)
Tamil Nadu	Kharif	3.0	3.2	3.5	3.7
Tamil Nadu	Rabi	3.2	3.0	3.0	3.7
Karnataka	Kharif	1.7	2.0	2.7	2.0
Karnataka	Rabi	3.9	4.9	4.4	3.9
Orissa	Kharif	1.7	1.7	2.2	1.5
Orissa	Rabi	4.9	5.7	6.2	5.9
West Bengal	Kharif	3.5	3.5	3.5	3.7
West Bengal	Rabi	5.9	6.7	6.9	6.7
Kerala	Kharif	3.2	3.9	2.0	2.2
Kerala	Rabi	3.2	1.9	2.2	3.2
Andhra Pradesh	Kharif	3.7	3.5	3.9	3.9
Andhra Pradesh	Rabi	3.9	5.2	4.4	2.9
Uttar Pradesh	Kharif	3.2	3.2	3.5	4.4
Punjab	Kharif	—	3.9	2.9	2.9
Bihar	Kharif	1.9	2.2	2.2	1.9
Maharashtra	Kharif	1.2	1.2	1.2	1.9
Average	—	3.2	3.5	3.4	3.4

Source: B. Lockwood, P. K. Mukherjee, and R. T. Shand (1971).

and Indonesia, large farms had higher yields than small farms, but in five out of thirteen villages in Malaysia, the Philippines, and Thailand, the average yield was higher on small farms than on large.

Data from other studies in eighteen locations, summarized in table 10.11, give the same picture. Each study uses a different number of farm size groups, so for this comparison the yields of the two smallest size groups of farms are contrasted with the yields of the two largest size groups, and, again, the average yields differed no more than 0.2 mt/ha. In eight cases, the yields on the smallest farms were 0.2 mt/ha lower than yields on the largest farms, but in seven cases yields on the smallest farms exceeded yields on the largest farms, confirming that there is no clear evidence of large farms benefiting at the expense of small ones.

Labor Use with New Varieties

Because of the high and growing labor-to-land ratio in the rice-producing countries of Asia and because of the limited possibilities for nonfarm employment, it is important that increases in agricultural output be generated, as far as possible, through increasing labor inputs. Because the modern varieties respond to higher rates of fertilizer than do traditional varieties, it is sometimes assumed that they are more capital intensive. Some observers argue that the new varieties form a "package" with machinery, and because machinery replaces labor, the new technology as a whole is labor displacing.

The question that really should be asked is: Has the use of the new varieties absorbed more labor than continued use of the existing varieties would have? This question cannot be answered with finality by appeal to empirical data because it is hypothetical. Still, the comparison of labor use with the two technologies that is presented in chapter 9 is indicative of the effect of technical change. Clearly, successful development of an economy requires that people be productively employed at a wage that provides an adequate level of living. Technologies that increase the demand for labor will contribute to that goal, even if the total performance of the economy falls short. Early observers reported that use of the new varieties resulted in high labor use. Frankel found that both permanent and casual laborers find work for six or seven months a year from all sources—agricultural as well as construction and other odd jobs in the slack season. This represents an improvement over the previous years, when, with less intensive cropping, work was available only about four months of the year.[24] The empirical studies that compare labor use with modern and traditional rice varieties summarized in chapter 9 show the same tendency. The modern varieties used on average 20 percent more labor per hectare than the local varieties, with hired labor showing a greater increase than family labor.

Some claim that new varieties are part of a larger labor-displacing complex, that "modernization in the Philippines and the rest of contemporary Asia is a

Table 10.11. Paddy Yields Reported by Farmers in Various Size Groupings in Farm Management Studies in Asia

	No. of groups	Yield (mt/ha)			Source
		Second smallest	Second largest	Largest	
West Bengal, India	6	3.4	3.5	3.6	Mandal and Ghosh, 1976
Orissa, India	8	3.5	3.3	4.1	Mandal and Ghosh, 1976
Tamil Nadu, India	7	5.0	4.3	4.1	Shanmugasundram, 1973
Assam, India	4	2.7	2.6	3.2	Mukhopadhyay, 1980
Andhra Pradesh, India	4	2.6	2.5	2.4	Mukhopadhyay, 1980
Bihar, India	4	1.7	1.8	1.9	Mukhopadhyay, 1980
Karnataka, India	4	2.4	2.9	2.3	Mukhopadhyay, 1980
Tamil Nadu, India	4	2.5	3.1	2.1	Mukhopadhyay, 1980
Assam, India	7	4.8	5.2	4.8	AERC, Jorhat, 1970
Haryana, India	6	2.4	3.1	3.0	AERC, Delhi, 1970
Orissa, India	5	4.5	3.7	4.9	AERC, Visva Bharti, 1969
Punjab, India	3	—	—	1.5	Khalon and Singh, 1973
C. Java, Indonesia	7	2.8	3.2	3.1	Palmer, 1977
Punjab, Pakistan	4	2.7	2.8	3.1	Khan, 1975
Sind, Pakistan	4	2.6	2.5	2.5	Khan, 1975
C. Java, Indonesia	5	2.9	2.8	3.3	Soejono, 1976
C. Java, Indonesia	5	3.6	4.3	5.0	Soejono, 1976
Laguna, Philippines	3	—	—	3.3	Kikuchi et al., 1982
Average		3.1	3.2	3.1	

seamless web; tractors and modern varieties are politically linked. . . . It is naive to imagine that one can readily eliminate the undesirable aspect of modernization, e.g., tractorization, while retaining the desirable aspects of modernization, e.g., the modern varieties."[25] However, in many places modern varieties have become widely used without tractors, although in some places the two changes have occurred over the same period.

Data from the study of thirty-six villages show little relationship between adoption of varieties and mechanization (figure 10.2). In his review of the links between mechanization and new varieties, Duff found little evidence to indicate a strong causal relationship between adoption of MVs and mechanization, particularly tractors.[26] He did find evidence of complementarity between varieties and some types of mechanization, such as water pumping capacity, but this does not displace labor.

The Effect of New Technology at the Village Level

The empirical studies reviewed above show that, in most cases, the adoption of new technology has generated benefits to consumers, has helped small subsistence farmers who consume a large fraction of their output (where they are able to adopt the technology), has been widely used by small farmers (although in some cases they have lagged behind their larger neighbors by a season or two), and has generally resulted in an increased use of labor in production. Although these benefits are not uniformly observed by research in every location, they are typical of most sites. One cannot, however, unambiguously state how the benefits of the new technology have been spread among various groups.

The introduction of new technology into the rice production systems of Asia has changed those systems in a major way. With the traditional technology, only land, labor, and some traditional forms of capital, such as draft animals and land improvement, were used in production. The income generated from rice production was distributed to those inputs. In many cases, a landlord received the income generated by land, and the tenant operator who owned and provided the other inputs received the rest of the income. If the operator hired labor for transplanting and harvesting, that labor received some of the income. Output growth depended on the rate at which land could be improved or brought into production, or the ability to intensify labor inputs.

The introduction of irrigation and modern rice technology provided an additional source of growth by making it possible to convert inputs produced outside the agricultural sector into food. The highly fertilizer-responsive rices of the IR8 type generate more food from a given amount of land, labor, and fertilizer than the traditional rices. Short duration and nonphotoperiod-sensitive characteristics make it possible for farmers to use their land throughout the year, but require irrigation systems to deliver water to the plants when natural rainfall is scarce.

The new production possibilities created additional opportunities for distribution. Irrigation may double production, but there are no traditional rules about allocating that additional production, especially when, as is the case in most countries, the irrigation is publicly financed. New technology may further augment production, but the way new factors of production claim part of production is a new socioeconomic issue.

The changes in technology imply that the total income generated by rice production, while larger than it would otherwise have been, is also divided differently than under traditional agriculture. One result is unambiguous—a larger absolute amount of income is transferred outside the agricultural sector to pay for production inputs. Hence, by its very nature, the new technology results in a reduction in the percentage share of earnings going to traditional factors of production. The increased output and the need to pay for nonagricultural inputs lead to pressures for adjustments in the traditional proportions of output received by each input supplier. These pressures may be reflected in changes in the sharing arrangements for harvesting, the emergence of subleasing, the creation of large farms managed and cultivated by former landlords, and the fragmentation of farms into tiny units, or other phenomena.[27]

There is no uniform experience. Each village responds to changes as determined by existing conditions. The handful of empirical village studies comparing the distribution of income before and after introduction of new technology that are reviewed below show that little direct effect was generated.

The rice variety, ADT27, was introduced into Thanjavur District in Tamil Nadu, India, in 1965, and by 1970, it and other modern varieties were being widely cultivated. A careful comparison of the distribution of income between farmers and landless laborers and among farmers of various size groups showed that the distribution was highly skewed in both periods, but that it did not change between the two periods. "The landless laborers had about the same gains in relative terms as most farm operator groups. While all farm operators as a group increased their total income by 12 percent, the landless laborers increased their real income by 13%. . . . The very large farm operator group had an 18% increase in total real income. All other farm size-groups had less relative increase in total real income than the landless laborers."[28]

In another study, eight villages in irrigated areas of Central Java were studied in 1968/69 and again in 1973/74.[29] Thirty farmers were chosen in a stratified random sampling procedure that overrepresented large farmers and those participating in government programs. In 1968/69, 32 percent of the sample farmers used fertilizer and local varieties, and 38 percent used fertilizer and modern varieties. By 1973/74, all farmers in six of the villages were using fertilizer and modern varieties. The modern varieties were not suited to the other two villages, where 75 percent of the farmers were using fertilizer on the old varieties in 1973/74. The income distribution was highly skewed in the initial period, with the lowest quintile receiving 1.1 percent of the net returns from rice and the highest quintile receiving 66.3 percent. In the second period, the share of the lowest income group was 2.7 percent while that of the largest group was 61.8 percent. The author of the study concluded "that while paddy farm incomes have increased in the sample areas mainly due to extension of HYV technology, incomes also became more evenly distributed among farmers."[30]

Another study comparing two villages in the Subang regency of West Java provides useful insights into the dynamic forces at work in somewhat similar villages.[31] Survey data are available for periods before and after the introduction of modern rice varieties in each village. Both villages are located in rice-dominated areas. The South Subang village had an older history of settlement and a more cohesive structure, while the North village, which was nearer the coast, was newer, had a higher land–man ratio, and had a greater diversity of social classes, with at least one large landowner. The South Subang village had a long-established irrigation system, while the North village did not become irrigated until 1972.

The intensification of double cropping was made possible with the availability of short-duration modern varieties and irrigation. In the North Subang village, the modern varieties spread rapidly. In 1968–71, about 7 percent of farmers grew MVs, and this increased to 100 percent in 1978/79. Fertilizer application increased, and yields went from 2.3 to 3.2 mt/ha (average of wet and dry seasons). This yield improvement, along with an increase in cropping intensity from 1.5 to 2.0, resulted in an 80 percent rise in output per hectare over the decade. The South Subang village had a cropping intensity of 1.9 before the introduction of new technology, and it remained constant. In that village, farmers tried the MVs but found them unsuited to production under their conditions. By 1978, only 14 percent were growing MVs. Fertilizer was in common use even in 1968–71. Thus, over the study period, little technological change occurred in the South village, and yields increased a modest 300 kg/ha, compared with three times that increase in the North village. Table 10.12 shows the distribution of income from rice output to capital, family, and hired labor, with the residual surplus assigned to the farm operator. In the North Subang village, where the introduction of technology was successful, there were significant gains to all claimants, and in particular to hired labor. In the South Subang village, the little gain in yield and income that did occur was captured by the farm operator.

Periodic surveys by the International Rice Research Institute of two samples of farmers in the Philippines provided data on the distribution of earnings before and after the new varieties were introduced. Modern varieties were not available during the first period (1966), while 98 percent of the first sample (Laguna)

Table 10.12. Income from Rice Production and its Distribution Among Participants in the Production Process, Two Villages in West Java, Indonesia

Period	Percent area in MVs	Yield (kg/ha)	Income (kg/ha) distributed to				Operator's surplus
			Current inputs	Capital	Family labor	Hired labor	
North Subang village							
1968–71	7	2,342	151	47	117	830	1,197
1978–79	100	3,237	334	154	252	1,070	1,427
Percent change	—	38	21	70	15	29	19
South Subang village							
1968–71	n.a.	2,600	345	136	427	830	862
1978–79	14	2,956	307	125	438	863	1,223
Percent change	—	14	−11	−8	3	4	42

Note: n.a. = not available.
Source: Y. Hayami and M. Kikuchi (1981) pp. 192, 207.

Table 10.13. Average Real Earnings and Distribution Among Participants in Production, Two Samples of Rice Producers in the Philippines, 1966 and 1970

Period	Percent area in MV	Wage rate ($US/day)	Yield (kg/ha)	Income (kg/ha) distributed to			
				Landlord	Hired labor	Farmer	Current inputs
Laguna sample							
1966	0	0.61	2,374	831	570	807	166
1970	98	0.84	3,349	971	871	1,072	436
Percent change	—	48	41	17	53	29	162
Central Luzon/Laguna sample							
1966	0	0.58	2,288	664	435	961	229
1970	65	0.84	2,589	777	569	880	362
Percent change	—	37	13	17	31	−8	58

Source: C. Ranade and R. W. Herdt (1978).

and 65 percent of the second sample (Central Luzon and Laguna) grew them in 1970.[32] The data in table 10.13 summarize the income changes over the period expressed in kilograms of rice per hectare. As in the North Subang village, the successful introduction of new technology brought gains to hired labor that were greater than to other claimants, with the exception of capital or current inputs. After 1970, land reform was instituted; and irrigation from a large-scale system spread widely through the second sample, so subsequent changes have multiple causes, although the pattern and direction of change continues much as shown in table 10.13.

Variability in Adoption at the Regional Level

Most of the discussion about the benefits and biases of the new technology has dealt with the farm and village. Much less attention has been given to regional differences in rates of adoption and benefits, but it is clear that the differences across regions have been very striking. Table 10.14 clearly shows that the regional bias is closely associated with differences in systems. With the exception of Vietnam, the mainland delta regions have a low percentage of irrigated rice area and a low level of adoption of modern varieties. By contrast, the insular countries tend to have a high percentage of irrigated rice area and a high level of adoption.

There are notable exceptions to the close association between irrigated area and the adoption of new technology. In Pakistan, despite the fact that essentially all of the rice area is irrigated, less than half is planted to modern varieties. This is because the government encourages the production of high-quality, local Basmati rice. It is an important source of foreign exchange earnings, and the government adjusts farm prices to be sure that there is an adequate supply to meet the export demand.

Table 10.14. Percentage of Rice Crop Area Irrigated, and in MVs, late 1970s

	Irrigated	Modern varieties
Mainland delta regions		
Bangladesh	12	15
Burma	17	6
Thailand	14	11
Vietnam	40	—
Eastern India	27	21
Mainland regions of India		
India		
North	89	82
Central	18	34
West	23	51
South	83	66
Insular countries		
Indonesia	63	53
Malaysia	66	54
Philippines	43	70
Sri Lanka	63	68

In the Philippines, by contrast, modern varieties have been widely adopted in the rainfed areas, which suggests that there is potential for developing new technology in the less favorable growing environments. This is discussed in chapter 14.

Technological Change—Who Benefits?

It would be naïve to believe that this literature review would change the opinions of anyone who firmly believes that the technological change embodied in modern rice varieties is not in the best interests of society. However, for those who are less certain, or who believe that one should examine the impact of the technology in each place where it has been introduced, the evidence may provide food for thought. The benefits from the lowered price of rice due to new technology clearly spread to all rice consumers. There is variation in adoption and hence the direct benefits that are obtained by producers. In some areas, there has been no adoption of new technology. In a few studies, the appropriation of technology by big farmers is evident; some studies show less labor used with new than old varieties; and some studies show that landlords get the bulk of the increases in the net income. On the other hand, most studies show that the new rice technology is widely adopted by farmers irrespective of size, and that it has had rather broad benefits to laborers and farm operators as well as landlords. The main group that has not gained are farmers in areas for which the new technology is poorly suited.

It is difficult for us to conclude that either the dependency model or the induced innovation model described the entire course of events. Certainly, there are situations in which the introduction of new technology has caused increased dependency. But there are many cases in which technology, accompanied by other changes (perhaps induced, perhaps not) has resulted in greater real incomes for the poor.

We believe that it would have been difficult to ensure the needed increase in rice production without the new technology. Even appeal to the experience in China, in which institutional and sociopolitical changes were made the keystone of agricultural development, illustrates the key role of technological change. In the early 1960s, China developed its own version of semidwarf, fertilizer-responsive rices. Modern varieties spread as rapidly in China as in any country and generated a high internal demand for fertilizer. Although China uses the highest level of organic fertilizer in the world, the Chinese found it advantageous to produce chemical fertilizers, first through small-scale factories and later through the importation of large-scale, urea-ammonia complexes from the West.[33]

What is important about the Chinese experience is that broad distribution of the direct gains of new technology has been ensured through collective ownership of resources. In market-oriented economies, owners of resources receive the earnings of those resources. Scarce resources receive relatively higher returns than plentiful resources, and since land and capital are scarce, they receive higher returns. If the ownership of these resources is concentrated in a few hands, then their earnings will likewise be concentrated. Technology provides ways to substitute one resource for another, and certainly any technology that substitutes capital for labor will tend to increase the earnings of capital. Technology that generates an increased demand for labor will do the opposite.

However, the effect of resource ownership on the distribution of earnings is so great that any effect caused by technological change is marginal. In fact, econometric studies attempting to measure the direction of bias in technological change in rice have failed to show any departure from neutrality.[34] That does not say that when incomes are increased because of a technological change, all participants benefit equally. On the contrary, they benefit in proportion to their ownership of resources and the earnings of the resources. Technology in one industry, rice, has relatively little influence on earnings for resources used in many industries (labor and capital), although the relative abundance of inputs has a large effect. The important factor determining who receives the direct income benefits is the ownership of resources. One cannot expect technological innovations introduced over a period of five years to modify a pattern of resource ownership derived from hundreds of years of history. Technology is a tool; to meet society's needs, its use must be determined by society acting in its own interest.

Notes

1. Yujiro Hayami and Vernon W. Ruttan, *Agricultural Development: An International Perspective* 2d ed., rev. and updated (Baltimore, Johns Hopkins University Press, 1985).

2. Concern over the failure of farmers to adopt the new technology is reflected in the statement of the former director of the International Rice Research Institute, who wrote in 1975, "On retiring from IRRI in 1972, the only real disappointment I felt was that somehow we did not understand sufficiently why the Asian farmer who had adopted the new varieties was not doing better. Somehow, I felt that rice scientists who obtained yields of 5 to 10 metric tons per hectare on the IRRI farm still could not explain why so many Filipino farmers (for example), obtained, on the average, less than one metric ton per hectare increase in yield after shifting from traditional to high-yielding varieties. All of us were a bit mystified as to why not more than 25 percent of the rice land in the less developed countries was planted to the new varieties." Robert F. Chandler, Jr., "Case History of IRRI's Research Management During the Period 1960 to 1972" (Taiwan, Asian Vegetable Research and Development Center, 1975).

3. These contrasting positions are reflected in many of the arti-

cles contained in Robert S. Anderson, Paul R. Brass, Edwin Levy, and Barrie M. Morris, eds., *Science, Politics and the Agricultural Revolution in Asia* (Boulder, Colorado, Westview Press, 1982).

4. The development of the induced innovation model can be traced to the works of Vernon W. Ruttan, Yujiro Hayami, and Hans P. Binswanger, with perhaps its most complete formal statement in Hans P. Binswanger and Vernon W. Ruttan et al., *Induced Innovation, Technology, Institutions, and Development* (Baltimore, Johns Hopkins University Press, 1978). Yujiro Hayami and Masao Kikuchi provide a somewhat more intimate view of the process in their book, *Asian Village Economy at the Crossroads* (Tokyo, Tokyo University Press, 1981).

5. The negative effect of price distortions is discussed in a number of articles in Theodore W. Schultz, ed., *Distribution of Agricultural Incentives* (Bloomington, Ind., Indiana University Press, 1978). An argument for transferring resources to the poor is presented in Wyn T. Owen, *Two Rural Sectors: Their Characteristics and Roles in the Development Process* (Ottawa, International Development Research Center, 1971).

6. John L. Dillon, "Board Structural Review of the Small-Farmer Technology Problems" in Alberto Valdez, Grant M. Scobie, and John L. Dillon, eds., *Economics and the Design of Small-Farmer Technology* (Ames, Iowa, Iowa State University Press, 1979).

7. A set of maps of the rice environments of Asia is in Robert E. Huke, *Rice Area by Type of Culture: Southeast and East Asia* (Los Banos, Philippines, International Rice Research Institute, 1982). The effort to focus biological research attention on the less favored areas is reflected in a number of IRRI conferences and publications including *Rainfed Lowland Rice* (Los Banos, Philippines, International Rice Research Institute, 1979), and *International Deepwater Rice Workshop* (Los Banos, Philippines, International Rice Research Institute, 1979).

8. An excellent description of research based on farming systems is contained in H. G. Zandstra, E. C. Price, J. A. Litsinger, and R. A. Morris, *A Methodology for On-Farm Cropping Systems Research* (Los Banos, Philippines, International Rice Research Institute, 1981).

9. D. G. Dalrymple, *Measuring the Green Revolution: The Impact of Research on Wheat and Rice Production*, Foreign Agricultural Economic Report No. 106 (Washington, D.C., USDA, 1975).

10. The yield-reducing effect on poor dry season water control is illustrated in chapters 7 and 9.

11. R. E. Evenson and P. Flores, "Social Returns to Rice Research," in *Economic Consequences of the New Rice Technology* (Los Banos, Philippines, International Rice Research Institute, 1978).

12. R. E. Evenson and P. Flores, "Social Returns to Rice Research."

13. Grant M. Scobie and Rafael T. Posada, "The Impact of Technical Change on Income and Distribution: The Case of Rice in Colombia," *American Journal of Agricultural Economics* vol. 60, no. 1 (February 1978) pp. 85–92.

14. Yujiro Hayami and Robert W. Herdt, "Market Price Effects of Technological Change on Income Distribution in Semi-Subsistence Agriculture," *American Journal of Agricultural Economics* vol. 59, no. 2 (May 1977) pp. 245–256.

15. Carl Gotsch, "Technological Change and the Distribution of Income in Rural Areas," *American Journal of Agricultural Economics* vol. 54, no. 2 (May 1972) pp. 326–341.

16. F. Frankel, *India's Green Revolution: Economic Gains and Political Costs* (Princeton, N.J., Princeton University Press, 1971).

17. Ibid.

18. B. Lockwood, P. K. Mukherjee, and R. T. Shand, *The High-Yielding Varieties Programme in India*, Part I (Canberra, Planning Commission, Government of India and the Australian National University, 1971).

19. M. Schluter, "Differential Rates of Adoption of the New Seed Varieties in India: The Problem of the Small Farm," USAID Research Project, Occasional Paper No. 47 (Ithaca, N.Y., Department of Agricultural Economics, Cornell University, 1971) p. 37.

20. India, National Council of Applied Economic Research, *Fertilizer Demand Study: Interim Report* (in 6 volumes) (New Delhi, National Council of Applied Economic Research, 1978).

21. Unfortunately, the results published at the time of this writing do not include data on MVs; however, the researchers made those results available to us, and we thank Dr. Surjit Sidhu for his willingness to share the information. The study is described and many results presented in Bangladesh Agriculture Research Council and International Fertilizer Development Center, "Agricultural Production, Fertilizer Use, and Equity Considerations—Results and Analysis of Farm Survey Data, 1979/80, Bangladesh" (Muscle Shoals, Alabama, International Fertilizer Development Center, 1982).

22. B. Lockwood, P. K. Mukherjee, and R. T. Shand, *The High-Yielding Varieties Programme in India*.

23. G. C. Mandal and M. G. Ghosh, *Economics of the Green Revolution: A Study in East India* (New Delhi, Asia Publishers, 1976).

24. F. Frankel, *India's Green Revolution: Economic Gains and Political Costs*, p. 177.

25. K. Griffen, "Comments on Labor Utilization in Rice Production," in *Economic Consequences of the New Rice Technology* (Los Banos, Philippines, International Rice Research Institute, 1978) p. 140.

26. Bart Duff, "Mechanization and Use of Modern Varieties," in *Economic Consequences of the New Rice Technology* (Los Banos, Philippines, International Rice Research Institute, 1978).

27. The changes brought about by population growth interacting with technological change are discussed and illustrated by Y. Hayami and M. Kikuchi, *Asian Village Economy at the Crossroads*.

28. C. G. Swenson, "The Distribution of Benefits from Increased Rice Production in Thanjavur District, South India," *Indian Journal of Agricultural Economics* vol. 31 (January–March 1976) pp. 1–12.

29. I. Soejono, "Growth and Distributional Change in Income in Paddy Farms in Central Java, 1968–74," *Bulletin of Indonesian Economic Studies* vol. 12, no. 2 (July 1976) pp. 80–89.

30. Ibid.

31. Y. Hayami and M. Kikuchi, *Asian Village Economy at the Crossroads*.

32. Chandra G. Ranade and Robert W. Herdt, "Shares of Farm Earnings from Rice Production," in *Economic Consequences of the New Rice Technology* (Los Banos, Philippines, International Rice Research Institute, 1978) pp. 87–104.

33. Anthony M. Tang and Bruce Stone, "Food Production in the People's Republic of China," IFPRI Research Report No. 15 (Washington, D.C., International Food Policy Research Institute, 1980).

34. C. Ranade and R. W. Herdt, "Shares of Farm Earnings from Rice Production"; Pranab K. Bardhan, "Size, Productivity, and Returns to Scale: An Analysis of Farm-Land Data in Indian Agriculture," *Journal of Political Economy* vol. 81 (July–December 1973) pp. 1370–1386.

11

Rice Consumption Patterns

Cereal grains account for two-thirds of the calories in the average Asian diet, with rice alone providing 40 percent of the total and wheat another 15 percent. Other cereal grains such as maize, millets, and barley are important in some parts of Asia, as are several noncereal staples, primarily cassava and sweet potatoes.

Per capita consumption of cereal grains has been rising in the developing countries of South and Southeast Asia, while in the more advanced economies of East Asia, cereal consumption has reached a plateau, and in Japan it is declining. Since the end of World War II, increasing investment in irrigation, some growth in rice land, and the adoption of new cereal grain technology helped to increase yields and production and keep prices low. The question of whether this trend can continue depends on sustained improvements in cereal production and mediated future increases in demand for cereals.

Trends in Cereal Grain Consumption

Patterns of cereal grain consumption differ throughout Asia, but as shown in figure 11.1, rice and wheat are clearly dominant. With the exception of Bangladesh, the rice-dependent countries were historically the leading world source of rice exports, although a number of factors have caused a reduction in surpluses (chapter 13). Rice in this area is produced in Asia's major river deltas including the Ganges-Brahmaputra in Bangladesh, the Irrawaddy in Burma, the Chao Phraya in Thailand, and the Mekong in Laos, Kampuchea, and Vietnam.[1] Wheat has been relatively unimportant, although recently Bangladesh and Vietnam have imported more wheat under foreign aid agreements, and domestic wheat production in Bangladesh has grown at a remarkable rate over the past two decades (from 36,000 metric tons in 1961–65 to 681,000 metric tons in 1978–80).

Countries in the rice-wheat producer group consume significant quantities of both cereal grains. With the exception of Pakistan, rice is still the dominant foodgrain. The third group of countries, rice-wheat importers, includes insular (peninsular) Asia from Japan and South Korea in East Asia to Sri Lanka in South Asia. Although wheat is not grown there in significant quantities, it is gradually gaining in importance relative to rice, particularly in the more affluent East Asian countries. Wheat needs in these countries are met almost exclusively by imports.

Because wheat is an important substitute for rice in some areas of Asia, one cannot examine rice consumption separately from that of wheat, at least in the aggregate. Imports will also be considered because they are an important caloric source in many Asian diets (see chapter 13 for an expanded discussion of trade).

Accurate data on the quantity of rice consumed by people in various countries are more elusive than one might think. Total national consumption does not equal total national production because rice is traded, stored, fed to livestock, and used for seed. Most countries do not have annual consumption surveys, and even when surveys are taken, they do not always account for seasonal variations or reflect

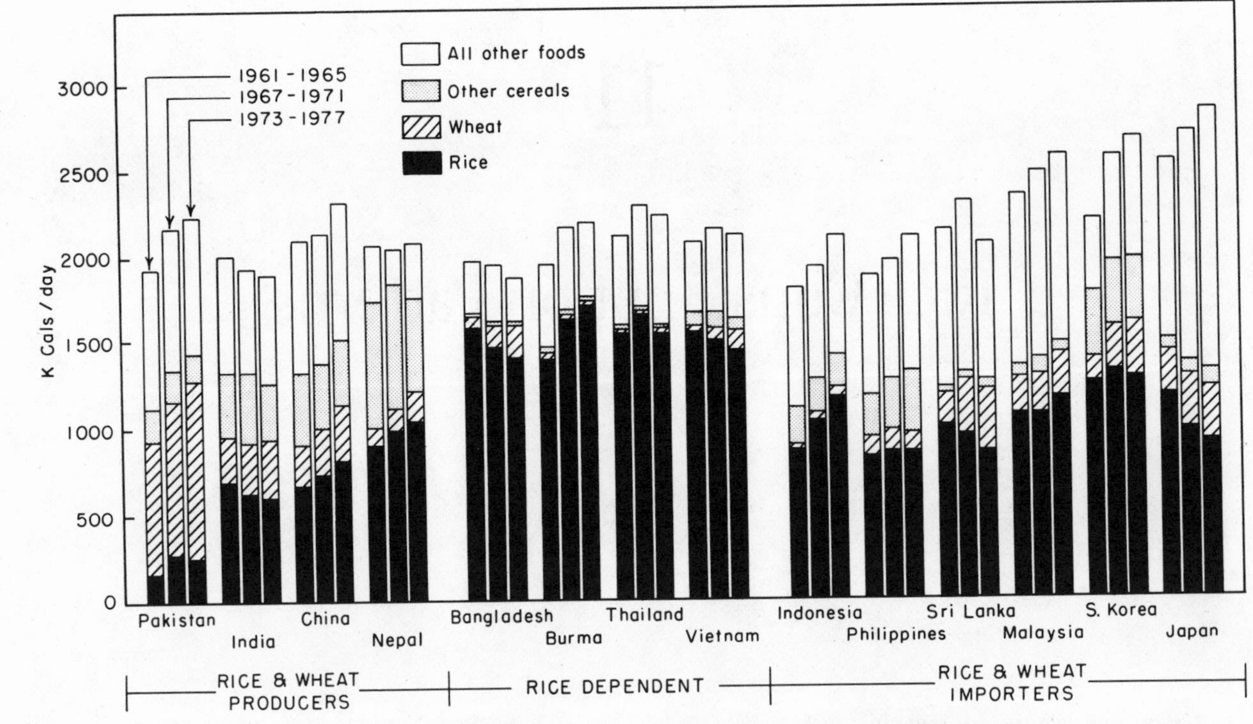

Figure 11.1. Apparent per capita daily calorie consumption, 1961–65, 1967–71, and 1973–77

waste. The method most often used to determine changes in average consumption over time is the food balance sheet. Supplementing that, various kinds of consumption surveys are used to obtain more accurate consumption estimates for subsectors of the population, or to measure relationships between consumption and income or between health and food consumption patterns.

Food Balance Sheets

All countries annually collect production data for rice and other major crops as well as for international trade. This information, together with population estimates, changes in storage grain, and amounts used for nonfood purposes, allows us to estimate net availability of commodities using the formula:

$$\text{per capita availability} = \frac{\text{production} + \text{imports} - \text{exports} - \text{feed} - \text{seed} - \text{waste} - \text{gains in stocks}}{\text{population}}$$

Data to calculate per capita availability estimates suffer from two primary problems—stock and feed components. Most countries have only vague data on the total amount of rice in storage at any time, and most storage data omit private stocks and hence underestimate the total. If there are significant changes in private stocks during a particular year, the food balance sheet probably will not reflect these changes. Many countries also do not have good estimates of the amount of rice consumed by livestock. Fortunately, the amount of rice fed to livestock is quite negligible in the lowest income countries, and in the higher income countries, where rates are somewhat higher, data are more comprehensive. By taking an average of several years' data, the effect of fluctuations in stocks from one year to another can be smoothed out. Finally, if one is interested in changes in consumption levels over time, and if livestock uses are not significantly different in each of two periods, the difference in estimated consumption should reflect changes that have occurred.

The Food and Agriculture Organization's (FAO's) estimates of consumption for the various regions of the world emphasize the degree to which rice production and consumption are concentrated in Asia.[2] In 1972–74, worldwide rice consumption averaged 76 kilograms per person (rough rice), but in the developed countries per capita consumption was only 5 kg/person/year. In Latin America, consumption was about five times as large; however, it was still only one-fourth the average 100 kg/person/year consumption in Asia.

The availability of more than 2,000 calories/capita/

day in most countries is not far below the FAO's estimated requirements of 2,200 for Asia (figure 11.1). Nonetheless, averages underestimate the seriousness of the food problem in most countries because food availability is skewed in line with income distribution. A large proportion of low-income Asian households have levels of food intake far below the national averages. Food self-sufficiency has little meaning if a large portion of the population cannot afford an adequate diet.

Consumption Trends by Country

The contributions of rice, wheat, other cereals, and other foods to total caloric consumption for the three groupings of countries already discussed are shown in table 11.1. The data reflect largely stagnant consumption levels in the rice-wheat and rice-dependent areas, while the rice-wheat importers of Southeast and East Asia show distinct upward trends in food availability. Table 11.2 shows wheat and rice consumption for each country separately.

Table 11.1. Daily per Capita Calorie Supply from Cereals and Cereal Substitutes, 1961–65 to 1973–77

	Kcal/capita/day[a]		
	1961–65	1967–71	1973–77
	Rice-wheat producers[b]		
Rice	664	678	702
Wheat	256	293	360
Other cereals	381	357	337
Roots and tubers	169	152	138
Total of above	1,472	1,478	1,626
All food	2,054	2,055	2,143
	Rice-dependent[c]		
Rice	1,532	1,536	1,486
Wheat	39	66	103
Other cereals	26	29	26
Roots and tubers	50	53	58
Total of above	1,647	1,683	1,673
All food	2,023	2,103	2,058
	Rice-wheat importers[d]		
Rice	1,014	1,022	1,052
Wheat	125	163	175
Other cereals	180	178	185
Roots and tubers	184	154	133
Total of above	1,503	1,525	1,136
All food	2,127	2,287	2,434

Source: Food and Agriculture Organization of the United Nations, *Food Balance Sheets 1975–77 and Per Caput Food Supplies, 1967 to 1977.*

[a] National averages are weighted by national 1970 populations, so the data reflect regional per capita availabilities rather than simple averages of the national numbers in table 11.2.

[b] Includes Pakistan, China, India, Sri Lanka, and Nepal, with a total population of 1,398 million in 1970.

[c] Includes Bangladesh, Burma, Thailand, and Vietnam with a total population of 173 million in 1970.

[d] Includes Indonesia, Philippines, Malaysia (Peninsular), South Korea, and Japan, with a total population of 301 million in 1970.

The rice-wheat importing countries increased consumption by about 300 calories per capita between 1961 and 1977. In all five countries, in the third period average per capita consumption levels exceeded 2,000 calories (figure 11.1). In most of the countries, rice provided a substantial share of the additional calories, but in Japan, rice consumption declined precipitously throughout the period because of a rapidly rising average income level. Interestingly, although it includes the highest income countries, this group consumes roots and tubers at as high a level as the lower income rice-wheat producer group, an indication of dietetic diversity. Sri Lankan consumption levels were especially erratic, with a sharp drop in the late 1970s.

Caloric intake in the rice-wheat producer countries increased by about 90 calories per day over the period, mainly because of improvements in China's consumption levels. In China, wheat consumption increased by 104 calories per day while that of roots and tubers decreased by about 30. India and Bangladesh both had a decline of about 5 percent in average caloric consumption, largely caused by a drop in the number of rice calories consumed.

Table 11.2 shows per capita availability of rice and wheat for individual countries.[3] There are four separate patterns of change. In three of the lowest income countries (China, Indonesia, and Nepal), per capita consumption of both rice and wheat appears to have risen. In five other countries (India, Sri Lanka, Bangladesh, Vietnam, and Japan), per capita consumption of wheat has risen and per capita consumption of rice has fallen, but for different reasons. In India and Sri Lanka, price and trade policy have shifted in favor of wheat over rice. In Bangladesh and Vietnam, wheat imports, largely in the form of food aid, have made up for the shortage of rice. In Japan, shifts in consumption have been strongly influenced by changes in consumer preference toward wheat products because of rising incomes.

The per capita consumption of wheat has been rising while the consumption of rice has remained fairly steady in four countries (South Korea, Malaysia, Pakistan, and the Philippines). Finally, in two traditional rice-exporting countries (Burma and Thailand), per capita rice consumption has been steady to rising, but wheat consumption is so small as to be insignificant. Overall, there has been little, if any, significant change in the total caloric consumption among Asian countries over the period. The exception is the higher income countries of East Asia where the consumption of staples has declined.

Table 11.2. Apparent per Capita Daily Caloric Consumption of Wheat and Rice in Asia, 1961–65, 1967–71, and 1973–77
(Kcal)

Region/country	Wheat			Rice		
	1961–65	1967–71	1973–77	1961–65	1967–71	1973–77
Rice-wheat producers						
India	247	284	321	696	627	603
Nepal	94	123	181	905	987	1,037
Pakistan	758	889	1,021	176	265	250
China[a]	228	256	339	672	737	800
Rice-dependent						
Bangladesh	71	111	180	1,567	1,471	1,403
Burma	20	15	16	1,400	1,617	1,692
Thailand	9	13	17	1,547	1,645	1,543
Vietnam	27	94	113	1,552	1,495	1,434
Rice-wheat importers						
Indonesia	4	28	44	876	1,039	1,174
Japan	263	302	308	1,162	979	909
South Korea	136	259	308	1,244	1,316	1,278
West Malaysia	214	228	248	1,063	1,063	1,161
Philippines	94	120	103	835	852	851
Sri Lanka	181	311	357	1,011	958	858

Source: Food and Agriculture Organization of the United Nations, *Food Balance Sheets 1975–77 and Per Caput Food Supplies 1967 to 1977.*

[a] Includes Taiwan.

Factors Affecting Rice Consumption

Three main factors determine the level of rice consumption of any group at any time: tastes and preferences, incomes, and the price of rice relative to the price of substitutes. These three factors are interactive and reflect a satellite of contributing elements. Tastes and preferences of urban people are generally distinct from those of rural people because of differences in life-style and level of physical activity. The tastes of one ethnic group within a country may differ from those of another group. Changes in the price of rice will have an effect on consumption because rice becomes more or less expensive relative to other alternatives. Where rice absorbs a large fraction of total expenditures, a change in its price has substantial effects on real income.

The effect of preferences, income (as in the previous two sections), and prices can be examined by comparing aggregate country consumption data or by studying consumption patterns of various groups of consumers within countries. The national consumption estimates are normally based on the food balance sheet approach, described earlier, which estimates rice available for consumption using production, stocks, and trade data. The data for particular groups are usually obtained through surveys carried out by nutritionists or social scientists. Nutritionists often conduct intensive surveys in which they may stay with families for up to a week, weighing portions of raw food prior to its preparation. Social scientists, on the other hand, may ask respondents what they consumed during a reference period (usually the previous week). Food balance sheets and consumption surveys obviously have a distinct potential for bias, and it is not surprising that where the results for identical populations can be directly compared, they show variant levels of consumption. Thus, the absolute levels of consumption reported here vary somewhat from those given earlier.

Another source of confusion about the relationship of consumption to various factors arises because some consumption data are obtained from expenditure surveys that treat households as units of observation. Typically, data from such studies obtain the household income, the amount spent for rice and various other commodities, and the quantity of commodities consumed by the household. Casual examination of such data may lead to erroneous conclusions. For example, 1978 Japanese data show that the higher the household income, the more the household spends on rice. The reason for this is not, however, that higher income earners consume more per person, as is shown by the per capita data. Rather, high-income households have more members and pay a higher rice price because they buy better quality rice. In Thailand, household *expenditures* for rice are the same in Bangkok as in other urban areas, but per capita *quantity* consumed outside Bangkok is much higher, reflecting a lower price paid by urban people outside Bangkok. Village-level consumption per capita is almost twice that of Bangkok, while expenditures on rice per household are only 20 percent higher in villages than in Bangkok, reflecting both price and household size differences.

Consumption Patterns

Real personal income is clearly linked to levels and types of cereal grain consumption. In the 1940s, Wickizer and Bennett prepared a graphic illustration of this relationship based on hypothetical data (figure 11.2)[4] They suggested that as incomes rise, grain consumption will also increase to a maximum of about 1,700 kcal/day. As consumption levels improve, preferred cereals, chiefly rice and sometimes wheat, will be substituted for inferior cereals or roots and tubers. Beyond the 1,700 calorie level, grain consumption would begin to decline in favor of preferred noncereal items—fruits, vegetables, and animal products.

Wickizer and Bennett's model fits quite well with most country data for consumption. Figure 11.3 shows per capita consumption of rice and other cereals in Japan. Japan is chosen for analysis because consumption trends reflect a long-term change in dietary composition from a grain base to a diet similar to that consumed in other developed countries. The data are also very complete and quite detailed. The long-term series for urban groups shows a steady increase in rice consumption from 1875 to about 1925 and a decrease in consumption of other cereals beginning about 1895. Over the period, per capita disposable incomes (as represented by total consumption) rose about 1.4 percent per year. Real incomes fell drastically during World War II, and rice was in extremely short supply, reducing rice consumption levels in 1947 to half the prewar level and pushing up the consumption of other cereals far above their 1935 levels. As postwar economic development proceeded, the consumption of rice and rice products increased to about 110 kg/capita/year in 1956. Thereafter, as disposable incomes grew, the consumption of noncereal foods grew at the expense of rice. From about 1900 to the 1930s, rice was gradually substituted for the less-preferred barley, millets, and root crops. The substitution of other cereals for rice during World War II reflects a push back to a more primitive consumption pattern forced by war-caused deprivation. However, the post-1960s substitution reflects a reduction in rice consumption because of an increase in more preferred commodities. Wheat consumption increased slightly over the period, but the most pronounced change was the rapid decline in total cereal consumption beginning about 1950.

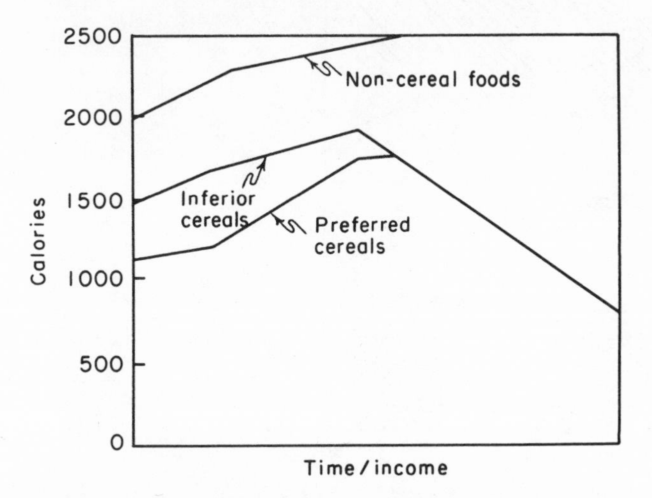

Figure 11.2. Hypothetical effect of a long-term increase in level of per capita real income on the composition of Asian diets (Source: Adapted from Wickizer and Bennett, *The Rice Economy of Monsoon Asia*)

Analysis of household consumption patterns in two Southeast Asian countries reveals the extent to which rice is substituted for other staple foods as incomes increase, even in locations where an alternative is considered the major staple. Figure 11.4 shows calories contributed by four foods in households of varying income levels in the Eastern Visayas in the Philippines, a "maize-eating" area.[5] The tendency for maize and root crop consumption to decline in favor of rice and other foods as incomes rise is striking. Total consumption also increases up to nearly the top income group. A similar tendency is evident in figure 11.5 for a maize-consuming area in East Java, Indonesia.[6] These data illustrate the pattern that Wickizer and Bennett hypothesized would hold as incomes increased, whether over time or across income groups at a given time.

Poleman has proposed that income-consumption behavior, if properly monitored to delineate the quantity and quality of changes, may contain sufficient information to develop a better understanding of the nutritional problem.[7] His argument is that the point at which consumers begin to substitute quality for quantity in other diets would reflect a threshold of dietary adequacy as perceived by the consumer. This might be a more adequate yardstick for measuring the degree of malnourishment than a threshold based upon nutritional standards. It is the view of many observers that the standards have been set too high and tend to exaggerate the nutritional problem.

As far as Poleman's behavioral hypothesis is

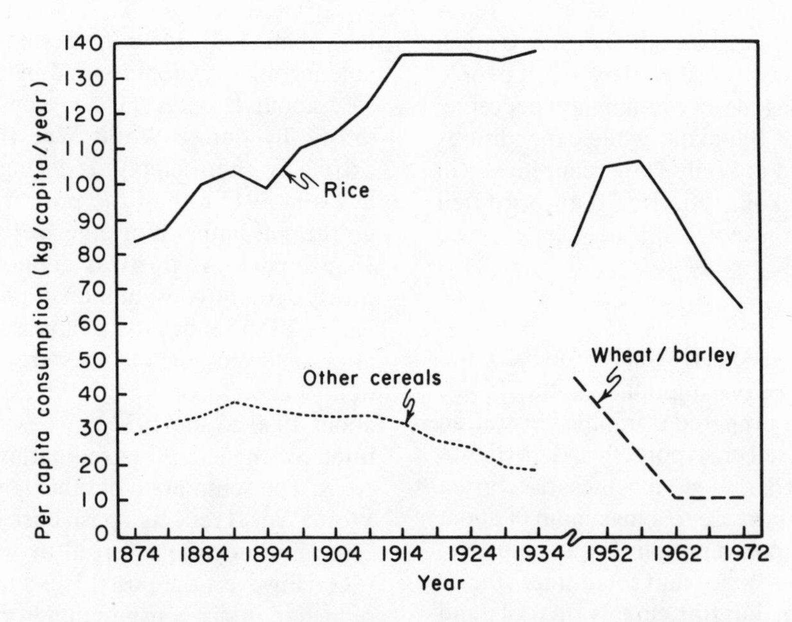

Figure 11.3. Per capita consumption of rice and other cereals by urban people, Japan (Sources: 1874 to 1934 from K. Ohkawa, M. Shinohara, and M. Umemura, eds., *Estimates of Long-Term Economic Statistics of Japan Since 1868;* post–World War II data from Japan, Statistics Bureau, *Annual Report on the Family Income and Expenditure Survey*)

concerned, it is clear that for the Philippines and Indonesia, the substitution of quality for quantity begins at the lowest income group. For the lowest income group in the Eastern Visayas, Philippines (below 100 pesos/capita/year), the caloric consumption level per person per day was 1,735. For the lowest income group in the Indonesian maize area (below Rp 60/person/day), the caloric consumption level was 1,541.

These "threshold" levels of substitution are

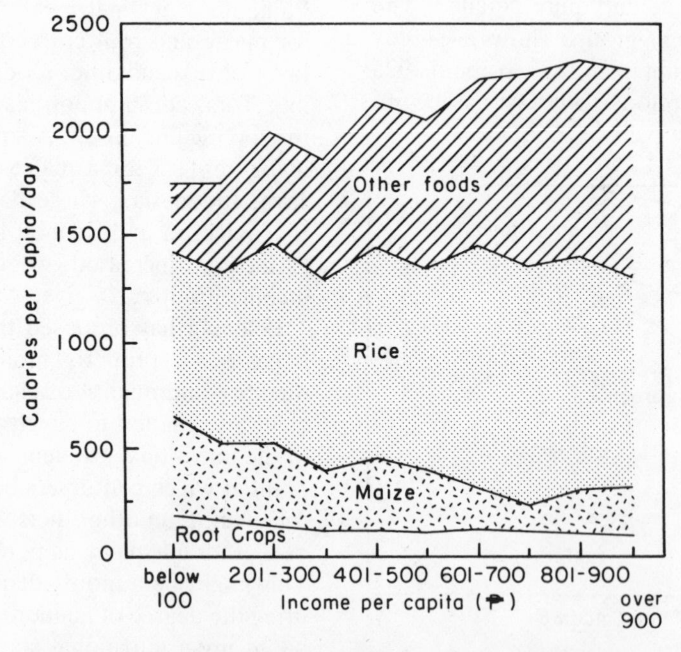

Figure 11.4. Average daily per capita calorie intake by income group, Eastern Visayas, Philippines, 1973–76 (Source: Adapted from Cynthia L. G. Santos, "Identifying Nutritionally Vulnerable Households in the Philippines," pp. 58 and 65)

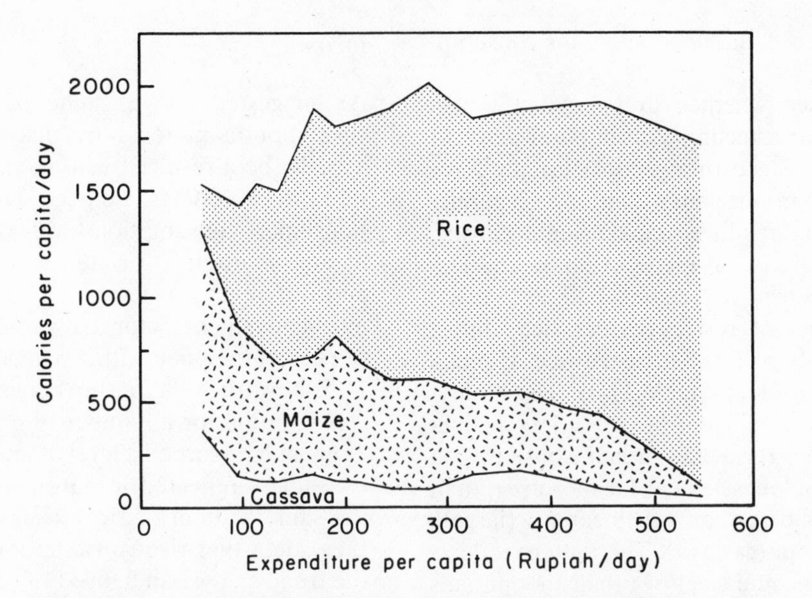

Figure 11.5. Average daily per capita calorie intake from three staple foods by expenditure class, maize-consuming area, East Java, Indonesia, 1978 (Source: Adapted from P. Surbakti, "Identifying the Nutritionally Vulnerable Urban and Rural Groups in Indonesia," p. 102)

considerably below the recent standards of nutritional adequacy used by agencies such as the Food and Agriculture Organization (FAO), the U.S. Department of Agriculture (USDA), and the World Bank in determining the percentage of the world's population that is malnourished. (These estimates have ranged from 2,200 to 2,700 kCal/capita/day.) These estimates are somewhat above the 1,500 calorie level that results when applying the formula, 1.2 × basal metabolic rate.[8]

In summary, the results of recent surveys suggest that, particularly in those areas where rice is not the predominant crop, low-income households have typically consumed large quantities of cheaper staples such as maize and root crops. Maize and cassava are typically grown in rainfed areas that have not significantly benefited from the introduction of modern technology and hence have shown very little growth in productivity and income. However, as incomes rise in these areas, there is a strong income elasticity of demand for rice.

Preference

National consumption data reflect the effect of tastes and preferences. People in some countries eat mostly rice, in other countries they eat both rice and wheat, while in others they may eat mostly rice and maize. Within-country differences may be as great or greater than between-country differences. In China and India, people in the north consume mainly wheat, maize, and other cereals like millets. In the southern rice-producing areas, they eat mainly rice, some other cereals, and very little wheat (table 11.3). In the Visayan Islands of the Philippines, people consume far more maize than rice, but in the central part of Luzon, maize is insignificant, wheat is relatively unimportant, and rice dominates. Migrants usually carry their food preferences when they move from one country to another, but after a few generations,

Table 11.3. Consumption of Rice, Wheat, and Other Cereals by Subgroups

	Consumption (kg/capita/yr)		
Country, subgroup	Rice	Wheat	Other cereals
India (rural)			
Punjab (north)	12	130	37
Tamil Nadu (south)	126	1	49
Philippines			
Central Luzon	135	7	1
Central Visayas	44	10	84
Malaysia			
Malays	120	15	[a]
Chinese	104	8	—
Indians	109	16	—

Sources: India: India, National Sample Survey Organization, *Tables on Consumer Expenditures, Twenty-eighth Round, October 1973–June 1979* (1977). Philippines: E. F. Aviguetero, F. V. San Antonio, I. G. Serrano, H. A. del Castillo and C. K. Cabilangan (1976). Malaysia: Malaysia, Department of Statistics, *Household Budget Survey of the Federation of Malaya 1957–58.*

[a] Amount was so small that data were not separately collected.

local food preferences influence migrant consumption patterns.

Place of residence, whether rural or urban, is also an important factor affecting consumption patterns in most countries. Since rural people usually have lower incomes than urbanites and are generally engaged in farming or other time-consuming work, they are more likely to eat meals at home and less likely to purchase expensive prepared foods like bread.

Thais, regardless of residence, eat rice almost exclusively, but village residents eat twice as much of it as Bangkok residents (table 11.4). The Malaysian diet is more diversified, but with a less distinct difference between rural and urban consumption patterns. Urban Indonesians eat far more rice than rural residents, who rely on other cereals (largely maize) and root crops (cassava). This pattern reflects the much higher income level of urban Indonesians and the effect of the Indonesian government policy of sale and distribution of rice at a controlled price in the cities. Consumption levels of all cereals are lower in urban Punjab than in rural Punjab, but in Tamil, Nadu, urban and rural residents consume the same level of rice, while rural residents eat far more other types of cereal.

Table 11.4. Per Capita Rural and Urban Consumption of Rice and Other Cereals

	Consumption (kg/capita/yr)			
Country	Rice	Wheat	Other cereals	Root crops
Thailand				
Bangkok	98			
Other urban areas	134			
Villages	178			
Malaysia				
Urban Malays	98	18	—	9
Rural Malays	129	9	—	7
Urban Chinese	92	8	—	6
Rural Chinese	108	21	—	5
Indonesia				
Urban	115	—	13	19
Rural	77	—	93	48
Punjab, India				
Urban	8	117	7	0
Rural	12	130	37	0
Tamil Nadu, India				
Urban	127	2	8	6
Rural	126	1	49	4

Sources: Thailand: Thailand, National Statistical Office, *Household Expenditures Survey BE 2506 (1962)*. Malaysia: Malaysia, Department of Statistics, *Household Budget Survey of the Federation of Malaya 1957–58*. India: India, National Sample Survey Organization, *Tables on Consumer Expenditures, Twenty-eighth Round October 1973–June 1974*. Indonesia: Indonesia, Biro Pusat Statistik, *Survey Social Ekonomi Nasional October 1969–April 1970*, vol. 4-5.

Income

As suggested above, some of the differences in consumption patterns by place of residence may actually be a result of income differences. To separate the two effects, we need data showing the relationship of consumption levels to incomes. Because income is often associated with place of residence and differences in tastes and preferences, it is the most convenient factor to use in differentiating levels of rice consumption within populations in Asia. Table 11.5 shows the relationship between incomes and rice consumption for a number of groups.

Japan has reached a level of income where all urban consumers reflected in national statistics eat about the same amount of rice. There is no indication in these data that rice consumption will increase, and the trend analysis in figure 11.3 suggests, if anything, that consumption will decline. The data from South Korea and Sri Lanka show increasing consumption up to the 3rd and 4th income group, with a reduction beyond that for the highest group. In Malaysia, the lowest income level had the highest consumption level, with constant rice consumption by the other three groups. Rice consumption in the other group is, to a greater or lesser extent, positively related to incomes. The Philippine data in figure 11.4 and the Indonesian data in figure 11.5 reflect similar patterns.

Cross-country studies The relationship between changes in income and cereal grain consumption is shown for selected Asian countries for three periods from 1963 to 1974 in table 11.6. Some higher income countries (Japan, Malaysia) had rapid per capita income increases but relatively constant or declining per capita grain consumption during the period. In Pakistan and Indonesia, both per capita income and consumption rose rapidly, but per capita incomes in 1965 were substantially below the first group, which showed no increase in grain consumption. A number of countries had only slowly changing per capita income and cereal consumption. Some of the poorest countries in South Asia had periods of declining per capita grain consumption, along with very slowly growing or declining incomes.

One can plot the data from table 11.5 to show that as incomes rise to about $US300/capita, direct grain consumption increases sharply. Beyond that, grain consumption levels off, and at about $US800/capita the level of grain consumption declines. There are large differences in the absolute level of consumption among countries so the specific pattern is unique for each country. This is caused, in part, by differences in tastes. However, even more important than this

Table 11.5. Per Capita Rice Consumption by Income Level in Asia

Country	Year	Per capita rice consumption by income (kg/capita/yr) Lowest	2nd	3rd	4th	Highest	References
Southeast and East Asia							
Japan, urban	1978	49	44	47	49	47	Japan, Statistics Bureau, 1979
South Korea, Suwon	1967	166	186	195	199	170	Shim, 1968
Philippines	1974–79	95	102	—	110	114	Aviguetero, et al., 1978
Malaysia	1957–58	153	130	—	130	125	Malaysia, Dept. of Statistics, n.d.
Indonesia	1969–70	22	72	111	138	158	Indonesia, Biro Pusat Statistik, 1970
South Asia							
Bangladesh, rural	1975–76	182	186	198	214	210	Bangladesh, Inst. of Nutrition and Food Service, 1977
India, rural	1973–74	71	76	87	98	105	India, National Sample
India, urban	1973–74	53	67	69	72	68	India, Survey Org., 1977
Sri Lanka	1969–70	89	99	105	104	97	Perera, et al., 1972

Notes: Bangladesh: Groups are those with 0–24; 25–49; 50–74; 75–99; 100; and over 100 taka/capita/month in 1975–76. Data are for all cereals by income groups, not rice by income groups. Rice makes up 95 percent of total cereal consumption.

India: Rural groups had 0–34; 34–43; 43–55; 55–75; and over 75 Rs/capita/month total expenditures and represent 23 percent, 19 percent, 20 percent, 19 percent, and 19 percent of households, respectively. Urban groups had 0–43; 43–55; 55–75; 75–100; and over 100 Rs/capita/month total expenditures and represent 21 percent, 16 percent, 21 percent, 16 percent, and 26 percent of households, respectively.

Indonesia: Income groups are those with 0–500; 501–1,000; 1,001–1,500; 1,501–2,500; 2,501; and over 2,501 rupiah/capita/month.

Japan: Data are grouped by quintiles.

Malaysia: Data are averages of separate tables showing consumption by residence and ethnic groups for household income levels of 1–150; 151–300; 301–500; and 501–1,000 $M/month.

Philippines: Data are for per capita income groups of less than 400; 4–799; 8–1,499; 1,500; and over 1,500 peso/capita/year.

South Korea: Data are for households with less than 6,000; 6–9,999; 10–19,999; 20–29,999; 30,000; and over 30,000 won/household/month.

Sri Lanka: Groups had less than 200; 200–399; 400–799; 800–999; 1,000 and over Rs/household/month.

are the differences in the way the data are collected and reported among countries. For example, in the Philippines, there is strong evidence that data from the food balance sheets, based on farm interviews, underestimate actual consumption by as much as 20 percent. In Thailand, the shift to crop cutting in 1969 resulted in a 15 percent increase in estimated crop production.

The relationship between changes in rice consumption and income are measured by the income elasticity of demand—the percentage change in rice demand that results from a 1 percent change in income. The income elasticity is the central parameter for understanding how demand for a commodity will change over the long run. It is clear that cereal demand expands very sharply with income growth at low income levels, but then is constant and eventually declines slowly. One, therefore, forecasts different income elasticities for different countries and expects them to change over time.

A U.S. Department of Agriculture study reports both price and income elasticities of demand (table 11.7)[9] for rice and wheat for three groups of countries in Asia. Income elasticities of demand for wheat are slightly higher than those for rice. This reflects the expectation that because rice is the principal staple in most of the area, wheat will tend to enter the diet in larger quantities as incomes rise.

An econometric analysis was made of the consumption data in table 11.6 to obtain the estimated income elasticities of demand shown in table 11.8. All the data were pooled, and logarithm (ln) of per capita consumption of cereals was made a function of the logarithm per capita income using the following equation:

$$\ln Q = a + b \ \ln Y + c\,(\ln Y)^2$$

where Q is grain consumption and Y is gross domestic product per capita. The equation used in the analysis makes the income elasticity itself a function of the level of per capita income. Thus, although only one equation was estimated, it gave all the income elasticities of demand shown in table 11.8 using the following calculation:

$$E = b + 2\,c \ln Y$$

The estimated income elasticities for 1980 cover a range from 0.23 to −0.25. Because these elasticities are for total grains, they may under- or over-estimate

Table 11.6. Growth in per Capita Grain Consumption and Gross Domestic Product per Capita at 1975 Constant Prices, 1963–74[a]

	Gross domestic product ($US/capita)					Grain[b] consumption (kg/capita/yr)				
				Growth rate					Growth rate	
Countries	1963	1969	1974	1963–69	1969–74	1963	1969	1974	1963–69	1969–74
South Asia, rice-wheat producers										
India	133	141	141	1.0	0.0	139	136	131	−0.5	−0.7
Pakistan	139	177	188	4.1	1.2	116	139	143	3.0	0.6
Nepal	133	134	135	0.1	0.1	174	174	174	0.0	0.0
South and Southeast Asia, rice-dependent										
Bangladesh	116	116	109	0.0	−1.2	166	162	174	−0.5	1.5
Burma	91	91	95	0.0	0.9	147	169	173	2.4	0.4
Thailand	219	297	340	5.2	2.7	155	167	167	1.2	0.0
Vietnam	189	189	196	0.0	0.7	166	173	182	0.7	1.0
Southeast and East Asia, rice-wheat importers										
Indonesia	123	133	175	1.3	5.6	115	124	141	1.4	2.4
W. Malaysia	551	670	801	3.3	3.6	144	145	151	0.1	0.8
Philippines	238	270	305	2.1	2.5	122	125	128	0.4	0.4
Sri Lanka	108	125	129	2.5	0.6	123	132	117	1.2	−2.5
S. Korea	196	303	451	7.5	8.3	180	208	209	2.4	0.1
Taiwan	456	659	909	6.3	6.6	160	163	151	0.2	−1.4
Japan	1,919	3,286	4,374	9.4	5.9	148	132	123	−1.9	−1.3

Sources: For GDP: FAO Commodities and Trade Division, General Studies Group, "Gross Domestic Product, Private Consumption Expenditure and Agricultural GDP at 1975 Constant Prices, Historical Series 1960–1975 and Projections, 1975–1990." For Taiwan: Taiwan, Economic Planning Council, *Taiwan Statistics Data Book, 1977.* For grain consumption per capita: FAO computer printouts, October 1975–February 1976, provided by the Statistics Unit of the Asian Development Bank. For Japan: Japan, Ministry of Agriculture and Forestry, *Abstract of Statistics on Agriculture, Forestry and Fisheries.*

[a] Five-year averages centered on the years shown except for 1974, which is a three-year average centered on 1974.

[b] Rice, wheat, corn, and other cereals.

the income elasticity for rice. The latter is much more difficult to obtain from cross-national data because of the substitution possibilities that exist. The elasticities in table 11.8 are used to project future demand for rice in chapter 18 by assuming that they reflect the income effects on both rice and wheat.

Price elasticities The significance of shortfalls in the efforts to increase production at the same rate as demand can be evaluated from "own" price elasticities of demand (that is, the elasticity of demand of a commodity with respect to its own price). These show the relative amount by which consumers adjust their consumption of a commodity in relation to a given change in price. The more inelastic the demand for a commodity, the less willing consumers are to substitute for it, and the greater will be the price increase from any shortfall. The price elasticities of −0.1 to −0.5 shown in table 11.7 are highly inelastic and indicate that, for example, a shortage of 10 percent in the availability of rice in South Asia will lead to a 30 percent rise in the price of rice, assuming the availability of substitutes like wheat is constant. In Southeast Asia, a 10 percent shortfall will lead to a 100 percent increase in the price of rice.

In this context, the importance of the cross-price elasticities of demand becomes evident. These provide a measure of the degree to which consumption patterns and hence needed level of imports can be altered through price policy. Although the price and cross-elasticity of demand are inelastic, small changes in relative wheat and rice prices can have a significant effect on consumption under certain circumstances. Imported wheat can be substituted when there are shortages of domestic rice in many Asian countries.

Table 11.7. Price and Income Elasticity of Demand for Wheat and Rice

	Elasticity with respect to:		
	Price of		
Region and product	wheat	rice	Income
South Asia			
Wheat	−0.50	0.16	0.40
Rice	0.09	−0.30	0.30
Southeast Asia			
Wheat	−0.04	0.25	0.30
Rice	0.00	−0.10	0.00
East Asia and Pacific Islands			
Wheat	−0.50	0.37	0.40
Rice	0.02	−0.30	0.20

Source: U.S. Department of Agriculture, *World Demand Prospects for Grain in 1980* (1971) table II.

Table 11.8. Income Elasticities of Demand for Cereal Grains, Selected Countries of Asia

Countries	Income elasticity at[a]	
	1980 income	1990 income[b]
Rice–wheat-producing countries		
India	0.18	0.16
Pakistan	0.14	0.09
Nepal	0.19	0.17
Rice-dependent countries		
Bangladesh	0.22	0.21
Burma	0.23	0.22
Thailand	0.07	0.03
Rice–wheat-importing countries		
Japan	−0.25	−0.32
South Korea	0.02	−0.05
Philippines	0.08	0.06
Malaysia	−0.03	−0.08
Indonesia	0.13	0.08
Sri Lanka	0.19	0.17

[a] Elasticity = 0.7806 − 0.1184 in income. The estimated regression coefficients each had t values exceeding 5.2, and the equation had an $R^2 = 0.80$.

[b] Assuming a growth rate of gross domestic product approximately equal to that achieved over the past 20 years. See table 11.6.

For example, based on table 11.7, in South Asia if the rice price were to rise by 10 percent and the wheat price remain constant, rice consumption would decline by 3 percent (more than 2 million metric tons), and wheat consumption would increase by 1.6 percent (over 0.5 million metric tons), given adequate supplies.

Nutrition of Rice Consumers

Interest in food consumption naturally leads to concern with the nutritional status of individuals who consume the food. What is the nutritional adequacy among Asian rice consumers? To what extent does it vary? How does it vary? And how does rice contribute to nutritional adequacy?

Sukhatme has summarized the data on availability and requirements of calories and protein in Asian diets. When grouped into regions,

> all the less developed sub-regions of Asia show a deficit in calories and only one shows a marginal deficit in protein. However, a country-wide study shows that as many as 18 countries have deficits in calories and 9 have deficits in protein. Furthermore, all the 9 countries which show a protein deficit have deficits in calories as well. It is likely that this is a particular feature of Asia where the diet is predominantly based on cereals and pulses. It shows that the protein problem is probably an indirect result of inadequate quantity of diet.[10]

If the average quantities of food consumed in the nine countries were increased, holding their present proportions constant so as to eliminate the caloric gap, then "the only country that will still have a deficit in protein is Indonesia which is the only country in the region that draws a substantial part of total calorie supply from starchy roots."[11] According to Sukhatme, an average of only a 15 percent increase in the production of cereals would be needed to wipe out the present overall deficit in calories in the countries with shortages. However, such an increase would not, in fact, wipe out the calorie deficit in many countries because of various factors that contribute to an unequal distribution of food.

Recognizing that incomes are a major factor causing inequality in food consumption, Reutlinger and Selowsky have used income distribution data to derive the distribution of food on a broad basis.[12] Their analysis generally agrees with Sukhatme's estimate that between a quarter and a third of the population in the less-developed countries of the region are undernourished, largely because of inadequate total calorie consumption.

Some observers persist in believing that the presence of a larger number of undernourished people in rice-dependent countries suggests that rice is a causal factor in the problem. However, the fact that undernourishment would disappear with a proportional increase in all food suggests that there is no problem with a rice-based diet. Detailed considerations of nutrition are beyond our scope, but a comparison of a few nutritional characteristics of rice with its substitutes is instructive. Table 11.9 shows the nutritional composition of rice and other cereals consumed in Asia, along with egg as a reference for protein quality.

Most cereals have about the same caloric-supplying value. The protein content of rice is somewhat lower than that of wheat and maize, but offsetting that is its higher "biological value," defined by Whyte as "the proportion of absorbed nitrogen that is retained in the body for maintenance and growth."[13] When both factors are considered, there is little difference

Table 11.9. Major Nutrients in Rice and Other Selected Foods

	Calories per 100 gm	Percent protein	Biological value of protein	Percent fat
Egg	173	13.3	100	13.3
Rice, milled	345	6.8	67	0.5
Wheat flour	348	11.0	52	0.9
Maize	342	11.1	56	3.6
Sweet potato	120	1.2	n.a.	0.3

Sources: R. O. Whyte (1974). L. N. Perera, W. S. M. Fernando, B. V. de Mel and T. T. Poleman (1973).

in the protein per 100 grams of cereals. This is also the view of Payne, who says that the crude protein in cereal diets cannot be compared directly with the requirement scale, which is expressed in terms of completely utilized protein, but must first be corrected for quality. His review of studies with rats, which were fed diets based on a number of Asian staples, shows that "with the exception of cassava, sago, and plantain, all provide sufficient utilizable protein to meet requirements after the age of 1 year. . . . Several years of careful and exacting work by Swaminathan and his associates at Mysore . . . shows that . . . very simple diets based upon cereals with minimal additions of pulses and vegetables have NDpCAL (net digestible dietary protein expressed in calories) percent values measured with eight to twelve year old children which are in excellent general agreement with the rat assays."[14]

In other words, Asians who eat rice-based diets will consume sufficient protein if their caloric levels are adequate. Calorie consumption depends on income levels. The reason so many people in Asia are undernourished is not that they eat rice, but that they have incomes that are inadequate to purchase the quantity of rice (or calories) they need.

Conclusion

Rice consumption in Asia has been increasing rapidly as a result of population growth and rising incomes. However, on a per capita basis consumption has improved in some countries, remained stable in others, and declined in still others. The factors behind these trends vary from one country to another.

Total calorie consumption has been stable or rising in most Asian countries, with the sharpest rises being recorded in countries with the highest incomes. Even so, the average per capita caloric intake is only slightly above 2,000 kcal per day. This means that by any standard of nutritional adequacy, malnourishment is a major problem in Asia. In part because of the difficulty in choosing appropriate standards, and in part because of the inadequacy of the data in most countries, it has been difficult to determine whether the nutrition problem is improving or becoming significantly more severe over time. National governments would like to claim the former, although it would appear that the absolute numbers of malnourished have been increasing.

The demand for rice will continue to rise rapidly in Asia. Only Japan has reached the point where there is a significant substitution of other foods for cereal grains in the diet. Although the income elasticity of demand for wheat is slightly higher than for rice, rice is the dominant staple in essentially all Asian countries, except Pakistan. Therefore, rising income will be a major factor in the future growth in demand for rice and is likely to compensate for any reduction caused by a decline in population growth rates. We would thus expect a steady growth in demand for rice at previous levels of 3 percent or more per annum, but a rise in per capita rice consumption where incomes show significant gains, particularly in those countries where inferior cereal grains and staples make up a substantial portion of the diet.

Notes

1. Those countries characterized as "rice dependent" typically have relatively little irrigated rice (chapter 3 and chapter 7, table 7.3), while the "rice-wheat" countries tend to have a high proportion of irrigated rice area.

2. Food and Agriculture Organization of the United Nations, "Report on the 21st Session of the Intergovernmental Working Group on Rice" (Rome, 1978).

3. The reader should be cautious in interpreting this data, as differences in consumption levels among countries may be caused as much by differences in data collection procedures as differences in other factors. For example, in the Philippines, production estimates are based on farm surveys and are believed to underestimate actual production by as much as 20 percent.

4. V. D. Wickizer and M. K. Bennett, *The Rice Economy of Monsoon Asia* (Stanford, Calif., Stanford University Press, 1941).

5. Cynthia Lina G. Santos, "Identifying the Nutritionally Vulnerable Households in the Philippines" (Ph.D. dissertation, Cornell University, 1983). These data are based on quarterly consumption surveys conducted by the Special Studies Division of the Ministry of Agriculture from 1973 to 1976.

6. Pajung Surbakti, "Identifying the Nutritionally Vulnerable Urban and Rural Groups in Indonesia" (Ph.D. dissertation, Cornell University, 1983). These data are based on the 1978 Susenas socioeconomic survey conducted in quarterly rounds by the Indonesian Central Bureau of Statistics. Surbakti divided the twenty-nine districts (Kabupatens) of East Java into three groups based on the relative importance of rice, maize, and cassava in the diet. Figure 11.5 is based on the maize group.

7. Thomas T. Poleman, "A Reappraisal of the Extent of World Hunger," *Food Policy* vol. 6, no. 4 (November 1981) p. 250.

8. An informative and comprehensive discussion of those problems associated with measuring nutritional adequacy is found in Thomas T. Poleman, "Quantifying the Nutritional Situation in Developing Countries," *Food Research Institute Studies* vol. 18

no. 1 (1981) pp. 1–58. Poleman argues that the use of estimates of nutritional adequacy of 2,200 calories or greater have provided an unduly pessimistic picture of the world food problems. The more recent estimates based on basal metabolic rate (BMR) may be more reasonable, but perhaps on the low side. Thus, a careful look at consumption behavior in terms of thresholds between quantity and quality seems warranted.

9. U.S. Department of Agriculture, *World Demand Prospects for Grain in 1980 with Emphasis on Trade in the Less Developed Countries*, Foreign Agricultural Economic Report No. 75 (Washington, D.C., USDA, December 1971).

10. P. V. Sukhatme, "The Present Pattern of Production and Availability of Food in Asia," in *Three Papers on Food and Nutrition: The Problem and the Means of Its Solution* (Brighton, England, Institute of Development Studies, University of Sussex, 1971) pp. 1–17.

11. Ibid., p. 10.

12. S. Reutlinger and M. Selowsky, *Malnutrition and Poverty*, World Bank Staff Occasional Paper No. 23 (Baltimore, Md., Johns Hopkins University Press, 1976).

13. R. O. Whyte, *Rural Nutrition in Monsoon Asia* (Kuala Lumpur, Malaysia, Oxford University Press, 1974) p. 96.

14. R. R. Payne, "The Nutritive Value of Asian Dietaries in Relation to the Protein and Energy Needs of Man," in *Three Papers on Food and Nutrition: The Problem and the Means of its Solution* (Brighton, England, Institute of Development Studies, University of Sussex, 1971) pp. 23–24.

12

Rice Marketing

To keep pace with the rapid urbanization that has occurred in Asia, the volume of rice traded through market channels has grown at 4 percent per year or more over the past three decades. By the 1980s, more than half the crop was marketed, and traditional premarketing practices such as hand pounding were obsolete in all but a few areas. The governments of many countries were deeply involved in rice marketing, in some cases handling up to 25 percent of total rice consumption. This reflects the common belief that it is necessary to control the marketing system in order to ensure that adequate supplies reach consumers at acceptable prices. However, careful economic research suggests that private marketing channels are relatively efficient given the transportation and communication facilities available. Improvements can be made in the technical efficiency of drying, milling, and storage, but only by incurring higher costs.

Marketed Surplus

There are few official statistics on the proportion of Asia's rice production that is market directed; that is, neither the quantity nor the trend in the proportion of rice sold has been estimated with much accuracy. Farmers use paddy (unhusked rice) for home consumption, for seed, as payments in kind, and sell it as a source of cash income. In general, rice consumed by the farm family does not enter the commercial marketing channel, although farmers may sell at harvest and buy rice later. Much of the portion used as payment in kind does find its way into the marketing channel, but the main source of market supply is the rice sold directly by farmers.

Urban population has been increasing 50 to 100 percent more rapidly than total population throughout Asia. Growth in urban consumption needs would normally be supplied from increases in marketed surplus. These could be induced by raising prices to encourage production, or by imports, or both. Exporting countries can reduce exports in favor of domestic urban consumers. In Burma, for example, rice exports and market surplus declined sharply in the early 1970s as a consequence of slow growth in rice production.

Only a few empirical studies provide direct estimates about the marketed surplus relationship: the price response of farm households in allocating their rice production between home consumption and the market.[1] In these, a number of indirect inferences were drawn from earlier studies that support the hypothesis that the response of marketed surplus to a change in current prices is positive.[2]

All of the results based on direct measurement have shown that the output elasticity of marketed surplus (proportional change in marketed surplus per unit change in production) for foodgrain is greater than unity. By implication, this means that cultivators' income elasticity of demand for foodgrains is less than unity. Therefore, as output increases, farmers retain a smaller percentage of output for home consumption. The output elasticities of marketed surplus for foodgrains were estimated to be 1.8 and 1.6 by Bardhan,[3] 2.0 and 1.8 by Haessel,[4] and 1.4 by Toquero et al.[5]

On the basis of the above, it is reasonable to conclude that, on average, marketed rice surplus for Asia has grown faster than production. Nevertheless, the areas of most rapid growth have been those where production increased most quickly. In areas where production stagnated, marketed surplus may have declined. One might speculate that if marketed surplus were around 30 percent in 1950, it would be close to 50 percent today.

Organization of the Marketing System

Rice sold in the market passes through a series of private or government channels, changing hands (and usually ownership) several times as it is conveyed from the farm to the consumer. Figure 12.1 shows a highly simplified diagram of the principal marketing channels, although rice normally crosses between channels more freely than is suggested by the diagram. Local channels are important in all countries, accounting for at least 30 to 40 percent of the rice, much of which is used for home consumption. The importance of government cooperative channels varies considerably among countries. In general, the cooperatives have been effective only when they have been given special government support. Their lack of success is one indication of the competitiveness of the market.

There are a number of common elements in the structure of Asian rice markets. First, paddy is collected from many producers who often market only a small quantity; it is then transported, stored, and processed by a relatively small number of middlemen; and finally it is distributed to a large number of consumers.

Second, as paddy moves through the marketing channels, costs are incurred for the basic services—transportation, storage, and milling. Services such as packaging, which are a large part of the rice marketing costs in highly developed countries, are

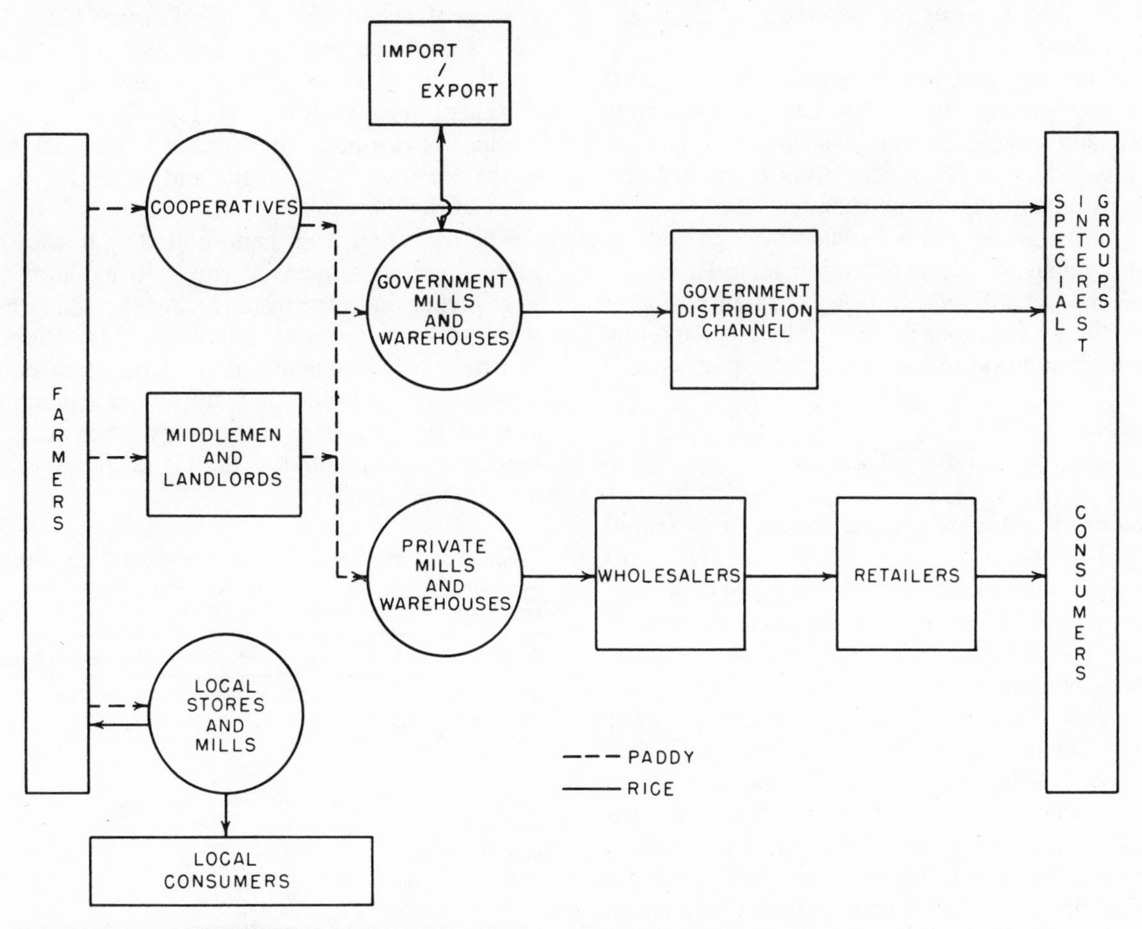

Figure 12.1. Traditional rice marketing channels (Source: Adapted from L. Mears, M. H. Agabin, T. L. Anden, and R. C. Marquez, *The Rice Economy of the Philippines*, p. 86)

not important for most of Asia. For example, until recently, in Indonesia harvested rice was commonly sold in stalk paddy form. In Japan and Korea, on the other hand, the farmer normally removes the husk and sells brown rice. The degree of milling and the amount of polishing differ substantially, as does the percentage of broken rice, reflecting the high variability in quality demanded in different Asian markets. Rice is parboiled in much of South Asia before it is milled, thereby preserving or salvaging grain that has been exposed to unfavorable weather conditions after harvest.[6]

A third common element in the structure is the differentiation between marketing channels that service rural consumers and the primary wholesale and terminal markets that serve urban consumers. In village markets, it is common to find paddy being milled by a small single-stage huller, but paddy for the urban market is processed in a more complex mill consisting of several separators, whiteners, and polishers. Governments in many Asian countries procure domestic rice for food security reserves, urban consumers, special urban consumer groups such as civil servants, the army, or the urban poor, and the export market.

In summary, there are three major flows in most marketing systems: the local market, which is largely private and services the rural consumer; the private marketing channel, serving the urban consumer; and the government marketing channels. Both exports and imports are normally handled through government channels, a notable exception being Thailand's export market. Milling, storage, and transportation are the major functions performed by the marketing system, in addition to importing and exporting.

Rice Milling

Unlike most grains, after rice is threshed to separate it from the straw, it is still covered by a hard inedible coating called the hull or husk. The husk is largely composed of silica and has no feed value. It must be removed by machine milling or, historically, by hand pounding. The product resulting after the removal of the husk is brown rice, the brown color coming from the remaining bran and other thin layers. Brown rice can be consumed directly, but most Asians prefer milled rice from which the bran has been removed. Technically advanced mills can separate bran from the starchy portion quite gently, but the most common type of village mill in many Asian countries results in a high percentage of broken grains. High-income consumers generally are willing to pay a premium price for white unbroken grains known as "head" rice, so milling equipment and rice-handling systems that can produce such rice are associated with a mature marketing system.

The development of the rice milling industry in Asia has passed through a number of distinct phases. Prior to World War II, most rice was milled by hand pounding, an operation typically performed by household members. However, in surplus-producing and export-oriented areas, large rice mills were established in the later part of the nineteenth century near major port cities such as Bangkok, Rangoon, Saigon, and Madras.

As transportation and communication improved, and as European gave way to indigenous ownership and management after the turn of the century, small, up-country mills grew rapidly. Location of the mills near the source of supply reduced both transportation and labor costs because during the height of the milling season rural labor was normally abundant.

For the local market and home consumption, hand pounding gave way to the single huller—a recent transition in areas of surplus labor and deficit foodgrain production such as Java and eastern India. Table 12.1 shows the extraordinary contrast in milling production practices among regions in India and Burma prior to World War II. The officially recorded number of rice mills increased in India from 20,000 in the 1950s to 50,000 at the end of the 1960s, with most of the increase being single hullers.[7]

The transition from hand pounding to the single huller, and subsequent attempts to modernize the new milling industry by replacing the single huller with a more complete mill, have caused considerable controversy. Opponents of modernization emphasized the labor-displacing features and capital intensity of the modern technology. Advocates of modernization pointed to the higher recovery rates

Table 12.1. Percentage of Rice Production That Was Machine Milled and Hand Pounded in Selected States of India and Burma Prior to World War II

State	Machine milled	Hand pounded
Assam	3	97
Bengal (India and Bangladesh)	16	84
Bihar and Orissa	10	90
Hyderabad (Andhra Pradesh)	62	38
Punjab (India and Pakistan)	90	10
Sind (Pakistan)	90	10
Average India	27	73
Average Burma	92	8

Source: India, Directorate of Marketing and Inspection, *Report on the Marketing of Rice in India* (1954).

that could be achieved not only through the milling itself, but through a more integrated approach toward processing, including more scientific drying, storage, parboiling, and milling of paddy.

Milling Equipment

The earliest and crudest form of rice milling was by hand pestle with a mortar or hollowed log, but the steel huller is the most widely used milling machine in Asia today. The equipment is designed to remove the husk and the bran layers of rice in a single operation. The most popular type, the Engelberg huller, was originally manufactured in Syracuse, New York, but the design was copied and modified in Indian foundries long before World War II.[8] Hullers are simple in design and easy to maintain, and are now locally manufactured throughout most of South and Southeast Asia. Their capacity is from 200 to 1,000 kg of paddy per hour. The major shortcomings of the steel huller are its low recovery rate and large percentage of broken grain, while its advantage is its low cost and small size; it can be located near the site of production to minimize transportation costs. Thus, most of the single hullers are found in villages where the service area is small, transportation facilities are poor, and quality standards are not assumed to be critical.[9]

In much of South Asia, large mills selling to the commercial market use as many as five hullers set in tandem, with the rice passed from one huller to the next. The hullers can be adjusted to reduce the percentage of broken grains and improve the recovery rate.

A second type of mill commonly found in Asia uses the disc-sheller frequently known as the *cono* system. It has two stages: the first removes the husk, producing brown rice, and the second whitens or polishes the brown rice. In the first stage, the paddy passes through one or more sheller units consisting of two cast iron discs coated with an abrasive material. The upper disc is stationary, and the lower one rotates. The friction on the grain as it passes between the two discs removes the hulls from the kernel. The brown rice is then passed through two cone polishers.

In comparison with the steel hullers, the cono mills have a higher milling recovery, and a higher percentage of head rice. These mills range in capacity from less than one-half ton per hour to more than 4 tons per hour. In South Asia, the most common large commercial mills employ both hullers and disc-shellers, reflecting an apparent gradual transition in technology.

A third general category of mill uses a rubber roller technology developed in Europe and Japan. The rubber rollers replace the discs in the hulling operation, permitting a higher rate of recovery and higher percentage of head rice. The comparatively large initial capital cost and the necessity for frequent replacement of rubber rollers discouraged their widespread adoption in South and Southeast Asia. However, the evidence in India and Sri Lanka suggests that existing rubber roller rice milling technology compares favorably with the most modern alternatives in terms of both market and social prices.[10]

Comparison of Alternative Milling Systems

Factors to be considered in comparing systems include: the input requirements of capital and labor, the cost of milling, the milling outturn, the quality of the milled rice, and the value of by-products. Table 12.2 shows the equipment found in typical mills. There are many variations. For example, in Thailand, disc-sheller mills typically have grading equipment to satisfy the demands of the export market.

In a technical sense, milling efficiency is measured by the recovery rate. The higher the recovery rate, the smaller the difference in weight and volume between rice and the paddy from which it is derived. In the transition from hand pounding and steel hullers to disc-shellers and rubber rollers, both the total rice recovered and the percentage of head rice increase (table 12.3), although there is a wide variability in performance of mills of the same type under actual operating conditions, particularly with respect to percentage of head rice.[11]

There is a sizable difference in the capital and labor requirements for mills typical of the types described above. Based on a study by Timmer[12] of choice of technology in rice milling in Java, and subsequent suggested revisions by Collier et al.,[13] estimated capi-

Table 12.2. Milling Equipment in Typical Huller, Disc-Sheller, and Modern Rice Mills

	Mill type		
Equipment	Huller	Disc-sheller	Modern
Precleaning			x
Paddy grading			x
Hulling	x	x	x
Husk separator		x	x
Paddy separator		x	x
Whitening		x	x
Polishing			x
Rice grading			x
Conveying			x

Source: M. Esmay, Soemangat, Eriyatno, and Allan Phillips (1979).

Table 12.3. Percentage of Milling Recovery for Four Types of Rice Milling

Type of milling	Husk and bran	Milled rice	Percent of milled rice	
			Head	Broken
		Indonesia		
Hand pounding	37.5	62.5	66.7	33.3
		South India		
Steel huller	36.6	63.4	73.3	26.7
Disc-sheller	32.5	67.5	82.8	17.2
Rubber roller	30.0	70.0	88.6	11.4
		Bicol, Philippines		
Steel huller	33.8	66.2	41.7	58.3
Disc-sheller	31.0	69.0	75.8	24.2
Rubber roller (multipass)	31.0	69.0	77.1	22.9

Sources: Indonesia—C. P. Timmer (1973). India—Central Food Technological Research Institute (n.d.). Philippines—International Rice Research Institute, "The Technical and Economic Characteristics of Rice Post Production Systems in the Bicol River Basin" (1978).

tal and labor requirements for the principal technologies are shown in table 12.4. Timmer concluded that, in Java, single hullers were clearly superior to larger modern mills, even allowing for a wide range of variation in existing capital and labor costs. At a time when Java was moving rapidly away from the traditional practice of hand pounding, Timmer sought to convince the Indonesian government that locally manufactured hullers were superior to larger and more costly imported Japanese mills. His analysis shows that the latter would lead to a substantial reduction in employment without visible economic benefit.

Collier et al. criticized Timmer for underestimating the effect of the demise of hand pounding on employment by overestimating the efficiency of hand pounding. The labor efficiency of the huller over hand pounding is greater than 20 to 1. The fact that, until 1970, more than 70 percent of Indonesian rice was hand pounded reflects the extraordinarily low opportunity cost of labor. However, even at the low wage rates that prevailed in Java, the cost of hand pounding was nearly three times that of the huller.

By contrast, Lele criticized the Indian government for neglecting the matter of modernization "while emphasizing the merits of hand pounding, if not in deeds at least in words."[14] She contended that government policies indirectly encouraged the expansion of single hullers by hampering the growth of organized mills.

There is, in fact, a general recognition that both single hullers and larger mills have a distinct but important role to play, the former servicing the local village community and the latter the market-oriented producers. Thus, the proportion of single hullers and larger mills should change over time with the increase in proportion of marketed surplus. In the Philippines in 1968, 80 percent of the rice mills were single hullers, accounting for 53 percent of rated capacity; the remainder were disc-shellers.[15] In India in the mid-1970s, about 85 percent of the mills were single hullers. Thus, a considerable amount of Indian rice destined for commercial channels was still processed by the single huller. Until the early 1970s in Indonesia, a sizable portion of marketed surplus was hand pounded.

Utilization

A common feature of the rice milling industry in Asia is that its technical potential or installed capacity is not fully used (table 12.5). The installed capacity of a rice mill is usually calculated by multiplying the per hour capacity by the number of hours the mill can technically operate during the year—a hypothetical or ideal potential or operating capacity.[16] The underutilization observed in table 12.5 is often referred to as the problem of *excess capacity*. It would be incorrect, however, to conclude that such excess capacity is a sign of inefficiency.

Table 12.4. Investment Costs and Labor Requirements for Milling Facilities in the Early 1970s

Type of milling	Milling (kg/hr)	Investment cost ($US)	Laborers per mill (number)	Labor input (hours/mt)
Hand pounding	4	0	2	500
Steel huller	200	3,000	5	25
Disc-sheller (1 mt/hr)	1,000	10,000	27	27
Modern (10 mt/hr)	10,000	100,000	39	4

Sources: Based on C. P. Timmer (1973) and W. L. Collier et al., "Choice of Technique in Rice Milling in Java: A Comment."

Table 12.5. Average Degree of Capacity Utilization in Rice Mills for Selected Countries, 1964

Country	Mills (no.)	Installed milling capacity (1,000 mt/yr)	Amount milled (1,000 mt/yr)	Degree of utilization (percent)
India	44,057	28,485	8,605	30
Japan	39,128	18,510	5,968	32
Sri Lanka	1,100	935	462	49
Thailand	14,099	42,297	5,092	12
USA	55	3,115	2,223	71

Source: FAO, *FAO Rice Report* (1965).

Seasonal variation in production and in demand for processing services is one obvious reason for excess capacity. This is particularly true in the case of the small, single-huller service mills whose utilization is closely tied to the seasonal pattern of production and marketing. Because variable costs (fuel, labor, etc.) per unit of rice milled do not differ greatly among the various types of mills, those mills with larger fixed costs must have a higher utilization rate to achieve a breakeven point between costs and returns. In the Philippine study cited above, the steel huller mills surveyed were operating at approximately 20 percent of capacity (that is, 3,500 hours/year), the small cono mills (1 mt/hr) at 35 percent of capacity, and the large cono mills (10 mt/hr) at 65 percent of capacity.[17] Evidence from India and Sri Lanka suggests that, unless the large-scale modern mills with associated complexes of silos and driers are operated at a long-term average exceeding about 90 percent of the engineering capacity, they are unlikely to break even at projected market prices.[18]

The commercial mills traditionally purchase and store paddy. The purchase of paddy requires a substantial amount of working capital. However, periods of high interest rates discourage rice millers from storing, making it even more difficult to operate mills near capacity.

A substantial part of calculated underutilized capacity (table 12.5) reflects apparent rather than real excess capacity. Still, excess capacity does arise as a result of the changing structure and location of the rice milling industry. As the rural areas of Asia have developed, commercial rice mills have gradually moved nearer their sources of supply. Excess capacity tends to occur in the areas of traditional milling near the urban centers.

Most Asian governments have attempted to control the expansion, location, and operation of rice mills through licensing. Such controls do not seem to have been particularly effective in promoting the orderly development of the rice milling industry.

Efficiency and Competition

The notion of efficiency in the marketing of rice is related to the cost of performing services—transportation, storage, processing—and to the degree of competitiveness in the market. There are several potential interrelated sources of inefficiency that will increase marketing margins. These include: (1) inadequate investment in transportation, storage and processing facilities, and information; (2) poor management in the performance of services; (3) collusion among buyers; and (4) underutilization of capacity.

In much of South and Southeast Asia, the private sector has traditionally played a dominant role in rice marketing—Burma and Sri Lanka being notable exceptions in the post–World War II period. In contrast, in East Asia, and increasingly in Indonesia and the Philippines, the government plays a dominant role. However, even in those areas where the private sector predominates, public sector investment in such matters as transportation and infrastructural services significantly influences the efficiency of the system. Either directly or through the establishment of cooperatives, government intervention in rice procurement activities has often been undertaken on the grounds that there is collusion or monopoly power in the market and that farmers lack bargaining power. The domination of the rice trade in many countries by particular groups—the Chinese in Southeast Asia, and specific castes in South Asia—tends to reinforce the belief that collusion exists, that farmers lack bargaining power and are squeezed by middlemen.

There are two approaches to investigating the issue of competitiveness in the market, and many studies combine them. The first is to establish whether the structure of the market tends to conform with the general criteria for competitiveness: (1) a large number of marketing firms; (2) freedom of entry to establish a business in marketing; (3) access by suppliers to information on prices. The second approach is to determine whether price movements and price and cost relationships reflect a state of competitiveness in the market. The investigation can, of course, take place at several points in the marketing channel—and can concern several functions.

Marketing Structure

Ruttan notes that much of what passes for market analysis represents little more than a repetition of conventional wisdom regarding the behavior of middlemen.[19] Few studies have examined the question of whether the rice market conforms to the conditions of competition defined in economic theory. However, studies by Lele[20] for India, Farruk[21] for Bangledesh, and Siamwalla[22] for Thailand have addressed this issue more rigorously than most.

Lele systematically examined data on the number of traders in village and primary markets, the number and capacity of rice mills, and the degree of concentration of the market in Tamil Nadu and West Bengal. She concluded that there is widespread evidence of (1) a good private network of market intelligence on prices; (2) free entry in the grain trade; and (3) extreme self-interest of traders, which generally discourages collusion.

Farruk observed that intermediate and terminal markets in Bangledesh were dominated by small groups of *aratdaris*, who handled 70 to 80 percent of supplies. While Farruk found no direct evidence of excessive commission rates, he argued that there were elements in the aratdari system, such as the strong barrier to entry of new firms, that could weaken competitive forces.

Siamwalla argued that the degree of competitiveness in the marketing of a crop in Thailand is related to the cost of shifting from one buyer to another. This cost is related to the nature of the commodity itself and to the technology employed in both production and marketing. The rice milling industry in Thailand consists of a number of small-scale firms that are well dispersed. The farmers and the various intervening middlemen thus face a relatively low *shifting cost* and do, in fact, sell to many different buyers over the years. Unfortunately, this observation is not well documented either in Thailand or elsewhere. Few studies have asked farmers how many different buyers they have sold to over a given period. The conventional view is that, regardless of the potential number of buyers, the effective number is reduced to one by the indebtedness of the farmers to the middlemen. However, available evidence suggest that loans from middlemen and rice millers form only a small portion of rural credit.[23]

Marketing Margins

Many studies have investigated rice marketing margins. The general procedure is to compare the price differential with actual costs for the farm-to-retail margin or for selected functions such as transportation, storage, or milling to determine whether there are signs of excess profit. Figure 12.2 represents a highly simplified schematic view of the rice marketing system from farm to retail, ignoring the various channels depicted in figure 12.1. The vertical dimensions of figure 12.2 reflect changes in space and form, while the horizontal axis relates to changes in time. The transportation and milling costs associated with the vertical axis are absolute charges (fixed per kg). The storage costs associated with the horizontal axis are proportional to the time the commodity is stored. Brokerage fees and profits of market intermediates are proportional to the value of rice. Thus, although the marketing margin is normally expressed as the ratio of the difference between the retail and farm price divided by the retail price, there is no reason to believe that this proportion should remain constant as rice prices fluctuate over time. In

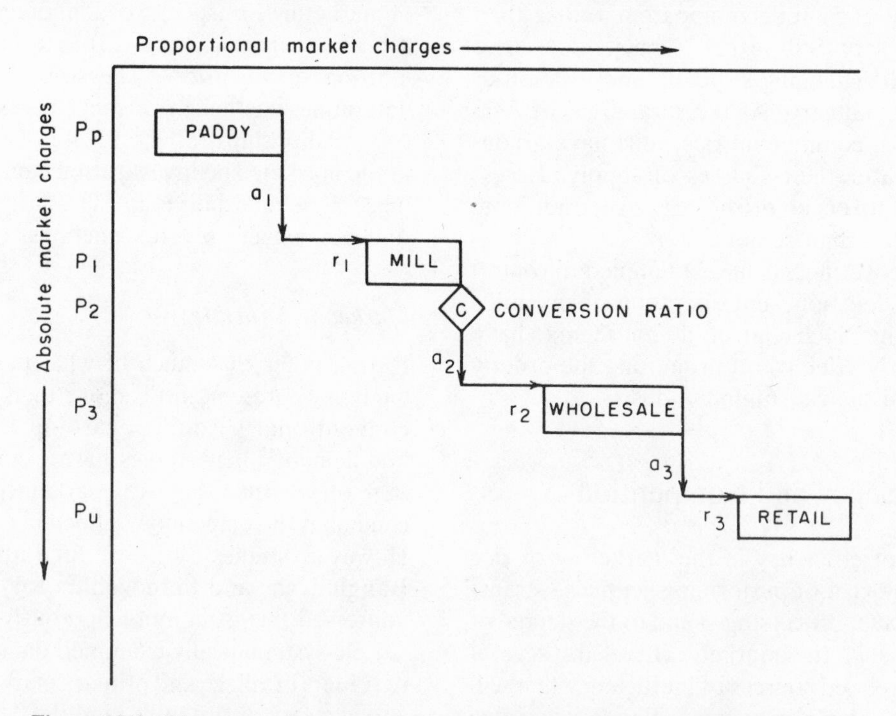

Figure 12.2. Rice marketing systems reflecting changes due to change in ownership and time (proportional) and in form and space (absolute change) (Source: Adapted from P. Timmer, "A Model of Rice Market Margins in Indonesia," p. 159)

fact, the bulk of marketing costs consists of transportation and milling, which are determined as charges per kg.

Table 12.6 summarizes the findings of selected studies of rice marketing margins. For each site, two or more margins are reported—estimated for different years and usually by different researchers. The producers' share of the retail ranges from 58 to 83 percent. Except in Taiwan, where the retail traders' margin seems to have risen sharply, marketing margins in more recent years tend to have decreased. The margin between farm and retail is, of course, low compared with developed countries such as the United States, where farmers receive less than 40 percent of the retail price. The high margins in the developed countries reflect the demand for more services. In fact, the figures shown in table 12.6 tell us nothing about marketing efficiency either within or among countries. Marketing margins are often judged as high or low only in comparison with costs similarly defined and measured under an alternative marketing system.[24]

Few studies have drawn comparisons between alternative systems. Mears et al. compared marketing margins for private, cooperative, and government channels in the Philippines.[25] They drew no strong conclusion about relative efficiency, suggesting that there would be a slight advantage for the private sector under an assumed condition of no windfall profits.

Ruttan interpreted evidence from regional analysis of Philippine price series as indicative of relatively competitive markets.[26] His conclusion was based on the high level of correlation between farm and retail prices. These correlations were not significantly different than 1.0 (that is, the margins were constant). Nicholls, however, argued that market inflexibility is consistent with tight oligopolistic control.[27] In the short run, margin flexibility (increasing margins as demand increases relative to supply) is consistent with rational maximizing behavior under perfect competition.

Blyn also commented on price series correlation as a measure of market integration.[28] Considering the imperfections of competition and the possible range of local market price fluctuations, it is not implausible that the overall correlation coefficient for a group of markets could be much lower than has been indicated in some studies and still be consistent with the integrated market.

In Indonesia, marketing margins vary appreciably over the year, reflecting the fact that rice flows from rural to urban areas during periods of relatively abundant domestic supply, but that this flow is reversed during the lean months of heavy importation. During these months, rural prices can at times exceed urban prices.

The Thai farm price is closely linked to the Bangkok wholesale price which, in turn, is affected by export demand. Thus, in Thailand, the price-making

Table 12.6. Marketing Margins for Rice in Asia, 1953–74

		Site				Portion of total cost				
Location	Year	Production	Market	Distance	Producers' shares (percent)	Processing, storage, and transportation costs (percent)	Transportation costs only (percent)	Traders margin (percent)	Marketing margin (percent)	References
Taiwan	1953	Pingtung	Taipei	400	67.9	30.6	3.6	1.5	32.1	Yeh, 1955
Taiwan	1974	Pingtung	Taipei	400	58.3	26.6	3.2	15.1	47.1	APO, 1976
Indonesia	1955	Krawang	Jakarta	100	60.0	—	—	—	40.0	Mears, 1961
Indonesia	1955	Krawang (Channel I)	Jakarta	100	79.8	11.5	—	8.7	20.2	Adiratma, 1969
		Krawang (Channel II)	Jakarta	100	72.5	20.2	—	7.3	27.5	
Indonesia	1978	Krawang	Jakarta	100	83.9	—	7.8	—	16.1	Mears, 1982
India	pre-WWII	W. Godavari	Madras	500	63.2	25.2	9.6	11.6	36.8	India, Directorate of Marketing and Inspection, 1954
India	1970	W. Godavari	Madras	500	82.9	13.8	8.7	3.3	17.1	George and Choudikar, 1973
Philippines	1966	Central Luzon	Manila	100	80.3	12.9	3.3	6.8	19.7	Mears et al., 1974
Philippines	1970	Central Luzon	Manila	100	78.5	12.6	4.4	8.9	21.5	Mears et al., 1974

force flows from urban to rural areas rather than rural to urban as in the case of most rice exporters. In both the Philippines and Thailand, marketing margins are appreciably higher at harvest than at other times of the year because of a shortage of milling capacity and the inability of producers to finance their own storage operations. Marketing margins varied by about 30 percent from peak to trough in the Philippines[29] and Thailand.[30] Increased farm storage and finance could raise farm prices, narrowing the harvest margin.

Market Integration and Transportation

The analysis of price differences between regions, as between stages of marketing, is based on the same methodology—the price spread between markets being compared with the cost of handling and transport. A price spread between markets approximating transportation costs suggests a competitive situation.

Transportation systems in Asia have undergone a complete change in the past three decades. Ingram's description of transportation in Thailand immediately after World War II typifies much of Asia:

> Prior to 1950 no natural road system existed in Thailand. Such roads as did exist were feeder roads for the railway, unpaved provincial roads largely used by bullock carts and short stretches of road around Bangkok. Interregional transport of goods and people scarcely existed. Such transport took place by rail and water and was therefore limited to the area served by these means.[31]

As road networks developed throughout Asia, there was a substantial shift from rail and water to truck transportation. For hauls in excess of 100 to 200 km, trucks are more expensive than either rail or boat.[32] However, the greater flexibility of trucks favors their use, even when rates for alternative modes of transportation are less expensive.

Data from both Mears et al.[33] and Pinthong[34] indicate that, as hauling distances extend beyond 200 km, the cost of transportation rapidly exceeds the cost of milling. Transportation costs ranging from 3 to 10 percent of total marketing costs for selected areas are shown in table 12.6.

The rapid improvement in transportation and communications has increased the general level of market integration in Asia. In separate studies, Lele[35] for two states of India, and Pinthong[36] for Thailand concluded that there is a high degree of market integration. Studies of the island economies of the Philippines and Indonesia imply a much lower degree of integration. In Mindanao, marketing margins are twice the level of Central Luzon's because of poor transportation and few larger mills, suggesting a greater opportunity for inefficiency and collusion.[37] Marketing margins in the outer islands of Indonesia are higher than on Java for the same reasons.[38]

Storage

A final note on efficiency and competition concerns storage. There are issues related to both technical efficiency and the loss of rice, and to economic efficiency and the cost of storage. Rice is stored as paddy or in milled form both in bulk or in sacks. Estimates of loss in storage range from less than 1 percent to as high as 10 percent. There is general agreement in studies investigating the profitability of storage that storing rice is extremely risky.[39] The financial riskiness of storage may be the main reason why cultivators market heavily soon after harvest. By selling at harvest, the farmer transfers risk to the trader. The volume of rice purchased by traders and the profit rate earned varies from year to year. Because of different trading strategies, all traders do not earn a high rate of profit in one year and a low rate in another. There is no basis for accurately predicting whether profit or loss will result from holding stocks during any specific year. The high loss probabilities found in research studies present quite a contrast to the picture of windfall profits inferred by the stereotype.

Government Intervention

Government intervention in the Asian rice market increased after World War II for social and economic reasons including: (1) stimulation of production through farm price guarantees; (2) supply of subsidized foodgrains to low-income consumers or other special interest consumer groups; (3) control of unhealthy competition or of monopolistic power; (4) redistribution of marketing activity among different social classes; and (5) the promotion of modernization or improvements in marketing. Methods of intervention included regulatory measures; direct or indirect financial support in the construction of transportation, storage, and marketing facilities; and direct procurement of some portion of the rice stock. Some measures clearly were more efficient than others in achieving government goals.

Regulatory Measures

Government pricing of rice is an important regulatory measure, and nearly all governments engage in it. Among the less developed countries of the region, pricing policy appears to have had a consumer bias, as central governments have sought to ensure a cheap supply of rice from the rural areas to meet urban demands. Gradually, as countries develop, there appears to be a pronounced shift toward a producer bias in rice policy. This is reflected in the levels of farm and retail prices that prevail in each country (table 12.7). Philippine and Indonesian prices are close to the world market level. Although isolated from world price gyrations, the prices of Malaysia and the East Asian countries are significantly higher, reflecting a strong producer-oriented policy. Those of Thailand and Burma are much lower, reflecting the fact that exporting countries tax producers to obtain government revenue. A ratio of retail price to farm paddy price of about 2 reflects a reasonable marketing margin, but does not necessarily indicate the absence of excessive middlemen profits. When the ratios are significantly lower, governments may be subsidizing marketing operations.

Another popular form of regulation is licensing firms and dealers, usually with the objective of controlling entry. As noted previously, government licensing in India seems to have encouraged the proliferation of single huller mills and discouraged the entry of larger and more modern mills.[40] In the Philippines, the Rice and Corn Board was established in 1958 to license rice traders, with the specific purpose of reducing the control of Philippine rice marketing activities by Chinese middlemen.

Table 12.7. Relationship of Farm to Retail Prices, 1976–78

Location	Farm paddy price ($US/kg)	Retail milled rice price ($US/kg)	Kg. milled/ kg. paddy
Consumer-biased policies			
Burma (Rangoon)	0.06	0.13	2.2
China (Guangdong)	0.10	0.14	1.4
Thailand (Suphan Buri)	0.10	0.20	2.0
Bangladesh (Joydebpur)	0.12	0.21	1.8
India (Cuttack)	0.11	0.21	1.9
India (Waltair)	0.10	0.21	2.1
India (Coimbatore)	0.11	0.22	2.0
Pakistan (Punjab)	0.09	0.30	3.3
Neutral policies			
Nepal (Kathmandu)	0.18	0.26	1.4
Sri Lanka (Kurunegala)	0.19	0.27	1.4
Philippines (Central Luzon)	0.15	0.28	1.9
Indonesia (Central Java)	0.17	0.32	1.9
Producer-biased policies			
Malaysia (Selangor)	0.19	0.40	2.1
China (Taichung)	0.24	0.37	1.5
Korea (Suweon)	0.51	0.67	1.3
Japan (Ibaraki)	0.78	1.15	1.5

Source: A. Palacpac (1982).

Most Asian governments have established a system of grades for the purchase and sale of rice to improve quality and reduce loss or spoilage. However, except in the more export-oriented countries of the region, government grading standards are not implemented at the farm level. Traders have tended to develop their own standards of quality, and it is frequently argued that this is one mechanism through which they exploit producers. Siamwalla notes that interest in quality on the part of middlemen is not a constant factor, with the buyers more quality conscious in times of falling prices.[41] It is difficult to discern how much grading acts to depress prices received by farmers.

Other steps have been taken in an apparent effort to reduce middlemen profits in the rice market. These include the promotion of cooperatives as an alternative marketing channel. However, the failure of rice marketing cooperatives in countries such as Malaysia and the Philippines is another indication that Asian rice marketing tends to be highly competitive. By contrast, the earlier success of the cooperatives in Taiwan must be explained largely in terms of the monopolistic control bestowed on them by the government, particularly for the purchase of rice and sale of fertilizers through the now defunct rice-fertilizer barter system.

Transportation and Communication

Improvements in transportation and communication automatically enhance the efficiency of marketing.[42] Price differentials between markets are reduced because supplies can be more cheaply transferred from low-price to high-price markets. Isolated producing regions are opened to the extent that as marketing margins are reduced, the production incentive is increased. Recognizing the importace of keeping farmers and traders appraised of demand–supply and price conditions, many governments disseminate daily information on rice prices.

Government Procurement

Government procurement operations vary widely among countries and within a given country over time. In some exporting countries, such as Burma, and exporting regions, such as the Punjab of India, the government has for some years procured most of the marketed surplus and has thus retained the traders profit. This is in contrast to Thailand, where exports pass through private channels, and the government extracts revenue through taxes. The subsidization of production and marketing activities in countries such as Japan and Sri Lanka and, more recently, in Taiwan and South Korea, has led to a high level of government procurement. In the Philippines and Indonesia, the government has historically procured only a small portion of the total rice crop, its effort in part being directed toward bolstering rural prices in years of surplus production.

In periods of shortages, governments typically attempt to defend price ceilings through importation. In periods of surplus, they defend price floors through stepped-up purchases of domestic supplies. During both types of crises, the private marketing system comes under attack, and middlemen are accused of reaping excess profits through hoarding or through purchase of paddy at low and discriminating prices. Governments frequently become more active in procurement during these periods. For example, government procurement as a share of total production rose dramatically in India during a period of severe shortage in the mid-1960s and again during a period of significant surplus in the mid-1970s. Philippine government purchases rose sharply in the mid-1970s, also during a situation of relative surplus.

Periods of rapid increase in government procurement do not necessarily imply an expansion in facilities. Government operations are frequently carried on by leasing private facilities, or by direct contract with private firms. For example, in Sri Lanka despite the fact that a large portion of the rice crop is procured for distribution under the government ration program, the bulk of the rice is milled in private facilities. However, both the Indonesian and the Philippine government agencies invested large amounts in new storage facilities in the late 1970s to give them greater leverage over the market.

There is much debate regarding government's role in direct procurement of rice stocks. Ostensibly, governments intervene to protect consumer and producer interests. However, evidence suggests that private marketing is relatively efficient, and that middlemen profits are not excessive. Any loss in efficiency through government operations should be offset by an equivalent social gain. Where the government has limited resources and controls only a small portion of the crop, government buying and selling activities often serve the interests of only a few privileged producers and consumers. This appears to have been the experience in countries such as Indonesia and the Philippines.

A few government activities, such as the improvement of transportation and communications, can be seen as improving marketing efficiency. Other activities have social objectives. Whether the social benefits of many of these activities exceed social costs is a subject of continuous debate and little serious research. However, governments are often more effective than middlemen in the exploitation of producers.

Modernization

There is evidence that the rice milling industry is competitive. At the same time, there is still a high level of *technical* inefficiency, which leads to considerable loss between the time the rice is harvested and the time it finally reaches the consumer. Low milling recovery rates and high losses in storage are evidence of technical inefficiency. Since the 1960s, there has been increasing talk of modernizing the processing industry, although steps taken in this direction have frequently met with failure. To understand why, we need to consider the concept of modernization as it is commonly described and understood.

Modernization of processing implies an improvement in the recovery rate of rice from paddy and in the quality of the final product. Both the recovery and the quality can be increased by improving the technical efficiency of post-harvest operations, including drying, handling, storage, parboiling, milling, and transport. For example, the use of mechanical driers instead of traditional sun drying of the grain can reduce the number of broken grains; bulk storage in well-constructed sites can reduce losses caused by rodents, insects, and moisture damage; and rubber roller mills can give the highest recovery. All these improvements require investments that need to be repaid through higher marketing costs. As long as governments maintain low marketing margins, the private sector will not find it attractive to make such investments, and governments will have to subsidize their own marketing activities if they make such investments to achieve substantially higher technical effectiveness.

A particularly vexing problem is that of grain drying. With traditional photoperiod-sensitive varieties, the crop was harvested after the monsoon rains had ended, but modern varieties are often harvested while it is still raining. As a result, much of the paddy is inadequately dried. Mechanical dryers can turn out paddy that results in a much higher percentage of high-quality head rice, but dryers that are economical for small quantities of paddy are not available. Farmers thus find it gives them a higher return to sell wet or poorly dried paddy to millers and government buyers rather than invest in dryers. Large-scale attempts to modernize are often plagued by a similar lack of economic returns to investments in drying facilities.

In the late 1950s, four large milling complexes with on-site drying and storage facilities were constructed in the Philippines, one of the first attempts to introduce modern facilities. In Sri Lanka, four modern complexes constructed by the government in the mid-1970s are virtually idle. The reasons given for failure of these and many similar projects are numerous—faulty design and construction, wrong location, poor management, and inadequate supply of uniform paddy to run the mill at capacity for an extended period.

A second type of problem is reflected in the recent experience of Indonesia discussed earlier. Although technically successful, the introduction of modern Japanese milling equipment in the 1970s appears to have led to an excessive substitution of capital for labor, and hence a socially inefficient outcome. However, hand pounding methods were very high cost, even with Java's low wage rates, which explains in large measure the rapid conversion to milling machines.

Less radical, but more successful attempts to modernize have involved the modification of existing mills. For example, in a large number of mills in Thailand, rubber rollers have replaced the second of two disc huskers in the hulling operation.[43]

The most comprehensive attempt to research the technical efficiencies of modernization involved the comparison of several modern rice mills in India with traditional mills.[44] The recovery rates from this study are reported in table 12.3, and are widely recognized as providing a reasonably accurate picture of rate of recovery among milling systems.

In recent years, research workers have come to recognize rice processing as a system that, like the farm production system, must be viewed holistically. Whether this new focus on the rice processing industry as a system can assist in the identification of technology appropriate to achieving social and economic gains as opposed to technical efficiency remains to be seen. In the eyes of most policy planners and technicians, modernization continues to be associated with technical efficiency.

Notes

1. See Kalpana Bardhan, "Price and Output Response of Marketed Surplus of Food Grains: A Cross-Sectional Study of Some North Indian Villages," *American Journal of Agricultural Economics* vol. 52 (1970) pp. 51–61; Pranab Bardham and Kalpana Bardham, "Price Response to Marketed Surplus of Food Grains," *Oxford Economic Papers* vol. 23 (1971) pp. 255–267; Walter Haessel, "The Price Elasticity of Home Consumption and Marketed Surplus of Food Grains," *American Journal of Agricultural Economics* vol. 57 (1975) pp. 111–115; and Z. Toquero, B. Duff, A. Lacsina, and Y. Hayami, "Marketable Surplus Functions for a Subsistence Crop: Rice in the Philippines," *American Journal of Agricultural Economics* vol. 57 (1975) pp. 705–713.

2. J. R. Behrman, "Price Elasticity of Marketed Surplus of a Subsistence Crop," *Journal of Farm Economics* vol. 48 (1966) pp. 875–893; Raj Krishna, "A Note on the Elasticity of the Marketable Surplus of a Subsistence Crop," *Indian Journal of Agricultural Economics* vol. 17 (1962) pp. 79–84; Raj Krishna, "The Marketed Surplus of Food Grains: Is It Inversely Related to Price?" *Economic and Political Weekly* vol. 17 (1965) pp. 325–328; M. Mangahas, A. E. Recto, and V. W. Ruttan, *Production and Marketing Relationships for Rice and Corn in the Philippines* Technical Bulletin No. 9 (Los Banos, Philippines, International Rice Research Institute, n.d.); and Mubyarto and L. B. Fletcher, "The Marketable Surplus of Rice in Indonesia: A Study of Java-Madura," Monograph No. 4 (Ames, Iowa, Iowa State University, International Studies in Economics, 1966).

3. Kalpana Bardhan, "Price and Output Response of Marketed Surplus of Food Grains."

4. Walter Haessel, "The Price Elasticity of Home Consumption and Marketed Surplus of Food Grains."

5. Z. Toquero et al., "Marketable Surplus Functions for a Subsistence Crop: Rice in the Philippines."

6. Parboiling is a hydrothermal treatment of rough rice prior to milling. In the process, rough rice is soaked, steamed, and redried before milling. In the United States and Europe, parboiling is done with modern equipment, but various methods of parboiling are used in Asia. Parboiled rice is more commonly eaten in South Asia than in Southeast Asia. For more details see Surajit K. De Datta, *Principles and Practices of Rice Production* (New York, Wiley, 1981).

7. Uma J. Lele, *Food Grain Marketing in India: Private Performance and Public Policy* (Ithaca, New York, Cornell University Press, 1971).

8. A. Aykrod, "The Rice Production in India," *Indian Medical Research Memoirs* vol. 32 (1940) p. 184.

9. Merle Esmay, Soemangat, Eriyatno, and Allan Phillips, *Rice Post-Production Technology in the Tropics* (Honolulu, Hawaii, East-West Center, 1979).

10. Barbara Harriss, "Allocation, Location, and Dislocation in Non-Market Rice Distribution," *Journal of Development Studies* vol. 15 (1978) pp. 87–105.

11. Both the Indian research conducted in the 1960s and the Philippine research in the 1970s are based on investigations in a limited number of mills: Central Food Technological Research Institute, *Report on the Evaluation of Modern Rice Mills in Comparison with Existing Mills* (Mysore, India, n.d.); International Rice Research Institute, "The Technical and Economic Characteristics of Rice Post-Production Systems in the Bicol River Basin" (Los Banos, Philippines, IRRI and University of the Philippines at Los Banos, Nov. 1978).

12. C. Peter Timmer, "Choice of Technique in Rice Milling in Java," *Bulletin of Indonesian Economic Studies* vol. 9 (1973) pp. 121–126.

13. William A. Collier, A. J. Colter, d'A. Sinarhadi, and A. Shaw, "Notes: Choice of Technique in Rice Milling in Java," *Bulletin of Indonesian Economic Studies* vol. 10 (1974) pp. 106–120.

14. Uma J. Lele, *Food Grain Marketing in India*.

15. E. V. Araullo, D. B. de Padua, and M. Graham, eds., *Rice Post-Harvest Technology* (Ottawa, Canada, International Development Research Center, 1976).

16. Food and Agriculture Organization of the United Nations, *FAO Rice Report* (Rome, FAO, 1965).

17. E. V. Araullo, D. B. de Padua, and M. Graham, eds., *Rice Post-Harvest Technology*.

18. Barbara Harriss, "Paddy and Rice Marketing in Northern Tamil Nadu," in *Studies of Marketed Surplus, Market Efficiency, Technology, and Livelihoods* (Madras, Sangram Publishers, 1977).

19. Vernon W. Ruttan, "Agricultural Product and Factor Markets in Southeast Asia," *Economic Development and Cultural Change* vol. 17 (1969) pp. 501–519.

20. Uma J. Lele, *Food Grain Marketing in India*.

21. Muhammad O. Farruk, "Structure and Performance of the Rice Marketing System in East Pakistan," Occasional Paper: USAID Employment and Income Distribution Project, No. 31 (Ithaca, N.Y., Department of Agricultural Economics, Cornell Univeristy, 1970).

22. Ammar Siamwalla, "Farmers and Middlemen: Aspects of Agricultural Marketing in Thailand" Agricultural Marketing Case Study No. 2 (Bangkok, U.N. Asian Development Institute, 1975).

23. Ammar Siamwalla, "Farmers and Middlemen: Aspects of Agricultural Marketing in Thailand."

24. Raj Krishna, "The Role of the Government in Agricultural Marketing Reform," *Review of Agricultural Economics: Malaysia* vol. 1 (1967) p. 115.

25. Leon Mears, M. H. Agabin, T. L. Anden, and R. C. Marquez, *The Rice Economy of the Philippines* (Quezon City, Philippines, University of the Philippines Press, 1974).

26. Vernon W. Ruttan, "Agricultural Product and Factor Markets in Southeast Asia."

27. William H. Nicholls, *Imperfect Competition in Agricultural Industries* (Ames, Iowa, Iowa State College Press, 1941).

28. George Blyn, "Price Series Correlations as a Measure of Market Integration," *Indian Journal of Agricultural Economics* vol. 28 (1973) pp. 56–58.

29. Leon Mears, et al., *The Rice Economy of the Philippines*.

30. C. Pinthong, "A Price Analysis of the Thai Rice Marketing System" (Ph.D. dissertation, Stanford University, 1977).

31. James C. Ingram, *Economic Change in Thailand, 1850–1970* (Stanford, Calif., Stanford University Press, 1971).

32. Leon Mears, et al., *The Rice Economy of the Philippines*; and Leon Mears, *The New Rice Economy of Indonesia* (Yogyakarta, Indonesia, Gadja Mada University Press, 1982).

33. Leon Mears, et al., *The Rice Economy of the Philippines*.

34. C. Pinthong, "A Price Analysis of the Thai Rice Marketing System."

35. Uma J. Lele, *Food Grain Marketing in India*.

36. C. Pinthong, "A Price Analysis of the Thai Rice Marketing System."

37. Leon Mears, *The Rice Economy of the Philippines*.

38. C. Peter Timmer, "A Model of Rice Marketing Margins in Indonesia," *Food Research Institute Studies* vol. 13, no. 2 (1974) pp. 145–167.

39. Uma J. Lele, *Food Grain Marketing in India*; Leon Mears, *The Rice Economy of the Philippines*; C. Pinthong, "The Price Analysis of the Thai Rice Marketing System"; and S. Tubpun, "The Price Analysis and the Rate of Return on Holding Rice and Paddy in Thailand" (MS thesis, Thammasat University, 1974). The intervention of governments in the rice marketing system to "stabilize" prices reduces even further the possible margin of profitability from storage, as shown by L. J. Unnevehr, "The Impact of Philippine Government Intervention in Rice Markets," IRRI Agricultural Economics Department Paper 82-24 (Los Banos, Philippines, International Rice Research Institute, 1982).

40. Uma J. Lele, *Food Grain Marketing in India*.

41. Ammar Siamwalla, "Farmers and Middlemen: Aspects of Agricultural Marketing in Thailand" Agricultural Marketing Case Study No. 2 (Bangkok, U.N. Asian Development Institute, 1975).

42. Raj Krishna, "The Role of Government in Agricultural Marketing Reform."

43. N. Soomboonsup, "Rice Milling Technology and Some Economic Implications: The Case of Nakorn Pathom, Thailand, 1974" (MS thesis, Thammasat University, 1976).

44. Central Food Technology Research Institute, *Report on the Evaluation of Modern Rice Mills in Comparison with Existing Mills* (Mysore, India, n.d.).

13

International Trade in Rice

The political and economic importance of international trade in rice is far greater than would appear from the 5 percent of world production that moves internationally, and this has long been true. Rice trade helped motivate the settlement of the major river deltas of mainland Southeast Asia between the 1850s and the 1930s. In the 1940s, Japanese imperial interests in the rice bowl area of Southeast Asia contributed to the expansion of World War II into the Pacific. More recently, with the domination of the world rice market by only a few exporters and the attendant volatility in rice supplies and prices, many countries have begun to place heavy emphasis on self-sufficiency and domestic price stability.

The world rice market is not easily described in terms of structure or function. An aura of mystery surrounds it, and elements of instability and uncertainty set it apart from the world markets for wheat and maize.

This chapter begins with a discussion of the patterns of rice trade. The development of rice trade can be divided into two distinct phases based on market function and participant objectives. The first period, extending from the 1860s to World War II, saw the emergence of a rice export market centered in three delta areas of mainland Southeast Asia—the Irrawaddy in Burma, the Chao Phraya in Thailand, and the Mekong in Vietnam. During this period, there was a minimum of government intervention in the market, but considerable government investment in opening new land to increase surplus production for export.

During the second period, extending from World War II to the present, Asian rice-producing countries continued to play a dominant role in world rice trade. Still, areas outside of Asia gained an increasing share of the market: the United States became a major exporter, and the Middle East and Africa major importers. Within Asia, the pattern of trade changed markedly: Vietnam and Burma lost their position as major exporters, and China and Japan shifted from import to export positions.

Domestic rice policies have had a marked effect on the pattern of trade. Domestic price stability and rice self-sufficiency have been primary policy goals in many Asian countries. Government control of imports and exports to achieve price stability has led to a greater volatility in world prices. The introduction of new rice technology, coupled with government programs to increase rice production and a slow but steady substitution of wheat for rice among many importers, have allowed Asia to maintain a fairly steady volume of imports while shipping an increasing share of its exportable surplus to the Middle East and Africa. As a result of these new markets, particularly among the OPEC (Organization of Petroleum Exporting Countries) countries, the volume of world trade, which had stabilized at about 8 million metric tons (mt) in the 1960s, increased in the late 1970s to 12 million mt, but still remains at less than 5 percent of total world production.

In the latter part of the chapter, we examine the factors related to the short-run instability in the market and to the long-term trends in supply of exports and demand for imports. Finally, we discuss the performance of the market.

Growth in Rice Trade: 1860s to World War II

Most discussions of rice trade prior to World War II center on the role of the exporters. There are many excellent accounts of the development of rice production and the expansion of exports in the major river deltas of mainland Southeast Asia during this period.[1] European demand for rice was a critical factor in the early development of rice trade. By World War I, however, export growth was being supported by a rising demand among Asian importers, particularly India and Japan.

The Development of Rice Exports in Delta Areas

Southeast Asia, particularly the mainland, has always been a demographic anomaly because the fertile and potentially most productive land has been significantly less densely populated than similar areas in India and China.[2] There are exceptions of course, such as Java in Indonesia or the Red River Delta in northern Vietnam. However, other major deltas in Southeast Asia, including the Mekong in southern Vietnam, the Irrawaddy in Burma, and to a lesser extent the Chao Phraya in Thailand, were sparsely populated in the mid-nineteenth century.[3] Mainland Southeast Asia exported small amounts of rice prior to 1850, but southern Vietnam, Burma, and Thailand restricted rice exports apparently because of fears of domestic shortages.[4] Prior to the 1850s, rice production was inhibited by widespread malaria and poor transportation facilities, but lack of profit was clearly the most significant obstacle to increased production. Subsequent events show that the rise in world prices after 1850 because of the growing demand in Europe and the Indian subcontinent provided sufficient incentive to overcome these constraints.

There were no significant technological breakthroughs in this period to stimulate production, but rice trade was spurred by a number of important political developments.[5] In 1852, the British annexed the Irrawaddy Delta, opening Rangoon to trade. The concessions made to the British under the Bowring Treaty of 1855 permitted the export of rice from Thailand at a fixed duty. In 1859, the French seized Saigon, paving the way for exports from the Mekong Delta. In less than a decade, all trade restrictions had been removed in the three delta countries.

During the next fifty years, the Delta population grew rapidly as immigrants cleared land for rice farming or developed business activities associated with rice trade. The rapid development of the Delta regions resulted in a clear division of labor according to ethnic background. The indigenous populations of Burma, Thailand, and Vietnam became farmers while Indians migrated to the Irrawaddy Delta to become laborers, moneylenders, and small businessmen, as did the Chinese in the Chao Phraya and Mekong deltas. The Europeans provided professional and technical support and foreign capital. The country governments promoted this ethnic division to encourage more rapid export development. Important social repercussions were not felt until several decades later.

A major role of the governments during this period was to develop infrastructure, particularly transportation, and to a much lesser degree water management and flood control. Canals were built principally to provide transportation, and irrigation did not become a serious undertaking until after World War I. Government infrastructural investment was greatest in Burma, despite the fact that there was no export duty on rice to provide funds, and least in Thailand. This may explain in part the rapid development of Burma, which until World War II accounted for almost two-thirds of Southeast Asian rice exports (table 13.1).

Export Markets

The development of shipping was critical to the growth of export markets. The Suez Canal opened in 1869, and it, together with developments in steam power, allowed shipping costs and shipping time to decline, making it more profitable to ship bulky and perishable products such as milled rice to distant ports in Europe.[6] During the decade of the 1870s, the three delta countries exported about 1.4 million mt of rice annually. By the turn of the century, their exports had reached over 4 million mt per year.

Initially, Europe imported more than 50 percent of the surplus rice from the deltas. However, the three countries did not equally share the European market, and the destination of exports was often dictated by political and personal connections.[7] British millers in Rangoon exported principally to England. Chinese millers in Bangkok and Saigon exported to Singapore and to Hong Kong. By the turn of the century, an increasing portion of Burmese rice exports was being shipped to India and Sri Lanka while the Vietnamese were expanding trade in the Dutch East Indies (Indonesia) and the Philippines.

Asian markets grew more rapidly than those in Europe (table 13.1), and at the end of World War I,

Table 13.1. Rice Exports from Burma, Thailand, and South Vietnam to Asia and Other Parts of the World, 1872–81 to 1936/37

Date	Burma		Thailand (Siam)		S. Vietnam (Cochinchine)		Three-country total			Asia as percent of total
	Asia	Other	Asia	Other	Asia	Other	Asia	Other	Total	
	(thousand metric tons)									
1872–81	204	703	161	37	292	23	657	763	1,420	46
1882–91	281	814	268	61	459	37	1,008	912	1,920	52
1892–1901	783	862	519	53	482	164	1,784	1,079	2,868	62
1902–1911	1,277	1,134	855	99	572	221	2,704	1,454	4,158	65
1911–1914	2,391	1,566	564	137	639	228	3,594	1,931	5,525	65
1917/18–1928/29	3,530	1,057	814	104	1,019	230	5,363	1,391	6,754	79
1929/30–1936/37	4,265	1,443	1,143	220	782	517	6,190	2,180	8,370	74

Sources: 1872–81 to 1902–11—N. Owen (1971) table II-A. 1911–14 to 1929/30–1936/37—V. Wickizer and M. Bennett (1941) appendix table VI.

nearly 80 percent of rice exports from Southeast Asia were retained in Asia. Between the 1870s and the 1930s, Burma's volume of exports to Europe remained fairly steady, but by the 1930s this volume represented about a quarter of total exports while India alone accounted for more than 50 percent of Burmese exports.

Rice Trade in the Japanese Empire

Prior to World War I, Japanese rice imports were relatively small. However by the end of the war, Japan faced an acute rice shortage. In 1918 and again in 1919, it was forced to import more than a million mt of rice. Although domestic production expanded steadily thereafter, it failed to keep up with demand. Initially, the bulk of Japanese imports came from Southeast Asia, but by the 1920s, Japan had begun to seriously develop rice production in its colonies Korea (Chosen) and Taiwan (Formosa).[8] By the 1930s, nearly all of Japan's import requirements were met by its colonies (table 13.2).

Impact of the Depression on Asian Trade

In terms of economic growth, development of rice trade in the delta regions of Southeast Asia appeared to be a major success. Throughout most of the period under discussion, export earnings accounted for over two-thirds of total export value in all three countries.[9] The growth in exports, rice area, and population for the three regions is summarized in table 13.3. At the yield levels prevalent in this period, approximately 0.15 hectares (ha) per person were required to supply an annual requirement of 150 kilograms (kg) of milled rice. Thus, there was clearly more than enough land to produce a surplus. Large Burmese exports were a reflection of the substan-

Table 13.2. Average Annual Japanese Imports from Korea (Chosen), Taiwan (Formosa), and the Rest of Asia, 1911–20 to 1931–38

Dates	Korea (Chosen)	Taiwan (Formosa)	Total imports	Percent of Japanese imports from Korea and Taiwan
	(thousand metric tons)			
1911–20	172	108	573	48.1
1921–30	659	242	1,253	71.9
1931–38	1,107	556	1,736	95.8

Source: V. Wickizer and M. Bennett (1941) appendix table 5.

Table 13.3. Growth in Rice Exports, Rice Area, and Population in the Delta Areas of Southeast Asia Prior to World War II

Year	Rice exports (thousand mt)	Rice area (thousand ha)	Population (thousand persons)	Rice area per capita
		Lower Burma		
1856	284	402[a]	1,318	0.3
1891	1,141	1,882[b]	4,408	0.4
1931	3,348	3,874	6,842	0.6
		Central Plain, Thailand		
1880	209	—	—	—
1911	851	1,152[c]	3,267	0.3
1937	1,330	1,920[d]	5,748	0.3
		Cochinchine		
1880	284	522	1,679	0.3
1890	747	1,175	2,937	0.4
1937	1,548	2,200	4,484	0.5

Sources: Burma—S. Cheng (1968) appendix tables I-A and II-A; and V. Wickizer and M. Bennett p. 226 (1941) table IV. Thailand—J. Ingram (1971) tables IV and V; and Ingram (1964) appendix A. Cochinchine—C. Robequain (1940) p. 220.

[a] 1855
[b] 1862/63
[c] 1910–1914
[d] 1935–1939

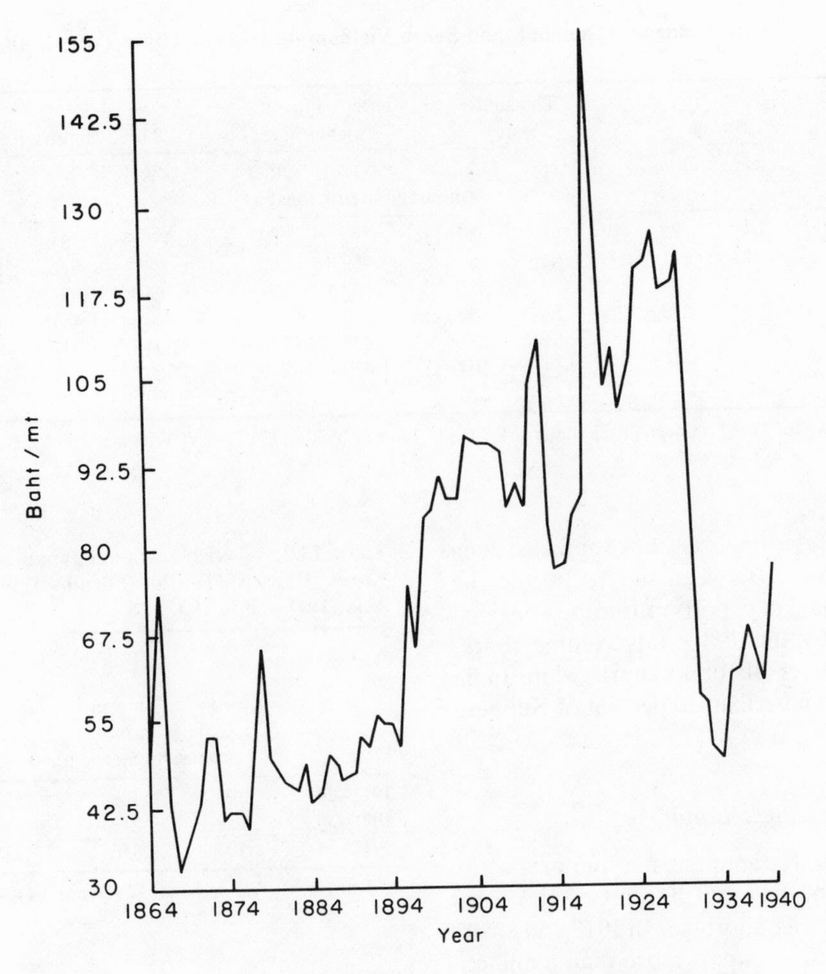

Figure 13.1. Average Thai export milled rice price 1864–1940 (Bhat/mt)

tially larger cultivated area. Rice area per capita was also somewhat larger in Burma in the 1930s than in the rest of Asia.

Throughout most of the period under study, rice prices rose gradually (figure 13.1). Prices in figure 13.1 are uncorrected for inflation and thus are difficult to interpret in real terms, but the trend is clearly positive. Except for two brief periods, prices rose steadily from 1864 to 1930.[10] In contrast to many of the export crops, rice was simply not a "boom or bust" commodity. The favorable rice price-to-wage ratio makes it easier to understand why Thai farmers did not want to work as wage laborers.[11]

The long period of stable or rising rice prices provided a very favorable environment for growth in rice trade. It has been suggested that Thailand's bargaining power in the rice market declined relative to that of Burma and Indochina, whose markets were protected by England and France respectively.[12] However, the Thai relied to a much greater extent on indigenous capital and labor resources, and levels of indebtedness and tenancy were much lower than in the other two countries.[13] Through time, there was a considerable difference in the socioeconomic development among the three delta areas, the details of which are beyond the scope of our discussion.

World Rice Trade After World War II

One might have expected the Southeast Asian delta areas to retain their comparative advantage in rice exports following World War II. Yet, Thailand alone among the three succeeded in maintaining a dominant role. Its share of exports never fell below 20 percent and rose at times to close to 30 percent. Part of the reason for this can be ascribed to the social and political problems in Burma and Vietnam resulting from the ethnic division of labor and colonial intervention. Although Burmese exports exceeded those of Thailand for more than a decade after World War II, Burmese domestic policies, particularly the maintenance of domestic rice prices well below world market levels, led to a gradual collapse of rice exports in the late 1960s and early 1970s (table 13.4).

Technological change also played a role in the

Table 13.4. Rice Exports for Selected Countries 1934–80[a]
(thousand mt milled rice)

Country	1934–36	1954–56	1959–61	1964–66	1969–71	1974–76	1978–80	Percent annual growth 1959–61 to 1978–80
Burma	3,118	1,654	1,668	1,281	664	345	535	−5.6
China	13	671	1,049	892	930	2,648	1,483	3.8
Pakistan	—	141	87	157	324	619	959	11.6
Thailand	1,558	1,266	1,290	1,763	1,221	1,289	2,383	2.9
Asia	8,892	4,087	3,778	4,981	4,445	5,456	6,826	2.0
U.S.	41	662	791	1,410	1,712	1,990	2,546	5.0
Other[b]	607	834	869	1,140	1,610	1,542	2,046	4.7
World	9,540	5,583	5,439	7,531	7,767	8,988	11,418	4.0
Asia as percent of world	93	73	69	66	57	61	60	—
Five leading exporters as percent of world	50	79	68	73	62	77	70	—

Source: FAO, *Trade Yearbook*, various years.
[a] Three-year averages.
[b] All other rice exports, excluding Asia and the U.S.

declining position of the Southeast Asian delta regions. The modern, short-stemmed, fertilizer-responsive varieties developed in the mid-1960s perform best under conditions of good water control and high solar energy, conditions that the delta areas lack. There is some evidence that more recent modern varieties will outperform traditional varieties under these adverse conditions and may help redress the imbalance. Burmese rice production and yields, for example, have shown remarkable growth since the mid-1970s.

In the postwar period, three other countries have emerged as leading exporters of rice—the United States, China, and, to a lesser extent, Pakistan. The five major exporting countries as a group (Burma, China, Pakistan, Thailand, and the United States) have, on average, accounted for about 70 percent of world exports since World War II (table 13.4). With the recent rise in Burmese exports, all five are likely to continue to play a dominant role in export trade for the next decade or two. The growth of U.S. exports, however, has significantly reduced Asia's share of world exports from 90 percent prior to World War II to 60 percent in recent years.

There has also been a substantial shift in the pattern of world imports (table 13.5). Prior to World War II, Asia and Europe together accounted for more than 90 percent of imports and, as recently as 1960, Asia and Europe accounted for three-quarters of all imports. In the last two decades, however, Middle Eastern, African, and Latin American rice imports have grown rapidly, particularly in comparison with those of Asia and Europe. Asia's share of total world imports in the late 1970s and early 1980s was well below 50 percent, but in absolute terms Asian imports remained fairly stable between 1960 and 1980 at around

Table 13.5. Rice Imports by Major Regions of the World, 1934–80[a]
(thousand mt milled rice)

Region	1934–36	1954–56	1959–61	1964–66	1969–71	1974–76	1978–80	Percent annual growth 1959–80
Asia[b]	6,774	3,601	3,896	5,196	4,561	5,674	4,538	0.9
Africa	348	317	534	774	889	928	2,095	7.0
Middle East	50	147	343	402	461	958	1,726	8.6
Western Hemisphere[c]	433	297	358	564	506	678	1,023	5.1
Europe	1,433	647	966	909	928	1,213	1,661	3.4
U.S.S.R.	40	385	403	292	327	266	580	3.7
World	9,078	5,394	5,419	7,735	7,672	8,764	11,623	3.2
Asia as percent of world	75	67	72	67	59	65	39	—

Source: FAO, *Trade Yearbook*, various years.
[a] Three-year average.
[b] Includes Oceania.
[c] Includes North, Central, and South America.

4 to 5 million mt, rising significantly above this level only after the period of widespread domestic shortfall in production in 1972.

There is a wide difference in the quality of rice sold on the world market. The bulk of the rice traded is medium and long grain, usually from indica-type varieties, which tend to be flaky (as opposed to sticky) when cooked. These rices are popular throughout South and Southeast Asia. There is a much smaller market for short-grain, japonica-type varieties grown widely in East Asia (Japan, Korea, Taiwan, and China) north of the Yangtze and in other temperate zones of the world. These rices are sticky when cooked. Indica and japonica rices accounted for about 75 percent of all rice traded in 1980.[14] Other rice types tend to be confined to particular areas or regions and are generally traded as specialty items. Low-quality, parboiled rice is produced and consumed largely in South Asia.[15] High-quality, parboiled rice is produced by major exporters, such as the United States and Thailand, and shipped to Africa and the Middle East. Parboiled rice accounts for about 10 percent of trade. Glutinous rice is consumed as a staple food in the territory stretching from the Shan areas of northern Burma to north and northeastern Thailand, Laos, and the mountain areas of Vietnam. It is also consumed in small quantities in sweets, snacks, and dessert dishes, usually on festive occasions, in many countries in East and Southeast Asia. Only a very small portion of glutinous rice enters world trade. *Basmati* rices are long-grained, scented or aromatic varieties grown largely in the Punjab of Pakistan and in India. In 1980, about 300 thousand mt were exported from this area to the Middle East, where basmati rices are commonly consumed. There are also a number of other minor specialty types.

Rice entering the world market is further graded according to the percentage of broken grains. A low-grade rice with 25 percent brokens will sell for a quarter to a third less than the price of high-grade rice with 5 percent or less brokens. The price differential among rice grades is much greater than for wheat. The different grades and qualities tend to further fragment an already thin market. However, the recent research work of Petzel and Monke indicates that there is a high level of integration in the long- and medium-grain indica market, which makes up the bulk of international trade and the bulk of exports from the leading exporters, Thailand and the United States.[16] There is much less evidence of integration between the indica medium- to long-grain (flaky type) and japonica short-grain (sticky type) markets.

Partly as a result of the wide differentiation in product and lack of market integration, there is no average world price for rice as there is for wheat. Rice traders meet weekly in Bangkok to agree on a "price" for different grades based on recent sales. In Thailand, a few Chinese traders and millers nicknamed the "six tigers" dominate the Asian rice trade.[17] What does not pass through their hands is generally tied up in government contracts. More than 50 percent of the international rice trade is handled under government-to-government contracts.

Exporters typically ship a wide range of grades of varying quality; consequently, average export values of rice vary less than prices for individual grades. However, there are significant differences among the exporters in average unit value and in quality traded (table 13.6). For example, the unit value of U.S. rice has consistently been well above the world average owing to higher quality. In contrast, Burma's average export values are much lower reflecting the lower grades commonly shipped from Burma.

Bangladesh, Sri Lanka, and Sénégal stand out as examples of importers of low-quality rice. The bulk of their import requirements are met by Burma, China, Pakistan, and Thailand. At the other end of the quality spectrum are the OPEC countries—Iran, Saudi Arabia, and Nigeria. The United States is a major supplier of these markets. As noted previously, Pakistan has also been a traditional supplier of high-qual-

Table 13.6. Average Unit Values of Rice Imports and Exports
($US/mt milled rice)

	Average annual unit value of rice	
Country	1970–73	1978–80
Imports		
World	166	399
Bangladesh	106	202
Hong Kong	200	389
Indonesia	203	325
Iran	221	530
Nigeria	257	578
Saudia Arabia	237	608
Sengal	128	239
Singapore	154	370
South Korea	134	327
Sri Lanka	108	257
Exports		
World	158	367
Australia	149	300
Burma	91	238
China	140	407
Egypt	140	347
Italy	163	479
Pakistan	176	352
Thailand	128	311
United States	218	400

Source: FAO, *Trade Yearbook*, various years.

ity basmati rice to the Middle East, and Thailand has been increasing its exports to the Middle East and Africa, serving both the high- and low-quality markets.

Instability

Instability in the world rice market is reflected in short-run price fluctuations and more broadly in the uncertainty that traders face in negotiating contracts. Weather is a major cause of fluctuations in supply, and technological change appears to have contributed to variability in production in the long run. The thinness of the market is another contributing factor to instability, and government policies have also been destabilizing. This section examines the various sources of instability in more detail.

Weather

It is commonly observed that since most rice production occurs in monsoon Asia, the volume of rice traded is particularly vulnerable to weather fluctuations. This factor, plus the desire of countries to maintain a relatively stable domestic price by importing in years of shortage, results in a very volatile world price. However, as noted in chapter 3, weather patterns in monsoon Asia are by no means homogeneous. Correlations in year-to-year shifts in production tend to be positive but not high. However, at least once every ten years unfavorable weather conditions in the region result in a drop in Asian rice production of 5 percent or more. Historically, eastern India has had the highest variability in production of any region in Asia because the failure of the monsoons has a catastrophic effect on production, and a large portion of the area is unirrigated. Yet the impact of production shortfalls in this region on world rice prices is not as great as might be expected because Indian foodgrain imports are largely in the form of wheat.

The effect of weather on prices can be seen in figure 13.2, which shows the long-term trend in the world price of Thai 5 percent brokens and American No. 2 hard red winter wheat. The world rice price rose in the mid-1960s as a result of the failure of the monsoons, which brought about a sharp drop in production in India, Bangladesh, Burma, and Sri Lanka. The extraordinary rise in prices in 1972 was touched off by unfavorable weather that lowered production throughout most of Asia. The coincidental poor wheat harvest, particularly in the Soviet Union, led to a sharp rise in world wheat prices, thus reinforcing the upward pressure on rice prices. The rise in rice prices in 1978 was largely a consequence of poor harvests by the largest exporter, Thailand, and the largest importer, Indonesia. The rise in prices in 1979 and 1980 appears to have been caused by an upsurge in demand in the Middle East and Africa, rather than by weather-related supply shifts in Asia. Total world trade increased approximately 50 percent from 9 to about 12 million mt (table 13.5). While weather-related shortfalls in production initiate the upward movement in price, other factors clearly contribute to the magnitude of the fluctuations.

Technology

There are elements of the new technology that can lead to greater stability in production, such as expanded and improved irrigation and development of disease- and insect-resistant varieties. There are other elements that potentially may increase instability, such as a growing dependency on nonfarm inputs, changing disease and insect patterns caused by the intensification of monoculture rice production (two or three crops of rice each year), and the narrowing of the genetic base through the use of a small number of modern varieties. Available evidence suggests that, on balance, the introduction of new rice technology has increased year-to-year fluctuations in production.[18] For example, in 1976 and 1977 in Indonesia, severe damage to the rice crop caused by brown planthopper infestations was largely responsible for record annual imports of close to 2 million mt from 1977 to 1980. In South Korea in 1980, damage to new varieties of rice caused by cold weather and disease led to imports of over 2 million mt in 1981.

The Thin Market

The Asian rice market can be characterized as "thin" in terms of the small volume of trade relative to the variability in supply or, more specifically, in production. Thus, the effect of year-to-year fluctuations in production is likely to be reflected in substantial price variability. Siamwalla and Haykin argue that the most important consequence of the thin market is the rise in transaction costs:

> The main problem with the rice market in our view is not instability in the sense usually understood, i.e., exhibiting large fluctuations in prices, nor particularly that these fluctuations are the consequence of "thinness"; . . . Rather the main problem lies in the fact that the transaction costs involved are very large. When a country enters the world market (either as exporter or importer), or even when it is staying put and buys or sells the same volume as before,

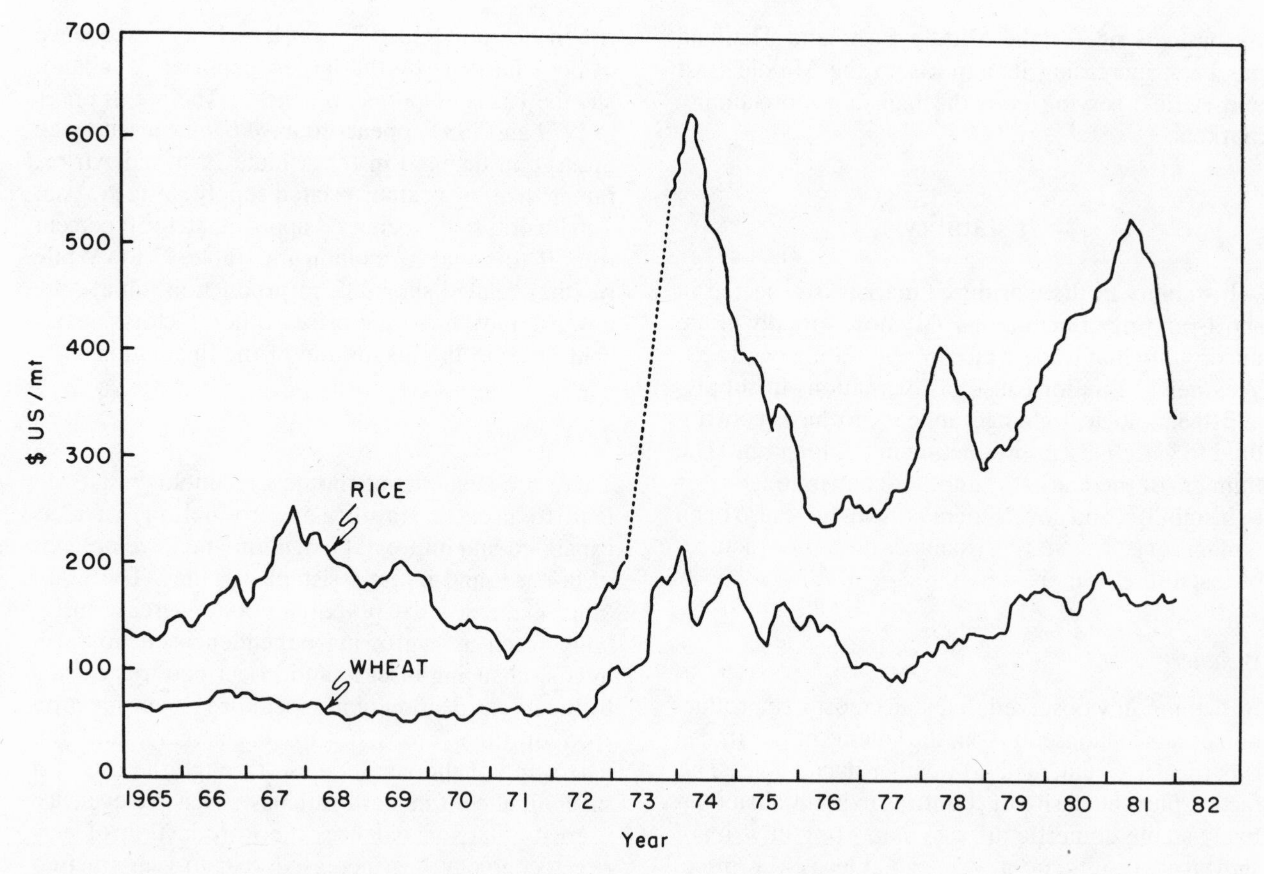

Figure 13.2. The export price of rice (Thai 5 percent brokens, FOB, Bangkok) and the export price of wheat (American No. 2 hard winter ordinary protein, FOB), 1965–82

it has to search for trading partners. There is no rice "supermarket" as there is in the United States for wheat.[19]

Transmitting Instability to the World Market

Government policies substantially contribute to the instability of the rice market. Most net importing countries are making concerted efforts to stabilize domestic prices and to transmit fluctuations in price to the world market. At the same time, there appears to be a general unwillingness among trading nations to absorb (reduce) the fluctuations through price and storage policies. We discuss this latter issue in the following section.

Government rice trade policies for Asia and the United States are summarized in table 13.7. The dominant mechanisms of interference or control are the government monopolization of trade and the imposition of import and export quotas.[20] As noted previously, more than 50 percent of the international trade is handled under government-to-government contracts. The proportion of rice handled in this manner rises significantly during periods of shortages such as occurred in 1973. The Asian countries rely exclusively on quantity rather than price control as the most effective means of isolating producers and consumers from the world market. That is to say, governments prefer quantitative trade restrictions rather than tariffs. This leads to a separation of domestic from world prices.

Governments in Asia have attempted to control grain imports and exports because of the high priority they have placed on domestic price stability. Through such control, they have been able to shift a major share of price instability to the international market. The stability of domestic versus international prices is illustrated in figure 13.3 for Thailand, the major Asian exporter, and Indonesia, the major Asian importer. The Indonesian import price has shown more stability than the Thai export price, but, in both Thailand and Indonesia, domestic rice prices have been more stable than export prices. Asian rice importers and exporters are relatively unresponsive to changes in the international price in the short-run. Each country decides how much to import and export on the basis of amounts needed to stabilize the domestic price.

Table 13.8 shows the degree to which countries

Table 13.7. Government Policy and Rice Trade in Asia and the United States

Country	Trade policy
Bangladesh	Government monopoly
Burma	Government monopoly (Myanma Export-Import Corporation)
China	Government monopoly (China National Cereals, Oils, Foodstuffs Import-Export Corporation)
Hong Kong	Importers are licensed and given quotas determined quarterly by the government.
India	Government monopoly (Food Corporation of India)
Indonesia	Government monopoly (BULOG)
Japan	Government monopoly
Korea	Government control (Ministry of Agriculture and Forestry)
Malaysia	Government control (National Padi and Rice Authority). Private importers are licensed, granted quotas, and required to purchase a portion of government-owned domestic rice.
Pakistan	Government monopoly of high-grade basmati rice (Trading Corporation of Pakistan). Government control of lower-grade rice exports through licensing of private traders; export taxes (since 1972).
Sri Lanka	Government monopoly
Thailand	Government control. Export permits required for private trade; use of rice premium and quotas dependent on domestic and world market conditions.
United States	No control over private trade. Before 1973, the Commodity Credit Corporation provided export subsidies when world prices fell below support prices plus marketing costs. Currently, intervention is limited to offering rice under favorable credit terms or providing rice as a part of international aid in some instances.
Vietnam	Government monopoly

Source: Food and Agriculture Organization (1977).

have been successful in achieving price stability. In all but three of the fourteen Asian countries, the coefficient of variation (standard deviation of deflated price divided by the mean) is well below that for world prices. In Burma, Sri Lanka, and Indonesia, there have been substantial changes in domestic price policy during the period that have contributed to the variability. Furthermore, Burma and Sri Lanka have extensively subsidized distribution systems and have undergone significant changes in subsidy programs. Thus, the coefficient of variation for Burma and Sri Lanka exaggerates the variability that consumers face.[21]

Econometric Studies

There have been numerous empirical econometric studies of international trade in rice.[22] This is a difficult area in which to conduct quantitative research because the rice market responds as much to political as to economic forces, and the political variables are hard to quantify. A related issue, raised previously, is the degree to which the market can be said to be integrated. Petzel and Monke conclude that, in the absence of complete data on prices for imports and exports of a given country, the price of the dominantly traded indica variety (such as Thai 5 percent brokens) can be used for analysis, not with complete confidence but with an expectation that the relationships discussed will be indicative of the time factors under study.[23] Essentially all of the econometric models of price response in the world market have been developed with the assumption, either implicit or explicit, that there is sufficient market integration to render meaningful results.

The general conclusion of these studies is that Asian countries have tended to isolate domestic from international markets through the use of quantitative trade controls. In the short run, they have not been very responsive to changes in rice price.[24] This factor, coupled with the thinness of the market, has led to highly volatile prices and an unstable international market. This contrasts with studies of domestic supply response, which show that Asian rice farmers are responsive to changes in rice prices.[25]

Siamwalla and Haykin fit individual country equations based on a time series (1961–80) of net traded quantities regressed against the world price of rice, a measure of production shortfall, and other exogenous variables.[26] They obtained statistically significant coefficients of price variables in only seventeen out of fifty-five countries or country groups for which regressions were fit. The coefficients of the price variables are shown for ten countries with high variance in production in column 3 of table 13.9. The country with the highest coefficient is China, but the coefficients of monsoon Asia as a whole account for about two-thirds of the variability due to price. Next to China, the United States is the most price-responsive country. However, U.S. response has been largely through acreage adjustments. As a consequence, export increases have been lagged by one to two

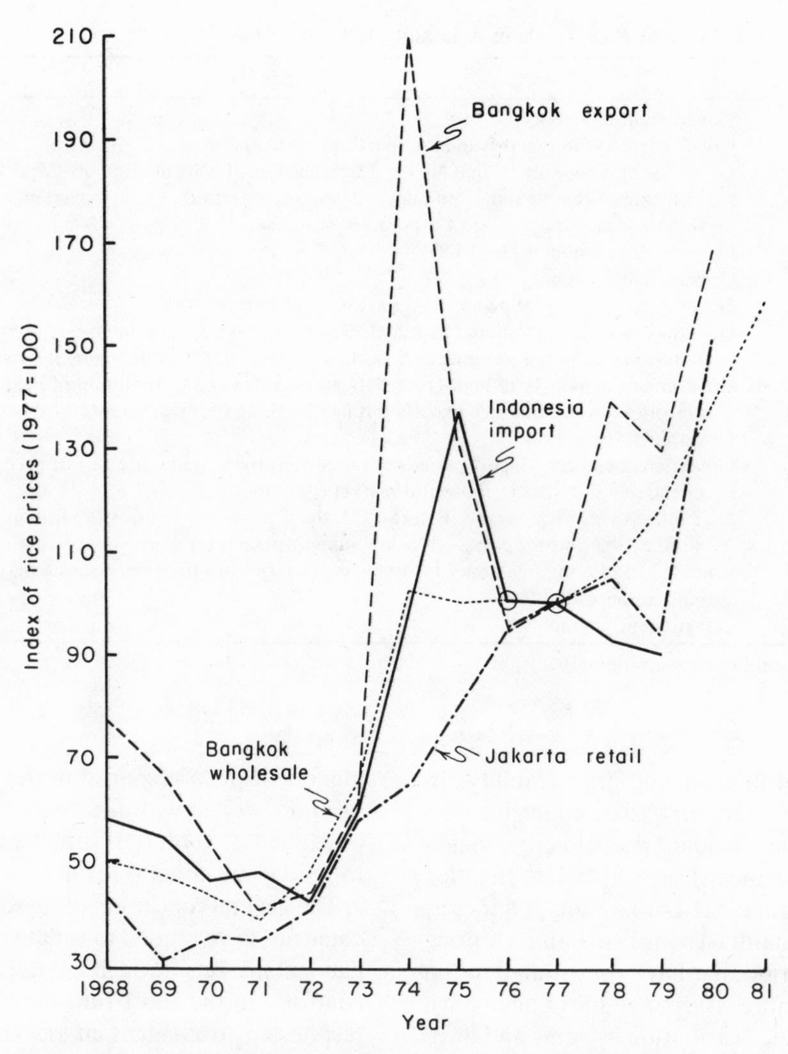

Figure 13.3. Index of 25 percent broken export rice, FOB Bangkok; medium quality retail, Jakarta; Bangkok Wholesale 25 percent broken rice; and Indonesia average CIF value import rice (1977 = 100)

years. For example, in the crisis years of 1973 and 1974, U.S. exports were below the 1972 level.

Falcon and Monke, summarizing the results of the previous econometric analyses of world trade, note that the linkage between price and quantity in these models is illusive.[27] They argue that given the importance of government policy, the appropriate functional form for a model of rice trade should include quantity as an independent variable and price as a dependent variable, a reversal of the form used in many previous studies. That is to say, short-run fluctuations in trade of a small group of countries are caused by, rather than result from world price movements. They identify eleven countries with the highest degree of variance in trade (the ten countries shown in table 13.9 plus Vietnam), arguing that the change in quantity traded by these countries as a group would tend to have the greatest effect on world prices.[28] They estimated a demand curve by regressing the eleven countries with the highest variance as a group for the period 1961–80. The demand curve is kinked around its long-run level, becoming less elastic as prices rise and more elastic as prices fall. As prices have risen reflecting a shortage, Asian countries as a whole have tended to be very rigid in their attempts to meet domestic import requirements. When price falls to a low level, a given change will have considerably more effect on the quantity traded. Siamwalla argues that, during these periods, the Thai government has been slow to reduce the premium (export tax) to meet changing conditions.[29]

In summary, the above studies show that most countries have been price takers. Only a few have been price responsive, but there has been no country that has been willing to accumulate stocks of rice as the United States stored wheat. In recent years, China

Table 13.8. Coefficient of Variation for Real Domestic Rice Prices and World Rice Prices, Selected Periods, Ranked by Coefficient of Variation for Domestic Price[a]

Country	Price	Coefficient of variation Domestic	(percent) World
Burma	1969–79	38.73	30.24
Sri Lanka	1961–80	35.65	30.29
Indonesia	1961–80	30.25	30.29
Indonesia	1971–80	11.37	28.82
Bangladesh	1961–80	17.67	30.29
Malaysia	1961–76	13.07	28.82
Pakistan	1961–80	12.80	30.29
Thailand	1961–80	12.25	30.29
Philippines	1961–80	12.22	30.29
South Korea	1961–80	11.91	30.29
Japan	1961–80	10.81	30.29
Taiwan	1961–80	10.77	30.29
Nepal	1961–80	10.57	30.29
India	1961–80	8.27	30.29

Source: A. Siamwalla and S. Haykin (1983).

[a] Retail domestic prices are based on domestic sources. The world price is the wholesale price of 5 percent brokens, Bangkok. For the price deflator, see World Bank, *Commodity Trade and Price Trends*, various years.

and Japan have acted as sellers of last resort. In a buyer's market, on the other hand, there is no price-responsive importer to soak up any excess supply. The reason for this is in part because of the uncertain long-term market prospects, particularly for the rice exporters. Long-term trends are discussed in the following section.

Table 13.9. Mean Volume of Trade, Variation of Trade, and Coefficient of Production Response to Change in Price, Ten High-Variance Countries, 1961–80, Ranked by Standard Error of Estimate

Country	Mean annual trade (thousand mt)	Standard error of estimate[a] (thousand mt)	Coefficients of price variables[b] (thousand mt/$US1.00 change)
Thailand	1,624	571	n.s.
China	1,257	503	1.70
Indonesia	−1,046	450	n.s.
Japan	16	408	0.79
Burma	806	344	−0.74[c]
South Korea	−247	269	0.78
Bangladesh	420	217	0.49
United States	1,695	214	1.54[d]
Pakistan	422	244	n.s.
India	−242	185	n.s.

Note: n.s. = not significantly different from zero

Source: Synthesized from A. Siamwalla and S. Haykin (1983) tables 7 and 18.

[a] Based on time series of net traded quantities regressed against time.

[b] Based on time series of net traded quantities regressed against the world price of rice and other exogenous variables.

[c] 1966–80.

[d] 1980, separate estimates for 1970 = 1.51, and for 1961 = 1.19.

Long-term Trends in Supply and Demand

Long-term trends in supply and demand have been influenced by the development of new rice technology and the adoption of a rice self-sufficiency policy among Asian importers, the rise in the price of rice relative to wheat, and the growing demand for rice outside of Asia.

New Rice Technology and Government Policy

As noted previously (chapters 7 and 10), the new rice technology performed best in irrigated areas with good water control rather than in the major river deltas, the traditional source for export rice. The major beneficiaries of the new rice technology included the insular countries of Asia (Indonesia, Korea, Malaysia, Philippines, and Sri Lanka) and the rice and wheat producers (China, India, and Pakistan). The remaining countries largely produce rice under rainfed conditions and lie along an axis running from Saigon northwestward to Kathmandu. They include the three traditional exporters, Burma, Thailand, and Vietnam, plus Kampuchea, Laos, Bangladesh, Nepal, and most of eastern India.

When the technology proved successful, the importing countries instigated rice programs designed to promote self-sufficiency, the ultimate security against an unstable market. The exporters, on the other hand, saw little to be gained from promoting the new technology (which initially performed relatively poorly in their environment) or increasing production, since many of the traditional importers appeared to be moving toward self-sufficiency. Thailand, for example, was more concerned with maintaining stable domestic prices than with promoting production for exports. As shown in table 13.10, however, up to 1980 the progress of the importers in achieving self-sufficiency was mixed, with Indonesia and Korea losing ground.

The efforts of countries to promote domestic rice production through policy formation, thereby changing their volume of imports or exports, is discussed in depth in chapters 16 and 17. Here, we note briefly the difference in pricing policy among the trading nations, and the consequences this has for trade.

In table 13.11, we classify the major rice traders in Asia and the United States according to whether they are importers or exporters and where their average farm price falls relative to international (border) prices. The right-hand column in the table shows the

Table 13.10. Net Rice Trade for Major Asian Importers, 1934–36 to 1978–80
(thousand mt milled rice)

Country	1934–36	1959–61	1969–71	1978–80	Net change 1959–80
East Asia					
China	−779	564	923	1,381	817
Hong Kong	−170	−350	−340	−350	0
Japan	−1,779	−193	594	428	621
South Korea	−24[a]	11	−803	−394	−405
Southeast Asia					
Indonesia	−278	−877	−685	−1,925	−1,048
Malaysia	−534[b]	−550[b]	−411[c]	−283[c]	267
Philippines	−28	−64	−145	159	223
Singapore	—[d]	—[d]	−218	−177	
Vietnam[e]	—	277	−312	−231	−508
South Asia					
Bangladesh	—[f]	—[f]	366	361	5
India	−240[g]	−466	−615	30	496
Sri Lanka	−529	−524	−376	−186	338

Source: FAO, *Trade Yearbook*, various years.
[a] Includes both Koreas.
[b] Malaysia includes West Malaysia, Sabah, Sarawak, and Singapore.
[c] Malaysia includes West Malaysia, Sabah, and Sarawak.
[d] Singapore included under Malaysia.
[e] Vietnam includes North and South Vietnam.
[f] Bangladesh included under India and Pakistan.
[g] India includes trade for Pakistan and Bangladesh.

degree to which the domestic currency is overvalued, as this would tend to lower the level of exports and raise the level of imports. In 1979–81, the world price of milled rice was close to $US 0.40 per kilogram (or $US 400 per metric ton, see table 13.6), and the equivalent farm price was approximately $US 0.20 per kilogram (or $US 200 per metric ton).

There are three major groupings of countries in terms of price policy. At one extreme are those advanced economies with prices well above the world market (Japan, South Korea, and Taiwan). At the other extreme are the traditional exporters with prices well below the world market (Burma, Pakistan, and Thailand). Other countries maintain prices reasonably close to the world market, including the traditional Asian importers.

When domestic prices are maintained at levels above the world market price, domestic supply is expanded, exerting a downward pressure on the international price of rice.[30] Conversely, maintaining prices below the world market prices suppresses domestic supply and increases domestic demand, placing an upward pressure on world prices. On balance, it appears that the high domestic prices of several of the more developed economies (including the European Economic Community) and the bias of the new technology in favor of the Asian importers (who have mounted rice self-sufficiency programs) have worked against the traditional exporters by reducing the volume of trade. It is more difficult, however, to determine the net effect on world prices. At least one study has concluded that those countries which have subsidized rice producers have been more than offset by those which have taxed rice producers, resulting in a net upward pressure on world prices.[31]

Rice Versus Wheat

Since 1960, wheat imports have grown at close to 4 percent per year in monsoon Asia (table 13.12). On a regional basis growth has been less even. East Asian wheat imports grew by 5 percent per annum, Southeast Asian imports increased by 8.9 percent, and South Asian imports declined by 1.3 percent over a twenty year period. Very little wheat is imported by the traditional rice exporters, Thailand and Burma. However, wheat imports have increased substantially in almost all rice-importing countries of Asia.

One of the largest wheat importers is Japan, where consumer preference is growing for wheat products. Nevertheless, compared to rice, wheat is still a relatively unimportant dietary component. China, another major wheat importer in Asia, has taken advantage of the price differential between wheat and rice to export significant quantities of rice when world rice prices are high, even though it is a major wheat

Table 13.11. Classification of Trading Nations According to Internal Price Policy and Import-Export Position Based on 1979/80 Paddy Prices

Country & location	Farm price[a] ($US/kg)	Domestic currency overvalue[b] (percent)
	Internal prices above world prices	
Importers		
South Korea (Hwaseong Pref.)[c]	0.66	*
Malaysia (Selangor)[d]	0.26	*
Exporters		
Japan (Ibaraki Pref.)	1.47	*
Taiwan (Taichung)[d]	0.36	*
United States (price received by farmers)	0.25	*
	Internal prices below world prices	
Importers		
Bangladesh (Joydebpur)[e]	0.22	230
Indonesia (Central Java)	0.17	*
Sri Lanka (Kuranegala)	0.13	*
Exporters		
Burma (Rangoon)	0.07	700
China (Hangzhou)	0.13[f]	—
India (Cuttack/Waltair/ Coimbatore)[c]	0.15	*
Pakistan (Punjab)	0.08	11
Philippines (Central Luzon)	0.16	*
Thailand (Suphan Buri)	0.11	*

Source: Asia—A. Palacpac (1982). United States—USDA, Crop Reporting Board, *Agricultural Policies, Annual Summary*. Prices are for selected locations and times. They should be viewed as indicative rather than a true reflection of the average price for the time period.

* Currency is overvalued at 10 percent or less.

[a] Based on conversion from local currency at official exchange rate.

[b] For 1980, based on official currency exchange rate and on foreign bank note selling rate of Deak and Co. (Far East) Ltd., Hong Kong and other sources.

[c] Korea has been an importer since 1977, and India since 1978.

[d] Malaysia has had a farmer support policy since 1970, and Taiwan since 1972.

[e] Based on currency overvaluation, Bangladesh farm price is well below world market equivalent.

[f] Quota price. Above quota price is 50 percent higher or close to world market prices for 1980.

Table 13.12. Net Wheat and Wheat Flour Imports for Selected Asian Countries, 1959–61 to 1978–80[a]
(wheat-equivalent thousand mt)

Country	1959–61	1969–71	1978–80	Percent growth in net imports to Asia 1959–61 to 1978–80
East Asia				
China[b]	3,790	4,836	10,002	
Hong Kong[c]	81	139	174	
Japan[c]	2,641	4,578	5,620	
North Korea	n.a.	271	443	
South Korea	273	1,554	1,754	
Subtotal	6,785	11,378	17,993	5.0
Southeast Asia				
Burma	40	31	13	
Indonesia	159	680	1,051	
Kampuchea	17	22	19	
Malaysia[cd]	255	481	588	
Philippines	147	568	739	
Thailand	35	70	169	
Vietnam[e]	81	691	1,481	
Subtotal	734	2,543	4,060	8.9
South Asia				
Bangladesh	[f]	924	1,564	
India	3,660	2,867	131	
Nepal	2	1	15	
Pakistan	1,036	187	1,298	
Sri Lanka	278	574	821	
Subtotal	4,976	4,553	3,829	−1.3
Total	12,495	18,474	25,882	3.7

Note: n.a. = not available.

Source: FAO, *Trade Yearbook*, 1963, 1972, and 1980.

[a] Based on 72 percent recovery rate from wheat to wheat flour.

[b] Includes Taiwan.

[c] Re-exports are netted out of imports to Japan, Malaysia, and Hong Kong. These countries re-export a portion of their wheat imports as milled flour.

[d] Includes West Malaysia, Sabah, Sarawak, and Singapore.

[e] Vietnam includes North and South Vietnam.

[f] Bangladesh included under Pakistan.

importer.[32] India exports rice in years of surplus foodgrain production and imports wheat in shortage years. Bangladesh and Vietnam rely heavily on donor agencies for financing grain imports and, over the years, have met an increasing share of their import requirements with wheat.

Other countries that have shown a remarkable growth in wheat imports are Indonesia, Malaysia, the Philippines, South Korea, and Sri Lanka, all of whom have had programs designed to achieve rice self-sufficiency. There is a strong reluctance among these countries to increase their dependency on wheat as a primary food source, a crop that none of them produces in any significant quantity. Yet they have shown a clear response to price in the long run as reflected by the very rapid growth in wheat imports. Over the same period, the level of rice imports has remained unchanged (table 13.13). The ratio of rice imports to wheat imports has switched in two decades from almost two-to-one in favor of rice to two-to-one in favor of wheat.

Before discussing the reasons for the remarkable rise in wheat imports, we must consider the differences between wheat and rice. Rice is milled before

Table 13.13. Wheat and Rice Imports of Five Asian Rice Producers, 1959–61 to 1978–80[a]

Crop	1959–61	1969–79	1978–80
Wheat imports (thousand mt)	1,129	3,676	4,853
Rice imports (thousand mt)	1,854	2,420	2,629
Ratio of wheat/rice	0.6	1.5	1.8

Source: FAO, *Trade Yearbook*, various years.
[a] Indonesia, Malaysia, Philippines, South Korea, and Sri Lanka.

being exported while wheat is milled after it is exported. Thus, we are not comparing the price of comparable items. Even after wheat is milled into flour, it will normally cost more than milled rice to process into edible form.

Table 13.14 shows the ratio of the retail price of rice to wheat flour for selected locations in Asian countries. Countries such as the Philippines have priced flour high and made a profit on wheat imports. At the other extreme are Pakistan, a wheat-producing country, and Japan, where the retail price of rice is high relative to wheat flour. The high price of rice in Japan reflects an income transfer to producers. For several countries, however, the price of rice at the retail level is from 1.1 to 1.4 times greater than the price of flour. The difference in the retail cost of rice and the final wheat product per kilogram or per calorie is probably not very great.

The relationship between processed products notwithstanding, the fact that the bulk of wheat is unmilled and the bulk of rice is milled before export does have a major effect on import demand. By importing wheat instead of rice, countries can satisfy domestic requirements for foodgrain with a much lower expenditure of foreign exchange, provided they have wheat milling facilities.

The major stimulant to growth in wheat imports has been the relative rise in the price of rice relative to the price of wheat after World War II. The ratio of the price of rice to wheat rose from approximately 1-to-1 in the prewar period to 2-to-1 in the postwar period (figure 13.4). Substantial fluctuations in the ratio are caused largely by the greater volatility of rice prices compared to wheat prices.

Despite the obvious importance of this price relationship, we are aware of only one study that has provided an analysis of the causal factors. Siamwalla and Haykin analyzed three factors that could explain the relative increase in the price of rice compared to wheat.[33] First, the supply of wheat has been increasing faster than that of rice (3.0 percent per annum versus 2.5 percent per annum from 1952 to 1978); second, the population in the predominantly rice-consuming areas has been growing more quickly than in the predominantly wheat-consuming areas (2.2 percent in rice-consuming countries compared to 1.4 percent in the wheat-consuming areas); and third, the income elasticities among rice consumers are higher than among wheat consumers. Based on the results of their quantitative analysis, they concluded that these three factors contributed about equally to the relative rise in the rice-to-wheat price ratio.

The rates of growth in rice and wheat production, area, and yield are shown for the world and for Asia (excluding China) in table 13.15. The growth rates for rice are essentially the same for the world and for Asia, where 90 percent of the world's rice is produced. However, growth in wheat production has

Table 13.14. Ratio of Retail Price of Rice to Retail Price of Wheat Flour for Selected Locations in Asia, 1979–81

Country & Location	Retail price rice ($US/kg)	Retail price wheat flour ($US/kg)	Ratio of the price of rice/price of wheat flour
Bangladesh (Joydebpur)	0.35	0.24	1.4
Burma (Rangoon)	0.13	0.56	0.2
Taiwan (Taichung)	0.58	0.42	1.4
India (Cuttack/Waltair/ Coimbatore)	0.27	0.25	1.1
Indonesia (Central Java)	0.32	0.28	1.1
Japan (Ibaraki Pref.)	1.54	0.45	3.4
Malaysia (Selangor)	0.50	0.35	1.4
Pakistan (Punjab)	0.38	0.09	4.2
Philippines (Central Luzon)	0.30	0.00	0.3
South Korea (Hwaseong Pref.)	0.92	0.31	3.0
Sri Lanka (Kuranegala)	0.31	0.22	1.4

Source: A. Palacpac (1982). Prices are for selected locations and times within the three year period. Prices are not average prices for the entire country under consideration, but are indicative of general trends.

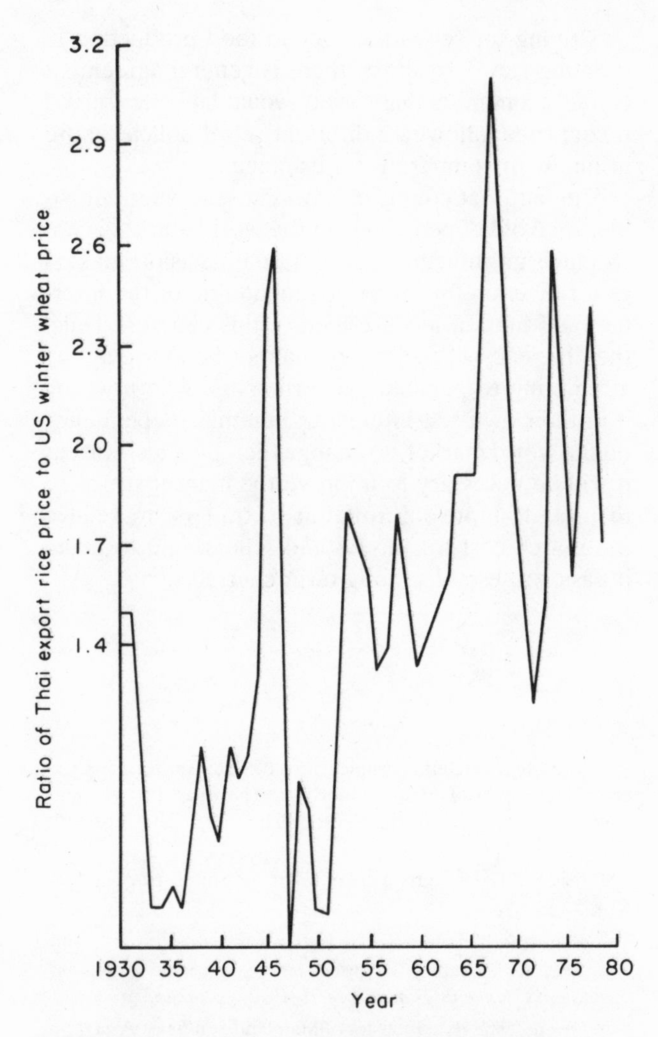

Figure 13.4. Ratio of average Thai export rice price to the U.S. No. 2 winter wheat price, FOB Kansas City, 1930–79

been much more rapid in the Indo-Gangetic Plain, which represents about 10 percent of world wheat production. The growth of wheat production in Asia is almost double that for the world as a whole (5.7 versus 3.0 percent), with both area expansion and yield contributing strongly to the growth. Although the figures are less accurate, the same basic picture emerges in China, with growth of wheat production (largely in the Central Plain) far outstripping rice production. Thus, improvements in wheat production in Asia that are the result of technological changes have largely been responsible for the decline in world wheat prices relative to world rice prices.

In summary, both the absolute price differential between rice and wheat and the relative price changes as well as population trends and differing income elasticities have favored the growth of wheat imports in most Asian countries. Wheat imports have grown not only in response to domestic consumer demand,

Table 13.15. Growth in Rice and Wheat Production, Area, and Yield from 1960 to 1980
(percent per year)

	World		Asia[a]	
	Rice	Wheat	Rice	Wheat
Production	2.5	3.0	2.5	5.7
Area	0.9	0.8	0.9	2.5
Yield	1.6	2.2	1.6	3.2

Source: FAO, *Production Yearbook*, various years.
[a] Excluding China.

but also in response to government demand. Wheat has also been the grain of choice among the aid donors, many of whom are important wheat producers themselves.

Non-Asian Demand for Rice

As noted previously (table 13.5), Asian imports as a percentage of total rice imports have dropped from about 65 to 70 percent to 40 percent. Growth in the rice export market is the direct result of increased demand among countries outside of monsoon Asia that produce very little of their own domestic supply (table 13.16).

The question arises as to whether this long-term growth in demand will be any more stable than the highly volatile demand of the monsoon Asian importers. Factors such as changing oil prices, interest rates, and foreign exchange rates (which are becoming increasingly volatile) are likely to have a significant impact on the import demands of these countries.

Improving Performance

In the previous sections of this chapter, we described a market characterized by short-run instability and long-run uncertainty. The short-run instability is related to fluctuations in production, to a lack of market integration and the resultant high transaction costs in the search for trading partners, and to national

Table 13.16. Rice Imports as a Percentage of Production, 1950–79

	Imports as percent of domestic production	
Period	Monsoon Asia	Other
1950–54	7.3	47.5
1955–59	7.1	71.5
1960–64	6.2	55.8
1965–69	4.7	64.8
1970–74	4.7	69.0
1975–79	3.8	85.7

Source: A. Siamwalla and S. Haykin (1983) table 3.

government policies designed to stabilize domestic prices thereby transmitting price instability to the world market. The long-run uncertainty is related to the introduction of new technology (which has favored the importers), to national government policies designed to take advantage of the technology to achieve national self-sufficiency, and to changes in the demand for rice relative to other staples, particularly wheat. Government policies can be seen as a response to the market environment, but these policies have led in turn to even greater instability and uncertainty.

Several studies have pointed to inefficiency in the world rice market.[34] Furthermore, in recent studies of food security issues, there seems to be general agreement that the food-deficit developing countries should store money in the form of foreign exchange reserves rather than stockpiling food in the short run, or striving for self-sufficiency in food production in the long run.[35] In short, there is general agreement among economists that society would be better served if countries followed a different set of policies more in line with comparative advantage.

The fact that countries have chosen, where possible, to avoid dependency on the world market is not so much an indictment of national decision makers as a reflection on the poor functioning of the international market as described in this chapter. Thus, the efficiency of the market cannot be improved by attempting to persuade governments, who now see it in their own self-interest to minimize dependency on the world market, to change their policies. Rather it will be necessary to improve the market structure to provide more information to traders, to reduce the search cost for buyers and sellers, and to make it easier to establish long-term contracts.[36]

Notes

1. Norman G. Owen, "The Rice Industry of Mainland Southeast Asia 1850–1914," *Journal of the Siam Society* vol. 59 (July 1971) pp. 78–142 presents an excellent account of the period prior to World War I as well as detailed comparative statistics on trade and prices. V. D. Wickizer and M. K. Bennett, *The Rice Economy of Monsoon Asia* (Stanford, Calif., Stanford University Press, 1941) contains basic comparative statistics for the period from World War I to World War II. Two excellent references on Burma's rice trade are: Soik-Hwa Cheng, *The Rice Industry of Burma, 1852–1940* (Singapore, University of Malaya Press, 1968); and Michael Adas, *The Burma Delta Economic Development and Social Change on an Asian Rice Frontier, 1852–1941* (Madison, Wisc., University of Wisconsin Press, 1974). An account of the Thai export industry is in J. C. Ingram, *Economic Change in Thailand, 1850–1970* (Stanford, Calif., Stanford University Press, 1971) and J. C. Ingram, "Thailand's Rice Trade and Resources," in C. D. Cowan, ed., *The Economic Development of Southeast Asia—Studies in Economic History and Political Economy* (New York, Praeger, 1964). The latter article contains a rich set of historical statistics and is one of the few resources that provides a continuous series on rice prices extending up to and beyond the World War II period. Charles Robequain, *The Economic Development of French Indochina* (New York, Oxford University Press, 1944) (translated from the 1939 French original) and Pierre Gorou, *Land Utilization in French Indochina* (Washington, D.C., Institute of Pacific Relations, 1945) touch on rice trade in the Mekong Delta, but Indochinese data are not as comprehensive as those for Burma and Thailand.

2. Charles A. Fisher, "Some Comments on Population Growth in Southeast Asia with Specific Reference to the Period Since 1830," in *The Economic Development of Southeast Asia—Studies in Economic History and Political Economy* (New York, Praeger, 1964) pp. 48–71 and Ammar Siamwalla, "Land, Labor and Capital in Three Rice Growing Deltas of Southeast Asia, 1800–1940," Economic Growth Center Discussion Paper No. 150 (New Haven, Yale University, July 1972).

3. The Mekong Delta occupies most of what was formerly Cochinchine, a part of French Indochina. The Chao Phraya Delta covers the lower part of the Central Plain of Thailand (formerly Siam).

4. Owen, "The Rice Industry of Mainland Southeast Asia," pp. 83–85.

5. Ibid., pp. 85–86.

6. See, for example, the discussion of this issue in Ingram, *Economic Change in Thailand 1850–1970*, pp. 42–43.

7. Owen, "The Rice Industry of Mainland Southeast Asia," pp. 110–111.

8. Wickizer and Bennett, *The Rice Economy of Monsoon Asia*, pp. 90–91.

9. Owen, "The Rice Industry of Mainland Southeast Asia," p. 92.

10. Ibid., pp. 90–91. The extremely poor harvest in India in 1918–19, coupled with a sharp rise in Japanese import demand, led to a tripling of rice prices. The 50 percent fall in rice prices from 1929 to 1934 had a more serious long-term impact on rice trade.

11. Ingram, "Thailand Rice Trade," pp. 112–113.

12. Erich H. Jacoby, *Agrarian Unrest in Southeast Asia* (New York, Columbia University Press, 1949) pp. 226–227.

13. This issue is discussed in some depth in Siamwalla, "Land Labor and Capital in Three Rice Growing Deltas of Southeast Asia, 1800–1940." The situation in Burma is documented in detail in Adas, *The Burma Delta*, and Cheng, *The Rice Industry of Burma, 1852–1940*. In the late 1930s, over half of the rice land in Lower Burma was owned by moneylenders and absentee landlords.

14. Norman Efferson, former Chancellor of Louisiana State University to the authors.

15. Parboiling is a hydrothermal treatment of rough rice prior to milling. Asian parboiled rice tends to be poor quality because

of the low quality of the rice used and rather primitive parboiling facilities. In contrast, U.S. parboiled rice is of uniformly high quality (e.g., Uncle Ben's).

16. Todd E. Petzel and Eric A. Monke, "The Integration of the International Rice Market," *Food Research Institute Studies* vol. 17 (1979–80) pp. 307–325.

17. Dan Morgan, *Merchants of Grain* (New York, The Viking Press, 1979) pp. 382–383.

18. For evidence of this see Randolph Barker, Eric C. Gabler, and Donald Winkelmann, "Long-Term Consequences of Technological Change on Crop Yield Stability: The Case of Cereal Grain," in Alberto Valdez, ed., *Food Security for Developing Countries* (Boulder, Colo., Westview Press, 1981); Shakuntla Mehra, *Instability in Indian Agriculture in the Context of New Technology*, Research Report No. 25 (Washington, D.C., International Food Policy Research Institute, 1981); and Peter B. R. Hazell, "Instability in Indian Foodgrain Production," Research Report No. 30 (Washington, D.C., International Food Policy Research Institute, 1982).

19. Personal communication from Ammar Siamwalla and Stephen Haykin.

20. Falcon and Monke, "International Trade in Rice," pp. 283–284.

21. Siamwalla and Haykin, *The World Rice Market: Structure, Conduct, and Performance,* Research Report No. 39 (Washington, D.C., International Food Policy Research Institute, June 1983) p. 55.

22. An important empirical study of international rice trade is F. G. Adams and J. R. Behrman, *Econometric Models of World Agricultural Commodity Markets* (Cambridge, Mass., Ballinger, 1976), which compares the response of developed, developing, and planned-economy countries. Other references that estimate separate equations for individual countries include Olin Chipravat and S. Pariwat, "An Econometric Model of World Rice Markets" (Bangkok, Department of Economic Research, Bank of Thailand, 1976); Warren R. Grant, T. Mullins, and W. F. Morrison, *World Rice Study: Disappearance, Production and Price Relationships Used to Develop the Model*, Economic Research Service No. 608 (Washington, D.C., U.S. Department of Agriculture, 1975); and Siamwalla and Haykin, "The World Rice Market." Four additional studies take an intermediate approach toward the aggregation problem and concentrate on important trading countries. These are: Virach Arromdee, "Economics of Rice Trade Among Countries of Southeast Asia" (Ph.D. dissertation, University of Minnesota, 1968); Hiren Sarkar, "A Simulation Model of the World Rice Economy with Special Reference to Thailand," DAE-CARD Sector Analysis Series No. 14, Center for Agricultural and Rural Development (Ames, Iowa, Iowa State University, 1978); Walter P. Falcon and Eric A. Monke, "International Trade in Rice," *Food Research Institute Studies* vol. 18 (1979–80) pp. 297–306; and Hiroshi Tsuji, "A Quantitative Model of the International Rice Market and Analysis of National Rice Policies, with Special Reference to Thailand, Indonesia, Japan, and the United States," in Max R. Langham and Ralph H. Retzlafeds, eds., *Agricultural Sector Analysis in Asia* (Bangkok, Agricultural Development Council, 1982).

23. Petzel and Monke, "The Integration of the International Rice Market," p. 323.

24. It should be emphasized that the lack of price response has been in terms of short-run supply and demand for rice in the international market. There can be little doubt that high world prices have a major impact on domestic rice production and policies and programs.

25. In the numerous studies of the response of rice supply to price conducted in the postwar period, the estimates of short-run elasticity of response fall primarily in the range of 0.1 to 0.3, and the estimates of long-run elasticity in the range of 0.4 to 0.6. See table 1 in J. F. Sison, Somsak Prakongtanapan, and Y. Hayami, "Structural Changes in Rice Supply Relations: Philippines and Thailand," in International Rice Research Institute, ed., *Economic Consequences of the New Rice Technology* (Los Banos, Philippines, IRRI, 1978).

26. Siamwalla and Haykin, "The World Rice Market," pp. 42–54.

27. Falcon and Monke, "International Trade in Rice," p. 287.

28. Ibid., pp. 288–295.

29. Ammar Siamwalla, "A History of Rice Policies in Thailand," *Food Research Institute Studies* vol. 14 (1975) pp. 233–249.

30. For a more detailed discussion of this issue, see James A. Roumassett and Arsenio M. Balisacan, "The Political Economy of Rice Policy and Trade in the Asian-Pacific Region" (Honolulu, East-West Center Resource Systems Institute, 1983).

31. Rodney Tyers, "Food Security in ASEAN: Potential Impacts of a Pacific Economic Community" (Canberra, Australian National University, ASEAN-Australian Economic Relations Research Project, May 1982).

32. There is little evidence of arbitrage in the strict sense of the word. Dennis L. Chinn, "A Calorie-Arbitrage Model of Chinese Grain Trade," Working Paper (Stanford, Calif., Food Research Institute, Stanford University 1979) suggests that the Chinese supply response occurs largely through trade flows, with rice exports and wheat imports increasing with a rise in the rice-to-wheat price ratio. His econometric analysis tends to confirm this relationship, but it is not clear how the trade decisions are made.

33. Siamwalla and Haykin, "The World Rice Market," pp. 29–33.

34. See for example: C. Peter Timmer and Walter P. Falcon, "The Impact of Price on Rice Trade in Asia," in G. S. Tolley and P. A. Zadrogny, eds., *Trade, Agriculture and Development* (New York, Ballinger, 1975); Eric A. Monke, Scott R. Peason, and Narongchai Akaransanee, "Comparative Advantage, Government Policies, and International Trade in Rice," *Food Research Institute Studies* vol. 9 (1976) pp. 257–283; James A. Roumasset and Arsenio M. Balisacan, "The Political Economy of Rice Policy and Trade."

35. This perspective is reflected in two recent volumes: Alberto Valdez, ed., *Food Security for the Developing Countries* (Boulder, Colo., Westview Press, 1981); and Anthony H. Chisholm and Rodney Tyers, eds., *Food Security: Theory, Policy and Perspectives from Asia and the Pacific Rim* (Lexington, Mass., Lexington Books, 1982).

36. Both Siamwalla and Haykin, "The World Rice Market" and Roumasset and Balisacan, "The Political Economy of Rice Policy and Trade" emphasize the need to remove the impediments to establishing long-term contracts. They suggest that improved efficiency could be achieved through long-term bilateral agreements rather than through future market operations as in the case of wheat.

14

Priorities for Rice Research and Extension in Asia

The success of the research that produced the Green Revolution has contributed to the widespread recognition of research as an essential component of agricultural development. Even so, it is not enough to recognize the need for investment in research. Funds must be allocated so that the parts of a research program logically relate to one another and the results of those research activities reach the farmers. This suggests an important role for extension work. This chapter concentrates mainly on research, but also treats extension activities—which are important where new technologies are so different that farmers cannot adopt them without assistance.[1]

The world rice research system consists of international, national, and, within some countries, regional institutions all of which are linked to varying degrees. The Food and Agriculture Organization of the United Nations (FAO) and the International Rice Research Institute (IRRI) provide mechanisms for international exchange of rice information and research. IRRI and sister international centers in Africa and Latin America are responsible for providing rice genetic material, developing research methodologies, and coordinating research activities worldwide. Scientists at national and regional institutions develop rice varieties needed for their specific environments and evaluate them under those conditions. Cooperation among researchers can speed the research process dramatically, especially in the case of varietal development where one year of testing at a number of locations can substitute for several years of testing at one location. All components must be working effectively, however, for the system to achieve its maximum productivity.

Unlike fertilizer distribution, machinery development, and rice marketing, there is a clear economic argument for direct involvement of government in agricultural research. Such research, particularly in the biological sciences, has many of the characteristics of a public good. The use of the product, that is, research findings, by one farmer or group of farmers, does not exclude its use by others. Furthermore, it normally is not possible to charge a user's fee for the products of agricultural research. This does not imply that information from research is equally usable by all. Certainly this has not been the case with the new rice technology, as illustrated by interregional differences discussed earlier. Still, private firms cannot readily internalize the benefits from research through such procedures as patents. Therefore, investment in biological agricultural research has traditionally been undertaken by the government rather than by the private sector.

While there is a clear role for the public sector in research, there are many difficult issues that come with implementing that responsibility. Administrative control, rather than control by market forces, means that there are no clear signals for deciding on allocation of research funds. Donor agencies in developed countries have supported the emerging world agricultural research system, but how much should they contribute to agricultural research? What types of research centers and activities should be supported? What should be the balance of support between national and international programs, between basic sciences and development of technology, between current research problems and development of research capacity? How can national governments be encouraged to provide more support for research activities and for training research workers? The above questions are of concern to the international centers as well as to national research systems.

The chapter is divided into four parts. The first section reviews the productivity of agricultural research investment and the structure of the rice research system in Asia. Then, methodologies for establishing research priorities are discussed. The third section presents the results of an analysis of the optimum allocation of research inputs and effort among the different rice growing environments. The fourth discusses the implications of these results for rice research in Asia.

The Organization of Rice Research

Throughout this century, the level of investment in rice research in East Asia (Japan, Korea, and Taiwan) has been severalfold that in South and Southeast Asia (table 14.1). Japan is, of course, the leader, with slightly more than half of the total world research investment in rice.

Beginning with the work of Griliches for hybrid corn in the United States, there have been a host of studies on returns to investment in agricultural research.[2] A significant number have dealt with rice (table 14.2). Hayami and Akino show that the internal rate of return to investment in research increased after 1930 when Japan shifted from a breeding program based on pure-line selection to an emphasis on the development of varieties by crossbreeding.[3] In tropical Asia, the internal rate of return for national research programs increased substantially in the period after 1966 as national programs were able to lay claim to a major share of the benefits of the new rice technology.[4] The consistently high internal rate of return found in all studies suggests a chronic state of underinvestment in rice research.

There has been considerable discussion about the accuracy and meaning of these findings.[5] We agree with those who believe it would be wrong to conclude that because returns are so high, it is unnecessary to worry about the allocation of funds. In fact, just the

Table 14.1. Annual Investment in Rice Research and Extension
(million 1970 constant $US)

	East Asia		Southeast Asia		South Asia		Other developing countries		Developed countries		
Period	R	E	R	E	R	E	R	E	R	E	IRRI Research
1900–20	0.9	—	—	—	0.1	—	0.5	1.0	1.0	0.5	—
1921–40	2.7	1.9	0.1	0.5	0.3	1.0	1.0	2.0	2.0	1.0	—
1951–55	10.0	3.0	2.1	3.0	1.7	2.7	1.5	3.0	5.0	3.0	—
1956–60	17.5	24.9	2.0	3.7	1.8	2.9	1.8	3.5	5.5	3.1	1.0
1961–65	32.0	7.1	2.7	5.7	3.0	4.8	3.0	6.0	7.0	3.5	1.8
1966–70	45.0	17.1	3.2	7.2	4.0	11.0	5.0	10.0	8.0	4.0	2.9
1971–75	48.2	18.3	3.1	7.2	4.4	11.7	7.1	12.0	11.2	5.5	4.0

Notes: Excludes China. R = research; E = extension.
Source: R. Evenson and P. Flores (1978).

Table 14.2. Summary of Studies of Rice Research Productivity

Study	Country/region	Time period[a]	Annual interest rate of return (percent)
Evenson and Flores, 1978	Asia-national	1950–65	32–39
Evenson and Flores, 1978	Asia-national	1966–75	73–78
Evenson and Flores, 1978	Asia-international	1966–75	74–105
Flores, Evenson, and Hayami, 1976	Philippines	1966–75	27
Flores, Evenson, and Hayami, 1976	Tropics	1966–75	46–71
Hayami and Akino, 1977	Japan	1915–50	25–27
Hayami and Akino, 1977	Japan	1930–61	73–75
Hertford et al., 1977	Colombia	1957–72	60–82
Scobie and Posada, 1976	Colombia	1957–74	79–96

[a] Refers to period for investments, but not for the stream of benefits.

opposite conclusion seems warranted. Given the scarcity of resources and high potential payoff, more time needs to be devoted to the question of allocation. But it is difficult to suggest how additional funds should be invested without a clear understanding of the structure, organization, and administration of rice research in Asia.

A typology of rice research systems can be drawn that relates research skills and institutional organization to the stage of development of the system.[6] A research system passes through three stages of development: (1) the *low-skilled stage*, dependent primarily on technical and engineering skills and characterized by widely diffused commodity-oriented experiment stations; (2) the *intermediate hierarchical stage*, with appreciable scientific skills and substantial economies of scale to be gained by the concentration of these skills in leading institutions; and (3) the *advanced science-based stage*, characterized by a large supply of conceptual scientific skills and emphasis by the most highly regarded centers on research that does not have a direct technological objective.

In the early stages of development, the low-skilled system depends heavily on the transfer and simple adaptation of technology. As a system matures, capacity to develop new technology, given the state of scientific knowledge, is added. In the final stage of the system's development, skills are appended that permit basic scientific breakthroughs.[7]

Japan is perhaps the only country in Asia where the rice research system has passed through all three stages and can today be characterized as advanced science-based. The shift from the low-skilled stages to the intermediate hierarchical stage occurred in the mid-1920s when the build-up of technical and scientific skills resulted in a major reorganization of agricultural research. Under the subsequent "assigned experiment station" system, the national experiment station was given the responsibility for conducting the initial crosses while breeding centers in each of the eight regions conducted further selections for different ecological conditions. The intermediate hierarchical system that emerged allowed Japan to capitalize on the development and dissemination of crossbred varieties.

Although the same scientific knowledge was potentially available to the experiment stations established throughout the tropics in the early part of the century, the scientific manpower needed to translate this knowledge into new technology did not emerge for many reasons. The handful of scientists in most Asian countries were often caught up in all phases of agricultural service—research, extension, and administration—and became "jacks of all trades."[8] Rice research started at a single experiment station in most countries, but the great diversity of existing rice varieties was considered to be inevitable because of their narrow adaptability, and therefore, in most countries, several stations were eventually established, each in a known ecological area. This further diffused the available scientific personnel.

In the decade after World War II, efforts to encourage agricultural development in the tropics still tended to ignore the potential of research in food crops. Extension received priority over research in part because the benefits promised to be more immediate, and in part because earlier experience had shown that higher production could be achieved with existing technology.[9]

Beginning in 1954, the extension model was superseded and incorporated into a more comprehensive organizational structure for agricultural development patterned after the land grant university in the United States.[10] The adoption and promotion of the land grant model was reflected in international aid agency funding of developing country research. This represented 40 to 50 percent of total investment in the 1950s and about one-third of the total in the mid-1960s.[11] In research, export crops continued to be favored over foodgrains. With one or two exceptions, such as India, the national research programs of tropical Asia could continue to be categorized as low-skilled systems. This lag in the development of research organization and scientific skills set the stage for the technological breakthrough that was to follow. The establishment of IRRI in 1962 as the "main station" in an international hierarchical system can be viewed as a temporary departure from the basic pattern of developing national research systems. On the other hand, countries with few human resources and limited geographic diversity may, for a long time, find it more advantageous to stick with low-skilled systems and look to an international center for backup. This suggests that the centers will be a feature of the global agricultural research system for some decades to come.

One could argue, of course, that following in the path of the Rockefeller Foundation country programs in Latin America and India, the aim should have been to develop national research main stations. However, the investment resources and the indigenous professional skills were not available and are still extremely limited in many countries in Asia today. In the creation of IRRI the founders gambled that the concentration of a "critical mass" of research skills could result in the development of technology with a high degree of transferability. The gamble paid off with a new plant type that proved to be widely transferable across the irrigated rice paddies in tropical Asia.

The slower rate of acceptance of the new varieties in rainfed environments outside of the Philippines suggests that the kind of diversity that is eliminated in well-controlled irrigated paddy fields needs to be served by research facilities located in similar environments. This clearly cannot be accomplished by the international centers alone—it requires strong national programs.

The creation of IRRI and the other international centers for biological research has been referred to as the *"big science" model*,[12] despite the fact that the centers typically have fewer doctoral level scientists than do many single academic departments in U.S. land grant universities. After 1965, international aid support for the land grant and extension activities declined as more and more funds were devoted to the establishment of the international agricultural research centers.[13] The main criticism of this approach is that the resulting new varieties tend to be adopted primarily by farmers who are located in relatively favorable environmental conditions (that is, with irrigation) and who can afford the purchased inputs that make the varieties productive. Furthermore, analysis of returns on research investment (table 14.2) shows that returns in national programs are very high. There is a strong degree of complementarity between the work of the national institutions and international centers.[14] A strong national program can facilitate the spread of new technology by adapting the exotic materials to local conditions. This capacity becomes increasingly important as the easy gains in productivity in the more favorable environments are fully exploited. The establishment of the International Agricultural Development Service (IADS) in 1975 and the International Service for National Agricultural Research (ISNAR) in 1978, both of which focus on strengthening national research systems, reflects a growing recognition of the need to achieve an appropriate balance of aid between the international and national programs.

In 1980, rice research in Asia had a number of components. The United States, Japan, and IRRI were conducting virtually all the advanced science-based rice research. India, China, and to a lesser extent Indonesia and Thailand, had intermediate hierarchical systems. All four had links to IRRI for germplasm and scientific exchange, but they also had their own domestic hierarchical organizations. Most other Asian countries were at the low-skilled level in 1980, with one or two effectively operating research stations. Most were in the process of developing hierarchical systems, but for all practical purposes, only their main stations were effectively operating, and those main stations were linked to the international institutes. Most of the research done in Asia, including that at the international centers, was of a very applied nature, with the basic biological research on which the technology of the future will be developed being carried out in laboratories scattered throughout the world. This basic research is neither confined to, nor necessarily related to, rice, but its fruits will find their way into the rice research system in time.

Methods of Establishing Research Priorities

Although the issue can be debated, there was undoubtedly more agreement among rice research workers in the 1960s than in the 1980s about the best research strategy for increasing rice production. The lag in technology development created a gap, but experience with small grains elsewhere suggested the potential to be gained from breeding a short-strawed fertilizer-responsive variety. When this objective was achieved, however, the subsequent steps to increase production were less obvious. Thus, a little more than a decade after the establishment of IRRI, the appropriate allocation of research resources was a matter of considerable debate. Scientific and management staff alike showed increasing concern for the need to develop a clearer perception of research priorities.

Agricultural research priorities have been evaluated by both ex post studies and ex ante models.[15] A primary objective of many ex post studies is to assess the economic returns to research investment (for example, see table 14.2). The various ex ante methods (that look to the future) are the most appropriate for establishing research priorities. The advantage of ex ante procedures is that they provide a formal means of using pooled judgment. In degree of methodological sophistication, the ex ante models range from simple scoring schemes to highly complex mathematical programming models. To a greater or lesser degree, all models depend on the judgment of researchers or other knowledgeable individuals concerning the outcome of future events. While the results may be sensitive to personal opinion, some of the most important findings are likely to hold under a wide range of sensitivity tests.

Rice researchers have applied some of these approaches to evaluate research allocation issues confronting the world rice research system. One analysis attempts to answer the question of whether the total investment currently made for rice research in Asia is adequate. Chapter 18 reports the details of that analysis. The growing gap between projected demand for food and the trend rate of growth in food supply reflects the need to achieve more rapid increases in production.[16] A preliminary study to determine the investments required in irrigation, fertilizer, and research to increase Asian rice supplies at a pace in keeping with projected demand showed that, without increases in fertilizer productivity, feasible increases in fertilizer and irrigation would not provide enough growth in output to keep up with demand.[17] The investigation looked on research as a method for bringing supply into balance with demand.

One method that has been used to determine appropriate allocations uses a productivity approach to examine the benefits expected from research in the different rice growing environments—irrigated, rainfed, upland, and deepwater—in the main rice growing countries of South and Southeast Asia. Scientists believe that many research findings are specific to a particular environment, and this is reflected in the more rapid adoption of new varieties in the irrigated areas. Thus, the analysis assumes that the four types of rice defined by different environmental conditions are, from a research input perspective, essentially different commodities.

This analysis of production potential in different rice growing environments has implications for income distribution. Many of the rural poor in Asia are located in the unirrigated rice-producing regions, particularly in eastern India and Bangladesh. The initial success of the new rice technology in the irrigated environment has tended to widen the disparity between irrigated and unirrigated regions. Successful technology specifically adapted to rainfed rice environments may redress that disparity.

Analyzing Allocation of Research Inputs to Rice Environments

In this section, we first present the conceptual framework of the productivity approach. Then the analysis and results are discussed under three headings: (1) gross benefits, (2) net benefits for irrigated vs. rainfed rice, and (3) contribution of research by country.

The classification of rice environments followed was discussed in chapter 2. Figure 14.1 provides a rough indication of the amount of rice crop area and production in each of these environments. Clearly, irrigated and shallow rainfed rice are the dominant categories.

Developing new rice technology for each of these environments involves consideration of a wide range of factors in addition to varietal type. The process that led to successful varieties for irrigated areas can, we believe, also produce modern varieties specifically suited for rainfed, floating rice, and dryland areas if rice researchers allocate enough resources for these environments. Such activities were getting under way in the early 1980s.

The Productivity Approach

Theoretically, to maximize the productivity of research resources, expenditures should be allocated so that the marginal productivity of research expenditures on each environment is equated. This means roughly that the increase in productivity expected from an additional dollar spent on research for each environment should be equated. A model can be formulated that takes into account the time required to obtain research results, the probability of success, the expected yield increase, expected changes in cropping intensity, area affected, direct cost of using the technology, and the investment cost.[18]

Given such a formulation and the necessary data for each type of rice environment, it would be optimal to allocate research resources so as to equate the net present value per dollar of research investment of potential new technology for each environment. The model was applied by IRRI to the problem of allocating research investments among the major types of rice growing environments. Because investments in building new irrigation capacity will continue, an additional effect on productivity resulting from the expansion of irrigation was also included.

Gross Benefits

A group of IRRI scientists estimated the expected increase in rice yield and cropping intensity that would be possible from "reasonable" research and extension inputs directed at each environment for South and Southeast Asia.[19] It was assumed that these yields would be realized over a twenty year period. At the outset, the probability of success, the direct cost of technology for each area, and the time required to achieve success, were assumed to be identical for all environments.

The first two columns of table 14.3 show how the land area in specific categories would be expected to change over the twenty year period because of the expansion of irrigation. Irrigated area was assumed to grow at 1.5 percent per year so the gross area in the irrigated environment would increase by 9 million hectares. The area in rainfed rice declines by 6 million

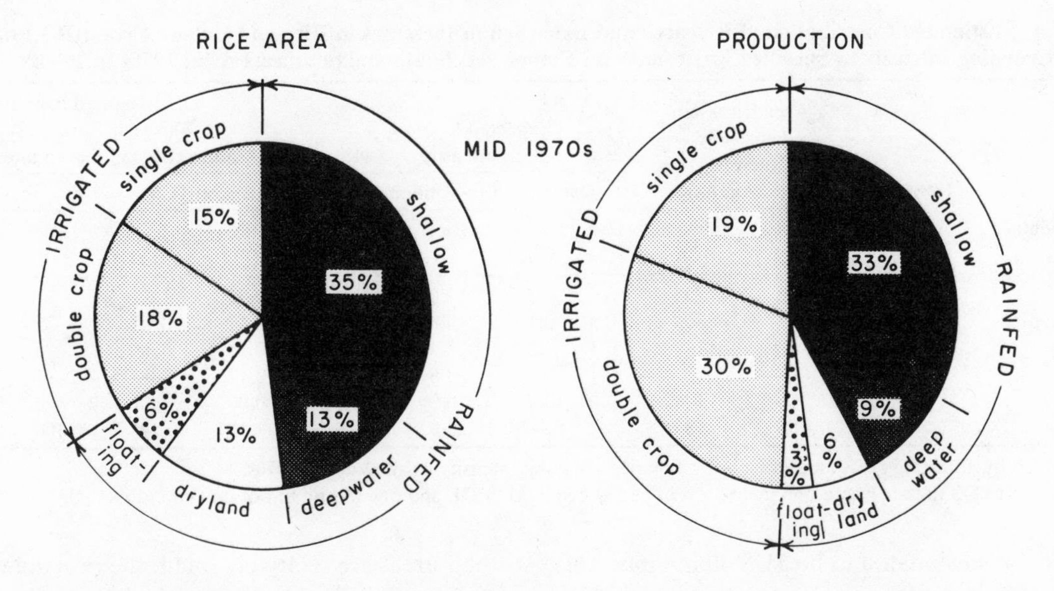

Figure 14.1. Estimate of the percentage of rice crop area and production by specific environmental complex in South and Southeast Asia, mid-1970s

as land is converted from rainfed to irrigated rice. The net increase of 3 million hectares is caused by increased double cropping in the irrigated area. The increase in rice production from expansion of irrigated land (assuming 1970s yield) was projected to be 17.8 million tons.

For each environment, the increase in average yield resulting from reasonable research inputs is shown in the fourth column, and the increase in production from expected yield gains is shown in the fifth column. The total is four times as great as the expected increase from irrigation. Irrigated rice accounts for 52 percent and rainfed rice for 38 percent of the total research and extension benefits expected to result from yield increases. Floating and dryland environments contribute the balance.

In addition to increasing yields, researchers are also expected to continue to discover new ways to intensify land use so that the number of harvests from each hectare of land will increase. The probable effect of land intensification is shown in table 14.4. Undiscounted benefits in the twentieth year are $US 8.10 billion from rice and $US 3.40 billion from upland crops. These benefits arise from increasing the number of rice crops harvested per year by an average of 0.4 on irrigated land, 0.3 on shallow rainfed, and the like (column 3), and increasing the number of upland crops harvested per year by an average of 0.2 on irrigated land, 0.1 on shallow rainfed, and so forth (column 4).

The benefits of research and extension investments are gradually felt over time. In the year 1976, the annual investment in research and extension in irri-

Table 14.3. Estimate of the Contribution of Irrigation Growth, and Research and Extension to the Increase in Rice Production in Specified Environmental Complexes, South and Southeast Asia, 1970s to 1990s

Environmental complex	Rice area (million ha)[a] 1970s (1)	1990s (2)	Undiscounted benefits from irrigation[b] (million mt) (3)	Expected yield increase (mt/ha) (4)	Undiscounted benefits in 20th year yield due to research and extension (million mt) (5)	(percent) (6)
Irrigated	27	36	27.0	1.1	39.6	52
Shallow rainfed	29	25	−7.2	0.8	20.0	26
Deepwater	13	11	−2.0	0.8	8.8	12
Floating	7	7	0.0	0.5	3.5	5
Dryland	8	8	0.0	0.5	4.0	5
Total	84	87	17.8	—	75.9	100

[a] Irrigated area is assumed to increase at 1.5 percent per annum and gross land area at 0.2 percent per annum. Land moves out of the rainfed and into the irrigated category.

[b] Col. 3 equals col. (2 − 1) × the estimated 1970s yield in each environment. Benefits are for the twentieth year.

Table 14.4. Estimated Contribution of Research and Extension to Increases in Rice and Upland Crop (UC) Production through Cropping Intensity in Specified Environmental Complexes, South and Southeast Asia, 1970s to 1990s

Environmental complexes	Area harvested at 1970s intensity[a] (million ha) 1970s (1)	1990s (2)	Expected increase in intensity: Ratio Rice (3)	UC (4)	(million ha) Rice (5)	UC (6)	Undiscounted benefits in 20th year from research and extension due to intensity: Billion $US[b] Rice (7)	UC (8)	(percent) Rice (9)	UC (10)
Irrigated	23	30	0.4	0.2	12.0	6.0	4.92	1.85	61	54
Shallow rainfed	41	36	0.3	0.1	10.8	3.6	2.81	0.70	35	21
Deepwater	16	14	0.1	0.2	1.4	2.8	0.25	0.38	3	11
Floating	8	8	0.1	0.1	0.8	0.8	0.12	0.09	1	3
Dryland	10	10	0.0	0.3	0.0	3.0	0.00	0.38	0	11
Total	98	98	—	—	25.0	16.2	8.10	3.40	100	100

[a] Computed by adjusting gross area in table 14.3 by rice cropping intensity existing in the 1970s.
[b] Converted to US dollars where one metric ton of paddy equals $US100, and one metric ton of upland crop equals $US75.

gated rice was estimated to be $US 40 million. This is assumed to increase at a rate of approximately $US 3 million per year, reaching $US 100 million in 1955. Its effect on output is assumed to begin in 1980, when yields increase by 55 kilograms per year on irrigated land (or a total of 1.1 metric tons per hectare by 2000) and yields increase 40 kg per year on rainfed land (or a total of 0.8 mt per ha) by 2000. The stream of net benefits remains constant after 2000 and is discounted from the year 2010.

The value of discounted benefits incorporating both the yield and crop intensification effect over the twenty year period is summarized in table 14.5. The largest share of benefits, 56 percent, is expected from the irrigated environment, both because the area will expand and because the absolute yield increases on shallow and medium rainfed areas are expected to be smaller. The shallow rainfed area is considerably larger than the medium deep rainfed so it generates more benefits. Dryland and floating rice account for only 3 and 4 percent of the total benefits because their areas are relatively small, the probable gains in yield are small, and the potential for land use intensification is relatively limited.

Net Benefits: Irrigated vs. Rainfed Rice

Excluding further consideration of dryland and floating rice, we now extend the analysis to compare the benefits, costs, and internal rate of return for: (1) investment in research and extension for the irrigated rice environment, (2) investment in research and extension for the rainfed rice environment, and (3) investment in new irrigation. Differences in farmers' costs for fertilizer and labor are also included.

The annual undiscounted investment, annual increased output, returns and costs for new irrigation and for research and extension investments in the twentieth year are shown in table 14.6.

By the year 2000, 120 kilograms per hectare of nitrogen-phosphorus-potassium fertilizer (NPK) is applied on all 48 million hectares of irrigated land,

Table 14.5. Present Value of the Stream of Added Benefits from Research and Extension due to Yield Increase and Cropping Intensity for Specified Environmental Complexes in South and Southeast Asia, 1970s to 1990s[a]

Environmental complex	Benefits from: Yield increase	Rice cropping intensity	Upland cropping intensity	Total benefits	Percent
	billion $US				
Irrigated	10.4	12.9	4.7	28.2	56
Shallow rainfed	5.2	7.4	1.8	14.4	29
Deepwater	2.3	0.6	1.0	3.9	8
Floating	0.9	0.3	0.2	1.4	3
Dryland	1.1	0.0	1.0	2.1	4
Total	19.9	21.2	8.9	50.0	100

[a] Discounted at 12 percent interest.

Table 14.6. Annual Increased Investments, Returns, and Costs Estimated for the Twentieth Year and Achieved from Reasonable Research and Extension Efforts in Irrigated and Rainfed Rice and Investments in New Irrigation, South and Southeast Asia, 1970s to 1990s

		From research and extension	
	From new irrigation	Irrigated rice	Rainfed rice
Annual capital investment ($US million)	675	100	100
New area irrigated (million ha)	9	—	—
Increased output per year (million metric tons)	17.8	88.8	59.4
Contribution to annual growth in production (percent)[a]	0.5	1.8	1.3
Value of output increase (million $US)	1,780	8,880	5,940
Current operating costs for new irrigation (million $US)[b]	100	—	—
Added fertilizer cost (million $US)[c]	180	1,896	632
Added labor cost (million $US)[d]	540	2,664	1,782
Net benefit (million $US)	960	4,320	3,526

[a] Output from irrigation includes the yield increase due to research on new irrigated area.

[b] $US 11/ha/yr.

[c] Fertilizer priced at $US 150/mt. Additional fertilizer applied at 60 kg/ha for newly irrigated area, plus 120 kg/ha on all irrigated area, and 40 kg/ha on all rainfed area.

[d] Added labor cost $US 30/mt of paddy rice.

and 40 kilograms per hectare is applied to 48 million hectares of rainfed land. The fertilizer cost is approximately $US 150 per kg or $US 0.33 per kg of NPK. Paddy is valued at $US 0.10 per kg, and the NPK-to-paddy price ratio is 3.3.

In both irrigated and rainfed environments, additional labor is a current input cost associated with increased production. Thirty days of labor are required to produce an additional ton of paddy, and labor is valued at $US 1 per day.

Beginning in 1976, irrigation is assumed to expand in South and Southeast Asia at a rate of 450,000 ha annually for twenty years, at which time there are 9 million ha of newly irrigated cropped area. However, the increase in physical area is only 6 million ha, half of which grows two crops of rice. The capital cost per hectare of physical area is $US 1,500. The maintenance cost per cropped area is $US 11 per ha per year.

Like the stream of benefits from research and extension, the stream of benefits from irrigation begins in 1980 and increases to 2000, remaining constant from 2000 to 2010. The investment stream, however, is constant at $US 450 million per year from 1976 to 1995.

The benefits and costs discounted at 12 percent and the internal rate of return for the three alternatives are shown in table 14.7. In these calculations the benefits that result from yield increases due to investment in research and extension on newly irrigated land are considered as part of the benefits derived from new irrigation.

The initial computations, shown in the lines marked "equal probability," assume a probability equal to one of achieving the production increases for investments in any of the three alternatives. The results suggest that returns to investment in new irrigation are relatively low, especially compared with the returns for investments in research and extension. Returns on the irrigated rice environment are high, and they appear to be even higher for rainfed rice.

If these calculations are anywhere near correct, why are the developing Asian economies placing so much emphasis on irrigation investments and so little on research for rainfed rice? The answer in part is related to the probability of success, an important element not yet incorporated into the model. Although the payoff is low, the greatest uncertainty is associ-

Table 14.7. Annual Discounted Investment Costs and Net Benefits, Benefit–Cost Ratio, and Internal Rate of Return for Alternative Rice Investments, South and Southeast Asia, 1970s to 1990s

		Research and extension	
	New irrigation[a]	Irrigated rice	Rainfed rice
Discounted investment costs (million $US)	5,800	2,000	850
Discounted net benefits (million $US)			
Equal probability	5,600	13,500	11,000
Unequal probability[b]	5,000	10,100	5,500
Benefit–cost ratio (12 percent interest)			
Equal probability	1.0	6.7	12.9
Unequal probability	0.9	5.0	6.5
Internal rate of return (percent)			
Equal probability	12	40	85
Unequal probability	11	35	40

[a] Benefits from new irrigation include the yield increase due to research on newly irrigated area.

[b] Assumed probability of achieving production gain from new irrigation = 100 percent, from research and extension on irrigated area = 75 percent, from research/extension on rainfed rice = 50 percent.

ated with increased productivity from expansion of irrigated land. At a minimum, it is known that the presently available technology will significantly boost yield on newly irrigated land. It is also clear from discussions with scientists that achieving a 1.1 mt/ha yield increase from further research on irrigated land has a higher probability of success than achieving 0.8 mt/ha yield increase on rainfed land.

The computations were modified to reflect these relative differences. In the modification, we assume that the probability of success in achieving returns on a hectare of newly irrigated land comparable to that for existing irrigated land is 100 percent, that the probability of achieving a 1.1 mt yield increase from investment in research for the irrigated environment is 75 percent, and that the probability of achieving a 0.8 mt increase from investment in research for the rainfed environment is 50 percent. Using these probabilities, the benefits, costs, and internal rates of returns are those given in the lines marked unequal probability in table 14.7. The differences among the alternatives are reduced, indicating roughly the same rate of return for research in rainfed as in irrigated environments and roughly half as large a discounted value of benefits in rainfed as in irrigated. Given these new calculations, one could argue that it made sense to concentrate research resources in irrigated rice in the early 1960s, when manpower was extremely scarce and the potential for increasing rice production was much less certain than today. However, given today's resources and potentials, a significant proportion of resources clearly ought to be directed at the rainfed environment.

Contribution of Research by Country

The importance of rainfed rice research in Asia is clear on a regional basis. However, guidance for national research programs needs to come from analysis of potential benefits from irrigated and rainfed rice research and extension on a country-by-country basis. Such an analysis was conducted for a number of important rice growing countries.

Questionnaires were sent to a panel of rice scientists from each country asking for their best estimates of the present and potential future yield increases that could be achieved from research in each rice environment. These unofficial data were used to compute the potential benefits from research and extension in each country. (Benefits due to cropping intensity were not considered.) The details of the results for each country are available elsewhere.[20] The percentage shares of the benefits attributable to each rice crop environment are summarized in table 14.8.

The results show two categories of countries: those with higher potential benefits from rainfed rice research and those with higher potential benefits from irrigated rice research. Among the rainfed-oriented countries are four traditional exporters: Burma, Thailand, Vietnam, and Nepal. The first three have large delta-based rice industries. The countries with a high proportion of potential benefits from research in irrigated areas include the island economies of Indonesia, the Philippines, and Sri Lanka. India does not clearly fall in either category. It shows a distribution of benefits similar to that for the whole of South and Southeast Asia, but if it were to be divided into regions, the same pattern of some areas with higher potential benefits from research on rainfed land and other areas with higher potential benefits from research on irrigated land would emerge.

These results do not imply that it would be unprofitable to expand irrigation in the rainfed-oriented countries. Rather, they indicate that for those countries, regardless of the rate at which irrigation expands, for some time in the future a substantial portion of total rice will be produced under rainfed conditions.

It is useful to compare the research expenditures of the rainfed and more export-oriented countries with those of the irrigated and more import-oriented countries. Table 14.9 shows the total and per 100-hectare research investment for selected countries. Both in absolute level and per unit of cultivated rice area, the research investment in general was lower in those countries with a high proportion of rainfed area. IRRI research to date has been of greater benefit to countries with a high proportion of irrigated area. Thus, both national and international rice research investment seems to have neglected the rainfed environments, at least prior to 1980.

Implications for Rice Research and Extension in Asia

Whether or not the potential benefits from rice research discussed in the previous section are realized in the next two decades will depend on the ability to (1) develop a suitable technology, (2) put the technology in the hands of farmers, and (3) provide the proper incentives to encourage farmers to use the technology. On all three counts, the task is more formidable for the rainfed areas than for irrigated areas, but it is becoming increasingly clear that additional research attention should be focused on the rainfed areas. This will have to include dryland and deepwater areas, but the major gains in the rainfed areas will come in the shallow-to-intermediate deepwater areas.

Table 14.8. Estimated Percentage Contribution of Research and Extension to Growth in Rice Production by Specified Environmental Complexes in South and Southeast Asia, 1970s to 1990s

	Share of benefits from research & extension[a]								
Environmental complexes	South and Southeast Asia	Nepal	Thailand	Bangladesh	Burma	India	Philippines	Indonesia	Sri Lanka
Irrigated	52	20	22	23	25	45	75	76	79
Shallow rainfed	26	67	49	41	53	10	11	10	8
Intermediate rainfed	12	—	20	17		19	1	—	5
Semideep rainfed	—	—	4	5	19	7	0.5	—	4
Floating	5	—	4	4		4	0.5	6	—
Dryland	5	13	1	10	3	15	12	8	4
Total	100	100	100	100	100	100	100	100	100
Implied growth in rice production	2.4	3.0	4.1	1.9	3.2	3.6	2.9	4.2	2.4

Source: R. Barker and R. Herdt (1982).
[a]Benefits are due to yield increase only.

Developing a Suitable Technology

There are three principal directions for rice research in the irrigated areas: (1) increasing yield potentials, (2) lowering input costs, and (3) reducing the growth duration of the plant. Basic scientific research will be searching for ways to break the current yield ceiling and to reduce farmer dependence on fossil-based fertilizers. But the time horizon for this research is probably twenty to fifty years. Much of the current effort will be directed toward "maintenance research" that will make a continuing stream of insect- and disease-resistant varieties available to replace varieties as the pests adapt to them. There is also substantial scope for more efficient application of fertilizer and chemicals that would make it possible to achieve the same yield level with significantly lower cash input. Whether it is possible to reduce the growth duration of rice to less than ninety days (from transplanting to harvest) without significant reduction in yield is questionable. But the potential of the ninety-day variety for promoting more intensive crop production has not yet been fully exploited.

Developing technology for the rainfed areas will also take time. Lack of water control in such areas results in a more heterogeneous environment than for irrigated rice. The present rice research effort on rainfed rice is exceedingly small (table 14.9). Furthermore, most major research stations are not located in typical rainfed environments. IRRI is certainly a case in point. Scientists there are forced to use deepwater tanks and greenhouses to simulate the flood and drought conditions of many rainfed areas. Even where stations are located in typical rainfed environments, most of the research plots at the stations are likely to be irrigated. Only recently has emphasis been given to varietal trials under rainfed conditions. During the 1970s, there was a considerable increase in our understanding of rice production under flood and drought conditions throughout Asia. There are already signs that the current modest investment will pay dividends, but there is not yet enough evidence to know whether the yield potentials for rainfed rice identified by the rice scientists are realistic targets.

The technology for rainfed areas should stabilize yields and require a minimum of purchased inputs. For these areas, it would be more appropriate to develop high-stability varieties (HSV) rather than varieties that respond to high input levels combined

Table 14.9. Rice Research Expenditures for Selected Countries in Asia, 1974

Country	Total research expenditures (thousand 1971 $US)	Expenditures per 100,000 ha (1971 $US)
	Rainfed—S & SE Asia	
Bangladesh	120	1.20
Burma	40	0.80
Nepal	100	8.10
Thailand	300	3.50
Vietnam	160	3.00
	Irrigated—S & SE Asia	
Indonesia	550	6.40
W. Malaysia	1,460	252.00
Pakistan	210	17.00
Philippines	500	14.00
Philippines (incl. IRRI)	2,900	80.60
	Other	
India	3,900	10.30
Japan	46,000	1,688.70
Taiwan	1,700	218.50
South Korea	250	35.70

Source: R. E. Evenson and P. M. Flores (1978).

with good weather, but fall to low yields under unfavorable weather conditions. Farmers should be given the choice of improved varieties that produce a minimum level of yield every year (HSV), as well as varieties that produce high yields under ideal conditions, but lower yields under unfavorable conditions. Farm-level surveys and investigations have shown that, even in the less progressive areas, farmers have considerable skill in selecting from among several varieties those that are most appropriate for their situation.

International trials for dryland and rainfed rice carried out through the International Rice Testing Program in cooperation with scientists in locations throughout South and Southeast Asia, have already begun to identify rices that are stable under the drought conditions that frequently occur in rainfed areas.[21] But if more serious attention is to be given to the rainfed areas, it will be necessary to identify more rainfed research sites within the rainfed environments. In our judgment, the shallow rainfed areas (0 to 30 centimeters maximum water depth) lying in the major flood plains of continental Asia should receive first priority, both because of the size of these areas and what appears to be the potential for raising yields. We already know that in the poorly drained (stagnant water) environment in eastern India improved varieties of medium height, such as Pankaj and Mahsuri, seem to perform better than other varieties. For social reasons also, research might initially concentrate in eastern India and Bangladesh since this is where population pressure problems are most serious.

Perhaps more than anything else, the problem of rainfed rice tends to emphasize the fact that the research establishment in South and Southeast Asia is a "top-down" organization. Until recently, there has been little attempt on the part of research workers to discover why farmers are doing what they are doing, and from this, to ascertain what technology is appropriate to increase production. Part of the reason is the false premise that it is the task of extension to provide the communication link between farmers and researchers. This may have arisen from an incomplete adaptation of the land grant model, with research, training, and extension functions separately identified. In Asia, the land grant model, incompletely interpreted, has resulted in a clear separation of function. In some cases, the ministry has extension responsibility while universities are expected to do research and teaching. This is very different from the United States where many state universities have responsibility for all three functions and where the private sector plays a major role in extension. The U.S. land grant model may be inappropriate for Asia at this time, given the limited resources of the developing countries.

About 1973, IRRI made a substantial shift in emphasis, with a major new focus devoted to farm-based research ("rice-based cropping systems" and "constraints on rice production"). A primary objective of this research is to develop a methodology to provide a link between the researcher and the farmer, to feed back information from the farm that would be useful in designing research.[22] Results of this research have consistently shown that many of the cultural practices appropriate for achieving high yields under experiment station conditions are not appropriate for achieving a profitable return under the farmer's environment.

Cultural practices as well as varieties will differ for the rainfed areas. For example, the wide variability in cropping patterns from one location to another, and even within a given location, reflects the extreme heterogeneity of conditions and attempts on the part of farmers to stabilize production by providing contingencies for variable weather conditions. Cropping patterns generally vary according to topography but are also adjusted to annual variation in weather conditions. The implications of these types of adjustments are important for technology development. Researchers can provide technology components, such as short-season varieties, that can be used to adjust cropping patterns and increase production in many different environments. But because of the extreme heterogeneity of the environmental conditions and the limited capacity of the research and extension network, normally it must be left to farmers to work out the most appropriate cropping patterns.

Getting Technology into the Hands of Farmers

In general, the input delivery system is not as well developed in the rainfed as in the irrigated areas. Seed multiplication and delivery are major weaknesses in many areas. For example, in eastern India from the time the first cross is made, it takes about ten years to develop and release a new variety of rice.[23] Even when the variety is released, there is no assurance that it will be made available to farmers. By contrast, varieties are typically available to farmers in the Philippines within five years of the first cross. The only modern varieties grown on a wide scale in eastern India in 1980, were Jaya, IR8, Pankaj, and Mahsuri, all of which were released more than a decade earlier. These varieties were grown almost exclusively in the irrigated areas, Jaya and IR8 in the dry season, and Pankaj and Mahsuri in the wet season.

Hargrove's research suggests that the problem begins with the breeding objectives.[24] Breeders had not been emphasizing those genetic characteristics required for the drought and flood conditions common to eastern India. The problem, however, is not only with research. Varieties are available in eastern India that seem to be more suitable than those farmers are using. But the mechanism for testing promising new varieties to ensure their rapid and wide dissemination is not adequate. The seed delivery system is so inadequately developed that it is difficult to determine whether superior varieties exist that are not being widely used by farmers.

Supply of other inputs must accompany the seed. The distribution and transportation system in rainfed areas is often such that farmers who live at some distance from commercial centers are at a considerable disadvantage in the prices they pay for inputs and receive for products. Governments surmise, perhaps correctly, that given the poor infrastructural development in the rainfed areas, the irrigated areas provide a higher return on investment. As these factors are overcome by general infrastructural development, new technology will become more attractive in rainfed areas.

Providing Incentives

Many observers associate farmer incentives with prices. Low prices and high input cost will discourage production. However, the disincentives for rice producers extend well beyond the pricing mechanism. Land fragmentation, small size of farms, inappropriate technology, and lack of knowledge on how to use inputs are often noted. A strong research and extension program and efficient input delivery system can overcome these problems and increase the incentives to farm production.

The expansion of irrigation is another important incentive to production. As an alternative to irrigation expansion, strengthening the research and extension system in the rainfed areas may have a higher economic benefit. Most countries cannot afford to ignore either alternative. It is a matter of finding the proper balance.

Ultimately, the strengthening of farmer incentives requires the strengthening of community leadership and organization. This is perhaps the greatest challenge for social science research.

Conclusion

The evolution of rice research systems in South and Southeast Asia has been accompanied by an extreme shortage of manpower and funds, leading to chronic underinvestment. Developed countries have played an important role in establishing the system that exists today. However, this has involved an extended learning process because efforts to transfer first technology and then institutions to the developing countries did not solve the production problem. IRRI's initial success in increasing rice production has been criticized on the grounds that it failed to give adequate attention to equity because it concentrated on irrigated rice technology. That strategy was followed because of the confidence that a technology could be developed for irrigated conditions.

The potential for increasing production in the nonirrigated areas is still in question, although it now seems appropriate for social as well as economic reasons to concentrate more research resources on the more promising rainfed areas. The success of such research will depend much more than in the past on an understanding of the clientele that the research is designed to serve. To design appropriate research for rainfed farmers, it is necessary to understand their present farming systems and the factors that constrain their production. There are already attempts to experiment with this new interactive model. Even so, increasing rice production in the rainfed areas will require a major research investment and a new philosophy in place of the drive for high yield that pervades most experiment stations. Given these obstacles, it may still take some convincing to persuade the national and international research establishment that the task is worth doing.

Notes

1. This chapter draws heavily on Randolph Barker and Robert W. Herdt, "Setting Priorities for Rice Research in Asia," in Robert S. Anderson, Paul R. Brass, Edwin Levy, and Barrie M. Morrison, eds., *Science, Politics and the Agricultural Revolution in Asia* (Boulder, Colo., Westview Press, 1982); and Randolph Barker and Robert W. Herdt, "Rainfed Lowland Rice as a Research Priority—An Economist's View," in *Rainfed Lowland Rice: Selected Papers from the 1978 International Rice Research Conference* (Los Banos, Philippines, IRRI, 1979).

2. The pioneering study was Zvi Griliches, "Research Costs and Social Returns: Hybrid Corn and Related Innovations," *Journal of Political Economy* vol. 66 (1958) pp. 419–431.

3. Yujiro Hayami and Masakatsu Akino, "Organization and

Productivity of Agricultural Research Systems in Japan," in Thomas M. Arndt, Dana G. Dalrymple, and Vernon W. Ruttan, eds., *Resource Allocation and Productivity in National and International Agricultural Research* (Minneapolis, Minn., University of Minnesota Press, 1977).

4. Robert E. Evenson and Pie M. Flores, "Social Returns to Rice Research," in International Rice Research Institute, *Economic Consequences of New Rice Technology* (Los Banos, Philippines, IRRI, 1978).

5. Reed Hertford, Jorge Ardila, Andres Rocha, and Carlos Trujillo, "Productivity in Agricultural Research in Columbia," T. Arndt, D. Dalrymple, and V. Ruttan, eds., *Resource Allocation and Productivity in National and International Agricultural Research* (Minneapolis, Minn., University of Minnesota Press, 1977).

6. This typology, covering more than just rice in Asia, is developed in Robert E. Evenson, "Comparative Evidence on Returns to Investment in National and International Research Institutions," in Arndt, Dalrymple, and Ruttan, eds., *Resource Allocation and Productivity in National and International Agricultural Research*.

7. Drawing a line between the various categories of research from very basic to applied is always difficult, somewhat arbitrary, and therefore confusing. Wortman and Cummings define five categories of research and suggest that "countries in a hurry to produce results" should give priority to the more applied levels. Evenson and Binswanger, on the other hand, conclude from recent studies that the complementarity between basic and applied research is so high that emphasizing applied at the cost of basic "indicate(s) that policymakers may be persisting in making the same errors as in an earlier period when the emphasis was on direct technology transfer." See Sterling Wortman and Ralph W. Cummings, Jr., *To Feed this World: The Challenge and the Strategy* (Baltimore, Md., Johns Hopkins University Press, 1978); and Robert Evenson and Hans P. Binswanger, "Technology Transfer and Resource Allocation," in Hans P. Binswanger, Vernon W. Ruttan, and others, eds., *Induced Innovation, Technology, Institutions and Development* (Baltimore, Md., Johns Hopkins University Press, 1978).

8. G. B. Masefield, *A History of the Colonial Agricultural Service* (Oxford, Clarendon Press, 1972).

9. Both in the United States and in Japan in the earlier part of this century prior to the perfection of the cross-breeding technique, an extension-oriented approach, with emphasis on transferring to others what the best farmers were doing, achieved considerable success.

10. Esman discusses the postwar research and development phase in terms of four models reflecting the particular emphasis of the period. These include: (1) the *extension model* covering the decade immediately after World War II when there was a strong belief that technology could be transferred from developed to developing countries; (2) the *land grant model* extending from the 1950s through the 1960s, reflecting the attempt to transfer the land grant system of research and extension to the tropics; (3) the *big science model* representing the establishment of the network of international agricultural research institutes in the late 1960s and early 1970s; and (4) the *interactive model*, as yet in the formative stage, in which researchers attempt to provide appropriate technology to those farmers bypassed by the Green Revolution by strengthening the link between the farmer and the research worker. See Milton J. Esman, "Research and Development Organizations: A Reevaluation," in William Foote Whyte and Damon Boynton, eds., *Higher Yielding Human Systems for Agriculture* (Ithaca, N.Y., Cornell University Press, 1983).

11. Robert E. Evenson, "Comparative Evidence on Returns to Investment in National and International Research Institutions."

12. Milton J. Esman, "Research and Development Organizations."

13. For an excellent discussion of issues relating to the development of the international agricultural research institutes, see Vernon W. Ruttan, "The International Agricultural Research Institutes as a Source of Agricultural Development," *Agricultural Administration* vol. 5 (1978) pp. 1–19.

14. The recent comprehensive evidence of the complementarity is provided in a paper by M. Ann Judd, James K. Boyce, and Robert E. Evenson, "Investing in Agricultural Supply" (New Haven, Conn., Economic Growth Center, Yale University, n.d.) mimeo.

15. Two useful review documents on this subject have been prepared at the request of the Technical Advisory Committee of the Consultative Group on International Agriculture Research, G. Edward Schuh and Helro Tollini, eds., "Cost and Benefits of Agricultural Research: State of the Art and Implications for CGIAR," (Washington, D.C., CGIAR Secretariat, World Bank, 1978) mimeo.; International Food Policy Research Institute, "Criteria and Approaches to the Analysis of Priorities for International Agricultural Research," IFPRI Working Paper No. 1 (Washington, D.C., IFPRI, 1978).

16. International Food Policy Research Institute, "Food Needs in the Developing Countries: Projections of Production and Consumption to 1990," IFPRI Research Paper No. 3 (Washington, D.C., IFPRI, 1977).

17. Robert W. Herdt, Amanda Te, and Randolph Barker, "The Prospects for Asian Rice Production," *Food Research Institute Studies* vol. 16 (1977) pp. 183–203.

18. Randolph Barker and Robert W. Herdt, "Setting Priorities for Rice Research in Asia."

19. The concept of "reasonable" research and extension is difficult to define, but it should be understood to include a wide range of technology improvements relating to cultural practices, use of inputs, and varietal improvement.

20. Randolph Barker and Robert W. Herdt, "Rainfed Lowland Rice as a Research Priority."

21. Data showing the yield stability of selected varieties under drought conditions are presented in Robert W. Herdt, "Focusing Research on Future Constraints to Rice Production," IRRI Research Paper Series 76 (Los Banos, Philippines, International Rice Research Institute, 1982) pp. 13–15.

22. Ruttan notes that another stated objective of cropping systems research at some international institutes is to develop more productive cropping systems, but that this is not a credible objective for these centers since the fine tuning of cropping systems is highly location specific. Vernon W. Ruttan, "The International Agricultural Research Institute as a Source of Agricultural Development."

23. Randolph Barker and T. K. Pal, "Barriers to Increased Rice Production in Eastern India," IRRI Research Paper Series No. 25 (Los Banos, Philippines, International Rice Research Institute, 1979).

24. Thomas R. Hargrove, "Genetic and Sociologic Aspects of Rice Breeding in India," IRRI Research Paper Series No. 10 (Los Banos, Philippines, International Rice Research Institute, September 1977).

15

Constraints to Increased Production

Constraints can be broadly defined as limiting factors. In the case of rice, constraints cover a broad range of influential factors from physiological limits in the biology of the rice plant to governmental intervention in the rice market.[1] Some constraints are under the control of human actors and may be directly influenced by policy or social change; others are determined by biological laws and, in the case of rice, are fixed norms that researchers must work around. Still other constraints may be removed or modified through the fruits of biological or technical research. A key fact in this discussion is that constraints cover a vast panoply of problems facing those interested in increasing rice production at all levels. Most researchers tend to view constraints on an individual basis according to their professional orientation. To biologists, constraints are forces that act on the rice plant to keep production below a maximum level. To economists, constraints can be artificially depressed prices or imperfectly functioning markets. To social scientists, the inability of people to organize themselves to exploit available technical potential may be a constraint. In the future, however, researchers and planners must take a more integrative approach as they attempt to overcome constraints. Constraints in one field act to modify conditions in another, and without a minimal consideration of all influential factors, improvement will be impossible.

Potential and Maximum Rice Yields

At a basic scientific level, a constraint is any factor that holds yields below the biologically determined maximum potential. As defined by plant physiologists, the yield potential is the maximum capacity of the rice plant to produce output given the nature of the cropland, moisture conditions, temperature level, and availability of solar energy. While the concept is clear, precise measurement is difficult because of the necessity of making several assumptions regarding solar energy uptake and other factors.[2]

Several methods are used to estimate yield potential. One method, depicted in figure 15.1, illustrates the potential yield (vertical axis) as a function of the daily incident solar radiation during the grain-filling period (horizontal axis) at two levels of photosynthetic efficiency ($E\mu$ = net gains of chemical energy ÷ total incident solar radiation) for temperate zone and tropical conditions. The temperate zone is distinguished from the tropics by a longer period of daily sunlight conditions, which allows for a lengthier grain-filling period. For example, assuming a daily incident solar radiation of 500 cal/cm^2, the maximum yield potential in the tropics ranges from 11 to 16 metric tons and in the temperate region from 16 to 22 metric tons. Yoshida suggests that this is a very crude measure of yield potential and that the relationship between

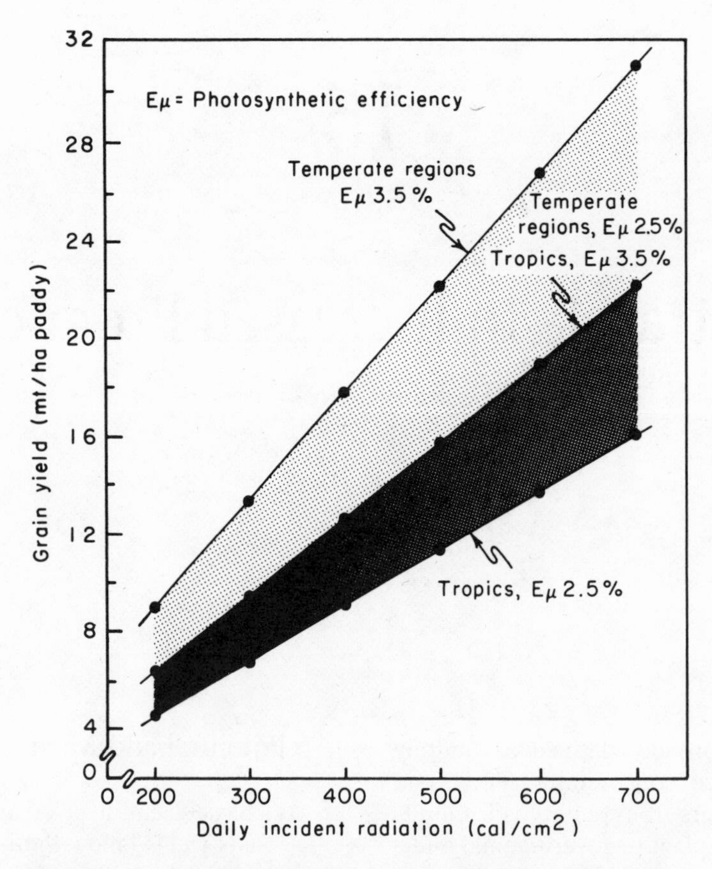

Figure 15.1. Relation between potential yield and incident solar radiation during the grain-filling period of rice (Source: S. Yoshida, *Fundamentals of Rice Crop Science* [Los Banos, Philippines, International Rice Research Institute, 1971] p. 93, reprinted by permission of the publisher)

incident solar radiation and potential yield is probably curvilinear, with the linear approximations underestimating potential yields at lower levels of incident solar radiation and overestimating at higher levels.[3]

Potential yields shown in figure 15.1 can be compared with maximum recorded yields of 13.2 mt/ha in Japan, with an estimated solar radiation of 400 cal/cm^2 per day and 11.0 mt/ha at the International Rice Research Institute (IRRI), with an estimated solar radiation of 550 cal/cm^2 per day.[4] At IRRI with good management, about 6 mt/ha is normally obtained in the wet season, with incident solar radiation at about 300 cal/cm^2 per day during the ripening period (see table 5.7). This compares with national yield levels of between 2 and 3 metric tons per hectare in tropical Asia.

The ten nations reporting the world's highest rice yields in 1978, 1979, and 1980 are shown in table 15.1. Only Spain's three year average exceeded 6 mt/ha paddy, and only three Asian countries had yields exceeding 5 mt/ha. These are far below the maximum recorded yields, which in turn are far below the physiological potential of the crop.

Of course, yield potential can also be measured in terms of total rice production per year in regions where more than a single crop is grown. In the tropics, if irrigation water is available, up to four harvests per year can be obtained from a single field using short-duration varieties. In a three year experiment in the Philippines, four crops per year gave an average of 23.7 mt/ha. At IRRI, a year-round continuous cropping system with sequential weekly planting and harvesting of small plots produced 23.6 mt/ha.[5]

Constraints

There are two main kinds of constraints that operate to keep rice yields significantly below their potential maximum. The first are technical constraints such as unfavorable soil conditions, diseases, or drought. The second consists of socioeconomic factors that prevent farmers from using more efficient technol-

Table 15.1. Rice Area and Yield for Selected Countries

Country	Area (thousand ha) 1980	Yield (mt/ha) 1978	1979	1980	Av.
Spain	68	5.9	6.2	6.4	6.2
North Korea	750	5.9	5.9	6.0	5.9
Australia	115	6.4	5.3	6.1	5.9
Japan	2,377	6.2	6.0	5.1	5.8
Italy	175	5.0	6.1	5.6	5.6
Egypt	437	5.4	5.7	5.5	5.5
South Korea	1,200	6.0	6.3	4.2	5.5
Greece	17	5.0	5.3	5.4	5.2
Yugoslavia	8	5.2	5.0	5.2	5.1
United States	1,354	5.0	5.1	4.9	5.0

Source: U.S. Dept. of Agriculture, Foreign Agriculture Circular, *Grains*, FG 36-80.

ogy—ignorance, poorly functioning markets, or land rental agreements that discourage the best use of inputs. Often the two constraints interact so that solving one of them does not necessarily result in increased yields. On the other hand, ameliorating conditions in one category may lessen constraints in the other. For example, the incentive to use fertilizer may be increased through a technical breakthrough, even if the price of fertilizer remains high.

Technical Constraints

There is no well-accepted set of measurements of the major technical constraints that limit rice production. Three main common parameters that restrict the world's rice areas are temperature, water, and soils. Anywhere these factors diverge from the ideal for rice production, yields are likely to be low, or rice production will be impractical.

Temperature

The rice plant is sensitive to temperature extremes. Temperatures that fall below 20°C or exceed 34°C frequently result in yield loss, sterility being one of the main yield-reducing factors.[6] However, cold-temperature stress is undoubtedly the more important problem, since a significant portion of the rice growing world is subject to temperatures below 20°C at some period during the growth of the rice plant.

The major regions of Asia subject to cold temperature stress include the temperate-zone areas—China, Japan, and Korea and the higher elevations (above about 300 meters) in the tropics and semitropics (below 25 degrees latitude). An estimated 10 percent (over 7 million ha) of the rice growing area below 25 degrees N latitude in Southeast and East Asia lies above 300 meters and is therefore potentially subject to cold-temperature damage.[7]

In the higher latitudes, significant progress has been made in developing varieties and cultural practices to withstand cold temperatures. As noted in chapter 2, the northern limits of rice production have been extended in this century from about 45 degrees to 53 degrees N latitude. By contrast, relatively little research has been done on developing technology for the higher elevations of the tropics, and in those areas, one usually finds traditional varieties and cultural practices. Moving rice into areas of temperature extremes will require intensive research and is likely to proceed only incrementally.

Water

Water is a major limiting factor in rice production in the tropics where the dominant monsoon weather patterns create pronounced wet and dry seasons. Water in appropriate amounts at the correct time in the growth cycle is crucial to high rice yields. Too little or too much, especially at critical stages of growth, stunts the plant and restricts yields. As with temperature, however, various kinds of rices are adapted to very different kinds of water regimes. Both dryland rice and much of the rainfed, wetland rice area suffer some stress from drought during growth. It is clear that drought severely limits yields, and research is underway to identify rices with superior drought resistance.

Excessive water also limits rice yield. In the broad valleys and plains that make up major rice areas, standing water in rice fields often is more than 30 cm deep for extended periods, and in extreme locations, it may exceed 100 cm. In other areas, floods may cover the growing rice plants for a week or more, then recede to leave the plant to grow with little or no standing water.

Research is underway to produce varieties that will give higher yields than the traditional rices grown under moderate and deepwater conditions, but they have not yet proven successful under field conditions. Some researchers believe it may be necessary to combine tolerance to drought with tolerance to excess water because these events often occur in the same areas. In many regions, modification of the environment through irrigation and drainage will continue to be the most efficient way to overcome the sensitivity of rice to appropriate water control (chapter 7).

Both regional and farm-level studies show that irrigation is closely associated with the adoption of modern varieties in most countries. Cross-country comparisons are moderately correlated between the percentage of area irrigated and the percentage of area in modern varieties (MVs) (table 15.2). Bangladesh and Burma have little area in MVs and very little irrigated area. Malaysia, Pakistan, and the Philippines, with a moderate or high proportion of irrigated rice land, were among the most rapid adopters of MVs in the 1965–69 period. In India, about 40 percent of all rice area was irrigated, and the same proportion was in modern varieties in 1975–79. However, state data for India (from the Directorate of Economics and Statistics, Ministry of Agriculture) show that while there is a correlation between irrigation and modern varieties, there are large nonirrigated areas planted to modern varieties. By 1975–79, an average of 68 percent of all rice in the Philippines was planted to modern varieties, while only 42 percent of the total rice area was irrigated (figure 15.2). In that period, over a million hectares were planted to rainfed modern varieties. Thus, the lack of irrigation has slowed the adoption of modern varieties, but has not prevented it.

Even countries which as a whole have lagged behind in adoption, such as Bangladesh, show different patterns in various rice growing environments. By 1978, the area planted to MVs in each of that country's three rice-producing seasons was about 500,000 ha, but the proportion in each season was quite different. Modern varieties occupied over 50 percent of the boro area, while in the aus and aman crops, both grown during the main wet season, MVs covered less than 10 percent of the area.[8] Much of the boro crop is irrigated, while the aman and aus are basically rainfed crops. The rainfed conditions include many areas where water depth exceeds 1 meter during the cropping season, making use of MVs impractical. Thus, Bangladesh seems to exemplify technical as well as adoption constraints traceable to water.

Table 15.2. Proportion of Rice Area Irrigated and Planted to Modern Varieties in Asia, 1965–79

	1965–69		1970–74		1975–79	
Country	Percent MVs	Percent irrig.	Percent MVs	Percent irrig.	Percent MVs	Percent irrig.
Pakistan	11	100	42	100	43	100
North India[a]	11	81	57	88	82	89
Indonesia	3	80	18	84	50	84
South India[b]	11	83	48	85	66	83
Malaysia	14	51	28	63	38	71
Sri Lanka	1	61	31	66	61	64
Philippines	22	40	57	42	68	42
East India[c]	3	28	10	28	21	28
West India[d]	6	22	20	26	50	23
Central India[e]	3	15	17	16	34	18
Burma	1	15	5	16	6	17
Bangladesh	1	7	11	11	15	12
Thailand	0	28	3	24	10	26

Sources: R. E. Huke (1982); R. W. Herdt and C. Capule (1983).

[a] North India: Jammu and Kashmir, Punjab, Haryana, Himachal Pradesh, Delhi.

[b] South India: Tamil Nadu, Andhra Pradesh, Karnataka, Kerala, and Pondicherry.

[c] East India: Assam, Bihar, Orissa, West Bengal, Manipur, Tripura, and Nagaland.

[d] West India: Gujarat, Maharashtra, Goa, Damon, and Diu.

[e] Central India: Madhya Pradesh, Maharasthra, Uttar Pradesh.

Soil-Related Constraints

"Soil-related factors are among the most significant environmental constraints on crop production in the developing countries. Through practical experience, farmers have favored areas with naturally highly productive soils and have shunned those with the less productive soils. Ignorance of improved technology and lack of inputs have constrained them from improving crop performance in areas with poor soils."[9] It is estimated that in densely populated South and Southeast Asia, "about 48 million hectares climatically, physiographically, and hydrologically suited to rice cultivation lie idle or are cultivated with poor results, largely because of soil salinity."[10] Much of this land has adequate (sometimes excessive) water for wetland rice cultivation, but with present technology, it would be impractical to make needed investments in land development and infrastructure. Some research indicates that areas where salinity and flooding are not severe could become productive without costly inputs if modern salt-tolerant varieties were grown. For much of the 48 million hectares where salinity is somewhat higher, there are not yet suitable varieties, although there are prospects for their development.

In addition to saline soils, it is estimated that there are 12.4 million ha of alkali soils, 5.4 million ha of acid sulfate soils, and 20.9 million ha of peat soils in tropical Asia.[11] The alkali soils are often located in extremely dry areas, making their potential for rice production questionable, although rice has been used in reclaiming such soils in the Punjab.

Other soil constraints are more intractable than salinity and alkalinity. According to experts, areas with severely acid sulfate soils and deep peats should be left untouched, although less extreme cases may eventually be productive given the right technology.[12]

There are many soils in tropical areas that are deficient in one or more nutrients necessary for high yields. Phosphorus, potassium, sulfur, zinc, and other micronutrients, as well as nitrogen, are all required

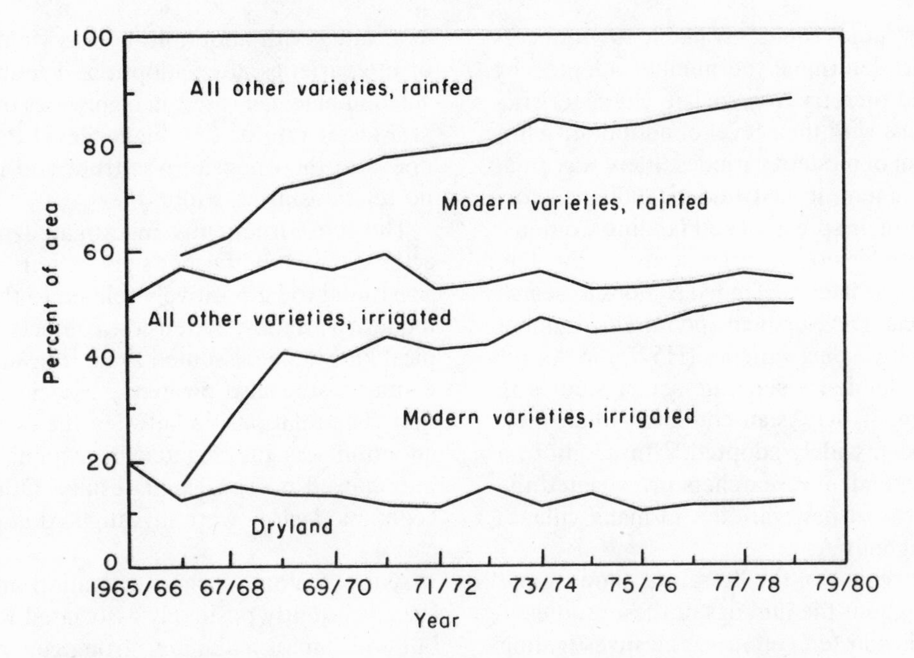

Figure 15.2. Percentage of rice area by variety and water regime, Philippines, 1965/66 to 1978/79 (Source: R. W. Herdt and C. Capule, *Adoption, Spread and Production Impact of Modern Rice Varieties in Asia)*

for healthy rice plant growth and high yields. In many areas, these elements are in short supply or are chemically bound with other elements so that they are not available to the plants. The submerged conditions under which rice is generally grown may enhance the availability of some elements and at the same time reduce that of others. Research to develop appropriate soil testing techniques to permit diagnosis of elemental soil imbalances is under way, but in 1980, most rice farmers in Asia could not gain access to anything other than standard, regional recommendations on the most appropriate fertilizer treatments for their soils. Thus, although soil deficiencies can be easily overcome by adding the right elements, most farmers do not have the necessary information to make the correct decisions about soil amendments. Thus, specific soil deficiencies can still be considered as technical constraints to improved yields.

Socioeconomic Constraints

Asian farmers have long been viewed as resistant to change, with limited aspirations. A common view in the 1950s was that "unless a man feels a desire to have more material wealth sufficiently to strive for it, he cannot be expected to have much interest in new techniques; there will be little attempt on his part to innovate."[13] More recently, most students of the development process have abandoned this view, although many social scientists still have a substantial distrust of technological change. Griffin interprets some data for a sample of rainfed farms as suggesting that "innovation almost certainly led to greater poverty on these farms and one could anticipate that in the future the farmers will revert to the traditional varieties of rice."[14] Subsequent studies of the same farms have disproved both points, but the general view that traditional techniques are preferred by farmers still enjoys popularity.

There is no question that new technology has been adopted more rapidly by some farmers than others and has benefited some groups more than others. Farmers located in marginal areas gained little from the first generation of modern varieties, but most studies by social scientists failed to recognize that the physical adaptability of the varieties to the growing conditions was a dominant factor associated with their use by farmers.

Adoption Studies

Socioeconomic constraints are generally considered to be the factors that slow the acceptance of innovations. A large literature attests to the belief that research on the pattern of adoption of innovations, the nature of adopters, and the institutions associated with adoption can give insights into the development process. In much of the research before 1965, little or no attention was paid to the innovations themselves. A typical approach was to consider a half dozen or dozen "innovations" available to producers in a study area, determine the number adopted by individuals, and then try to associate characteristics of the individuals with their level of adoption. After the development of modern varieties, there was great interest in documenting and understanding factors associated with their spread. The Planning Commission of India undertook a large study in the late 1960s.[15] A few years later the United Nations Research Institute for Social Development sponsored a number of studies of high-yielding varieties (HYVs) in Asia.[16] IRRI also coordinated a separate set of studies in thirty-six villages in six Asian countries where new varieties had been widely adopted.[17] In addition, a number of independent researchers investigated the adoption patterns of new varieties in many villages scattered throughout Asia.

An extensive review of the literature allows some generalizations about the findings of these studies.[18] Over sixty studies in ten countries, all investigating the adoption of modern rice varieties or the use of fertilizer on modern varieties, were examined. The studies fall into two types: those that describe the pattern of adoption and those that attempt to associate causal factors with adoption.

Some studies of the historical pattern of adoption trace the percentage of adopters in a sample over time, often hypothesizing that adoption follows a logistic or S-shaped curve. With innovations that have a quantitative dimension, individuals are classed as falling into one of five categories with respect to an innovation—awareness, interest, evaluation, trial, and adoption. Still other studies categorize farmers as adopters, nonadopters, readopters, and those who adopted but then stopped using the innovation.

A number of researchers have used regression or tabular analysis in an attempt to associate various factors with adoption or nonadoption. Causal factors may be determined by the characteristics of the farmer or farm. The most common among the former are: age, education, extension contacts, years of farming experience, family size, and contacts with groups promoting progressive agriculture. The farm charactertistics most frequently examined are farm size, tenure, and type of irrigation or degree of water control.

A review of the adoption studies shows that in eleven of twenty-five cases in which farm size was studied, it was positively associated with adoption of modern varieties (see chapter 10). In seven cases, a negative relationship was found, and in seven cases the relationship was indeterminate or not significant. Irrigation was the second most frequently studied variable, and it was generally found to be positively associated with adoption, as was yield performance of the varieties after adoption. Tenure was investigated in nineteen cases: in eight cases owners adopted at a faster rate or to a higher level than tenants; in one case the opposite was true; and in seven cases no relationship was found.

The most frequently investigated personal characteristic was the farmer's education, and education was found to be positively related to the adoption of modern varieties. Extension contacts and the technical knowledge assumed to be transmitted by those contacts were also positively associated with adoption. The relationship between the farmer's age and adoption was investigated in a number of studies, but showed no consistent results. Other social and economic factors were investigated in relatively few cases.

A few factors, such as irrigation and education, are consistently positively associated with adoption, but one cannot make any strong generalization that certain social or economic factors are constraints to adoption. Many others have been only occasionally investigated. Unfortunately, the available studies do not clearly define the variables, or describe the manner and circumstances under which these variables act as constraints to adoption. Thus, it is not possible to draw implications for policy.

Measurement of Constraints at the Farm Level

The lack of consistency in adoption studies and the obvious speed with which many Asian farmers adopted modern varieties and fertilizer seem to indicate that the initiation of the adoption process is less of a problem than it was thought to be in the 1950s. Despite rapid adoption, however, average rice yields in Asia are still low—many developing countries had average rice yields below 2 mt/ha in 1970, when MVs had yield potentials of over 6 mt/ha. This discrepancy created a desire to explain the yield gap between average yields and the potential yield of new varieties.[19] Limited satisfaction with national-level analysis led to an organized and intensive research effort to study yield constraints at the farm level in a number of countries.

National-Level Analysis, Philippines[20]

Data for 1980 show that national average rice yields in the Philippines were 2.1 mt/ha. Comparing this with the highest yield ever achieved at IRRI (11.0 mt/ha) gives a yield gap of 8.9 mt/ha. The reasons for this gap are as follows.

First, it is clear that national average yields should be compared with some kind of *average* maximum yield, not with a maximum reached only once. The average maximum yield in dry-season fertilizer response trials with good irrigation during 1977–79 at four Philippine experiment stations was 6.5 mt/ha. However, only a small part of the Philippines' rice is grown under the high-yield potential dry-season conditions. The corresponding wet-season maximum yield was 5 mt/ha. Even irrigated rice suffers from drought, which is estimated to reduce yields by an average of 1.6 mt/ha in the dry season and 0.3 mt/ha in the wet season under moderate seepage and percolation conditions.[21] About 45 percent of the Philippines' rice is rainfed wetland, which has a potential yield about 1 mt/ha lower than the wet-season irrigated land because the rainfed always suffers some drought problems. Another 15 percent is dryland rice, which has a potential yield of about 3.5 mt/ha. All these maximum yields can be achieved only with unrestricted fertilizer levels. It is estimated that an average of 29 kg/ha of fertilizer nutrients were applied to rice in the Philippines during 1976–79 (table 6.2).

Recognizing the different yield responses of modern and other varieties to fertilizer, one can calculate the level of yields implied by the available fertilizer, as was done in chapter 6. By using the above estimates, one can calculate that the weighted average maximum possible national yield for the Philippines under the 1976–79 distribution of seasonal area, the given water control conditions, and the fertilizer applied, was 4.4 mt/ha. The recorded yield was 2.1 mt/ha, implying a yield gap of 2.3 mt/ha. Part of this can be attributed to differences in the basis for the estimates—the national data were from interviews with farmers while the potential yields were based on experiments. We assume that 0.5 mt/ha of the yield gap is caused by the difference between cropcut and farm survey yield estimates. The remaining difference of 1.8 mt/ha is attributed to all the other unmeasurable differences between the average Philippine rice farm and the average experiment station. The experiment station typically has better soil conditions, better water control, and better pest management. Yields are not reduced by plant nutrient deficiencies, such as zinc, that are widespread in the Philippines.[22] Of course, the degree to which these factors contribute to the yield gap of between 4.4 mt/ha and 2.1 mt/ha is not known.

Farm-Level Analysis

The national analysis provides some insights, but questions still remain, especially when considering a specific farm location or farm. Two contrasting opinions are often voiced that have very different policy implications. Biologists are apt to suggest that farmers are not adopting high-yielding practices because of conservatism. Social scientists, on the other hand, are likely to say that farmers behave optimally considering the price structure and the productivity of the technology under local conditions. Therefore, they say, there must be something amiss with non-adopted technology in that farm location.

One clear way to resolve the issue is to test the technology in the farmers' fields and evaluate its economic return using local prices. This was the approach taken in the IRRI Constraints Project.[23] The project focused on yield and economic return as indicators of technology performance. The researchers measured the maximum or potential yield, the farmers' actual yields, and then explained the difference between the two.

As illustrated in figure 15.3, the gap can be explained either in terms of technical constraints or in terms of socioeconomic constraints. For example, the farmer may have achieved a lower yield in part because less fertilizer or insecticide was used. But the farmer may have chosen not to use the level needed to achieve the potential yield because credit was unavailable or it was unprofitable. The various reasons for not choosing the higher level of input are shown under socioeconomic constraints.

In the interdisciplinary IRRI team, agronomists attempted to determine which technical factors contributed to the yield gap, and economists tried to explain why farmers chose not to apply the high level of inputs. Economists further distinguished between those socioeconomic constraints that were motivated by economic considerations (it was not profitable or profits were too uncertain) and those caused by other factors. The economically recoverable yield gap is the yield increase that could be achieved if socioeconomic constraints other than economic behavior and risk aversion were removed.

The IRRI experiments were conducted by agronomists in farmers' fields. To determine yields, the levels of inputs and nature of practices followed by farmers were carefully monitored and experimentally simulated. Then, agronomists tried to produce the highest yields possible by changing important inputs or practices that are under the direct control of individuals. Economic researchers interviewed both the farmers whose fields were used for the experiments and a random sample of other farmers in the same villages to identify the economic, institutional, and social conditions in the area. IRRI researchers used this approach in selected areas in the Philippines

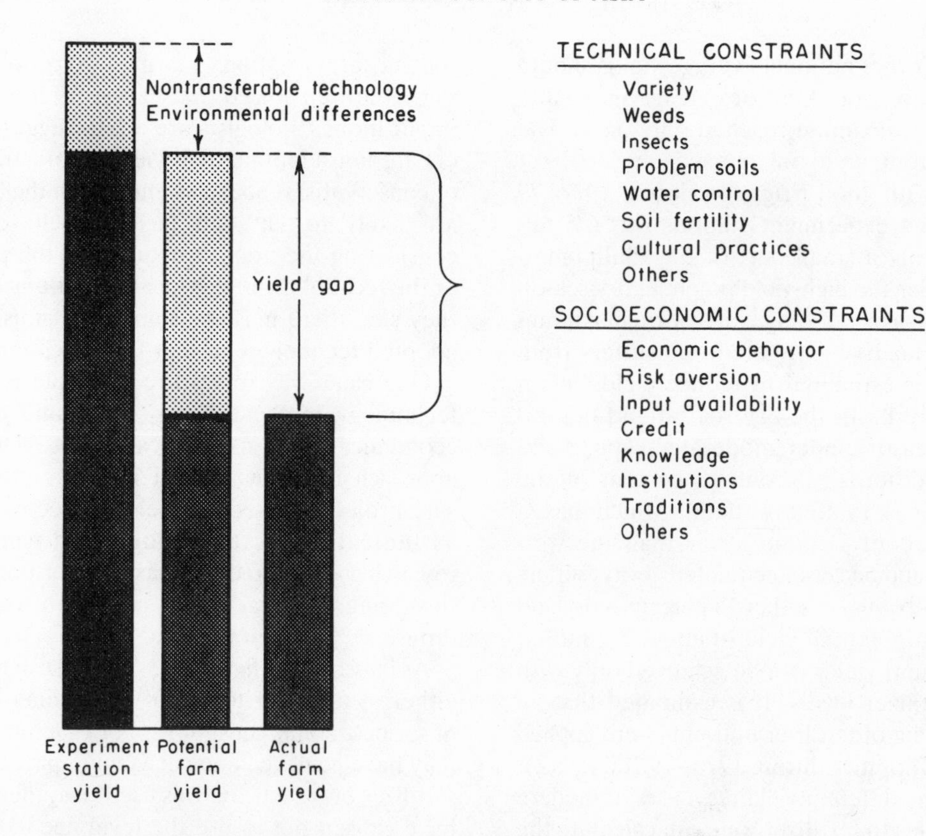

Figure 15.3. Conceptual model explaining the yield gap between experiment station yields and actual farm yield

in 1972 and 1973. Researchers in five other Asian rice-producing countries participated in the project beginning in 1974. The results summarized here are the product of a large collaborative effort by the researchers listed in appendix 15.A.[24]

Approximately 800 experiments were conducted in ten locations in six countries over the 1974–77 period. The wet-season experiments demonstrated a yield gap of 0.9 mt/ha, and the dry-season experiments demonstrated a yield gap of 1.3 mt/ha (table 15.3). Not only were the yield gaps higher, but both farmers' and researchers' yields were higher in the dry than in the wet season. There was considerable variation in the results from location to location, but in only one location was the wet-season yield gap larger than 1.2 mt/ha, while in the dry season the largest yield gap was 2.2 mt/ha.

Technical constraints In each location, researchers tried, ex ante, to identify critical factors not being used that would give higher yields. These were then used as the test factors in the experiments. Each participating group was urged to choose those factors that appeared to be most limiting in their situation. Fertilizer and weed and insect control were most frequently chosen, but others were also tested, including variety, plant spacing, land preparation, organic amendments, and various separate fertilizer elements. The experiments were designed to permit allocation of the total yield gap to each of the test factors, where interactions were not significant, and in most seasons and locations they were not.[25]

In the dry season, inadequate fertilizer was clearly the most important among the measured yield constraints, and insect control was second (table 15.3). In the wet season, the two factors were equally important, but both had smaller contributions than during the dry season. The contribution of the high

Table 15.3. Results from Yield Constraints Experiments in Farmers' Fields in Six Asian Countries, 1974–77

	Wet season	Dry season
Farmer's yield (mt/ha)	3.6	4.3
High yield (mt/ha)	4.5	5.6
Yield gap (mt/ha)	0.9	1.3
Contribution (mt/ha)		
Fertilizer	0.40	0.88
Weed control	0.13	0.21
Insect control	0.42	0.57
Number of trials measuring		
Yield gap	410	372
Fertilizer effect	272	187
Weed control effect	229	169
Insect control effect	254	193

Source: R. W. Herdt (1979).

level of insect control was more variable than was the contribution of high fertilizer. High insect control had a coefficient of variation (CV) of 73 in the dry season and 113 in the wet season, compared with a CV for fertilizer of 58 in the dry season and 80 in the wet season. The experiments showed that in many cases, it was possible to raise rice yields in farmers' fields by 1 mt/ha or more.

Socioeconomic constraints The study of socioeconomic constraints was designed to explain why farmers were not taking advantage of yield potential. The first method used was partial budgeting of the experiments. Two-hundred thirty-nine wet-season experiments were usable for budgeting analysis. In those trials, the farmers' yields averaged 3.5 mt/ha, and the yield gap averaged about 1 mt/ha. The value of increased rice output was calculated at local farm prices. Currencies were converted to $US at the rates prevailing in the 1974–77 period.

Figure 15.4 summarizes the yield and economic results obtained. On the left side of figure 15.4, the yield with high inputs is plotted against the yield with farmers' inputs. The vertical distance above the 45° line shows the yield gap. On the right side of the figure, the economics of the gap is shown. Points below the line are locations where the farmers' practices were more profitable than the "high" practices.

Economic analysis of the wet-season trials showed that, in four locations where the yield gap was small, the change to high inputs *reduced* profits (figure 15.4). In the Central Plain of Thailand, the yield gap amounted to 1 mt/ha, but because of the low price in Thailand, this was not enough yield gain to offset the increased costs. Increased profits from high inputs were large only in Bangladesh, and at one site in Indonesia. Thus, in most wet-season cases, the increase in yield was obtained with little or no economic gain.

In the dry season, the results were generally more favorable except in Sri Lanka (where the yield gap was about 0.4 mt/ha because of very erratic water supply). In most locations, the increased net returns from high inputs in the dry season exceeded $US 50/ha and in half of the sites exceeded $US 85/ha. Thus, the technology performed more economically in the dry season than in the wet.

One reason for the lack of additional profit in the wet season was the poor economic performance of the high level of insect control (table 15.4). This was partly because the expenditure on insect control needed to achieve maximum yields is high relative to that needed to achieve maximum yields from fertilizer. As a result, the high level of insect control added more to costs than it added to returns in six out of nine locations in the wet season. The high level of fertilizer decreased net returns in three out of ten locations, but on the average added $US 20/ha to net returns. High weed control had a very small and erratic contribution that was not economically analyzed.

In the dry season, a high level of fertilizer raised net returns by a substantial amount in all but one

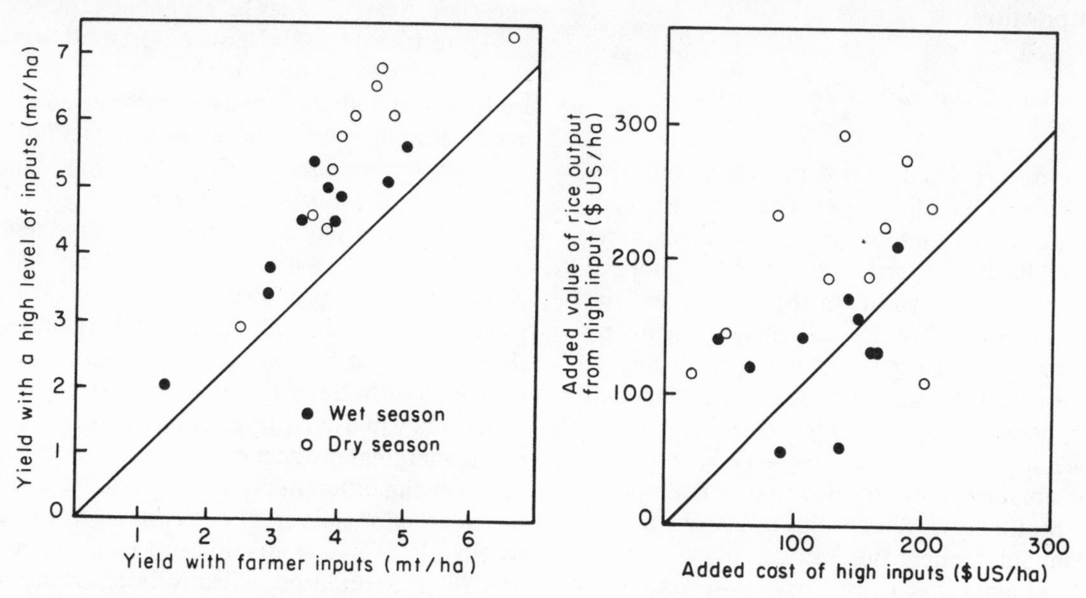

Figure 15.4. Yield gap and economics of the yield gap in wet and dry seasons. Each point represents the average results from one of ten locations in six Asian countries, 1974–77 (Source: R. Herdt, "An Overview of the Constraints Project Results," in International Rice Research Institute, *Farm Level Constraints to High Rice Yields in Asia: 1974–77* [Los Banos, Philippines, IRRI, 1979] p. 405, reprinted by permission of the publisher)

Table 15.4. Economic Performance of High Levels of Fertilizer and Insect Control Compared to Farmers' Levels in Six Asian Countries, 1974–77

Test factor	Fertilizer		Insect control	
	Wet season	Dry season	Wet season	Dry season
Percent of sites with increased profits	70	90	33	33
Average increase ($US/ha) in input cost	30	45	86	65
Average increase ($US/ha) in profit	20	71	−16	12
Average benefit-to-cost ratio	1.7	2.6	0.8	1.2

Source: R. W. Herdt (1979).

location, averaging $US 71/ha increase in profits. The average benefit-to-cost ratio of high fertilizer compared with the farmers' level was 2.6 (table 15.4). High insect control, on the other hand, reduced net returns in six out of nine dry seasons and barely covered its costs, on the average yielding a benefit-to-cost ratio of 1.2.

This disaggregation by input shows that, in the dry season, the application of higher levels of fertilizer appeared to give a sufficiently high return above added costs to generate a strong incentive for its use. By contrast, high insect control was not sufficiently attractive to encourage farmers to use it except in a few cases. Thus, the economic analysis indicated that the most profitable opportunity for increased yields was in the dry season through the use of improved fertilizer practices.

Socioeconomic Constraints in Nueva Ecija, Philippines[26]

The basic results that held for most locations also held in Nueva Ecija, Philippines. The wet season had a yield gap of about 1 mt/ha, largely because of insect control and fertilizer, but 1 mt/ha often cost more to produce than it was worth. In the dry season, the yield gap was about 2 mt/ha, contributed equally by insect control and fertilizer. Again, the insect control cost more than the value of rice it saved. The high fertilizer, however, cost much less than the value of added rice it produced in the dry season.

In this situation, there seemed to be a real opportunity to profitably increase yields by about 1 mt/ha. Why then did farmers not recover those profits? Obvious constraints were absent—credit was adequately available through the government's Masagana 99 program, and inputs were abundantly used. In fact, the average level of fertilizer used in Nueva Ecija in the 1977 dry season was 90 percent of the researcher's level, but at comparable levels, yields were lower and farmers' net returns were 15 percent lower than researchers' (figure 15.5). This observation suggests the need for an analysis of efficiency of input use.

Theoretical framework. Production efficiency has received considerable attention in economic literature in recent years. As pointed out by Shapiro and Muller,[27] producers may achieve different output-to-input ratios for several reasons: (1) they may actually face different technologies; (2) the differences may arise from random disturbances; or (3) some producers may be more successful than others in exploiting the same technology. In the first two cases, there is no necessary difference in efficiency; only in the third situation does efficiency play a role.

The failure of producers who face the same production function to achieve the same level of efficiency arises from two sources (1) technical inefficiency—failure to operate on the technically efficient production function, or (2) allocative inefficiency—failure to apply the level of inputs that maximizes profits (assuming identical prices). The production function is defined as the function that describes the greatest possible output from a given combination of inputs. Therefore, failure to operate on the production function *is* technical inefficiency. Farmers' failure to use the profit-maximizing level of inputs is defined as allocative or price inefficiency. The literature has characterized the combination of technical and allocative efficiency as economic efficiency.

Figure 15.6 shows the basic model used to conceptualize the constraints imposed on rice production by technical and allocative inefficiences in the one-variable inputs case. TPP_1 shows the yield response to input X when used with full efficiency. One may conceive of a whole series of technically inefficient response curves like TPP_2 that relate input to output at various levels of efficiency. Assuming that TPP_1 is available, any producer operating on TPP_2 or a similar curve is technically inefficient. Economic theory shows that given P_x, no restriction on input use, profit-maximizing behavior, and competitive markets, producers will use X_1 level of input where the value of the marginal product of X is equal to its price. If prices facing different producers vary, they will have different optimal levels of X. A producer with a technically inefficient production process, such as TPP_2 and MVP_2, would be allocatively efficient with X_2 input level, in spite of technical inefficiency.

IRRI researchers conducted agronomic experiments in farmers' fields to identify the production function for each field (that is, TPP_1). Because the

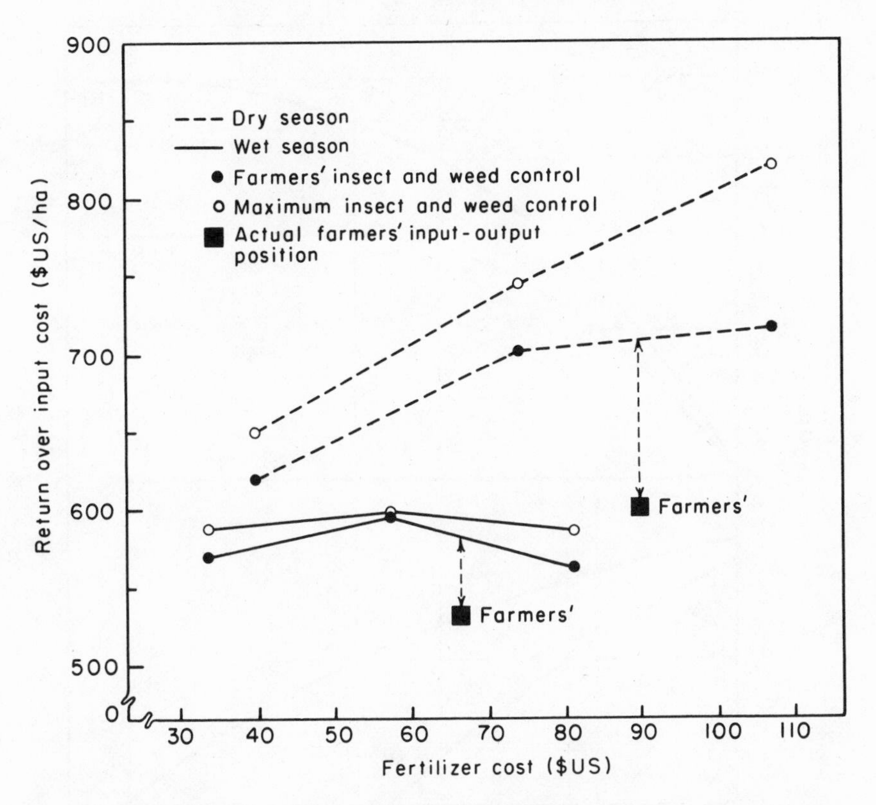

Figure 15.5. Return over cost of variable inputs for four levels of fertilizer inputs and two levels of weed and insect control inputs (average of nine experiments in farmers' fields, Nueva Ecija, 1977) (Source: International Rice Research Institute, *Annual Report for 1977* [Los Banos, Philippines, IRRI, 1978] p. 353, reprinted by permission of the publisher)

researchers by definition used the most technically efficient methods, the difference between their production function and that of each farmer (that is, TPP_2) was identified as technical inefficiency. The curves TPP_1 and TPP_2 were estimated from the agronomic experiments using a complex production function that included the important environmental variables, in addition to various levels of fertilizer and other managed inputs. Economic principles were used to determine if each farmer was allocatively efficient.[28]

The yield gap. The average yield gap as measured in all the Nueva Ecija experiments is shown in table 15.5 on the line marked "experiments." An estimate of the yield gap, reported on the line marked "function," was calculated by inserting the average value of environmental, insect, and weed control variables in the estimated production function. The maximum yield level of fertilizer was calculated with the researchers' efficiency and compared to the calculated yield with the average farmer's input level and efficiency. The total yield gap was then attributed to three factors: profit-seeking behavior, allocative efficiency, and technical efficiency.

The yield gap calculated from the function was the same as measured by the experiments in the wet season, although there was a slight overestimate of both the farmers' and high input yields by the function. Most of the yield gap of 0.9 mt/ha was attributed to technical inefficiency (67 percent). The remaining 0.3 mt/ha was attributed to profit-seeking behavior and allocative efficiency. In the dry season, the function gave a slightly lower yield gap estimate than the experiments. Of the 1.6 mt/ha dry-season yield difference, 56 percent was explained by technical efficiency and 38 percent by allocative efficiency. The contribution of profit-seeking behavior was rather small, 0.1 mt/ha each season.

Thus, the econometric analysis supported the implication of figure 15.5 that there was technical inefficiency in fertilizer use. An attempt was made to explain the variation in efficiency across farmers by using variables believed to be associated with management. Bhati listed technical knowledge, education, tenurial status, and access to external

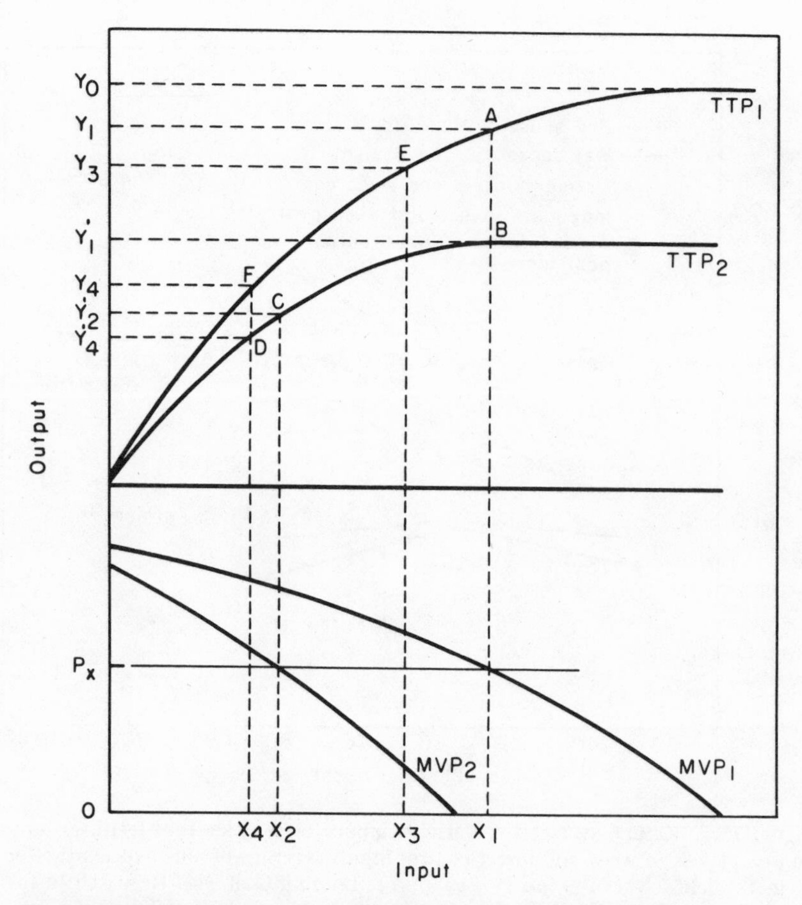

Figure 15.6. Theoretical model of technical and allocative efficiency in the one-variable input case (Source: R. Herdt and A. Mandac, "Modern Technology and Economic Efficiency of Philippine Rice Farms," *Economic Development and Cultural Change* vol. 29 [1981] p. 377, reprinted by permission of the publisher)

markets as affecting farmers' management ability.[29] Shapiro and Muller used technical knowledge and extent of adoption of a list of inputs and practices defined as "modern" by the authors to explain the differentials in technical efficiency among farmers.[30] In addition to these variables, irrigation availability, reported labor shortage, credit constraints, perceived risk, and the farm size were included in an attempt to explain efficiency differences across farmers.

The variable most highly associated with technical

Table 15.5. Actual (experiments) and Estimated (function) Yield Gaps and Estimated Contribution of Profit-Seeking Behavior, Allocative and Technical Inefficiencies, 1974 Wet Season to 1977 Dry Season, Nueva Ecija, Philippines

	Grain yield (mt/ha)			Average contribution (mt/ha)		
Source	Farmers' behavior[a]	High input and efficiency[b]	Yield gap	Profit-seeking behavior	Allocative inefficiency	Technical inefficiency
Wet season						
Experiments	2.4	3.3	0.9	—	—	—
Function	2.7	3.6	0.9	0.1	0.2	0.6
Dry season						
Experiments	4.4	6.4	2.0	—	—	—
Function	4.6	6.2	1.6	0.1	0.6	0.9

Source: R. W. Herdt and A. M. Mandac, "Modern Technology and Economic Efficiency of Philippine Rice Farms," *Economic Development and Cultural Change* vol. 29 (1981) p. 391, reprinted by permission of the publisher.

[a] For the experiments, the farmers' input plots; for the function, farmers' input level with farmers' efficiency.

[b] Maximum yield inputs, disregarding input costs, and with researchers' efficiency.

efficiency was area planted to rice per farm. The data indicate that, on the average for the study group of farmers, for every 1 hectare increase in farm size, technical efficiency in the use of fertilizer *decreased* by 0.88 points. To the extent that farm activities are spread out over time on large farms, it is more difficult for large farmers than for small farmers to conduct their operations at the optimal time, and rice is known to respond to intensive care. Thus, it is plausible that increased farm size would diminish the timeliness of certain farm operations. Information was positively associated with technical efficiency, as was expected.

About 50 percent of the variation in allocative efficiency was accounted for by the regression equation, but only three variables were significant at the 5 percent level or better. Total farm area was highly significant, with a negative coefficient, indicating that for every additional 1 hectare of rice area, a farmer's allocative efficiency rating dropped by 0.07 points. The number of days spent off the farm was positively associated with allocative efficiency, indicating better decisions by farmers with wider experience. Somewhat surprisingly, farmers who reported past significant loss from natural factors, such as flood or drought, were closer to the allocatively efficient level of fertilizer than farmers without such experience. Thus, the hypothesis that high risks led to underallocation on fertilizer was not sustained.

The failure of share tenure, labor shortage, and lack of credit to show significant effects on allocative efficiency indicates that most farmers make decisions that are relatively allocatively efficient when judged according to the actual production and economic conditions they face. Similarly, expressed lack of credit may seem to be important, but when actual prices paid and received and sharing arrangements for harvesting were included in the computation of the allocatively efficient point, those farmers who stated they had a lack of credit were as allocatively efficient as those who perceived no such lack.

The results of this analysis support the general picture that emerges from the review of adoption studies. Socioeconomic factors may be important in the adoption and efficient use of modern technology, but it is extremely difficult to generalize about which factors will be important without knowledge of the specific circumstances.

Constraints Beyond Farmers' Control

The experimental approach to the measurement of constraints is appropriate for evaluating the impact of factors that can be manipulated by farmers on their own fields, but there are many socioeconomic factors beyond farmers' control, and data from the constraints project can also be used to measure the impact of some of these.

Price-related constraints Prices of rough rice and urea fertilizer represent the relative incentive to use high levels of fertilizer, and the ratios of these prices varied across the locations of the constraints projects. In the Philippines, 1.7 kg of paddy were needed to purchase 1 kg of urea; in Thailand, 1.5 kg; in Indonesia, 1.2 kg; in Sri Lanka, 0.7 kg; and in Taiwan only 0.5 kg. These price ratios directly affect the yield gap (figure 15.7). In those areas where the price of urea relative to rice was high, the yield gap attributable to fertilizer was also high. The relationship is especially striking in the dry season. In the areas where it took less than 0.8 kg of rice to buy 1 kg of urea, the yield gap attributed to fertilizer was 0.5 mt/ha or less. Where it took over 1.5 kg of rice to buy 1 kg of urea, the yield gap generally exceeded 1 mt/ha. The correlation was much lower in the wet season ($r^2 = 0.04$) than in the dry ($r^2 = 0.61$), probably because the wet-season yield gaps embody a greater degree of variability caused by weather, insects, and diseases than do the dry-season gaps.

Environmental constraints Farmers recognize that some land is inherently better than other land for certain purposes. Some farmers are fortunate in being located where floods cannot reach, others plant at just the right time to avoid typhoons, others have productive soils, and so forth. The formulation of the constraints problem inherently assumes these site differences are minor and that the "appropriate" high level of technology will give high yields under all conditions. In fact, there is a great deal of variability in the performance of modern inputs in different environments. The comprehensive production function described above was used to identify the causes of variability in yields across farms in Nueva Ecija.

For analytical purposes, the experiments on the Nueva Ecija farms were classified according to farmers' yields.[31] The 25 percent of sites with the highest farmers' yields and the 25 percent of sites with the lowest farmers' yields were called high-yield and low-yield farms. To compute the predicted yield difference between these groups—the mean level of inputs, weather variables, soil characteristics, insect damage, and disease incidence of each group were substituted in the production function. The difference in predicted yields was attributed to the separate factors in proportion to the differences in the computed effect of each factor (table 15.6).

In the Nueva Ecija data, the differences in yields between the high- and low-yield groups based on the experiments were 2.36 mt/ha in the wet season and

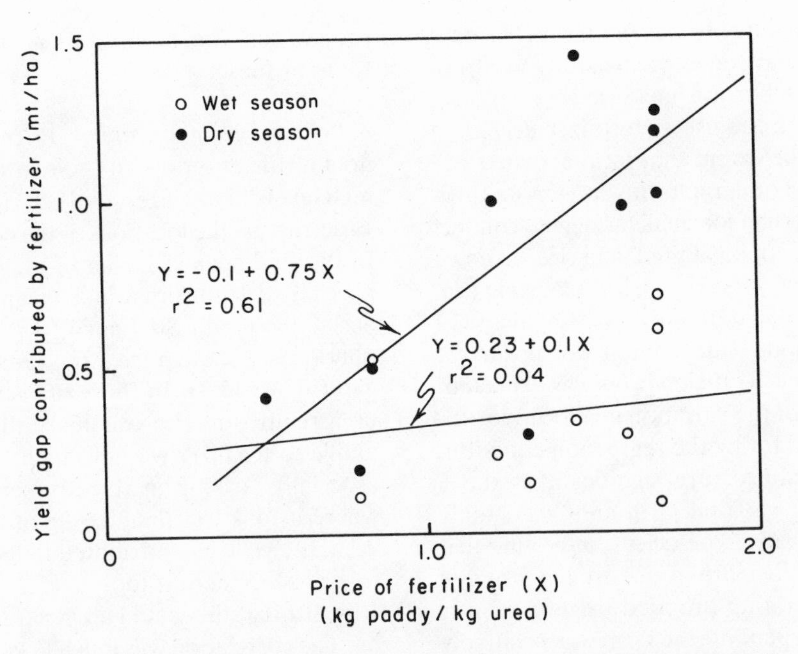

Figure 15.7. Relationship of real price of fertilizer to the yield gap contributed by fertilizer (Source: R. Herdt, "An Overview of the Constraints Project Results," in International Rice Research Institute, *Farm Level Constraints to High Rice Yields in Asia, 1974–77* [Los Banos, Philippines, IRRI, 1979] p. 409, reprinted by permission of the publisher)

1.32 mt/ha during the dry season. About 1.7 mt/ha or 71 percent of wet-season difference was explained by the variance in the measured variables in the wet season. For the dry season, differences in the factors included in the equation explained about 97 percent of the total yield difference (1.3 mt/ha).

Weather variables (solar radiation, typhoons, and water stress) accounted for 1.0 mt/ha, more than 40 percent of the yield difference during the wet season. Pests and diseases explained about 0.4 mt/ha or 15 percent, while overall interaction between managed inputs and the physical environment factors accounted for 13 percent.

Table 15.6. Accounting for Yield Differences Between Low- and High-Yield Groups, 1974–77, Nueva Ecija, Philippines

	Wet season		Dry season	
Class	mt/ha	percent	mt/ha	percent
Yield difference	2.36	100.0	1.32	100.0
Components of yield difference				
Inputs	0.07	3.0	0.34	25.8
Weather	0.99	41.9	0.46	34.8
Pest and disease	0.35	14.8	0.17	12.9
Soils	−0.05	−2.1	0.02	1.5
All interaction	0.31	13.1	0.29	22.0
Total explained difference	1.67	70.7	1.28	97.0
Unexplained difference	0.69	29.3	0.04	3.0

Source: A. M. Mandac and R. W. Herdt (1979) p. 19.

In the wet season, very little of the yield difference between the two groups was accounted for by managed inputs—fertilizer, insect control, weed control, and seedling age. This indicates that environmental factors were the major agents causing yield differences among farms during the wet season when the conditions for growing rice in the tropics are less favorable owing to low levels of solar radiation and frequent typhoons. These results support the general belief among farmers and the experience of researchers and extension workers in the field that only with favorable environmental conditions can the yield potential of the existing modern rice technology be realized.

In the dry season, environmental factors accounted for 35 percent of the yield difference, and pests and diseases accounted for about 13 percent. The increased effect of managed inputs during the dry season is notable, accounting for nearly 26 percent of the variability in the yield gap. Also, the proportion of the total attributed to interaction of managed inputs with the environment was greater in the dry than in the wet season.

Farmers' options to increase yields What are the possibilities for economically increasing yields in the low-yield farm environments? To answer this, the mean values of the environmental factors of the two groups were substituted into the production function to arrive at "collapsed equations for fertilizer response with farmers' efficiency and researchers' efficiency." From these equations, the optimal levels of fertilizer were calculated at prices prevailing during the study period. (The optimal levels assume that cash and fertilizer are available in any desired quantity.) The results are presented in table 15.7.

Except for the low-yield group during the wet season, the unconstrained economic optimum for fertilizer is reached at a higher level of application than farmers are currently using. This is true with both the farmers' and the researchers' efficiency levels. However, the yield and profit gains that would be achieved by raising fertilizer from the actual to the optimal level are small with both farmers' efficiency and researchers' efficiency in the wet season.

In contrast, yield gains would be more substantial by raising farmers' fertilizer use during the dry season—an increase of almost 0.4 mt/ha with farmers' efficiency and over 0.6 mt/ha with researchers' efficiency. This demonstrates that response to farm inputs such as fertilizer is greater under more favorable environments—as in the dry season—when there are more solar radiation, fewer typhoons, less pest damage, and lower disease incidence.

The results also demonstrate that, on the average, the farmers with low yields were correct in not applying much fertilizer during the wet season. In fact, the analysis implies that they would have been better off had they not applied any fertilizer. This was largely because of the direct yield-reducing effect of typhoons and lower solar radiation, and hence lower fertilizer response, which occurred on the low-yield, wet-season farms. The fact that the farmers in those conditions applied a level of fertilizer in excess of the ex post optimum indicates their willingness to take the risk incurred by wet-season conditions. This contrasts with the usual stereotype of farmers' unwillingness to take risks.

Table 15.7. Estimated Optimum and Actual Fertilizer Use of Different Yield Groups Based on Farmers' Treatments, 1974–77, Nueva Ecija, Philippines

Season	Yield group	Optimum fertilizer level (kg/ha)	Estimated yield (mt/ha)	Actual fertilizer level (kg/ha)	Actual yield (mt/ha)
			Farmers' efficiency		
Wet	Low	0	1.72	42	1.78
	High	116	3.43	81	3.45
Dry	Low	153	4.52	90	4.16
	High	211	5.80	147	5.44
			Researchers' efficiency		
Wet	Low	14	1.81	42	1.89
	High	150	3.89	81	3.68
Dry	Low	187	5.06	90	4.22
	High	245	6.47	147	5.81

Source: A. M. Mandac and R. W. Herdt (1979) p. 20.

Implications for Overcoming Constraints

The study of constraints measured the physical yield gap and the economic return that could be obtained by closing it in specific locations. The research results surprised both biological and social scientists. The biological scientists were surprised and disappointed at the *smallness* of the yield gap. Where farmers' yields are low, it is difficult for researchers to consistently obtain high yields, and where farmers' yields are high, it is difficult for researchers to get much higher yields.

There does, however, seem to have been some significant degree of inefficiency in farmers' use of fertilizer in the dry season. This is attributed by the biological scientists to differences between their timing and method of application and that of the farmers. To the surprise of social scientists, these inefficiencies could not be strongly associated with any specific characteristics of individual farmers.

The analysis of constraints data showed that physical differences in environments often explain a large part of the variability between farms with high and low yields. Yield gaps were larger in the dry season when only a small fraction of Asian farmers grow their rice. In the wet season, there was often no economically recoverable yield gap. The analysis also showed that, where prices give a strong incentive to use inputs, the yield gap was smaller, especially in the dry season.

All associated with constraint studies were surprised with the relatively small yield gap in most locations. One implication of this is that the economic return associated with overcoming the gap will be small. This conclusion is strengthened when one recognizes that overcoming the yield gap will, in itself, involve a cost—in diagnostic effort and in extension time. Given the difficulty in pinning down the exact nature and source of the constraints, these costs are likely to be high. Thus, research might better concentrate on raising the yield potential rather than worrying about the gap between actual and potential yields.

Notes

1. See for example, T. W. Schultz, "Constraints on Agricultural Production," and Sir C. Pereira, "The Changing Pattern of Constraints on Food Production in the Third World," in T. W. Schultz, ed., *Distortion of Agricultural Incentives* (Bloomington, Ind., Indiana University Press, 1978); ICRISAT, *Proceedings of the International Workshop on Socioeconomic Constraints to Development of Semi-Arid Tropical Agriculture* vol. 19, no. 23 (February 1979).

2. Shouichi Yoshida, *Fundamentals of Rice Crop Science* (Los Banos, Philippines, International Rice Research Institute, 1981). Chapter 2 in Yoshida, "Climatic Environment and Its Influence," presents a detailed scientific discussion of the estimation of potential rice yields.

3. Ibid., p. 93.

4. Ibid, pp. 93–94.

5. Y. Morooka, R. W. Herdt, and L. D. Haws, "An Analysis of a Labor-Intensive Continuous Rice Production System at IRRI," IRRI Research Paper Series 29 (Los Banos, Philippines, International Rice Research Institute, May 1979).

6. Yoshida, *Fundamentals of Rice Crop Science*, pp. 77–82.

7. Frederick W. Wackernagel III, "Rice for the Highlands: Cold Tolerant Varieties and Other Strategies for Increasing Rice Production in the Mountains of Southeast Asia" (Ph.D. dissertation, Cornell University, 1984).

8. Bangladesh, *1979 Statistical Yearbook of Bangladesh* (Dacca, Bangladesh Bureau of Statistics, Ministry of Planning, April 1979).

9. J. F. Metz and N. C. Brady, "Foreword" to *Priorities for Alleviating Soil Related Constraints to Food Production in the Tropics* (Los Banos, Philippines, International Rice Research Institute and New York State College of Agriculture and Life Sciences, Cornell University, 1980).

10. F. N. Ponnamperuma and A. K. Bandyopadhya, "Soil Salinity as a Constraint on Food Production in the Humid Tropics, J. F. Metz and N. C. Brady, eds., *Priorities for Alleviating Soil Related Constraints to Food Production in the Tropics* (Los Banos, Philippines, International Rice Research Institute and New York State College of Agriculture and Life Sciences, Cornell University, 1980) p. 203.

11. Ibid., p. 205.

12. N. van Breeman, "Acidity of Wetland Soils, Including Histosols, as a Constraint to Food Production," in J. F. Metz and N. C. Brady, eds., *Priorities for Alleviating Soil Related Constraints to Food Production in the Tropics* (Los Banos, Philippines, International Rice Research Institute and New York State College of Agriculture and Life Sciences, Cornell University, 1980) pp. 189–202.

13. Kusum Nair, *Blossoms in the Dust* (London, Gerald Duckworth and Co., 1961) p. 193.

14. Keith Griffin, *The Political Economy of Agrarian Change* (Cambridge, Mass., Harvard University Press, 1974) p. 58. Contrary to his speculation, by 1979 the farms to which Griffin was referring had almost adopted in toto modern varieties. See Violeta Cordova, Aida Papag, Sylvia Sardido, and Leonida Yambao, "Changes in Practices of Rice Farmers in Central Luzon: 1966–79," paper presented to the Crop Society of the Philippines, April 1981, (Bacnotan, La Union, Philippines).

15. Some of the findings of studies mentioned in this section are presented in chapter 10. The first broad study was reported by P. K. Mukherjee and B. Lockwood, "High Yielding Varieties Programme in India—An Assessment," paper presented at the 28th International Congress of Orientalists (Canberra, January 6–12, 1977).

16. The UNRISD studies include separate monographs on the Philippines, India, Sri Lanka, Indonesia, and Malaysia and a summary volume. Ingrid Palmer, *The New Rice in Asia: Conclusions from Four Country Studies* (Geneva, U.N. Research Institute for Social Development, 1976).

17. The IRRI studies are summarized in International Rice Research Institute, *Changes in Rice Farming in Selected Areas of Asia* (Los Banos, Philippines, IRRI, 1975) and in IRRI, *Interpretive Analysis of Selected Changes in Rice Farming in Selected Areas of Asia* (Los Banos, Philippines, IRRI, 1978).

18. R. W. Herdt and Celia Capule, *Adoption, Spread and Production Impact of Modern Rice Varieties in Asia* (Los Banos, Philippines, International Rice Research Institute, 1983).

19. R. W. Herdt and T. H. Wickham, "Exploring the Gap Between Potential and Actual Rice Yields in the Philippines," *Food Research Institute Studies* vol. 14 (1975) pp. 163–181.

20. Data presented in this section are drawn from official statistics of the Philippine Bureau of Agricultural Economics and various *Annual Reports* of IRRI.

21. T. H. Wickham, R. Barker, and M. W. Rosegrant, "Complementarities Among Irrigation, Fertilizer and Modern Rice Varieties," *Economic Consequences of the New Rice Technology* (Los Banos, Philippines, International Rice Research Institute, 1978) pp. 221–232.

22. A. C. Bautista, J. C. Bunoan, and R. Feuer, "Development of the Zinc Extension Component for Irrigated Rice in the Philippine Masagana 99 Production Program," *Proceedings of the Symposium on Paddy Soils* (Beijing, Science Press, 1981, and Berlin, Springer Verlag, 1981) p. 656.

23. The rationale for the project is spelled out in R. Barker, "Adoption and Production Impact of New Rice Technology—the Yield Constraints Problem," in International Rice Research Institute, *Farm Level Constraints to High Rice Yields in Asia: 1974–77* (Los Banos, Philippines, IRRI, 1979) pp. 1–26.

24. This summary draws on R. W. Herdt, "An Overview of the Constraints Project Results," in International Rice Research Institute, *Farm Level Constraints to High Rice Yields in Asia: 1974–77* (Los Banos, Philippines, IRRI, 1979) pp. 397–411.

25. The research methodology is described and illustrated in detail in S. K. De Datta, K. A. Gomez, R. W. Herdt, and R. Barker, *A Handbook on the Methodology for an Integrated Experiment-Survey on Rice Yield Constraints* (Los Banos, Philippines, International Rice Research Institute, 1978).

26. This section draws heavily on R. W. Herdt and A. M. Mandac, "Modern Technology and Economic Efficiency of Philippine Rice Farms," *Economic Development and Cultural Change* vol. 29 (1981) pp. 374–399.

27. K. H. Shapiro and J. Muller, "Sources of Technical Efficiency: The Role of Modernization and Information," *Economic Development and Cultural Change* vol. 25 (1977) pp. 239–310.

28. Herdt and Mandac, "Modern Technology and Economic Efficiency of Philippine Rice Farms."

29. U. N. Bhati, "Technical Knowledge as a Determinant of Farmers Income," in Agricultural Development Council, *Economic*

Theory and Practice in the Asian Setting (New Delhi, Wiley Eastern Ltd., 1972).

30. Shapiro and Muller, "Sources of Technical Efficiency," pp. 299–302.

31. A. M. Mandac and R. W. Herdt, "Environmental and Management Constraints to High Rice Yields in Nueva Ecija, Philippines," IRRI Agricultural Economics Department Paper 79–03 (Los Banos, Philippines, International Rice Research Institute, 1979).

Appendix 15.A

Principal Collaborators in the Rice Yield Constraints Project, 1974–77

Bangladesh Rice Research Institute, Joydebpur, Bangladesh

Ekramul Ahsan, agricultural economist
M. Zahidul Hoque, cropping systems specialist

Faculty of Agriculture, Gadja Mada University, Yogyakarta, Indonesia

Djoko Prajitno, agronomist
Sri Widodo, agricultural economist
Samino Wirjosuhardjo, dean, Faculty of Agriculture

Central Research Institute of Agriculture, Department of Agriculture, Bogor, Indonesia

Al Sri Bagyo, agricultural economist
Aten M. Hurun, agricultural economist
Hidajat Nataatmadja, agricultural economist
Richard Bernsten, agricultural economist

Faculty of Agriculture, University of Ceylon, Sri Lanka

H. P. M. Gunasena, agronomist
T. Jogaratnam, agricultural economist
A. P. Jinadasa, research assistant
V. Premakumar, research assistant

National Chung-Hsing University, Taiwan

Yi-Chung Kuo, agricultural economist
Cheng-Chang Li, agronomist
Carson K. H. Wu, agricultural economist

Kasetsart University, Bangkok, Thailand

Kamphol Adulavidhaya, agricultural economist
Tongruay Chungtes, agricultural economist
Somporn Isavilanonda, agricultural economist
Jongjat Janprasert, agricultural economist

Department of Agriculture, Bangkok, Thailand

Natavudh Bahsayavan, agricultural technician
Sanga Dangratana, statistician
Suchavadi Nakatat, statistician
Saowanee Pisithpun, statistician

International Rice Research Institute, Los Baños, Philippines

Randolph Barker, agricultural economist
Surajit K. De Datta, agronomist
Kwanchai A. Gomez, statistician
Robert W. Herdt, agricultural economist

16

Government Rice Policies

Much of the policy discussion in the literature focuses on the price both of rice and of major inputs such as fertilizer. However, there is a range of other policies used to manipulate the rice sector. These include investments in irrigation systems, roads and marketing facilities, agricultural research, taxation, and trade policy. Some of these policies have effects beyond the rice sector, and a complete discussion of all of them exceeds the scope of this chapter. It is nevertheless, important to recognize that all may influence behavior in the rice sector.

In this chapter, we focus on policies used by governments to manipulate rice production, consumption, and trade. The first part of the chapter compares the levels of economic development and resource endowments of the major Asian rice economies. Then we consider the national rice policy objectives and alternative policy instruments for achieving these goals. An important problem is that different objectives are often at odds. The section on the rice price dilemma discusses the conflict between efforts to maintain low prices for consumers and adequate price incentives to producers. Countries are classified on the basis of their price policies. This helps to identify the strategy being followed to achieve production objectives. We end the chapter by discussing the elements of a rational rice policy.

The Asian Rice Economies

A comparison of basic economic indicators illustrates the range of conditions existing in the Asian rice economies. Countries in table 16.1 have been ranked on the basis of GNP per capita. Japan stands alone as the one developed country in the group. South Korea and Malaysia have per capita incomes roughly twice the level of Thailand and the Philippines, which in turn have income levels two to three times those in China, Pakistan, and India, which in turn have incomes almost twice as high as in Burma, Nepal, and Bangladesh.

Growth rates of income in the 1960s and 1970s were strongly correlated with absolute income levels in these countries. Japan and Korea grew at more than 7 percent per capita per year. Malaysia, Thailand, and Indonesia grew at between 4 and 5 percent per capita annually, while the lower income countries grew at lower rates. Bangladesh, with a zero trend in per capita GNP, has the worst record, but it is likely that Vietnam and Laos would show similar trends if data were available—in all three countries poor economic conditions are the direct result of extended periods of war and civil disruption.

Agriculture plays a dominant role in the economies of the low-income countries, contributing 45 percent or more of gross domestic product (GDP) in Burma, Nepal, and Bangladesh (probably also in Vietnam, Laos, and Kampuchea). Agriculture contributes about one-third of GDP in Sri Lanka, Pakistan, and India. In Malaysia, Thailand, the Philippines, and Indonesia, agriculture contributes about 25 percent, while in Japan it contributes less than 5 percent. Thus, per capita income is inversely correlated with agriculture's share of labor force in agriculture.

There is a general, but by no means uniform, inverse

Table 16.1. Economic Indicators in the Asian Rice Economies

Country	GNP/capita (1980) 1980 $US	GNP/capita (1980) Growth 1960–80 (percent/yr)	Percent of GDP from agriculture 1980	Percent of labor force in agriculture 1978	Per capita milled rice consumption 1975–79 (kg/yr)	Population/ sq km of arable land
Japan	9,890	7.1	4	12	90	2,064
Malaysia	1,620	4.3	24	49	138	369
South Korea	1,520	7.0	16	41	148	1,615
Philippines	690	2.8	23	47	93	551
Thailand	670	4.7	25	76	181	255
Indonesia	430	4.0	26	60	127	833
Pakistan	300	2.8	31	54	27	371
China	290	n.a.	31	61	94	655
Sri Lanka	270	2.4	28	53	100	1,402
India	240	1.4	37	65	73	390
Burma	170	1.2	46	53	171	328
Nepal	140	0.2	57	93	105	570
Bangladesh	130	0.0	54	84	157	850

Sources: GDP from agriculture: World Bank, *World Development Report 1982*. Labor force and population/sq km: A. Palacpac, *World Rice Statistics* (1982). Rice consumption: U.S. Department of Agriculture, Foreign Agricultural Service, *Foreign Agriculture Circular: Grains*, FG-5-80. Population: Appendix tables.

correlation between income levels and per capita rice consumption. Rice consumption is at its highest in some of the lowest income countries: an estimated 179 kg/capita/year in Laos, 187 kg/capita/year in Burma, and 156 kg/capita/year in Bangladesh. Among the lower income countries, Indonesia and Nepal have rice consumption levels about equal to those in Japan, by far the highest income country. Conversely, Pakistan, India, and Sri Lanka have proportionately lower per capita rice consumption. This reflects a diversification of diets in South Asia, where both wheat and rice are important staple cereals. The inverse relationship between per capita income and rice consumption contrasts with the generally positive relationship between those variables within countries (chapter 11).

In 1980, national populations in the Asian rice economies ranged from 3 million in Laos to India with over 800 million and China with over 1 billion. North Korea, Malaysia, Sri Lanka, and Nepal were also relatively small with fewer than 20 million each in 1980. South Korea, Thailand, the Philippines, Burma, and Vietnam are medium-sized countries that had between 40 and 50 million people in 1980. Japan, Indonesia, Pakistan, and Bangladesh are large countries, with 75 to 150 million people in 1980.

Population densities, measured per square kilometer of cultivated area, are at the extremely high levels of 1,500 to 2,000 in Japan, Taiwan, and South Korea. Indonesia and Bangladesh, with about 850 people per square kilometer, have nearly as high a level of population pressure. China, Sri Lanka, the Philippines, and Nepal have about 600. Malaysia, Laos, Pakistan, and India have 350 to 400, while Thailand has the least population pressure, with fewer than 300 people per square kilometer of arable land (table 16.1). More than 75 percent of the labor force is agricultural in Thailand, Laos, Nepal, and Bangladesh, and about two-thirds is in agriculture in China, Indonesia, and India. Only in Japan is agriculture rather unimportant as a source of livelihood, with only 12 percent of the labor force engaged in agriculture in 1978.

Although rice is the dominant food crop throughout the region, substantial relative differences exist between countries. In Bangladesh and Vietnam, practically all of the cropped land is used for rice at least once a year, although low levels of modern inputs are used. In Korea, Thailand, the Philippines, Burma, Laos, and Nepal, one-half to two-thirds of cropland is devoted to rice. In Malaysia and Sri Lanka, rice occupies a minor fraction of cropland because of the importance of plantation crops. In Pakistan, rice covers 10 percent of the cropland and in India it covers 27 percent; in both countries nonrice cereal crops occupy a substantial area. There is generally a high correlation between income levels and the intensity of rice production across countries: irrigation, fertilizer use, and other indicators of technological development are higher in the high-income countries than in the low.

Policies for the Rice Sector

The diversity in the indicators examined in the previous section reflects differences in resource endowments, levels of economic development, and government policies. Roughly speaking, rice is much more important in the economies of countries at an early stage of development. Those countries generally have less capacity to adopt protectionist policies or invest in their rice sectors. Farmers have lower incomes, are closer to subsistence, and may therefore be less able to take advantage of new technologies. In low-income countries, there is usually a high income elasticity of demand for rice so that as incomes rise, the demand for rice grows rapidly.

National Rice Policy Objectives

The set of policy instruments used by a government reflects the desired goals within the limitations imposed by the agricultural, economic, political, and social environments of the country. The objectives themselves are determined by complex, long-term interactions among these environments and external, foreign influences.

The overriding aim of any society or government is survival. For governments, survival requires political stability, which is tied to the rice economy in Asia. A number of related objectives can be identified:[1]

1. maintenance of a low and stable rice price to protect consumers
2. generation or saving of scarce foreign exchange
3. generation or saving of government revenue
4. increasing rice supplies by means of production incentives
5. provision of adequate incomes for farmers
6. reduced dependency on foreign food supplies (self-sufficiency)
7. regional development and interregional equity
8. provision of adequate nutrition

In the short run, political stability is somewhat more closely tied to the maintenance of low and stable prices to urbanites and other privileged groups, but one could argue that all objectives are necessary in the long run. Judging from the behavior of many governments, short-run protection of consumers is often sought at the expense of long-run incentives for production. This is understandable in light of the overriding goal of governmental survival (government is used here in the sense of a political party retaining power over a period of time). Because governments have both limited time horizons and limited financial resources, several of the objectives enumerated above are in conflict with one another. Thus, rice policy in any given country at any given point in time reflects a compromise among the competing goals of production efficiency, greater equity, nutritional adequacy, and food security.

The most basic conflict is that between objective one—low consumer prices, and objectives four and five—the generation of adequate incomes and incentives for producers. However, adequate incentives can be provided without resorting to raising prices. A reduction in production costs generated through lower cost inputs, research-generated technological change, or socially financed infrastructure (roads, irrigation) can improve production incentives. But these instruments use rather than generate government revenue, and they conflict with objective three. In the event of a current or impending food shortage, most governments turn to imports, a tactic in conflict with the second and sixth objectives.

Policy Instruments

The instruments available to governments may be classified by the length of time required to implement each one. In a very short run of weeks or months, no instrument can affect rice production, but imports, exports, direct government purchases, and sales can affect both the farm and the retail price. Over a period of years, production can be enhanced or depressed by changing incentives. Some instruments operate on the supply side to increase or decrease supply, others operate on the demand side, and some operate on both sides of the market at the same time.

Short-run Short-run policy instruments (those that generate results within a year) are aimed at preventing a sharp rise or fall in price by fixing or allowing prices to fluctuate within a narrow range. Price fixing has little real effect unless backed by the ability to intervene in markets to ensure that the desired price is maintained, through either buying or selling from government-controlled stocks. Governments of importing countries have typically attempted to defend price ceilings by selling imported rice and to defend price floors through domestic purchases. However, historically in South and Southeast Asia, it has been politically more difficult to obtain funds to defend the price floors and prevent prices from falling below the government purchase price.

Government purchases often are most vigorously used in periods of scarcity and have some compulsory aspects. In India and Sri Lanka, compulsory sales to government agencies have been imposed along with restrictions on movements across boundaries. Theoretically, these restrictions allow governments to obtain rice at the official government price during periods of shortage. But most governments lack the administrative capacity to enforce controls, and "black market" trade develops at prices well above the official rate.

The emphasis on increased production and the drive for self-sufficiency have been accompanied by a strengthening of the power of government marketing agencies and enlargement of buffer stocks. For example, by the early 1980s BULOG in Indonesia and the National Grain Authority in the Philippines controlled adequate supplies of rice to maintain short-run price stability.

Medium- and long-run In the medium run of one to three years, governments can influence rice production by increasing the use of fertilizer or other technologies, or encouraging greater land use intensity through the construction of irrigation facilities. In some situations, water control can be improved within existing irrigation systems and on farms. On the demand side, programs can be set up to encourage substitution of other commodities for rice. In the medium run, developments in the nonagricultural economy can contribute through the construction of plants to produce fertilizer or irrigation equipment.

In the long run, government policy instruments can be focused on improving the productivity of the rice farming sector through construction of large-scale irrigation systems, or by enhancing water distribution through reshaping the topography, or by farm-level water distribution investments. Research investment is also important over the longer run, given sustained support and appropriate inputs.

Medium- and long-run policy instruments that improve the productivity of the farmers' land and labor inputs can provide the key to simultaneously achieving both low consumer prices and adequate incentives to farmers. Farmers respond to incentives, which are a function of prices, inputs, *and* yields. Thus, incentives can be provided even with low or stable prices if unit costs of production decline. Much of the increase in rice output that was achieved in Malaysia, India, and the Philippines during the 1970s when price policies were unfavorable can be traced to improved incentives generated through irrigation investments and new technology. One can even argue that the new technology and infrastructural investment made it possible to keep prices low.

In addition to investments in physical infrastructure and research, considerable government efforts have gone into supervised credit and input delivery programs in a number of countries. These are based on the recognition that use of modern technology requires the purchase of inputs from outside the agricultural sector and the argument that purchase of such inputs is beyond the financial capacity of small farmers. The experiences of several countries with programs of this type are discussed in the next chapter.

The Rice Price Dilemma

Rice prices perform a dual role. On the one hand, they determine consumption levels, especially among the poor who depend on rice as a main source of calories. On the other hand, prices influence the level of rice supplies by affecting the level of production incentives to farmers.[2] As previously noted, consumer pressure to maintain low prices can be at odds with the need to provide the necessary incentives to stimulate domestic production and investment in the rice sector. This results in a dilemma faced by most national governments that manipulate the price of rice.

The problem for policymakers is made more difficult by the considerable controversy about the actual as opposed to the theoretical effect of prices on consumption and production. Opposing views are described by Timmer:

> Two curiously inconsistent views of the role of food prices in the economic development process have dominated thinking in academic and decision-makers' circles since World War II. The first, or structuralist, position argues that food prices are irrelevant to the long-run development process since both producers and consumers are insensitive to changes in prices. Consequently, political leaders can feel free to manipulate food prices for whatever short-run political effect is desirable. Usually this manipulation takes the form of keeping urban food prices low to satisfy workers, politically active students, and the urban middle class.
>
> The second, or neoclassical view, holds that food prices are a critical factor in farmers' decisions about what crops to grow and how intensively to grow them, even in fairly traditional peasant economies. In the presence of new biological and chemical technologies that offer significantly higher yields for basic food crops when used properly in a package, price incentives become the major factor in determining what yields farmers will achieve. As empirical evidence has been gathered over the past decade demonstrating a dramatic long-run response to price, this neoclassical view has increasingly been pushed on third world leaders who are urged to get their prices right.[3]

Rigorous econometric analysis has failed to settle the question of exactly how much farmers respond to price incentives. The weight of empirical evidence shows that producers respond positively to a rise in price of one crop relative to all others by producing more of that crop. However, the more important issue of whether agricultural production in the aggregate can be improved by an increase in the price of all agricultural products relative to nonagricultural products is much more uncertain.[4]

Challenging the position of the price fundamentalists who urge governments to "get their prices right" or bear the consequences of stagnant production are those economists who argue for a more balanced view of price policy. Government investment in infrastructure, technical change, input distribution, and land reform can generate agricultural growth, even with prices that seem low by international standards. Krishna argues that these "technology policies" have a greater impact on output than price policies.[5] In a recent review of the book *Distortions of Agricultural Incentives* edited by T. W. Schultz, Hayami also takes issue with those who advocate higher foodgrain prices.

> Indeed, Taiwan represents a unique challenge to the Schultz thesis. The development of the Taiwanese economy was to a significant extent based on savings exploited from agriculture by means of the unfavorable terms of trade. Significant increases in agricultural output and productivity were achieved, despite the unfavorable price conditions, by means of government investments in research, extension, irrigation, and other forms of infrastructure. One may argue that, if there were no distortion such as the Rice-Fertilizer Barter System, Taiwan's economy could have developed faster. Such an argument should be supported by a major study including an investigation of intersectoral resource flows.[6]

It can be added that before economists began advocating price increases in Indonesia and the abolition of the Thai rice premium, it was common practice to recommend to the government of Taiwan that they abolish the "rice-fertilizer barter system."[7] This they eventually did in the early 1970s, but only when it became evident that they had exhausted all means of increasing production through further crop intensification and higher yields. At this point the demand for labor in the nonfarm sector lead to a decline in the agricultural labor force. The role of the rice economy on Taiwan shifted dramatically from the position of supporting nonagricultural development to being supported by the nonagricultural sector.

The controversy over price still continues. Price fundamentalists, for example, can find support for their argument in the remarkable growth of Chinese agriculture (including rice production) following the price increases in 1978, although these were accompanied by concomitant major structural changes in the agricultural economy.[8]

In a recent survey of agricultural food policy issues, McCalla puts this controversy in perspective:

> One of the most striking things emerging from the interviews (with policymakers) was the demise of the limited factor mentality. Seldom among policymakers at the national level does one hear that credit or risk, or extension or fertilizer or water management or improved varieties are single constraints which if removed would solve all problems. Unfortunately, the acceptance of the general equilibrium nature of the policy issues has not been as widely accepted by researchers, economists, and biologists, as it has by policymakers.[9]

Our observation of policies followed by individual countries in the sections and the chapter that follow seems to support McCalla's contention that policymakers do have a broad perspective of policy alternatives and tradeoffs, although there is still substantial room for improvement of policy choices.

Classifications of Countries

All countries have policies that affect the price of rice, and many countries also intervene to affect the price of inputs like fertilizer. As a result, prices of rice are very different across countries, and although they do not reflect the total thrust of rice policy for any single country, the differences are a useful indication of divergence in price incentives.

Table 16.2 shows the farm price of rough rice as a percentage of the prevailing world price since 1960 in the Asian rice economies.[10] Countries have been ranked according to their percentage in 1976–80.

Three groups of countries can be distinguished: (1) medium- to high-income countries (table 16.2) have supported the rice price well above the world market; (2) low-income traditional importers and China that have maintained domestic prices somewhat below world market prices; and (3) low-income traditional exporters that have extremely low domestic prices. The first group of countries includes Japan, South Korea, Taiwan, and Malaysia. In general, they have shown an upward trend in their support levels. In the second group, the domestic price relative to the world price seems to have been declining as these countries have moved toward rice self-sufficiency. The exception is Indonesia, one of the world's largest importers, where the 1976–80 price was increased to be about on par with the world market price. The traditional exporters, Burma, Thailand, and Pakistan, continue to maintain farm prices at an extremely low level and to tax the rice sector heavily to raise revenues for various government activities.

Table 16.3, which is arranged by country groups in the same order as table 16.2, shows important relationships between the farm price of paddy and (1) the retail price of rice, (2) the price of nitrogen, and (3) the price of labor.

A ratio of retail-to-farm price of 2 reflects the

Table 16.2. Farm-level Price of Rice as a Percentage of World Price, 1961–80[a]

Country	1961–65	1966–70	1971–75	1976–80
Japan	203	228	246	391[b]
South Korea	119	104	111	187
West Malaysia	—	—	149[c]	173
Taiwan	160	134	150	168
Indonesia	—	63	66	98
Bangladesh	127	126[d]	163[e]	93
Philippines	120	93	85	77
China[f]	109[g]	96	71	76
India	146	109	98	76
Sri Lanka	178	141	128	76[h]
Thailand	71	55	62	70
Pakistan	—	—	—	48[i]
Burma	56	42[f]	44[f]	37[f]

Sources: Farm-level prices: Appendix tables; A. Palacpac, *World Rice Statistics*. World rice prices: FAO, *FAO Trade Yearbook*.

[a] Ranked from high to low based on 1976–80 percentages. Farm-level price in "paddy-equivalent." World price based on quantity and value of total world exports and imports as reported by FAO. World price divided by 2 to obtain paddy equivalent. Paddy-equivalent farm price for each country divided by paddy-equivalent world prices and multipled by 100 to obtain percentage of world price.

[b] 1976–78 only.

[c] 1975 only.

[d] 1966/67 only.

[e] 1972/73–1975/76 only.

[f] Official procurement price used.

[g] 1965 only.

[h] 1976/77–1979/80 only.

[i] 1977–79 only.

approximate cost of marketing. Ratios that are less than 2 suggest that countries are subsidizing the marketing of rice, while those in excess of 2 suggest that excess profits are being made on marketing.

Wealthy countries—such as Japan, Taiwan, and South Korea—can afford to maintain producer prices that are above the retail price including marketing costs. China, which also subsidizes rice marketing costs, increased the farm price of paddy in the late 1970s without a corresponding increase in the retail price of rice. This placed a severe strain on the budget. Sri Lanka followed a similar policy for most of the period after independence, but by 1980 had changed its policy. With the exception of Pakistan, retail prices appeared to be in line with farm prices (allowing for marketing costs) in all other countries.

Nitrogen is the most important purchased input for most Asian rice farmers. The price of nitrogen, like the price of rice, is subject to control. Based on world prices in 1979–81, the nitrogen-to-paddy ratio in free market countries should be about 2.5 (table 16.3).

In Japan and South Korea, it took less than 1 kg of paddy to purchase 1 kg of nitrogen in 1980, not because nitrogen was subsidized but because of the high price supports for rice. A number of countries, including both importers and major exporters, have chosen to subsidize the price of nitrogen. If the supply of fertilizer is short, as has been true in Burma because of limited imports, or as was true in much of Asia in 1973, the subsidized price only leads to a black market for fertilizer.

Farm wages are less amenable to price control than fertilizer or paddy and tend to reflect the general stage of economic development. Wage rates measured in kilograms of paddy that can be purchased with a day's wage are high in the high-income countries, but are also high in the traditional exporting countries (Thailand, Pakistan, and Burma) because rice prices are low (table 16.3). The purchasing power of a day's wage in terms of the basic rice staple is lowest among the traditional importing countries of Asia, particularly Bangladesh, India, and Indonesia, countries we associate with high population pressure and widespread poverty.

Achieving Production Objectives

It can be argued that among less-developed regions, especially the larger countries of Asia, the central food policy issue for the medium run is not whether to strive for self-sufficiency, but whether means can be found to increase domestic food production at a rate high enough to keep pace with demand.[11] Although the principle of "food self-sufficiency" may have little economic logic per se, in practice, expanding agricultural output is consistent with the comparative advantage of many Asian nations.

The work carried out in the Stanford "Political Economy of Rice in Asia" project showed that comparative advantage in producing rice varies considerably among Asian countries.[12] While Thai producers have a comparative advantage at prices as low as $US 150/mt, Philippine rice production becomes inefficient below the $US 275 to $US 350/mt range. Indonesia's breakeven price falls between the Philippines and Thailand at $US 250 to $US 275 for Java and $US 175 to $US 200 for the outer islands. Malaysia, for which comparable analysis is not available, undoubtedly stands higher than the Philippines. Differences in comparative advantage do influence the choice of government policies because they determine to a large extent the cost of intervention. Countries like Thailand, with a strong comparative advantage, have the widest range of policy options. They can, as we have seen, tax output and force domestic prices below world levels. High-cost

Table 16.3. Farm, Retail, and Fertilizer Prices and Wage Rates Throughout Asia, 1979–81

	Farm-to-retail			Fertilizer		Labor	
Country/location	Farm paddy price ($US/kg)	Retail price paddy ($US/kg)	Ratio of retail-to-farm price	Nitrogen price ($US/kg)	kg nitrogen/ kg paddy	Daily male wage ($US/day)	kg paddy/ day
Japan (Ibaraki Pref.)	1.47	1.54	1.0	0.66	0.4	22.20	15
South Korea (Hwaseong Pref.)	0.66	0.92	1.4	0.62	0.9	10.67	16
West Malaysia (Selangor)	0.26	0.50	1.9	0.72	2.8	5.05	19
Taiwan (Taichung)	0.36	0.58	1.6	0.38	1.1	15.30	43
Indonesia (Central Java)	0.17	0.32	1.9	0.27	1.6	0.80	5
Bangladesh (Joydebpur)	0.22	0.35	1.7	0.37	1.7	1.05	5
Philippines (Central Luzon)	0.16	0.30	1.9	0.56	3.5	1.63	10
China (Hangzhou)	0.13	0.19	1.5	0.67	5.2	—	—
India (Cuttack/Waltair/ Coimbatore)	0.15	0.27	1.8	0.52	3.5	0.64	4
Sri Lanka (Kurunegala)	0.13	0.31	2.4	0.19	1.5	0.81	6
Thailand (Suphan Buri)	0.10	0.21	2.1	0.34	3.4	1.83	18
Pakistan (Punjab)	0.08	0.38	4.8	0.38	3.6	1.00	12
Burma (Rangoon)	0.07	0.13	1.9	0.12	1.7	0.98	14
World price	0.20[a]	0.40	2.0	0.50[b]	2.5	—	—

Sources: World price of rice: A. Palacpac, *World Rice Statistics* (1982); all other data: *FAO Trade Yearbook*.

[a] World price based on quantity and value of total world exports and imports, 1979–81, as reported by FAO divided by 2 to obtain paddy-equivalent.

[b] World price of nitrogen based on a world urea price of $US 220/mt.

producing countries, like Malaysia, on the other hand, must rely on a system of input and output subsidies and a high effective rate of protection to create the incentives for expanding output.

Maintaining production growth in the 2 to 4 percent range, given the constraints on bringing new land under cultivation, can be accomplished by (1) investment in research; (2) investment in land infrastructure (irrigation, drainage, and the like); (3) institutional changes like stronger extension services and land reform, and (4) price incentives or subsidies. The price differences shown in tables 16.2 and 16.3 seem to reflect the differences in comparative advantage associated with achieving production growth targets in the range of 2 to 4 percent annually. Contrast, for example, the price policies of Thailand and Taiwan.

It is important to distinguish carefully between the various types of policies for increasing production. Policy tends to be associated with prices and various production incentive programs initiated to affect a production response in keeping with short-run political objectives. However, the long-run consequences of these and other policies are the prime determinants of growth over time. Research and improvement of land quality are activities that loosen constraints. By investing in these areas, it is possible to raise production potential, shifting the "frontier" production function upward or creating "economic slack." (Here "economic slack" is defined as the difference between the present product of a sector and the product that could be realized if all resources were optimally used). Slack-reducing activities, such as improving extension services or developing local organizations for water management, become more profitable when production potential is increased. Price policy can affect the rate at which new technology and institutions are adopted and slack is reduced. We have argued elsewhere that with the rapid transition from land to labor surplus that has typically occurred in much of Asia over the last two decades, there has been a lag in institutional adjustments needed to make efficient use of the new land-saving technology.[13]

Complementary and Competitive Policies

Thus far we have stressed the heterogeneity of situations among Asian countries and have discussed the policy alternatives used to achieve objectives such as increased production and consumer protection. The most efficient mix of policies is likely to be time and location specific. In the Asian setting, price policy, infrastructural development, research, and institutional change compete for scarce government revenue even while their effects complement each other in terms of effect on production.

In most of Asia, the rate of investment in research and irrigation development is determined by government policy and by policies of the major international lending agencies, rather than by private enterprises. However, there are indications that the level of public investment may be subject to price inducements. That is to say, in years of high prices brought about by crop shortages, the political will to commit the needed agricultural sector investment resources is stronger. As a result larger irrigation investments are made.[14] The support for irrigation development in recent years suggests growing government awareness of the longer run implications for low and stable prices, although the international donors seem to be influenced by short-term surpluses and shortages.

Programs, such as Masagana 99 in the Philippines and BIMAS in Indonesia, were designed to stimulate rice production through credit and input subsidies and extension support. In terms of developing viable institutional credit and extension systems, these programs must be judged as failures. However, as short-term "pump priming" activities to extend modern inputs and technology, they seem to have been fairly successful. And they should be viewed, at least in part, as income transfer to the rural areas.

Welfare Implications

An increasing amount of interest is being shown in the welfare implications of various policies, and rightfully so. We have previously suggested that the policy objectives related to welfare are likely to have low priority over other objectives unless there is strong political pressure, but it is important to identify the beneficiaries of various policies. The welfare implications of technology change have been discussed in chapter 10. Thus, we refer here only to implications for price policy.

Price policy presents one of the obvious dilemmas in welfare. Low rice prices designed to serve special consumer interest groups, including civil servants and in some cases the military, also benefit the rural and urban poor who spend a major portion of their income for the purchase of rice. High rice prices benefit the larger rice farmers. Prices are therefore a blunt tool for welfare objectives.

One resolution to this dilemma, especially for Indonesia, is to hold prices of the nonpreferred staples (maize, cassava, sweet potatoes) low relative to rice to provide an alternative dietary source for the poor.[15] The problem is that the poor also prefer rice, and an adequate supply of the alternative commodities is needed. Income elasticities are positive for rice in Indonesia and negative for the inferior staples. Some professionals think that the low-income portion of the Indonesian population may decline over the next decade and that the option of increasing consumption of nonrice staples among this group should be encouraged. However, there is little evidence of this happening to date.

A Rational Rice Policy

No one looking at the wide disparity in domestic relative to international prices (tables 16.2 and 16.3) could argue that the rice policies followed by Asian countries as a group were rational on purely economic grounds. In fact, the economic literature has been replete with articles demonstrating the economic irrationality of many national rice policies. It is our contention, however, that rationality must be judged in terms of the effectiveness of achieving the broad spectrum of political and economic objectives set forth in the early part of this chapter. Rationality must also be judged in the context of the constraints imposed by the international market discussed in chapter 13.

The major exporting countries have been very discriminatory toward producers in their price policies. As modern technology more suited to their environment became available, exporting countries began to invest more heavily in "technology policies."

The price policies of the major South and Southeast Asian importers have been much more closely in line with world prices, but they too have relied heavily on technology policies in an effort to meet production targets. China, however, has been able to achieve significant gains in production and productivity since 1978 by raising prices to producers.

In East Asia, price policy appears to be almost completely devoid of economic rationality. The high price supports in Japan, South Korea, and Taiwan (and recent concerns over a shortfall in rice production in Japan) emphasize the political imperatives of rice policy. Outside observers are hard pressed to view these policies as rational.

In the light of both productivity and welfare objectives, there are two programs that should be at the foundation of any rational rice policy: (1) technical change in agricultural production, and (2) redistribution of assets in favor of the poor. As we have discussed in the previous section, investments to raise production potential are critical to long-run growth and development. The historical underinvestment in rice research has been discussed in chapter 14. Redistribution of assets can raise the employment and income potential of the poor and the demand for rice.

In their recent book, *Food Policy Analysis*, Timmer, Falcon, and Pearson point to a number of programs that are frequently adopted, but are not useful in solving the food problem.[16] These include: (1) eliminating the middlemen (chapter 12 discusses this issue); (2) crash programs (we agree that crash programs tend to crash, although as income transfers that also encourage adoption of new technology they may be useful); (3) subsidizing farm inputs; (4) direct food deliveries to the poorest of the poor; (5) nutritional intervention projects; and (6) food aid. These programs are largely attempts at short-term solutions to the problem and, therefore, tend to appeal to politicians. But their costs are high and long-run impact minimal or in some cases counterproductive.

In summary, we argue that the pragmatic balance of low prices, government subsidies for irrigation, investment in research and extension, subsidized credit, and "crash" production programs followed by most of the Asian countries may make more sense than academic observers are willing to concede. The relative success of a number of countries attests to the viability of that route where the potential for increasing production exists. Continued investment in technical change and growth in production potential should be a central element in rational rice policy. Because rice is a labor-intensive crop, some gains in employment have been achieved by emphasizing rice production. However policies to develop technology cannot be expected to solve the welfare problem. More emphasis should be placed on programs to ensure that the poor have greater access to productive assets.

Notes

1. C. P. Timmer, "The Political Economy of Rice in Asia: A Methodological Introduction," *Food Research Institute Studies* vol. 14, no.3 (1975) pp. 191–196.

2. Whether one refers to the rice price dilemma or in the broader context to the overall food price dilemma, the issue is basically the same. See C. Peter Timmer, Walter P. Falcon, and Scott R. Pearson, *Food Policy Analysis* (Baltimore, Md., The Johns Hopkins University Press, 1983). The food price dilemma is a major theme in this book.

3. C. Peter Timmer, "Food Prices and Economic Development in LDCs," a paper prepared for the World Food Policy Seminar, Harvard Business School, May 13–14, 1979.

4. "Although supply response is a heavily researched area, there are surprisingly few studies of the response of aggregate farm output to lagged terms of trade *inter alia*. Such studies are obviously crucial for measuring the marginal leverage of terms of trade as a means of stimulating agricultural growth. In recent survey papers of the World Bank tabulating about 100 single-crop price elasticities of acreage and supply for developing countries, only two aggregate supply elasticities are recorded." Raj Krishna, "Some Aspects of Agricultural Growth, Price Policy and Equity in Developing Countries," *Food Research Institute Studies* vol. 18 (1982) p. 234.

5. Ibid., pp. 219–260.

6. Yujiro Hayami, Review of T. W. Schultz, ed., "Distortions of Agricultural Incentives," *Economic Development and Cultural Change* vol. 29, no. 2 (January 1981) pp. 433–434.

7. The "rice-fertilizer barter system" was established in 1950 and abolished in 1972. The barter exchange ratio of rice for fertilizer was set by the government, which had a monopoly on fertilizer supplies. To obtain fertilizer, farmers had to provide rice to the government at the prescribed exchange rate. During most of the twenty-two year period the progam operated, the exchange rate, based on world prices, served as a tax on producers. Yet fertilizer use expanded rapidly during this period. See T. H. Lee, "Government Interference in the Rice Market in Taiwan," in International Rice Research Institute, *Viewpoints on Rice Policy in Asia* (Los Banos, Philippines, IRRI, 1971) chapter 4.

8. For a discussion of this issue see Thomas B. Wiens, "Chinese Economic Reforms: Price Adjustments, the Responsibility System, and Agricultural Productivity," *American Economic Review* vol. 73 (May 1983) pp. 314–324.

9. Alex F. McCalla, *Agricultural and Food Policy Issues Analysis: Some Thoughts from an International Perspective* (Washington, D.C., International Food Policy Research Institute, 1978) p. 21.

10. Comparison of domestic prices across countries is complicated by the problem of conversion to a single monetary unit and the divergence of official exchange rates from "market" exchange rates. Although we recognize these complications, there is no completely satisfactory way to overcome them. For the analysis reported in table 16.2, we converted all prices to $US at the official rates prevailing during the respective years. Farm-level prices found in the appendix for each country (printed separately from this book) were converted to $US and divided by the world price calculated from the *FAO Trade Yearbook* series on quantity and value of rice exports and imports.

11. This question is addressed directly in the final chapter of this book. It is also raised in Walter P. Falcon, "Food Self-Sufficiency: Lessons from Asia," in *International Food Policy Issues: A Proceedings*, U.S. Department of Agriculture Foreign Agricultural Economic Report 143 (Washington, D.C., 1978).

12. *Food Research Institute Studies* vol. 15 no. 3 (1976) contains various studies of the domestic resource costs of rice production for five Asian countries and the United States.

13. Randolph Barker, "Barriers to Efficient Capital Investment in Agriculture," in T. W. Schultz, ed., *Distortions of Agricultural Incentives* (Bloomington, Ind., Indiana University Press, 1978).

14. Yujiro Hayami and Masao Kikuchi, "Investment Inducements to Public Infrastructure: Irrigation in the Philippines," *The Review of Economics and Statistics* vol. 6, no. 1 (February 1978) pp. 70–77.

15. Timmer, "The Political Economy of Rice in Asia: A Methodological Introduction."

16. Timmer, Falcon, and Pearson, *Food Policy Analysis*, pp. 293–290.

17

National Rice Programs

Case Studies for Selected Countries

Different countries in Asia employ a wide range of rice-related policy instruments to increase production and provide a reliable, reasonably priced supply of rice for urbanites and other select consumer groups. As we have seen in chapters 12 and 16, much of the literature deals only with price policy, but price is only one element of the mix of policies employed at any given time. Furthermore, national policies change over time as incomes rise and other circumstances, such as the level of supply relative to demand, also change.

In this chapter, we trace the policy changes of six countries by examining their national rice programs in the post–World War II period. The countries were chosen to emphasize the plethora of approaches that are used. India and China, Asia's two largest rice producers, differ in the degree of government control and planning in the economy. Indonesia, the world's largest rice importer in most years, has a strong consumer bias in its policy, while Thailand, a high-volume exporter, has discriminated against producers in rice price policy. Malaysia, the country with the highest per capita income in South and Southeast Asia, is exhibiting a producer bias in policy formation similar to the high-income East Asian countries. The Philippines reflect some of the problems that occur when a traditional importer achieves national self-sufficiency and attempts to market a small rice surplus. In the final section of the chapter, we discuss some of the common elements in the various rice programs.

India: Programs to Increase Rice Production

Like China, India's agriculture is remarkably diverse, and rice plays a less overriding role than in many other countries of Asia. Wheat occupies about half as much land as rice, and other cereals and pulses take up a greater area than wheat and rice combined. The contribution of other foodgrains and non-foodgrains is so important to consumption and farm income in India that it is impossible to consider rice policy separately. Also, Indian government policy generally considers foodgrains as a group.

Policy instruments used by the Indian government include an array of tools ranging from price fixing to government monopoly of imports to broad scale extension efforts. India's general price policy has consistently favored consumers over producers. As indicated in tables 16.2 and 16.3, the Indian economy features rather low product and high input prices, although a series of programs has been undertaken to increase production, some of which have been aimed at helping producers.

The first national-scale programs were instituted in the early post–World War II period. The "grow more food" campaign, narrowly focused on technical agricultural recommendations, was organized to help India overcome the ravages of the 1944 Bengal famine. Government officials attempted to use the colonialists' rudimentary agricultural extension network to dispense improved technical information through posters and slogans, but the information was not widely dispersed and had little effect on food production.

After independence, it quickly became clear that a simple and narrow technical focus to national agricultural policy was not sufficient to overcome India's food production problems. The Community Development Program was formulated to provide an administrative framework through which the government could reach down to the district, subdistrict, and finally the village level with a range of development programs. India's 400-odd districts were divided into development blocks, with a block development officer (BDO) in charge of activities. Within each block, some 20 to 40 village-level workers (VLWs) had direct contact with the villagers in the 10 to 20 villages for which each was responsible. A nationwide structure of national, state, and regional training centers was set up to train the thousands of BDOs and VLWs needed to run the system. At the top, a Community Development Research Center was staffed by the best of India's academic world, and the Community Development Organization became a ministry within the government.

The originators intended the system to serve the broad needs of India's masses, whether for village roads, public health facilities, adult education, agricultural innovations, or assistance in developing cottage industries. But they also hoped that the Community Development Program would provide the rapid increases in agricultural output that the nation needed. However, the system rapidly became overburdened as the responsibility for developing projects, whether area or village-level, health or agriculture-related, was devolved to development program personnel.

Foodgrain production increased at a satisfactory rate between 1951 and 1956, and imports were reduced from the high 1951 level of about five million mt of foodgrains to less than half a million mt (figure 17.1). Grain output in 1956 and 1957 failed to increase substantially, however, and in 1957/58 production fell to pre-1954 levels. Large imports were needed, and dissatisfaction with the pace of agricultural development led the government of India to follow a new strategy developed by a team of foreign consultants in the late 1950s.[1] This approach called for a sharply focused, integrated agricultural program.

In 1961 with the assistance of the Ford Foundation, the government set up the Intensive Agricultural District Program (IADP) on a pilot basis in one district in each of seven states. The program was established with the goal of tripling the rate of increase in agricultural production and attaining Third Plan targets.[2]

The number of village-level workers and block agricultural development officers was doubled in IADP districts. Jeeps were provided to increase the mobility of the block development officers. Funds were made available for demonstrations and fairs. Nonagricultural responsibilities were deemphasized, and funds for agricultural credit were directed toward the project districts. There was even an attempt to get the government to implement special product price supports in project districts—an attempt that was unsuccessful because of the larger market forces at work.

The Intensive Agricultural District Program was based on a "package" approach to development. The package consisted of a combination of institutional, economic, and technical innovations to be implemented at the district, block, village, farm, and field level. Noncollateral production credit based on individual farm plans was intended as a key part of the program, although in some of the districts the state and local cooperative credit institutions could not be convinced of the wisdom of the practice. Field demonstrations contrasting farmers' methods with the "package of improved practices"—treated seeds, fertilizer, weeding, pesticides—were one of the main responsibilities of village-level workers under the program (along with farm planning). Soil testing laboratories were set up in the original seven IADP districts, and farmer-participants were supposed to apply fertilizer on the basis of soil tests. In a number of the districts, engineering workshops were established to demonstrate, modify, and test implements thought to be of use to the district's farmers. Plans for improving the water management of the irrigated districts were developed jointly with district agricultural and irrigation specialists.

The resources being poured into the seven pilot districts attracted the attention of states that had no package program district. The government agreed to expand the program so that, by 1963, seven new districts were added. Additional staff, funds, and transportation were provided to the new districts. In 1964–65, a new program, modeled on IADP but with far less financial backing, was announced for 100 districts. This Intensive Agricultural Areas Program (IAA) was a pale imitation of the original IADP, but it did draw the experienced district-level staff out of the original IADP district for the new national program and consequently weakened it.

During the 1950s, agricultural output in India grew at over 2.5 percent per year while population grew at about 2 percent. However, between 1958 and 1963, despite the IADP, foodgrain output stagnated around

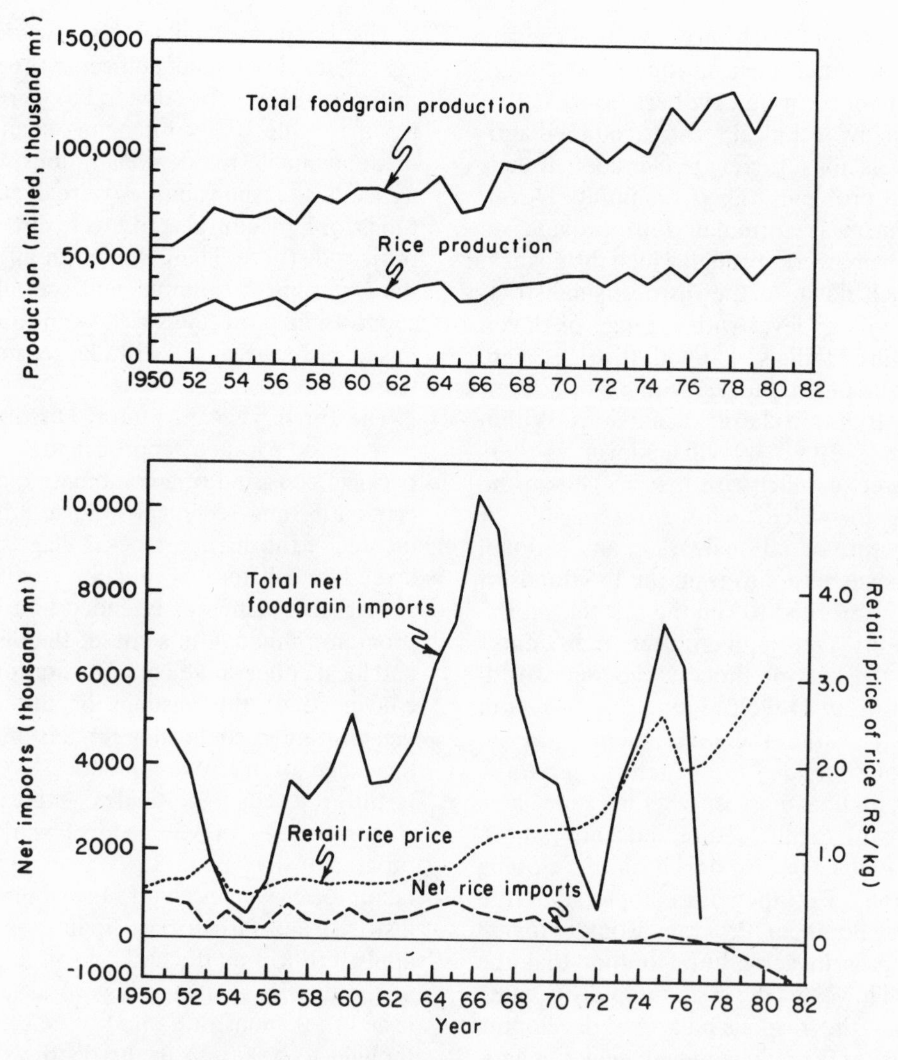

Figure 17.1. Production, imports, and prices of rice and foodgrains, India, 1950–81. Years in the upper graph are crop years.

the 80 million mt level (figure 17.1). Favorable weather in 1964 finally resulted in a rise to 89 million metric tons, but severe droughts in 1965 and 1966 cut production to 72 and 74 million mt, respectively. Imports, which had ranged from 3 to 5 million mt between 1957 and 1963, reached 10 mt in 1966. The long stagnation followed by the rapid rise and fall of such large magnitudes convinced planners and many observers that something more was needed. A number of careful studies of IADP failed to provide convincing evidence that IADP districts grew faster than similar nonprogram districts. And, after the fact, those closely associated with IADP acknowledged the same thing. The expert committee assessing IADP performance said in 1969 that: "One major obstacle that was not fairly recognized at the inception of the program was the low yield response to fertilizer and other inputs of the then recommended varieties of foodgrains. It was the IADP experience which highlighted the low yield response of the indigenous improved varieties. Only after about 6 years of the operation of the program, varieties responding to fertilizer use and capable of giving high yields became available."[3]

In 1966, dissatisfaction with the performance of IADP and the Community Development Program, together with bold assurance by agricultural leaders that the new semidwarf varieties of wheat and rice were capable of giving much higher yields if adequately fertilized and irrigated, led to the formulation of a new strategy of agricultural development in India's draft Fourth Five-Year Plan. The strategy, embodied in the High-Yielding Varieties Program (HYVP), involved deliberate concentration of seeds, fertilizer, and administrative talent in areas where irrigation was of high quality. Thus, resources were poured into the potentially productive areas at the expense of others, a turnaround from the equalitarianism under the IADP and IAA programs. This represented a deepening and narrowing of the "package" concept with less emphasis on developing institutional support in the credit and cooperative fields and more emphasis on the agronomic components of the package. Selected areas were given priority in receiving fertilizer shipments. Seeds of the semidwarf varieties, still in scarce supply despite imports of 18,000 mt of wheat seed from Mexico and shipments of rice seed from the Philippines, were also directed to the water-assured areas. At the same time, national agricultural programs designed to stabilize farm prices, boost the total availability of credit to agriculture, increase the area irrigated, improve the quality of irrigation, and expand the scope of research were implemented.

The new varieties, under the HYVP, took hold rapidly. In 1967 nearly 2 million hectares were planted; in 1968 6 million; in 1971 15 million; and by 1975 27 million hectares. It is estimated that the new varieties supplied 6 percent of total cereals output in 1967; 15 percent in 1968; 35 percent in 1971; and 62 percent in 1975.[4] Modern varieties are estimated to have occupied 88 percent of the irrigated cereal area in 1975 and to have received 47 percent of the fertilizer used on cereal crops that same year. Output of foodgrains increased from about 80 million mt in the early 1960s to 95 million mt in the late 1960s and further averaged 103 million mt for 1971 to 1975. Imports fell from 10 mt tons in 1966 to about 3 million mt between 1971 and 1973.

Beginning in 1970, with the more comfortable total food situation, attention in India turned to the problems of poverty and equity.[5] "Integrated rural development" became a byword among donors. Quantitative estimates of the number of people living in absolute and in relative poverty were generated. The economic gains of new technology were balanced against the political costs, and the new strategy was questioned. Interregional equity questions were raised as production increased in the resource-rich areas while others stagnated or made little progress.

Political response to these concerns was slow, but when momentum had built up, it became an important force. However in 1972, bad weather caused crop production to fall again. The need for imports during the world food crisis of 1973–74 diverted political attention from the distribution issues for several years. But, by the election of 1980, the poverty issue had become an important topic, with Mrs. Gandhi promising to "banish" it from the country. Government planning reflected these concerns, with schooling, health, food distribution, and family planning receiving renewed emphasis.

A renewed focus on teaching farmers how to use the components of technology was evident with the introduction of a new approach to agricultural extension called the Training and Visits System. This approach, pioneered in Israel and spread by the World Bank, involved agricultural extension agents (like village-level workers) in a regular two-week routine consisting of one day of training in a very specific skill followed by one-day visits to about ten villages where the skill is passed on to farmers. The cycle is repeated every two weeks. This method is extremely valuable for passing on specific techniques and skills. The program was initiated with World Bank assistance in the mid 1970s, and by 1980, half a dozen states had instituted the program using it in conjunction with the high-yielding varieties program that was already in progress.

During the entire period since independence, India has struggled to ensure that adequate foodgrains would be available to its population at affordable prices. Programs of direct distribution during famines were replaced by a system of fair-price shops and rationing in the early 1950s. Grain was made available to the fair-price shops by the government-run Food Corporation of India. The Food Corporation, in turn, obtained its supplies from imports (both commercial and concessional purchases through the U.S. PL 480 program) and domestic purchases.

Domestic procurement was facilitated through the imposition of "food zones" between which food could not move without explicit government permission. This policy had the effect of bringing down prices of foodgrains in zones that produced more than they needed and driving up prices in zones that were in deficit. The government took responsibility for buying up the excess in the surplus zones at the procurement price—which was usually below what the prevailing market price would be in the absence of zones. During periods when foodgrains were plentiful, it was easy for the Food Corporation to meet its procurement targets, and zones were unnecessary. Consequently, zones were imposed during short years and abolished during years of high grain production. Thus, the government believed that it could do a better job of obtaining food and transporting it to locations where needed than could the free market. In the process, the government reduced the levels of price incentives below what they would have been in areas with a relative advantage in production. At the same time, however, programs designed to provide physical inputs and capital investments such as irrigation were being concentrated in the same areas in an attempt to increase production.

One can trace the pattern of government policy emphasis in India through wide cycles, often coinciding with the worldwide concerns of the development community. This is not surprising in light of India's size and importance. The focus of technical agricultural issues prior to independence gave way to the broad concerns of community development in the 1950s. Dissatisfaction with agricultural performance led to the package program concepts of IADP, which were further narrowed to almost exclusive concern with technical issues in the High-Yielding Varieties Program. Relative success with that strategy permitted renewed concern with the broader issues of development in the middle and late 1970s. The diversity of instruments resulted in a relatively satisfactory production situation by the early 1980s, with modest imports or even exports each year from 1976 to 1980.

China: A Centrally Planned Economy

China, like India, is a large country with a diverse agricultural economy in which not only rice but other foodgrains play an important role. Broadly speaking, in the area south of the Yangtze River rice is the dominant crop, and in the area to the north, wheat and a wide range of other cereal grains are traditionally important. However, throughout this century, rice production has tended to move northward and wheat production southward. Today, rice accounts for 45 percent of China's grain production and wheat and maize, which are of about equal importance, another 45 percent.[6]

At the time of liberation in 1949, the new communist government found itself in a unique position with respect to agricultural development among the countries of Asia. China shared a long history of population pressure and an unfavorable population–land ratio with its East Asian neighbors, Japan and Korea. Moreover, China lacked important preconditions for agricultural development: (1) a formal agricultural research system, (2) an industrial sector capable of producing inputs such as chemical fertilizer, and (3) a transportation and communications network to ensure that inputs could be supplied to farmers at affordable prices. At the turn of the century, such a system of modern agriculture was gradually materializing in Japan. Following World War I, Japan extended this system to its colonies, Korea and Taiwan.

China's neighbors to the south also lacked these preconditions for agricultural modernization at the end of World War II. But unlike China, the countries of South and Southeast Asia had additional land to be developed and could depend for at least a decade on further exploitation of traditional inputs for a large share of agricultural growth.

With opportunities for the expansion of agriculture through traditional inputs essentially exhausted, and population growing at a rapid rate, modernization of agriculture was urgently needed in China to avoid a deterioration in the per capita production and standard of living and to provide a surplus for industrial development. In this context, it is interesting to review the development strategy that emerged.

China's historical focus on cropping intensity through emphasis on water control and on early maturing varieties continued after liberation, receiving even stronger emphasis under the new government. The objective was to maximize output per hectare per year under the assumptions (1) that there was no opportunity for expanding land area, (2) that there was a tight restraint on liquid capital for purchased inputs, and (3) that there was no constraint on the supply of labor.[7] Between 1952 and 1957, the multiple cropping index (number of crops harvested per year per unit area × 100) rose from 167 to 187 percent in the South China rice growing region. During the same period in Taiwan, it rose from 174 to 179 percent, essentially reaching a peak of 189 during the mid-1960s.

Despite lip service to "agriculture first" and to equity between rural and urban incomes, China's strategy of economic development has been heavily biased toward industry and urban consumers. As a consequence, while agricultural output has grown at 2.5 percent in the past three decades, industrial output has grown at 10 percent. The gap between rural and urban standards of living is still large.

A complex set of policy instruments was used to guarantee an adequate supply of basic foods such as rice, but at the same time to squeeze agriculture in favor of industrial development. Price policy, the collectivization of agriculture, and the development of agricultural technology will be discussed with respect to their independent and collective effect on the Chinese rice economy.

Low consumer rice prices were maintained by purchasing paddy rice under a two-price system. Production quotas were assigned to each team, and rice was purchased by the government at quota prices that were well below the world market price. Rice purchased above the quota, on the other hand, was purchased at 30 to 50 percent above the quota to encourage production above the target. Quotas were infrequently changed to give producers a further incentive. In addition to the low procurement price, further subsidy in the marketing and distribution of rice guaranteed an extremely low retail price to consumers. Rice and other staples handled in this manner were rationed to hold demand in check.

The collectivization of agriculture created both incentives and disincentives to agricultural development. Land reform, carried out in 1951 and 1952, destroyed the power of the rural elite by confiscating land and dividing it into small, private holdings, similar to land reforms throughout East Asia. This was followed by a Russian-style collectivization, which culminated in the creation of communes in 1958. Such a radical departure from the traditional social system initially resulted in a loss in producer incentives. These social changes, coupled with unfavorable weather, resulted in an extraordinary decline in grain production in the early 1960s (figure 17.2). A series of modifications in commune structure followed, with the responsibility for many day-to-day decisions and for the determination of income and sharing of profits being lowered first to the brigade and then to the production team, which typically consisted of about thirty families farming 10 hectares of land.

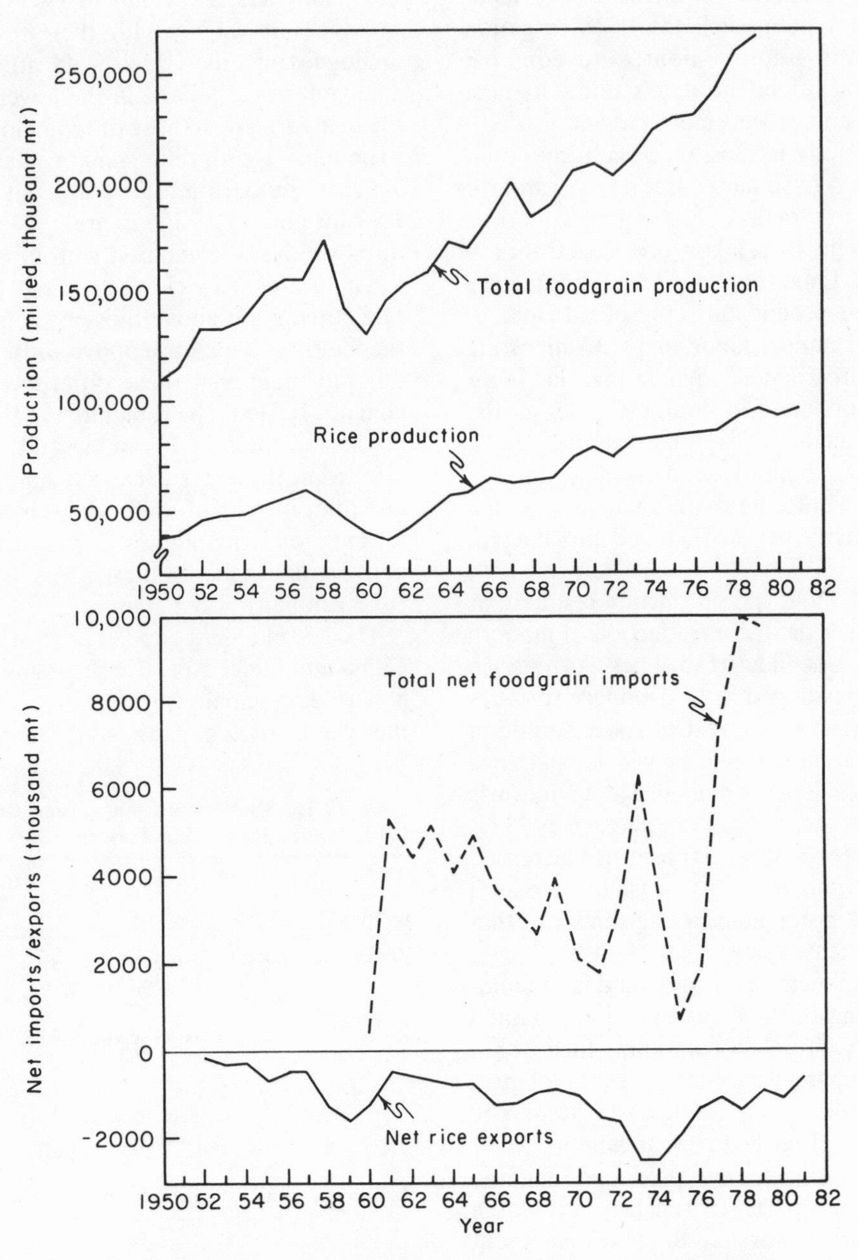

Figure 17.2. Production and international trade of rice and total foodgrains, China, 1950–81

Maoists made an attempt to restore the unit of accounting to the brigade level during the Cultural Revolution. However, since 1978, the trend has been in the opposite direction, with the adoption of the "responsibility system." The state contracts directly with individuals or families in what approximates a share rent or a cash arrangement in Western agriculture (the state being the landlord).

The degree to which the collectivization of agriculture encouraged or discouraged agricultural production is a matter of debate. It is clear that the commune structure facilitated the mobilization of labor for capital investment in such things as irrigation development, land improvement, and compost production, and facilitated the state's ability to meet grain production and procurement targets.

Despite the steadily increasing agricultural population working on a fixed land base, the demand for labor grew even more rapidly. The annual work days for participants in the rural labor force rose from 159 to well over 200.[8] Labor employed for manufacture of compost in rice production represented close to one-third of total annual labor input. Composting labor alone is approximately equal to the total labor input for rice production in countries such as the Philippines and Thailand. Given the massive growth of rural labor, as population expanded rapidly, productivity per worker-hour declined despite the fact that productivity per worker and per hectare increased.

Increases in rice yield and production resulted, as in the rest of Asia, from the introduction of modern fertilizer-responsive semidwarf varieties. As noted in chapter 5, China developed its own modern varieties (MVs) slightly ahead of the rest of Asia. Adoption of new varieties was more rapid and widespread since most of China's rice area is irrigated. The semidwarfs were grown on 80 percent of all rice acreage by 1977. From 1962 to 1977, chemical fertilizer use increased at 17 percent per annum, and rice yields increased at about 2 percent per annum, a rate similar to that in India and Southeast Asia.[9]

As the new grain technology took hold in the mid-1960s, the left wing of the communist party gained ascendancy. The Cultural Revolution, which began in 1966, was accompanied by a policy of self-reliance, which continued until after the death of Mao. The development of the "four-level research network"—county, commune, brigade, and production team—exemplified the concept of self-reliance in research and extension. This network led to the development of strong applied research and extension capacity at the local level, totally lacking in most other Asian developing countries.

Higher level education and research in general were undermined by the Cultural Revolution. However, certain rice research continued to receive priority, particularly in Hunan, the home province of both Mao and Hua Guofeng, where the first F_1 hybrid rice was developed in the early 1970s (see chapter 5). The F_1 hybrids were released in 1977 and spread to 15 percent of the rice area by 1979.[10]

Despite these advances in rice technology, by the mid-1970s, rice production was stagnating in much of China. The reasons for this are not clear. Table 17.1 compares rice yields in 1957 and 1978 for two provinces in China with their respective climatic analogues in other, high-yield areas in East Asia. The yields of Zhejiang in the lower Yangtze Valley are compared with those of Japan for the same years. (The most appropriate climatic analogue is Kyushu, Japan's southern-most island, but yield levels for Kyushu and all of Japan are very similar). Guangdong yields are compared with those of Taiwan.

There are two yield gaps. First, just as the yields of Zhejiang are above those of Guangdong, so also the yields of Japan are above those of Taiwan. It is our judgment that these differences are related to climate. It should be remembered that a much larger portion of the land is double and triple cropped in semitropical areas such as Guangdong and Taiwan and that, in spite of lower yields in a single cropping season, total rice and crop production per hectare per year are higher in these areas than in temperate-zone Zhejiang and Japan.

The second yield gap is between the provinces of China and their respective analogues. This yield gap is caused principally by differences in inputs. In 1978, the yields of Zhejiang and Guangdong were at

Table 17.1. Rice Yield Comparisons Between 1957 and 1978, China, Japan, and Taiwan

Location	Area	Year 1957	Year 1978	Percent increase
Zhejiang province	Total sown	3.7	4.7[a]	27
	1st crop	—	5.2	
Japan	Total sown	4.4	6.2	41
Guangdong province	Total sown	2.1	3.2[a]	52
	1st crop	—	3.7	
Taiwan province	Total sown	3.2	4.3	34
	1st crop			
China, Mainland	Total sown	2.7	4.0	30

Source: Perkins and Yusuf (1980).

[a] Assuming total yield 10 percent below first crop yields.

approximately the level of Japan and Taiwan in 1957. As we look back on the recent experiences of Japan and Taiwan, we are reminded of the substantial incentives that were needed to bring yields to their current levels.

Following the political upheaval upon the death of Mao, agricultural policies have undergone substantial revision. The Gang of Four was blamed for the "failure" of old agricultural policies as illustrated by the fact that per capita food consumption was no greater in the late 1970s than in 1957 or in prerevolutionary China.[11] Despite improvements in distribution, 10 to 20 percent of the Chinese population was said to have had an insufficient amount of foodgrain in 1972.[12]

The new policies called for giving greater incentives and more flexibility in decision making to producers. The quota price of rice and other grain crops was increased by 20 percent in 1979, and the surplus or above-quota price raised to 50 percent above the quota. In contrast to the past, it became permissible to sell surplus grain on the private market. With the shift away from self-reliance, hectarage in rice declined, particularly in those areas where producers chose to revert from triple- to double-cropping systems. However, rice yields and total rice production increased significantly. It is difficult to separate the effect of price incentives from that of structural changes caused by the "responsibility system" in bringing about these production gains. The important issue is to what degree these policies will result in short-term once-and-for-all change compared with a long-term sustained rate of growth.

Ultimately, growth in agricultural production and productivity depends on expanding the technical capacity of Chinese agriculture. Increases over the past three decades have largely come through the introduction of new technology and the exploitation of labor for capital investment. During and immediately after the Cultural Revolution, formal agricultural research and teaching were interrupted for eight years. To what degree this affected development of new agricultural technology is difficult to ascertain. Although there seems to be some potential for making improvements in the existing irrigation systems, a cardinal component in improved rice production, most of the easy investments have already been made. Further expansion of irrigation will require development of major projects with long gestation periods. It seems unlikely that the government will finance such investments in the near future in part because of the strain on financial resources as a result of high producer prices. For the immediate future, further gains in rice yields are likely to be achieved by increased application of chemical fertilizer, particularly phosphorus and potassium.

The Chinese record on labor absorption and mobilization for capital investment deserves special attention by governments confronted with a severe land constraint and a rapidly growing population of landless agricultural laborers. Emphasis on rice production has been a central part of that strategy.

Indonesia: Rice for Consumers[13]

In contrast to India and China where community development are important objectives, Indonesia's agricultural policies are more sharply focused on rice availability and prices.

After Indonesia's independence, the bureaucratic structure inherited from the Dutch was expanded. Rice was distributed to civil servants and the army beginning in 1951 and 1952 in an effort to retain their loyalty in the face of severe inflation. Rice production gradually increased, and rice prices stabilized from 1952 to 1954. Imports were modest, and distribution of rice rations to government employees ceased in 1953 and 1954. A drop in rice production in 1955 and 1956 caused a sharp increase in imports in 1956 (figure 17.3). Nineteen fifty-six marked the beginning of an eight-year period during which rice imports rose, peaking at nearly two million metric tons in 1964. Although in part a consequence of inflation caused by massive government deficits, rising domestic rice prices brought about a reinstatement of rice rations for civil servants and the army, and provincial governors began to procure rice supplies for the government. Provincial borders were closed to rice trade, increasing price differentials and driving prices down in surplus-producing provinces.

During the 1950s, rice imports drained considerable foreign exchange, but in 1959 the government formulated a new agricultural program based on village "padi centers." Each center was to stock seeds and fertilizer, teach farmers how to use them, provide credit, and buy back paddy to liquidate the credit (generally at somewhat below the prevailing market price). Each center was to service about 1,000 ha with a target of 1.5 million ha by 1964. These targets were not met, and production did not increase. There was 350,000 metric tons *less* production in 1963 than in 1959. Imports topped 1.7 million mt in 1964. Under the pressure of runaway inflation, rice prices skyrocketed from 200 rupiahs/kg to 1,800 rp/kg during 1965[14] and continued to rise even after demonitization and issuance of new rupiahs in 1966, contributing to the political instability in this period.

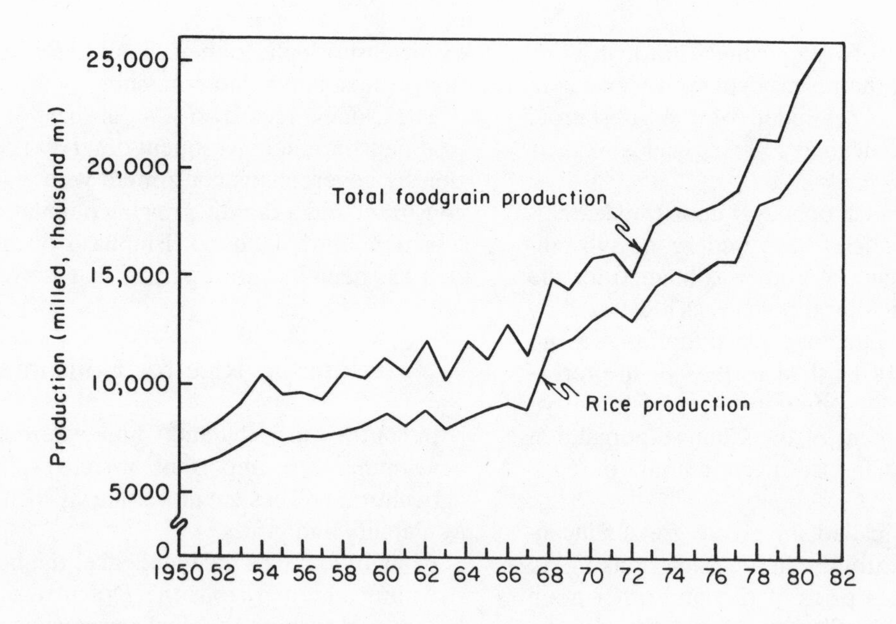

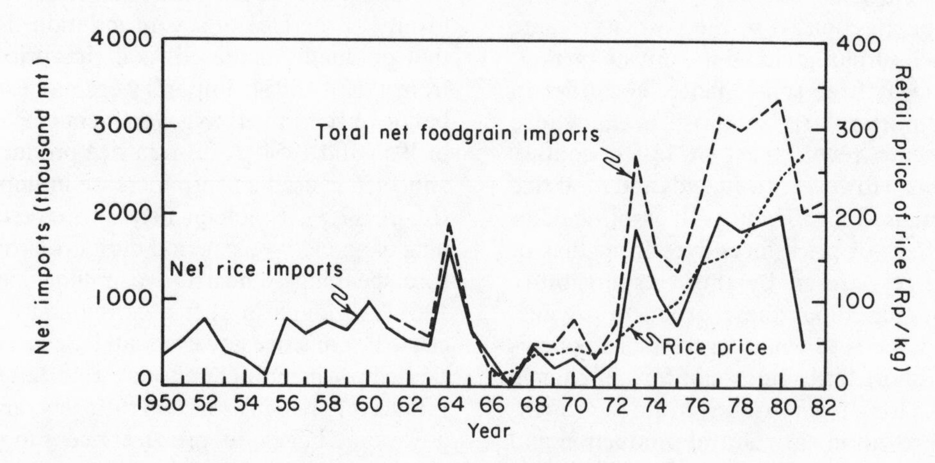

Figure 17.3. Production, imports, and prices of rice and foodgrains, Indonesia, 1950–82

The abortive coup of 1965 and its aftermath of reprisals against members and suspected members of the Communist Party diverted official attention away from agriculture. The 1,000 to 1 revaluation of the rupiah at the end of 1965 signaled the determination of the new government to tackle the economic problem, but the lack of rice within the country made continued dependence on massive imports necessary. Wheat was imported at a lower price than rice in small quantities, but did not become accepted by the general population until the 1970s.

The rate of inflation was reduced from 650 percent in 1966 to 120 percent in 1967, but the instability in rice price and production continued into 1968. The government, at this point, instituted a price incentive program for farmers. It held the price of milled rice and urea at about the same level and also greatly expanded the extension effort through the BIMAS program described below.

During the early 1960s the struggle to increase rice production attracted the attention of students and faculty at Bogor Agricultural University. They began an experimental program to increase rice production by first gaining the confidence of farmers and then attempting to teach them how to use selected seeds, chemical fertilizer, and insecticides. Efforts were concentrated on a few farmers, and they did not hesitate to pressure government officials, even up to provincial governors, to ensure timely delivery of inputs.[15] Yields on the guided farms increased up to 50 percent.

Initially the program was called DEMAS (Demonstrasi Massal, or Mass Demonstration). By 1965, the program had been institutionalized by the government under a new name, BIMAS (Bimbingan Massal, or Mass Guidance). It expanded rapidly from 10,000 demonstration hectares in 1965 to 500,000 ha in 1956 and over 1.5 million in 1968. The key component of BIMAS remained similar to those of the padi centers—seeds and fertilizer, technical know how, credit, and an assured market. Increased emphasis was also placed on improved water management.

The BIMAS program involved a heavy burden of supervision and guidance by the government. As the program matured, efforts were made to encourage the farmers to use seed, fertilizer, and improved technology with less direct guidance. This approach was institutionalized as INMAS (Intensifikasi Massal, or Mass Intensification), which permitted farmers to obtain inputs without credit and with less government involvement.

Rice production continued to increase, but a shortage of foreign exchange in 1968 prompted the goverment to invite foreign manufacturers of fertilizers and pesticides to participate directly in supplying inputs and management advice to rice farmers in certain areas. BIMAS GOTONG ROYONG (mutual help BIMAS) lasted only four seasons because of the limited capacity of each company to supply chemicals for specific problems and the tendency for the use of heavy-handed techniques, such as aerial spraying of large areas without farmers' consent.

In 1968, the semidwarf rice varieties were first introduced in limited quantity to Indonesia. Local breeding work quickly absorbed the genetic material and developed the variety Pelita, which was similar to the IRRI varieties but had a grain type and plant structure considered more desirable by most Indonesian farmers. Use of the new varieties became part of the BIMAS and INMAS programs.

By early 1970, widespread defaults in the credit program led to a termination of BIMAS GOTONG ROYONG and an abandonment of the fixed package approach. A new approach retained the BIMAS name, but provided individual farmers credit for a flexible package of inputs. The new BIMAS exerted less pressure for participation and emphasized timely delivery of inputs and somewhat more attractive rice prices. Perhaps the biggest change, however, was the revitalization of the padi center in the *unit desa* (village unit). Each center was composed of a representative of the government bank, two assistants from local villages, an extension worker, and a fertilizer retailer. These agents worked as a team in the delivery of credit, fertilizer, and extension advice over an area of 600 to 1,000 hectares. The new structure was impressive, not only in getting inputs delivered on time, but in getting loans repaid.

Rice production grew from 9 million mt in 1965–68 to 12 million in 1968–70 and increased by between 5 and 10 percent annually (except for 1972) between 1970 and 1981, reaching 21 million mt in 1981 (figure 17.3). Between 1968 and 1977, imports of rice rose from only half a million to nearly 2 million mt, despite the significant growth in production (figure 17.3). By the early 1980s, however, as production growth continued to accelerate, there was renewed confidence that the country would be able to keep up with basic food needs. But it required significant increases in wheat imports to meet total foodgrain demand.

There continues to be considerable difference of opinion about the appropriate strategy for Indonesia to follow to solve its food production problem. Two issues concern policymakers: the relative emphasis to be given to rice as compared to other crops and the relative emphasis on Java as compared to the outer islands. In the coming years, a balance between policy imperatives for each of these problems will have to be developed.

Thailand: The Position of the Exporter[16]

One of the three major river deltas of Southeast Asia, the Chao Phraya, forms the central plain of Thailand. It, along with the Irrawaddy in Burma and the Mekong in Vietnam, provided the major source of world rice exports from the latter part of the nineteenth century until World War II. Each area contributed more than a million metric tons of rice per year to the export trade in 1939. With continuing turmoil in Indochina (Vietnam), the rice export trade in the Mekong Delta collapsed after World War II. Because of internal problems in Burma, rice exports declined from more than a million mt as recently as 1966 to only 200 to 300 mt in the early 1970s. Alone among the Southeast Asian exporters, Thailand was able to maintain its position. Although there have been some predictions that growing domestic demand would absorb the export surplus, in fact Thai exports have gradually risen over the past two decades, along with exports of maize.

Until after World War II, there was little government intervention in Thai rice trade. After the war, government intercession was accomplished principally through an export duty, commonly referred to as the "rice premium," and rice export quotas.[17] Benefits included higher government revenues and lower domestic rice prices, but at the cost of depressed farm-harvest prices (table 16.2). Furthermore, until recently, efforts to stabilize rice prices through quantitative restrictions of exports have meant lowering prices when they were high, but there were no provisions for raising low rice prices. Low farm prices for paddy, coupled with an extraordinarily high price for fertilizer, have left Thai farmers in a very unfavorable position compared with farmers in neighboring importing countries. Until recently, the nitrogen price relative to rice was twice the level of other Asian countries (table 16.3). Burma and Pakistan have also tended to squeeze farmers on price, and in Burma, this policy contributed significantly to the 1960s decline in exports.

There have been numerous studies of Thai rice policy, many of them focusing on the rice premium. The central issue concerning the premium relates to the elasticity of demand for Thai rice on the export market and the elasticity of supply of domestic rice. In a recent study, Chung Ming Wong concludes that in the long run as opposed to the short run, both demand for Thai exports and domestic supply are likely to be elastic provided that the total export demand does not decline as a result of the self-sufficiency drive among importers.[18] Tsuji, almost alone among academics, is a strong proponent of the premium.[19] He argues, based on his research, that export demand for Thai rice is inelastic and that price increases will be translated into higher, not lower, revenue for Thailand. There is, unfortunately, no strong empirical evidence to support either of the above positions. Siamwalla takes a more pragmatic view in arguing that, until recently, the government has been slow to adjust the level of the premium to changing conditions. His position is that the short-run elasticity of demand for exports changes with fluctuations in export supplies and input demand, becoming more elastic as prices fall.[20] This view is supported by Falcon and Monke, who describe the export demand curve as being kinked around its long-run level, becoming very inelastic with higher prices and very elastic with lower prices.[21]

There has been little attempt on the part of the Thai government to promote rice production because of the uncertainty of the export market (chapter 13). Despite unfavorable farm incentives, a reasonably stable level of rice exports has been maintained largely owing to an expansion of dry-season production. Steps were taken in the middle 1970s to provide more incentives to farmers, not only through reduction of the premium to bolster farm prices, but also by lowering fertilizer prices. These measures can be seen more as a response to concern about issues of equity and growing peasant unrest, rather than as an attempt to encourage domestic production.

Malaysia: A Turn Toward Producers[22]

Malaysia's agricultural development policies under colonial rule emphasized export crops so that, in 1957, 40 percent of its rice requirements were imported. With independence, three major rice policy goals were formulated: (1) to reduce dependency on world markets, (2) to save foreign exchange, (3) and to increase the welfare of rice farmers. The emphasis on the welfare of farmers is an unusual feature.

Given a relatively developed economy and the somewhat smaller role played by rice in the agricultural economy than in most other Asian countries, the government had little difficulty in pursuing a producer-oriented policy by restraining rice imports and thereby maintaining rice prices above world market levels.

During the early 1950s, domestic production accounted for 55 to 60 percent of Malaysia's requirements. Government officials were unhappy with this situation, but it was not until independence in 1957 (when Malaya became Malaysia), that vigorous action was finally taken to achieve the goal of complete self-sufficiency in rice production. That goal was, however, linked to the objective of producer welfare, even as early as 1960 when the Drainage and Irrigation Department stated: "The department's irrigation planning had hitherto been concentrated on developing new land for rice cultivation and, although this objective is still important, it is only so against the background of the development of an economic farm unit. . . . More intensive use must be made of the land and the area of the family unit (now 3–5 acres or less) must be increased."[23]

To achieve self-sufficiency, Malaysia invested heavily in irrigation. A leading project was the Muda River Irrigation System undertaken with the assistance of the World Bank. Malaysia invested about $US 180 million in irrigation and drainage projects between 1956 and 1970, about 30 percent of its expenditures on agriculture and rural development during the period.[24] The main thrust of irrigation development was the construction of systems capable of supporting double cropping. Until 1960, the off-season crop was of minor importance, providing less than 5 percent of output. The investments had such a big impact that by 1971, the off-season area amounted to 45 percent of the main-season paddy area.

The rapid increase in double cropping made a major contribution to the doubling of paddy production between the early 1950s and the early 1970s, and to the reduction of imports from 302,000 to 153,000 mt (figure 17.4).

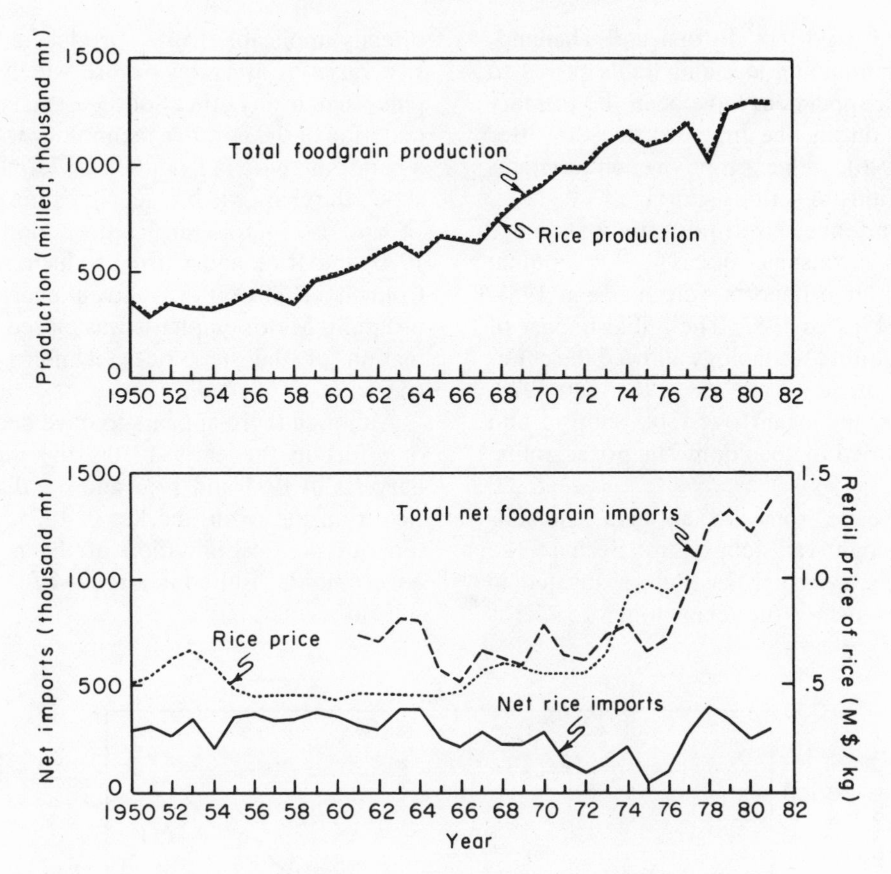

Figure 17.4. Production, imports, and prices of rice and total foodgrains, Malaysia, 1950–81

Goldman argues that World Bank assistance was the key to the improved technology and subsidies that were behind the success in raising production.[25] There is little doubt that the financial and technical assistance of the World Bank made a major contribution to the development of Malaysia's rice irrigation system, especially in the 100,000 ha Muda River project. The development, dissemination, and adoption of new rice varieties, some of which had their origin in the FAO (Food and Agriculture Organization) programs of the 1950s, also contributed to higher output. The use of fertilizer was a third significant element in the increased rice production. The potential for dry-season production probably contributed to the rapid adoption of the shorter season, nonphotoperiod-sensitive varieties as well as their response to fertilizer.

In 1970, a review of rice policies led to a decision to continue the objectives set forth at the beginning of this section but to give priority to increasing consumer incomes and welfare.[26] Price supports for rice were continued. Beginning in 1974, fertilizer was heavily subsidized. As in other Asian countries, the large-scale irrigation systems contained a substantial element of subsidy as well.

The Malaysian government, aware of the rising costs associated with further increases in production, and the already high cost of production relative to other countries (in part as a consequence of its pricing policy), adopted a target of 80-to-90 percent self-sufficiency in 1976. The goal has been pursued with uneven vigor, often with the welfare of producers the main concern. In 1980, Malaysia implemented a policy of providing enough fertilizer free to paddy farmers to plant up to 6 acres of land.[27] This policy was justified on welfare grounds rather than on its expected effect on production.

The Philippines: Problems of Self-Sufficiency[28]

Throughout most of the twentieth century, the Philippines imported about 5 to 10 percent of domestic rice needs. Rice policies have been consumer biased, but not to the same degree as in Indonesia or the traditional exporters, Burma and Thailand. Nevertheless, maintaining low and stable prices to urban consumers appears to have been the primary policy objective during the first two decades after independence (1946), with crash production programs organized in periods of serious shortage.

After independence, continuing imports were needed to maintain consumer prices at an acceptable level. Especially large imports were made in 1951, 1958, and from 1963 to 1967. The initial impact of the "green revolution" technology allowed the country to be "self-sufficient" in 1968, 1969, and 1970, but by 1971 demand again overtook supply, and imports were needed to keep domestic prices in line (figure 17.5).

During the decade from 1965 to 1975, the self-sufficiency campaign was stepped up. Because the International Rice Research Institute is located in the Philippines, much of the technology of IRRI is directly applicable to rice producers there. Modern rice varieties are grown more widely in the Philippines than in any other Southeast Asian country. The adoption of the new rice technology was coupled with a major increase in irrigation investment. Table 17.2 shows the change in budget allocation between 1961–65 and 1973–77. During the period from 1966 to 1970, the Rice and Corn Production Coordinating Council (RCPCC) was given charge of the rice program. Major emphasis was placed on the dissemination of the new rice technology and on the construction of feeder roads.

Although there appears to have been a slackening of effort in the early 1970s, the unfavorable rice harvests in 1971 and 1972 and the disappearance of rice from the world market in 1973, along with the tenuous political situation of the new martial law government, resulted in a renewed emphasis on rice self-sufficiency.

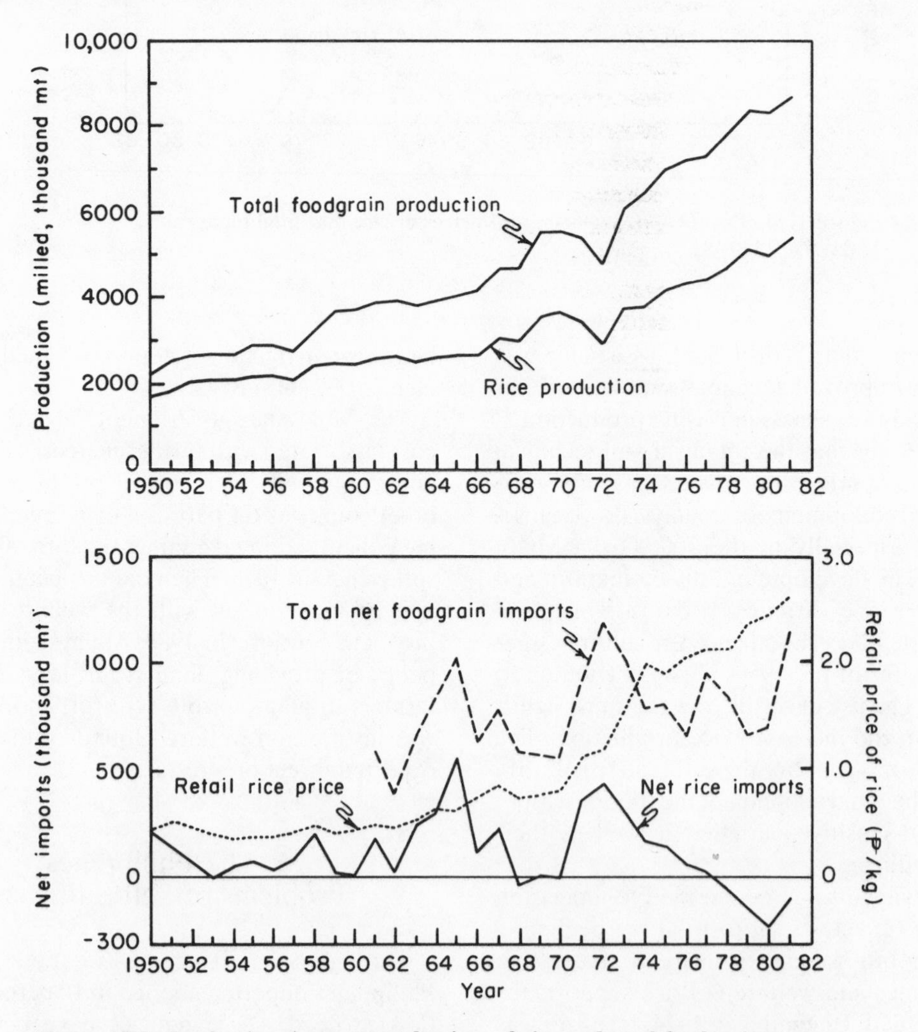

Figure 17.5. Production, imports, and prices of rice and total foodgrains, Philippines, 1950–81

Table 17.2. Average Annual Expenditures for Agricultural Development and Percentage Distribution of Expenditures, 1961–65, 1966–70, and 1973–77

Category	1961–65 (Pre-HYV)	1966–70 (RCPCC)	1973–77 (Masagana 99)
	million pesos[a]		
Price supports and subsidies	85.6	34.6	60.0
Irrigation	26.8	32.8	396.1
Feeder roads, community development[b]	0.0	70.8	115.0
Research and extension[c]	85.6	91.8	241.6
Agrarian reform	21.6	34.2	113.6
Environmental management, conservation	59.2	90.2	130.6
Total	278.8	354.4	1,056.8
	percent		
Price supports and subsidies	30.7	9.8	5.7
Irrigation	9.6	9.2	37.5
Feeder roads, community development	0.0	20.0	10.9
Research and extension	30.7	25.9	22.8
Agrarian reform	7.8	9.6	10.7
Environmental management, conservation	21.2	25.5	12.4
Total	100.0	100.0	100.0

Note: HYV = high-yielding varieties. RCPCC = Rice and Corn Production Coordinating Council.

Source: M.S.J. de Leon (1983) p. 27.

[a] Constant 1972 prices in million pesos.

[b] The major expenditure was for feeder roads.

[c] From 75 to 80 percent of the total was for extension in all three periods.

Beginning in 1973/74, the Philippine government supported a subsidized credit and fertilizer program, Masagana 99,[29] which combined low interest and noncollateral credit with recommended production practices. Fertilizer was heavily subsidized in 1973 and 1974, with a decreasing subsidy in 1976 and thereafter. At the same time, farmer welfare received more attention. Land reform was heralded as the "corner-stone of the New Society." The primary beneficiaries of land reform were the larger tenants (2 to 3 hectares) who held political power in the villages.

The Masagana 99 program notwithstanding, irrigation development was perhaps the single most important factor accounting for sustained growth in Philippine rice production throughout the 1970s. Between 1966–70 and 1973–77, investment in irrigation in constant pesos rose tenfold from 33 to 396 million pesos and increased from less than 10 percent to over 37 percent of the annual government expenditures for agricultural development.

The Philippine government established a floor price for paddy for many years, but this seldom had any effect because market prices were usually higher. When high production did push prices below the floor level, the government did not have the financial or managerial resources to guarantee the floor price to producers. On the other hand, government distribution of rice to consumers at the ceiling price has been maintained, except during the disaster year of 1973/74. The government weathered that crisis by distributing a mixture of rice and corn (obtained domestically). With the advent of Masagana 99, the price support system was more effective, and sharp downward movements were largely checked.

The data in tables 16.2 and 16.3 show that rice prices in the Philippines increased less than in most countries of the region in spite of high relative input prices. Despite this policy slant, production increased between 1975 and 1978. Rice production increased by an average of 7 percent per year, largely as a result of higher yields. Good weather, the expansion of irrigation, and the introduction of 100-day varieties have all been important contributing factors. The government has greatly expanded its own storage capacity to handle some of the surplus, but by the late 1970s, annual production in excess of needs had become a regular event and the question of whether the Philippines could economically export was raised.

The rather dramatic transformation of the Philippines from a chronic importer to a potential exporter has attracted the attention of many governments. However, despite the achievement of rice self-sufficiency, the total volume of cereal grain imports (rice, wheat, and maize) has increased (figure 17.5).

The official position of the Philippines on its success with rice is that Masagana 99, and even more specifically the subsidized credit and fertilizer portions of Masagana, is the primary factor associated with rice self-sufficiency.[30] A careful examination of the events of the 1970s raises some questions about the relative contributions of credit and input subsidies.

In 1974/75, over 40 percent of the Philippines' rice area was financed under the Masagana program, but by 1977/78, this proportion had fallen to 10 percent.[31] In all years from 1976 to 1981, domestic fertilizer prices were substantially *above* world prices, despite the "subsidy."[32] Thus, the Masagana program could have had an operational effect in the 1973 to 1976 period, but thereafter its operational relevance was minimal.

Commenting on the credit-based programs that have been extensively used in the Philippines, one perceptive observer said:

"interest rate subsidies have not significantly altered the unfavorable economic incentives in agriculture caused by government policies. . . . Relative prices as well as yields are major factors determining rates of return to most enterprises. Cheap credit will not make an unprofitable activity profitable! It is also clear that credit subsidies through low interest rates worsen income distribution because only a few, typically progressive farmers, receive the cheap credit."[33]

Philippine rice production continued to increase despite the reduced coverage of the Masagana program and the relatively high fertilizer prices through the late 1970s and early 1980s. This seems to indicate that production gains occurred in spite of the price and credit policies, apparently largely caused by government irrigation investments and technological change. Despite the rhetoric about Masagana, the contribution of irrigation is well understood by policymakers because of the proximity of Central Luzon to Manila: it is very easy for Manila government officials to personally witness the dry season transformation of the nearby countryside that irrigation has made possible. Continued growth in output will be needed into the future and will require a balancing of incentives. The extent to which past sources can provide growth in output depends on the commitment made to functioning investments. Complicating this is the need to divert resources to the rainfed areas where production potential is lower and a greater diversity of crops is grown, but poverty is also a serious problem.

Common Elements in National Rice Programs

In chapter 16, we observed the similarities and differences in national rice policies across countries. In this chapter, we have studied six national rice programs as they have developed over time. Despite the diversity in the programs followed in the six case studies, there are some common elements. It can be said, in general, that national programs have responded to opportunities and to crises. The opportunities presented by the new rice technology developed in the mid-1960s have resulted in major shifts in emphasis. These changes in budget allocations, shown in table 17.2 for the Philippines, undoubtedly took place in somewhat similar fashion during the same period in most other Asian countries, exceptions being countries, such as Thailand and Burma, where technology suited to their environmental conditions was not available until the late 1970s, and where their export status has caused them to set different priorities. For the majority of Asian countries, the new technology provided an opportunity to reduce dependence on rice imports. They seized this opportunity by greatly expanding investments in irrigation. Extension cum credit programs were developed, such as the High Yielding Variety (HYV) Program in India or BIMAS in Indonesia, to get the package of practices into the hands of farmers. Agrarian reform was attempted in some countries to provide greater security in tenure and property rights, although these programs have not been implemented in South and Southeast Asia with the same degree of success as in East Asia. Nearly all countries have attempted to control price and imports in order to stimulate production. Subsidies for fertilizer have been more popular than price supports for rice.

Major crises have occurred throughout the period since World War II. Weather, in particular drought, has had perhaps the most dramatic impact in terms of stimulating investment in rice programs. The extremely dry years in the mid-1960s in the Indian subcontinent spurred Indian efforts to look for a solution to the foodgrain problem. The 1972 drought, which affected most of Asia, coupled with the shortage of foodgrains and fertilizer in the world market, had a very profound effect on many countries. For example, the Philippine Masagana 99 program was launched in 1973, largely as a consequence of the disastrous 1972/73 crop year in which Philippine rice production fell by 17 percent from the level reached in 1970/71.

The rice programs have in general been successful in terms of stimulating growth in production and reducing dependency on world markets. However, it is very clear that these programs have been heavily subsidized. The subsidies have taken many forms including price of inputs, credit subsidies, nonrepayment of loans, and low or zero charges for irrigation water. Unfortunately, though, there has been no systematic evaluation of the effect of specific programs or policies on production and of the cost of additional rice programs in terms of overall economic growth. Malaysia declared 90 percent of self-sufficiency as a target because they sensed that the cost was getting too high. But rice continues to attract the bulk of the attention (and money) in most economies because the alternatives and opportunity costs are not clearly understood. The strategy for agricultural development beyond rice is not obvious in many Asian countries. This is an issue that countries that have achieved a sustained growth in rice production and have reduced or eliminated imports will have to address in the future.

Notes

1. India, Ministry of Food and Agriculture, "India's Food Crisis and Steps to Meet It" (New Delhi, Government of India, 1959).

2. Carl C. Taylor, Douglas Insminger, Helen W. Johnson, and Joyce Jean, *India's Roots of Democracy* (Calcutta, Orient Longmans, 1965).

3. India, Expert Committee on Assessment and Evaluation, Ministry of Food, Agriculture, Commodity Development and Cooperation, *Modernizing Indian Agriculture: Report on the Intensive Agricultural District Program 1960–68* (New Delhi, Government of India, 1969).

4. Susanta K. Ray, Ralph W. Cummings, Jr., and Robert W. Herdt, *Policy Planning for Agricultural Development* (New Delhi, Tata McGraw-Hill, 1979) p.116.

5. Leading examples of the literature illustrating the concerns discussed in this paragraph are: Uma Lele, *The Design of Rural Development* (Baltimore, Md., Johns Hopkins University Press, 1975); V. M. Dandekar and N. Rath, "Poverty in India," *Economic and Political Weekly* vol. 6, no. 1 (January 1971) pp. 25–48; Francine R. Frankel, *India's Green Revolution: Economic Gains and Political Costs* (Princeton, N.J., Princeton University Press, 1971); and Ashok Rudra, "Planning and the New Agricultural Strategy," *Economic and Political Weekly* vol. 6, no. 6 (February 1971) pp. 429–430.

6. A. M. Tang and B. Stone, *Food Production in the People's Republic of China*, Research Report No. 15 (Washington, D.C., International Food Policy Research Institute, 1980).

7. Thomas B. Wiens, "The Evolution of Policy Capabilities in China's Agricultural Technology," in *The Chinese Economy Post-Mao, Policy and Performance* vol. 1, U.S. Congress, Joint Economic Committee (1978) pp. 671–703.

8. Thomas G. Rawski, *Economic Growth and Employment in China* (New York, Oxford University Press, 1979).

9. Tang and Stone, *Food Production in the People's Republic of China*.

10. A joint study conducted by the Chinese Academy of Agricultural Sciences and the International Rice Research Institute suggests that while hybrids give significantly higher yields than conventional varieties, because of the extra costs involved, their economic profitability varies with conditions. See Kwei-ting He, Amanda Te, Shi-gang Zhu, S. Lee Travers, Hsui-fang Lai, and Robert W. Herdt, "The Economics of Hybrid Rice Production in China," IRRI Department Paper No. 83-22 (Los Banos, Philippines, International Rice Research Institute, 1983).

11. Nicholas Lardy, "Food Consumption in the People's Republic of China" (New Haven, Conn., Department of Economics, Yale University, 1980) mimeo.

12. This observation is based on a report of the Chinese Communist Party Central Committee, "Decisions on Some Problems in Accelerating Agricultural Development (Draft)." The Chinese text was published in Hong Kong *(Zhangwang)* no. 417 (June 26, 1979) pp. 21–24 and in Taiwan *(Zhonggong yanjiu)* vol. 13, no. 5, pp. 149–162. The quotation is from page 151. A garbled translation of the document appeared in Foreign Broadcast Information Service, *Daily Report China*, August 31, 1979, pp. 122–137. This report states that more than 100 million peasants have inadequate foodgrains. It has since appeared in the official Chinese press, Niu Ruofang, "Does 'Taking Grain as the Key Link' Suit Measures to Local Conditions?" *Guangming Daily*, December 8, 1979. See also Lin Shen, "The Inside Information on China's Economic Readjustment," *Zhengming*, May 1979, pp. 9–13. This communist-controlled Hong Kong periodical painted an even bleaker picture stating that "the annual foodgrain ration of two hundred million peasants in China is less than 300 jin (150 kilograms per year), that is to say, they are living in a state of semi-starvation."

13. This section draws heavily on C. P. Timmer, "The Political Economy of Rice in Asia: Indonesia," *Food Research Institute Studies* vol. 14, no. 3 (1975) pp. 197–231.

14. Richard Goldman, "The Formation of Seasonal Rice Prices in Indonesia" (Ph.D. dissertation, Stanford University, 1973).

15. Ace Partadiredja, "Mass Extension Among Farmers in Indonesia," paper presented at the 17th International Conference of Agricultural Economists, Banff, Canada, 1979.

16. One review of Thai rice policy is provided by Ammar Siamwalla, "A History of Rice Policies in Thailand," *Food Research Institute Studies* vol. 14, no. 3 (1975) pp. 233–249.

17. The "rice premium" is technically a fee to be paid by a private company to the government and not a tax that would require approval by the legislature. This has given the executive greater flexibility in changing premium levels and thus using the premium as a policy instrument. See Siamwalla, "Rice Policies in Thailand."

18. Chung Ming Wong, "A Model of the Rice Economy of Thailand" (Ph.D. dissertation, University of Chicago, 1976).

19. Hiroshi Tsuji, "An Economic and Institutional Analysis of the Rice Export Policy of Thailand with Special Reference to the Rice Premium Policy," *The Developing Economies* vol. 15 (1977) pp. 202–220.

20. Siamwalla, "A History of Rice Policies in Thailand."

21. Walter P. Falcon and Eric Monke, "International Trade in Rice," *Food Research Institute Studies* vol. 18 (1979–80) pp. 297–306.

22. This section draws heavily on Richard H. Goldman, "Staple Food Self-sufficiency and the Distributive Impact of Malaysian Rice Policy," *Food Research Institute Studies* vol. 14, no. 3 (1975) pp. 251–293.

23. Cited in Goldman, "Staple Food Self-sufficiency and the Distributive Impact of Malaysian Rice Policy," p. 263.

24. Donald C. Taylor, *The Economics of Malaysian Paddy Production and Irrigation* (Bangkok, The Agricultural Development Council, 1981) p. 33.

25. Goldman, "Staple Food Self-sufficiency and the Distributive Impact of Malaysian Rice Policy," p. 266.

26. Taylor, *The Economics of Malaysian Paddy Production and Irrigation*, p. 30.

27. Mohammad Ismail Ahmad and Zainal Abidin Mohamed, "Fertilizer Marketing in Malaysia," a paper presented at the IFPRI/IRRI/IFDC workshop on the Rice Policies in South East Asia Project, Jakarta, Indonesia, August 1982.

28. A review of Philippine rice policy is provided by Mahar Mangahas, "The Political Economy of Rice in the New Society," *Food Research Institute Studies* vol. 14, no. 3 (1975) pp. 295–309.

29. "Masagana" means "bountiful" in the local language. The number 99 signifies the nominal target of 100 cavans (23 mt) per hectare. In terms of scope and objectives, the Masagana 99 program was not unlike the BIMAS program in Indonesia discussed earlier.

30. Speech by Arturo R. Tanco, Jr., "Philippine Rice Production 1973–76," to the World Food Council, June 14, 1978.

31. Mark W. Rosegrant and R. W. Herdt, "Simulating the Impacts of Credit Policy and Fertilizer Subsidy on Central Luzon Rice Farms, the Philippines," *American Journal of Agricultural Economics* vol. 63, no. 4 (1981) pp. 655–665.

32. Amanda Te and Robert W. Herdt, "Fertilizer Prices, Subsidies and Rice Production," a paper presented to the 1982 Annual Convention of the Philippine Agricultural Economics and Development Association, June 4, 1982.

33. Cristina C. David, "Credit and Price Policies in Philippine Agriculture," Staff Paper Series 86-2 (Manila, Philippines, Philippine Institute for Development Studies, 1982) pp. 22–23.

18

Projecting the Asian Rice Situation

Will Asia be able to produce enough rice to feed itself in the coming decades? Although Asia is no longer seen as a major food crisis region of the world as it was during the 1960s (its food production has more than kept up with population, while in Africa per capita food production has declined for at least fifteen years), 75 percent of the developing world's population is in Asia, and any significant shortfall in food production would put a strain on the export capacity of other regions of the world. Because rice is such a dominant component of the Asian diet, rice supplies in Asia will determine future food availability in the region as a whole.

Many demand and supply projections for food or for particular commodities have been made.[1] The usual conclusion is that a deficit (or surplus) of a given amount will occur at some time in the future, even though most analysts recognize that such statements are not predictions of what will happen, but are projections of past trends. The term deficit is used to indicate the amount of food that would have to be supplied from some source other than domestic production of the countries included. This kind of projection does not suggest, however, that there will be a gap between the total quantity supplied and the total quantity consumed. Those two quantities, by definition, must always be equal. The interesting issue is how they are equated at any particular time and place. Most analyses recognize the interrelations between income growth and demand, between technological change, irrigation investments, fertilizer prices, government policies, and rice production. However, it has not been common practice to explicitly include such interrelationships in projection models. In our projections, however, we have attempted to see how future rice demand can reasonably be supplied—how irrigation, fertilizer, technological change, and imports can contribute to that goal.

The Projections Model

To include the interactions between the population and income factors that affect rice demand, and the investment, adoption, and technological factors that affect rice supply, we have constructed simplified models of the rice sectors of eight important rice-producing and consuming countries. These models permit us to integrate much of the information discussed in earlier chapters of the book into a consistent framework. The framework indicates, for example, how much production would increase with a given irrigation investment, what the resulting rice price would be if output does not keep up with demand, and, at the same time, how the level of fertilizer demand would respond to the changed price ratio of rice and fertilizer. We outline the model and how it works, explain the demand projections, discuss the supply projections, and, in the final section of the chapter, discuss our results.

The future rice situation in a country depends on the rate of growth in rice demand, the rate of growth in rice production, the quantity of imports or exports, and the acceptability of the resulting level of rice prices and trade. National agricultural policies are designed to achieve a desired combination of production and prices at acceptable government costs, but often policymakers try to achieve the incompatible goals of a low food price for consumers, a high product price to producers, and a low fiscal burden on government. The present quantitative projection model illustrates what tradeoffs must be made among these goals.

The model can be concisely described as demand and supply projections with a computation of the resulting equilibrium rice price. Demand is determined by population and income growth rates. Production is determined by the area of land planted to different types of rice, where rice types are defined as combinations of irrigation and varieties, and the extent of each is determined partly by government policies and partly by farmer behavior. Supply is equal to production plus imports, where imports are determined by a government policy decision. The equilibrium market price is the result of the fixed supply intersecting the downward-sloping demand curve. In addition to rice production and price, the model internally determines the demand for fertilizer based on the relative rice/fertilizer price.

The model is built around the ideas presented in chapter 16 about government rice policy instruments. The most important policy instruments that we include in the projections model are:

1. investments to determine the rate of growth of irrigated land
2. control over the price of fertilizer and the rate of growth of fertilizer availability (all available fertilizer need not be used)
3. the level of rice imports or exports that are used to control the consumer rice price
4. investment in research and extension to speed development and adoption of technology.

Variables that are not directly controlled by government but that are built into the model include the area of land available for rice, the rates of growth of population and per capita income, the milling ratio, and the marketing margin for rice. Target variables calculated by the model include the direct government financial cost of irrigation investment, fertilizer subsidy (or implicit tax), and rice imports. The model is used to determine what irrigation investment, fertilizer availability, and technology level is needed to achieve the desired level of rice supply and price while keeping the fiscal cost at a tolerable level. A five-year cycle is used. The system first simulates events for 1965, 1970, 1975, and 1980; then projections are made for 1985, 1995, and 2000. The models for individual countries mainly differ in the way projections are made for future rice area. These are discussed after the section on factors affecting supply and demand. Because of data limitations, the model is used for only eight of the Asian rice economies.

Future Demand for Rice

The demand for rice at a given time depends on (1) the current level of demand, (2) the rate of population growth, (3) the growth rate of per capita incomes, (4) the income elasticity of demand for rice, (5) the price of rice relative to prices of consumption substitutes for rice, and (6) the direct and cross-price elasticities of demand for rice.

Even in a simulation model, it is difficult to deal simultaneously with variation in all of these factors. Also, it is impossible to project the availability and prices of consumption substitutes without a full model of the food sector of each country. We assume that the availability and prices of consumption substitutes are constant, which eliminates the need to explicitly consider cross-price elasticities. The demand for rice is defined as

$$D = aTY^{n}P^{e}$$

where
D = the demand for rice
a = the constant term
T = the total population
Y = the per capita income
n = the income elasticity of demand for rice
P = the retail price of rice
e = the price elasticity of demand for rice.

The numerical value and parameter of each variable at any time is uncertain, as illustrated in the following discussion.

Population Growth

Several agencies continuously conduct population research.[2] The most widely used data come from the United Nations, which has formulated a range of projections. Since population growth is the single most important factor affecting future demand for rice, slight differences in growth rates will have a significant effect on demand. A simple projection of past population growth rates will likely overestimate future growth because rates can be affected by other variables such as economic development. For example, China and India had an average annual population growth of 2.1 percent between 1960 and 1970 and 1.9 percent between 1970 and 1979.[3] Bangladesh grew at 2.4 percent in the 1960s and at 3.0 percent in the 1970s; Indonesia grew at 2.0 percent in the 1960s and at 2.3 percent in the 1970s. Other countries experienced similar changes.

The United Nations assessed world population growth in 1980 separately for each country using information about changes occurring in fertility and mortality rates.[4] Because population growth is the net difference between the birth and death rates, and because death rates tend to decline with development while birth rates tend to rise and then decline, population projections that reflect fertility and mortality trends are better indicators of future changes than are past growth rates. The average projected rates of population are shown in the first column of table 18.1, although the rates actually used in the projections change every five years, following those in the UN projections.

Income Growth

The income concept that best reflects the factors influencing the demand for rice is personal disposable income, but such statistics are not available on a comparable basis for all countries. We used per capita personal consumption data from World Bank sources, even though, based as they are on national income concepts, such data are open to some questions for developing countries.[5] What is important for our purposes is not the absolute income level, but the trends.

Past rates of growth in national income are probably the best indicators of future growth rates, although, in many circumstances, divergences can be expected. Available data show that the two largest rice growing countries, India and China, had about constant GNP growth rates during the 1960s and the 1970s.[6] GNP growth rates accelerated during the 1970s in Malaysia, the Philippines, Indonesia, Burma, and South Korea while they slowed somewhat in Pakistan, Sri Lanka, Thailand, Japan, and North Korea. The growth for the 1970–79 period was used as the basis for projecting demand to the year 2000, so our projections reflect the rates of economic growth prevailing in the 1970s.

Income and Price Elasticities

There is an extensive literature on the impact of growth on the demand for food. The first internationally comprehensive set of income elasticities was generated by the Food and Agriculture Organization (FAO) for the "Indicative World Plan in the 1960s."[7] That project estimated income elasticities of the major food commodities separately for most countries, separating rice from wheat and coarse grains. At about the same time, the U.S. Department of Agriculture (USDA) projected world food demand using a set of price and income elasticities of demand for categories of countries and commodities.[8] The parameters included cross-elasticities between rice and wheat, which provided a basis for understanding the probability of substitution between the two commodities. More recently, the International Food Policy Research Institute (IFPRI) projected food demand for individual countries using income elasticities for cereals.[9] The income elasticities of each study are shown in table 18.1.

The U.S. Department of Agriculture report classifies Asia into three groups of countries—South,

Table 18.1. Annual Growth Rates of Population and Income, and Elasticities Demand with Respect to Income and Prices Used in the Projection Model

Country	Projected growth rate of		Income elasticities			Elasticities used in model	
	Population[a]	Income per capita	USDA	FAO	IFPRI[b]	Income elasticity	Price elasticity
China	1.2	2.0	[c]	—	—	0.45	−0.50
India	1.8	2.0	0.3	0.4	0.45	0.45	−0.50
Indonesia	1.5	5.0	0.0	0.7	0.39	0.50[b]	−0.60
Bangladesh	2.5	2.0	0.3	0.3	0.49	0.45	−0.50
Thailand	2.3	5.0	0.0	0.2	0.03	0.05	−0.30
Burma	2.4	2.0	0.0	0.4	0.49	0.30	−0.40
Philippines	2.7	3.5	0.0	0.2	0.25	0.25	−0.40
Sri Lanka	2.3	2.0	0.3	0.4	0.46	0.40	−0.60

[a] This is the value used for 1980–85. Because of the high rate of per capita income growth, the income elasticity is assumed to fall by .1 every subsequent five year period.

[b] The income elasticity for cereals, high-income growth variant.

[c] No estimates presented in the sources.

Southeast, and East. The income elasticity of demand for rice in South Asia is 0.3, for Southeast Asia it is zero, and for East Asia 0.2. The FAO provides separate estimates for each country, with income elasticities ranging from −0.1 for Japan, to 0.3 for Korea, 0.4 for Burma, and 0.7 for Indonesia. The IFPRI estimates also cover a wide range—from less than 0.1 for Thailand, Malaysia, Korea, and Nepal to about 0.45 for Indonesia, Burma, India, and Sri Lanka.

Our analysis of the demand for foodgrains showed a similar, although narrower set of income elasticities, ranging from −0.25 for Japan to 0.23 for Burma (chapter 11). These income elasticity estimates cannot be directly used in the projections model because they were estimated without including price effects. However, since we explicitly wish to reflect the price implications, we need to include the price elasticity of demand for rice in our projections. Because there is no good source of consistent price and income elasticities, we use a theoretical construct. The relationship between the price and income elasticities is reflected in the Slutsky equation, which a number of analysts have used to help generate reasonable price and income elasticities. Mellor argues "as a result of the deficiency of empirical data, reliance has generally been placed on the working assumption that the sum of the price elasticity, the income elasticity and the cross elasticities is equal to zero."[10] Using that working assumption and the available estimates, we have derived a set of income and price elasticities of demand for use in the model (table 18.1).

A conventional projections model would use the expected growth rates and income elasticities to compute the quantity demanded at some future date. Our model uses them to project the demand function, which together with estimated supply, results in equilibrium prices. Alternatively, one can use the model to achieve any desired rice price by assuming imports or exports of the needed quantity of rice.

Future Rice Production

Future production is usually projected by extending past trend rates of growth. This approach implies that past changes will continue in a linear fashion into the future, but we know that such an assumption is usually false. In the projections made here, the current level of adoption of modern varieties and fertilizer is used to determine the potential for additional growth from further adoption; the current proportion of irrigated rice land is used to determine the potential for future conversion of rainfed to irrigated land; and the availability of unused land is used to determine the potential for new land in rice.

Potential for Increasing Rice Area

Observations for most Asian countries lead many to conclude that the potential for increasing the area planted to crops is very limited. Nearly every analysis includes a statement such as "South Asia, unfortunately has the least amount of unused arable land remaining. . . . Land development costs, particularly in areas where investments must be made for extensive resettlement facilities and for irrigation and erosion control, greatly limit the feasible rate of expansion of cultivation into new areas." Data comparing cultivated land with potentially cultivable land in South Asia and China show the two are nearly equal.[11] Cultivable land per capita in Asia is far less than in other parts of the world. It would seem that the possibilities for any growth from land are practically nil.

However, the analysis of past sources of rice output growth shows that a significant part of the growth achieved during the 1960s and 1970s has in fact come from increasing area (table 4.15). Even in the period from 1972 to 1977, China achieved an annual growth rate of 1.3 percent from land, while India obtained 0.9 percent annually from increasing area planted to rice. Most of this area was gained through multiple cropping, especially in East and South Asia. But in Southeast Asia, much was obtained through increases in cultivated land. Irrigation contributes significantly to multiple cropping, especially in areas with a distinct dry season. Because of the lack of data, it is impossible to separate the cropping-intensity effect of irrigation from the increases in cultivated land, but it is clear that both effects may continue, although at a declining rate. Based on the examination of changes in past sources of output growth and the available data on multiple cropping and arable land, we have projected the growth in harvested rice area in each country. The county-specific definitions of land types and method of projecting their changes are discussed in a later section of this chapter.

Potential for Increasing Rice Yields

The possibility of increasing farmers' yields depends on the present level of yields and the technologically "potential" yield. Both of these depend on the quality of the rice production environment, especially the area irrigated and drained. The present yield level depends on the extent to which farmers are currently using fertilizer and fertilizer-responsive varieties. Fertilizer and modern variety adoption are, in turn, related to the extent and quality of irrigation, as well as other factors. The technologically potential yield can be pushed up with research, but there are physiological limits to the process that science has not yet overcome. The present technologically potential yield, represented by yields commonly achieved at experiment stations, is well above average national yields in the South and Southeast Asian countries, leaving an adequate exploitable gap. Only in East Asia are actual national yields approaching their potential level, and in China this may be a serious factor limiting future production growth.

Modern varieties (MVs) have a significantly higher yield potential than local varieties, and that advantage is most pronounced when they are grown on high quality (irrigated) land with fertilizer. Local varieties also benefit from irrigation and modest amounts of fertilizer. Much of the output growth achieved during the past two decades can be traced to the use of these inputs, and foreseeable continued growth in per hectare yields will depend, to a large extent, on the remaining potential to exploit these sources.

Irrigation Irrigated rice land is the result of deliberate government decisions to invest in irrigation construction and improvement, but creating irrigated land is expensive, and governments and their financiers, the development banks, seek to expand irrigated land no more rapidly than is needed to meet requirements. Irrigation is usually so advantageous for farmers that it is automatically used when made available.[12] Thus, the rate of growth of irrigated land depends on national policy decisions in most countries.

Fertilizer The use of fertilizer is much more dependent on individuals' decisions than is irrigation. Government can encourage fertilizer production, ensure that fertilizer is imported, and set fertilizer prices, but farmers decide how much fertilizer is actually applied. Fertilizer use has increased especially rapidly where irrigation is available and modern varieties are used.

The increase in fertilizer use reflects a process of gradual adoption; that is, farmers become aware, learn about, experiment with, and then habitually use it. When all farmers have gone through the process and arrived at an equilibrium application level, adoption is complete. Hence, one cannot simply project a linear rate of yield increases from fertilizer. There is a limit of 100 percent to adoption, and, as farmers approach optimal application levels, the yield response ratio falls. Better irrigation and more responsive varieties have the effect of maintaining high response ratios at higher levels of input, but at some level, there will be no more response to added fertilizer (that is, marginal returns diminish to zero). As a result, fertilizer use may increase very rapidly over an initial period until most farmers are applying the level that provides that initial high yield responses; after that, further output increases from fertilizer will be slow.

After farmers become familiar with fertilizer use through adoption, the price of fertilizer relative to the price of rice becomes a more important factor determining changes in use. That is, when low rates are being applied, the stage in the process of adoption determines use; one could compute the economically optimum rate, and the effect of price variation on that rate, but the computation would be of no value in predicting output changes because actual use is far below the optimal. After fertilizer has been used for some time, and assuming enough is available to meet market demand, farmers arrive at their own economic equilibrium levels. At that point, variations in prices will become important factors determining farmers' use of fertilizer. The projection model reflects these considerations.

The spread of modern varieties, like initial fertilizer use, is an adoption process. Farmers experiment with and evaluate the varieties under field conditions, and if suitable, the new varieties are adopted. The experience with available modern varieties has shown that they are more rapidly and widely adopted by farmers who have irrigation, and hence, one would expect them to spread as irrigation expands.

To reflect the diminishing marginal returns to fertilizer and the complementarity of modern varieties, fertilizer, and irrigation, fertilizer response curves like those illustrated in chapter 6 have been used in projecting future rice production. These functions define the yield response of modern varieties with and without irrigation. Differences in the proportion of area in the land types result in differences in the total fertilizer used and in the average rice yield in each country. Hence, increases in irrigated area or adoption of modern varieties will result in a greater total use of fertilizer, even with constant prices.

Country-Specific Data and Assumptions

While many of the interrelations are common to all countries, there are also many that are specific to individual countries. Definitions of land type and projections of area planted to each land type are prepared somewhat differently for each country, depending on the specific conditions.

China China is a challenge because its great agricultural diversity calls for a disaggregated model to adequately reflect future events; however, disaggregated time series data are not available. Until recently, complete time series have been limited to those pieced together by China scholars showing total rice area and production. There are also no data on irrigated rice area or fertilizer used on rice, only on the totals for the agricultural sector as a whole.

Drawing on the personal knowledge of Chinese rice scientists, data on the area of rice irrigated and grown in dryland and shallow rainfed conditions have been assembled.[13] Out of the 36 million ha, less than 2 percent is in upland, and about 5 percent is in nonirrigated shallow rainfed. In the projections, these are assumed to be a constant absolute area over time. In 1980, about 92 percent of China's rice area was irrigated; rice cropping intensity was 1.43 (computed by assuming one crop per year on area reported in "early rice crop" and japonica rice).

The Chinese developed and introduced their own semidwarf MVs during the 1960s. Those varieties spread to 6.7 million ha by 1973.[14] The International Rice Research Institute (IRRI) team that visited China in 1977 reported that dwarf varieties are now grown on more than 90 percent of the rice area.[15] The Chinese also have developed and rapidly disseminated true F_1 hybrid rice, claiming a yield advantage of 25 to 30 percent over conventional semidwarf rices. By 1981, about 6 million ha were planted to hybrids. In the future, hybrid areas will likely be limited by cropping patterns because currently available hybrids have a longer growth period than conventional semidwarfs.

Five rice land types are defined in the model for China: (1) irrigated hybrid, (2) irrigated modern varieties, (3) irrigated traditional varieties, (4) rainfed other varieties, and (5) dryland. For projection purposes, we assume that the area planted to rainfed and dryland rice will remain fixed and that the increases in rice area harvested have come about through increases in rice cropping intensity made possible through improvements in irrigation. Future area increases from the same source are dependent on irrigation policy, but the potential from this is very limited because nearly all of China's rice land is already well irrigated. In the basic projection, hybrid varieties spread at a medium rate, covering 70 percent of the rice area by the year 2000; in the fastest growth projection they spread 25 percent faster. Other MVs are projected to cover 90 percent of the remaining irrigated area.

Organic materials are the predominate fertilizer source, although the proportion of chemical fertilizer is increasing over time as domestic production capacity has improved. By 1977, China used 8.5 million mt of chemical fertilizer nutrients, and 23 million mt came from organic sources.[16]

The 1977 level of 31.5 million mt of nutrients provided an average of about 200 kg of nutrients per hectare on all crops. If we assume that fertilizer applications on rice were equal to its fraction of total output value (21 percent), then rice received over 187 kg/hectare. This may be more than the crop can biologically use. If we consider only chemical fertilizer, and again assume that rice received 21 percent of the total, 59 kg/ha of nutrients were applied in 1980.

Although organic fertilizer as a source of plant nutrients will likely remain stagnant or decline in numerical terms, investments in the chemical fertilizer industry will permit continued growth in the application of fertilizer nutrients.[17] The model assumes that the present organic nutrients provide a high "base" yield and that rice yields will respond to additional applications of chemical fertilizer.

India Like China, India's geographic diversity demands a disaggregated approach to projections. To keep the model manageable, five land types were identified: (1) irrigated modern varieties in the north, west, and south, (2) irrigated modern varieties in the other (eastern) states, (3) rainfed modern varieties in all states, (4) irrigated traditional varieties in all states, and (5) rainfed and dryland all varieties in all states.

Physical rice area is assumed to remain constant at 30.5 million ha in the eastern states and 10.5 million ha in the north, west, and south (Haryana, Punjab, Jammu and Kashmir, Andhra Pradesh, Karnataka, Kerala, and Tamil Nadu). With only about 40 percent of rice land irrigated in 1980, there is considerable scope for growth from additional irrigation.

Modern varieties covered approximately 85 percent of the irrigated paddy area in the north, and over 50 percent in the west and south in 1978. It is assumed that 15 percent of irrigated rice land will be planted to other varieties into the future. The model projects that modern varieties spread at 1, 2, or 3 percent of total rice area per year, depending on the assumptions made in a particular projection. This means that MVs continue to spread to rainfed areas after 1980. This trend started in the late 1970s in the eastern states, where MVs covered 35 percent of total rice area while irrigated rice covered only 25 percent. If irrigation is assumed to increase more rapidly, then the area in irrigated MVs also increases more rapidly. In the absence of irrigation growth, the MVs spread more slowly and are less productive.

Indonesia Historical data for rice area by water source are difficult to find for Indonesia, and the available data are open to several interpretations. In an extensive discussion of contemporary irrigation data, Nyberg and Prabowo are unable to make a conclusive statement on irrigated area planted to rice, even for the 1975–80 period.[18] Part of the difficulty is in distinguishing between irrigated rice and *sawah* (wetland rice), part in the distinction between irrigation service area and irrigated area of rice. Typically the irrigation service area or command area is larger than the area actually receiving the water. They do provide data on irrigated area classed as technically irrigated, semitechnically irrigated, and *sederhana* (simple) irrigation.

For projection purposes, we define four rice land types for lowland (sawah) areas: (1) technically irrigated modern varieties, (2) other irrigated modern varieties, (3) other irrigated traditional varieties, and (4) rainfed sawah. Dryland rice is a fifth category. Table 18.2 shows historical data on area in each type derived from available information.[19] The rice cropping intensity is 1.55 on technically irrigated sawah and 1.30 with other irrigation. Areas by type for 1980 are derived from the 1975 statistics by using information on irrigation under construction in 1977. Ninety-five percent of the area under construction is assumed to be complete by 1980. The rehabilitation projects are assumed to result in additional technically irrigated areas and the other irrigation projects are assumed to upgrade land presently in rainfed sawah. These imply a gross harvested rice area equal to the officially reported area of 9.3 million ha.

The adoption of modern varieties took place at a rapid rate in Indonesia. By 1970, over 10 percent of all rice area was planted to MVs, by 1975 over 45 percent, and by 1978 over 65 percent.[20] Our model assumes that the MVs are first planted on technically irrigated land, and then on all other irrigated land.

Projections of output for Indonesia are made using the following assumptions:

- Area technically irrigated has a rice cropping intensity of 1.55, and area with semitechnical or sederhana, irrigation has a rice cropping intensity of 1.30. These cropping intensities are assumed to remain constant through 2000.
- Physical area devoted to rice is constant at 6.8 million ha.
- Growth rate of irrigated area is a policy variable, and the proportion of new irrigated area with technical irrigation is a policy variable. Technical irrigation costs $US 4000/ha; other types of irrigation cost $US 2000/ha.
- Upland rice area declines from 17 to 14 percent of total rice area between 1975 and 2000.
- Rainfed area is computed as a residual.
- All area with technical irrigation is assumed to be planted to MVs after 1980.
- MVs are assumed to increase from 70 to 80 percent

Table 18.2. Estimated Area of Rice Land, Indonesia, 1965–80
(thousand ha)

	1965		1970		1975		1980	
Category	Physical	Cropped	Physical	Cropped	Physical[a]	Cropped	Physical	Cropped
Technically irrigated	1,214[b]	1,942	1,333[b]	2,133	1,434	2,359	1,933[c]	2,996
Other irrigated	2,145[b]	2,788	2,359[b]	3,066	2,278	3,267	3,046[c]	3,960
Rainfed sawah	1,465[d]	1,465	1,805[d]	1,805	1,712	1,710	1,210[d]	1,210
Dryland[e]	1,135	1,135	1,135	1,135	1,135	1,135	1,135	1,135
Total of above	6,800	7,330	6,800	8,140	6,800	8,471	6,800	9,301
Reported areas[f]	—	7,330	—	8,140	—	8,500	—	9,300
Modern varieties	—	0	—	902	—	3,757	—	6,045

[a] Dryland area from R. Huke (1982). Others from A. Nyberg and D. Prabowo (1982), table 6.

[b] Implied by backward interpolation from 1975 using the rate of increase in irrigated rice fields shown in A. Lains (1978), table 5.

[c] Derived from A. Nyberg and D. Prabowo (1982), table 25, assuming "rehabilitation" created net additions to technically irrigated area and that the total under construction is achieved by 1980, but assuming 1 percent/year depreciation out of irrigated status.

[d] Residual.

[e] From R. Huke (1982), assumed constant.

[f] Biro Pusat Statistic.

of total rice area between 1980 and 2000, and to be planted on the irrigated area.

Bangladesh Official statistical data from Bangladesh show six categories of rice—three seasons (aus, aman, and boro), local, and high-yielding varieties. Statistical information for irrigated area of local and high-yielding varieties is not reported. The relationship between the area with well-controlled water and the area of MVs is clear—the boro season (dry) has a much higher proportion of both than other seasons. Investments in many kinds of water-control devices (deep tubewells, shallow tubewells, and power pumps) have increased sharply in recent years. Projects to protect agricultural land from flooding have also been implemented in a number of areas, although these are costly. Total irrigated area increased 35 percent between 1974–75 and 1978–79. Despite these investments, the area of irrigated rice is rather small, reaching only about 12 percent of the total rice area in 1978–79.

The five rice land types used in the model are (1) irrigated MV in boro, (2) irrigated MV in aus and aman, (3) rainfed MV in all seasons, (4) irrigated traditional varieties all seasons, and (5) rainfed and dryland rice all seasons. The MVs are grown primarily under conditions of well-controlled water. In Bangladesh, where flooding is a significant problem, it is unlikely that MVs can spread any faster than irrigated area.

For the model, it is assumed that aus and aman crops respond to fertilizer and irrigation in similar ways, so they are grouped together. During the initial years of introduction, MVs were grown only on irrigated area—in 1970 Bangladesh had 1 million ha of irrigated rice area and less than 0.5 million ha of MVs. By 1975, Bangladesh had 1.5 million ha of MVs and only 1.1 million ha of irrigated area, and by 1980 there were 1.9 million ha planted to MVs and no more than 1.5 million ha of irrigated area. After 1970, MVs began to spread to nonirrigated areas, but the rate of adoption slowed significantly.

For the projections, we assume that all boro rice land is irrigated and planted to modern varieties. It is hypothesized that modern varieties are planted on 85 percent of the irrigated land in the aus and aman seasons. In addition, depending on the particular projection, modern varieties spread onto the nonirrigated land to some extent. As in all countries, increases in irrigation depend on the investments determined within each particular projection.

Thailand Data on rice in Thailand show the area harvested in two seasons, wet and dry. There are also data on irrigated rice area, but the official statistics do not show area planted to modern varieties. Farm-level research studies show that practically none of the wet-season area is planted to MVs, while almost all of the dry-season irrigated crop is in modern varieties.

Thailand is one of the few Asian countries with additional uncultivated land available for production. In Thailand, the use of MVs and fertilizer is not widespread, and a relatively small proportion of rice area is irrigated, so considerable scope exists for increasing output.

For the projections, four rice land types are defined: (1) irrigated, dry-season, modern varieties; (2) irrigated, wet-season, modern varieties; (3) irrigated, wet-season, traditional varieties; and (4) rainfed and dryland area. The area in each type is projected as follows:

- The physical area in rice is assumed to grow at 1 percent per year.
- The irrigated area is determined by the investments made under each projection. The percentage of irrigated area with two crops of rice per year is assumed to continue increasing from 14 percent in 1975 and 18 percent in 1980 to 34 percent by the year 2000.[21]
- MVs are assumed to cover the entire dry-season irrigated area and an area equal to 90 percent of the dry-season area during the wet season, based on their performance up to 1980.

Burma Burmese data are available from an annual official government report that show area planted and production for all rice and for selected "high-yield variety" rices.[22] Two types are included in the latter, modern semidwarf varieties and intermediate improved varieties. Total irrigated rice area is available, but irrigated area by variety is not, although only 16 percent of all rice was irrigated in 1980. The data indicate that fertilizer availability was low in the 1970s and the use of modern varieties started much later than in most other countries, but by 1980, the MVs and improved types covered about 50 percent of the total area.[23]

Five rice land types are defined for the Burma model: (1) irrigated, modern varieties; (2) irrigated, improved varieties; (3) rainfed modern and improved varieties; (4) rainfed other varieties; and (5) dryland. It is assumed that 90 percent of the irrigated land is planted to modern and improved varieties, that dryland area is constant at its 1975 level, and that modern and improved varieties continue to spread, either more or less rapidly, depending on the particular projection assumptions.

The Philippines Rice area data for the Philippines are well suited to the projection model. Beginning in the mid-1960s, area planted to modern and local varieties with and without irrigation was regularly reported. These are four of the rice land types used in the model. The fifth is dryland.

The area in each land type is projected as follows:

- According to official data, MVs were planted on 90 percent of the irrigated rice area in 1980. This percentage is assumed to increase to 95 percent by the year 2000.
- The total irrigated area is determined by investment for each projection, and irrigated local variety is a residual.
- According to official data, MVs were planted on 74 percent of the rainfed wetland area in 1980. This is assumed to increase to 84 percent by the year 2000.
- The proportion of rice in dryland is assumed to continue the trend of the past twenty years and decline to 8 percent by the year 2000.

Sri Lanka Statistics for Sri Lanka report the total area planted in the two major rice crop seasons, *maha* and *yala*, and the area planted to "new improved varieties" and "old improved varieties" in each of the seasons.[24] By 1979/80, the area in old improved varieties was sharply declining. The data on irrigated area do not distinguish between season or type of variety, so a number of assumptions were needed to synthesize the data on area used in the model.

Five rice land types are distinguished: (1) irrigated, modern varieties yala; (2) irrigated, modern varieties maha; (3) irrigated, other varieties both seasons; (4) rainfed all varieties both seasons; and (5) dryland. It is assumed that 80 percent of the yala crop is irrigated and that the balance of irrigated paddy area is planted in maha. Modern varieties (new improved varieties) covered 68 percent of the total area by 1980, and the area was assumed to gradually increase to cover all of the irrigated land in both seasons. Area in rainfed and dryland is assumed constant.

Results for Selected Countries

Results are presented for eight countries that produced 85 percent of Asia's rice on 88 percent of Asia's rice land. Table 18.3 shows the countries, their demand for rice in 1980, and three projections of demand. Although some important countries are omitted from the exercise, the range of conditions included is broad enough to reflect what is likely to happen in Asia as a whole.

If per capita income does not grow, population alone will generate an increased demand of about 37 percent between 1980 and 2000. This projection is dominated by the rather modest population growth rate expected in China (1.3 percent by 1985). If per capita incomes grow at about the rates they have over the past decade (medium), there will be a 58 percent increase in total rice demand by the year 2000. If incomes grow at a 50 percent faster rate, there will be about a 65 percent increase in demand, only slightly higher than the increase expected with the medium income growth rate. The bottom line shows projected demand if one assumes that demand growth in the other countries of Asia is similar to these eight. It indicates that the demand for rice in Asia in the year 2000 is likely to be about 570 million metric tons (unmilled).

Table 18.3. Projected Demand for Rice in the Year 2000
(thousand mt paddy)

Country	1980 demand		Projected demand with income growth rate[a]			Percent change with medium
	kg/capita	Total[b]	Zero	High	Medium	
China	97	138,007	183,367	212,432	206,478	50
India	83	79,552	110,430	129,451	125,147	57
Indonesia	146	33,105	43,430	68,495	59,488	114
Bangladesh	158	21,455	32,513	39,488	36,988	73
Thailand	192	14,119	20,632	21,487	22,341	39
Burma	173	9,166	13,353	15,228	14,578	39
Philippines	99	7,042	10,688	12,509	12,387	58
Sri Lanka	131	3,033	3,429	4,119	4,009	56
Total or average	135	305,389	417,842	503,099	481,526	58
Total Asia	—	360,765	493,609	594,326	570,008	58

[a] Income elasticities shown in table 18.1. Except in the case of Indonesia, the income elasticity was assumed to decline from 0.5 in 1980 to 0.1 in 2000.

[b] Production + imports − exports.

Table 18.4. Base Run Projections of Production, Consumption, and Prices, for Selected Asian Countries for the Year 2000

Country	Production[a] (mil. mt)	Fertilizer (kg/ha)	Percent area MVs	Percent area Irrigation	Annual irrigation investment (mil. $US)	With zero imports: Rice price index (1980 = 100)	With zero imports: Consumption (kg/capita)	With imports to hold price at 1980 level: Consumption (kg/capita)	With imports to hold price at 1980 level: Imports (mil. mt)[b]
China	196.1	148	65[c]	94	238	113	109	116	6.0
India	99.4	67	68	51	576	210	69	89	13.0
Indonesia	34.1	89	74	84	457	380	112	204	8.5
Bangladesh	28.7	32	63	24	86	171	144	188	5.4
Thailand	23.8	25	18	41	82	100[d]	201[d]	201	−0.3
Burma	14.7	71	56[e]	21	99	127	178	195	0.6
Philippines	9.6	61	89	52	37	225	82	114	1.7
Sri Lanka	3.1	102	73	66	48	207	99	141	0.5
Total or average	408.8	75	64	54	1,623	192	126	156	35.4

[a] Rough rice.
[b] Milled rice; negative sign indicates exports.
[c] Hybrid rice.
[d] Exporting nation, assumed to continue exports.
[e] Includes modern and improved varieties.

Basic Projections with Constant Technology

Table 18.4 shows the projected outcome if demand grows at the medium rate and output grows at an attainable rate given recent trends. In this projection, irrigated area grows at the historic rate, modern varieties continue to spread where possible, and fertilizer availability increases at 5 percent per year. All three components of modern technology reach a high level by the year 2000: fertilizer use averages 75 kg/ha, modern varieties cover an average of 65 percent of the rice area in all countries, and irrigation is extended to 54 percent of the rice area. A number of countries have a considerable area remaining unirrigated in the target year; in those countries MV adoption and fertilizer use remain low, but given what we believe about the capacity of these countries to increase their rate of irrigation construction, this projection of supply appears to be a "best guess" case. We calculate that this level of production will require an approximate $US 1.9 billion annual irrigation investment (1980 prices). Two sets of performance criteria are presented, one in which the deficit countries are assumed to impose self-sufficiency conditions, and one in which rice is imported to hold its real price at the 1980 level.

With zero imports, the real rice price almost doubles in most countries except China. Indonesia, which is projected to have a very buoyant demand, experiences the largest increase in real price. Consumption levels average $US 125 kg/capita, which is 95 percent of the 1980 levels, but this average hides increases in China, Thailand, and Burma, and no change in the Philippines. Other countries are projected to experience significant reductions in consumption if self-sufficiency is imposed.

There is a striking difference in per capita consumption levels between the zero-import, high-price case and the fixed-price, high-import case in most countries. On average, per capita consumption would be 30 kg per person higher in the case where rice prices are kept at their 1980 levels through imports.

If all countries decide to import sufficient rice to maintain a constant real price, about 33 million mt would be needed for the region in the year 2000, mostly for India, Indonesia, and China. The availability of such large quantities at unaltered prices is essentially nil.

An alternative to importing the 33 million metric tons of rice is to substitute imported wheat for some or all of the rice. As discussed in chapter 13, this strategy is already being practiced in many grain-importing countries in Asia. It is likely, however, that more than 1 ton of wheat will be needed to substitute for each ton of rice. This implies the need for something like 40 to 50 million mt of wheat. To meet this demand, wheat imports would have to grow at a 6 percent rate. While world wheat imports are large, demand of this magnitude would put significant upward pressure on wheat prices.

A second projection was made to reflect what might happen if the rate of growth of irrigated land could be doubled in all countries except China. The results are summarized in table 18.5. Irrigated area reaches

Table 18.5. Fast Output Growth Projections of Rice Production, Consumption, and Prices for Selected Asian Countries for the Year 2000

Country	Fertilizer (kg/ha)	Percent area: MVs	Percent area: Irrigation	Annual irrigation investment (mil. US$)	Production (mil. mt)	With zero imports: Rice price index (1980 = 100)	With zero imports: Consumption (kg/capita)	With constant price: Imports (mil. mt)	With constant price: Consumption (kg/capita)
China	138	82	94	495	207	97	118	0.0	115
India	80	82	61	1,293	115	156	79	4.3	88
Indonesia	108	75	86	934	46	211	152	7.0	204
Bangladesh	69	66	47	160	34	158	150	3.2	188
Thailand	31	22	49	186	25	100[a]	202	−0.9	202
Burma	81	83	33	177	17	100[a]	196	−0.1	196
Philippines	78	89	57	87	12	145	98	0.8	114
Sri Lanka	107	80	71	94	6	100	148	−0.3	148
Total or average	87	72	62	3,426	466	130	144	13.6	157

[a] The country is projected to have exports in 2000; price will depend on the amount exported.

62 percent of rice area on average, with substantially higher levels than in the base run in several countries, notably Bangladesh and Thailand. Modern varieties reach 72 percent of the total rice area, compared to 64 percent in the base projections. Average fertilizer rates are 12 kg/ha higher than in the base run because of the capacity of the MV irrigated area to productively absorb more fertilizer. Total output reaches 466 million mt of rough rice compared to 409 in the base run.

If the faster rate of output growth is achieved and importing countries were to impose self-sufficiency, real rice prices would increase by 30 percent on average, compared to the 84 percent increase in the base run. If countries chose to import enough to hold real prices constant, net imports for the region would total 13.6 million mt, compared with 35.4 million mt in the base run. If imported wheat substituted for part of the rice imports, the scenario might be feasible.

Although it is impossible to determine whether either of these projections is accurate, the base run seems possible for the eight countries because it requires investments similar to those currently being made. A possible problem is the requirement for very large imports. The fast rate of output growth demands substantially higher irrigation investments, but because of the extra output produced, lower import expenditures will result in subsequent years. This effect is illustrated in table 18.6. The estimated annual cost in 1985 for the irrigation investments, fertilizer, and imports (assuming they would be available) needed to hold real rice prices constant is $US 5.1 billion under the base run and $US 4.5 billion under the fast output case. In subsequent years, the cost of the base run case continues to exceed that of the fast output case, suggesting that the latter is superior. However, a real question exists as to whether it is possible to assemble the human resources needed to double the rate of growth of irrigated land. Also the gestation period for irrigation projects is such that they cannot be directly substituted for imports in a year of short production, and it is politically easier

Table 18.6. Costs Associated with Two Alternative Projections of the Future Rice Situation in Eight Asian Countries (million $US/yr)[a]

Year	Base run case: Irrigation	Base run case: Fertilizer	Base run case: Net imports	Base run case: Total	Fast output case: Irrigation	Fast output case: Fertilizer	Fast output case: Net Imports	Fast output case: Total
1985	1,741	1,410	1,903	5,054	3,224	1,500	−270	4,454
1990	1,815	1,720	6,195	9,730	3,458	1,906	1,755	7,119
1995	1,917	2,030	10,955	14,902	5,954	2,272	4,515	12,741
2000	2,051	2,407	15,972	20,430	7,199	2,818	6,000	16,017

[a] Irrigation costs are the annual investment costs; fertilizer costs are the value of fertilizer used in rice production at a price of $US 225/mt of urea; cost of imports are for the amounts shown in tables 18.4 and 18.5 at a price of $US 300/mt of milled rice.

to get funds for food imports than for irrigation investments.

Alternative Projections with Improved Technology

It is evident that even with the fast rate of output growth reflected in table 18.5, a substantial increase in imports will likely be required in the year 2000 if the level of technology remains as it was in the 1970s. The largest possible investments in irrigation, complete fertilizer availability, and the full spread of modern varieties will not be sufficient to meet rice demand except in a few countries. And if irrigation investments fall below these levels, then the contribution of fertilizer and modern varieties will be smaller.

The only foreseeable additional source of production is from productivity gains—if these can be obtained with the irrigation and fertilizer levels expected to be in use during the decades of the 1980s and 1990s. The analyses of the experience of the Philippines, South Korea, Burma, and Indonesia during the 1970s show that substantial growth was obtained from sources other than land and the measured inputs (table 4.16). The additional growth in output likely came from more productive use of inputs, which may have resulted from the intensive extension programs each of these countries mounted during the 1970s. While those programs were built around MVs and fertilizer, it is evident, both from the analysis in the earlier section and from that in chapter 4 that other forces were also at work.

Increases in productivity can be reflected in the projections model as follows. A gain in the productivity of irrigation can be shown by an increase in the constant term in the fertilizer response function; the effect of an increase in fertilizer productivity can be reflected in an increase in the coefficient of fertilizer in the response function; and an increase in both can be seen in their combined effects. To demonstrate these effects, a set of altered response curves for irrigated MVs in the Philippines is plotted in figure 18.1. During each five-year period, the constant term was increased by 250 kg/ha, and the coefficient of fertilizer was increased by 1.0. The maximum yield increases from 2.9 mt/ha at 105 kg fertilizer nutrients per ha in the first year to 4.4 mt/ha at 128 kg fertilizer per ha in the year 2000. Of course, a major question is how such a change in productivity can be obtained.

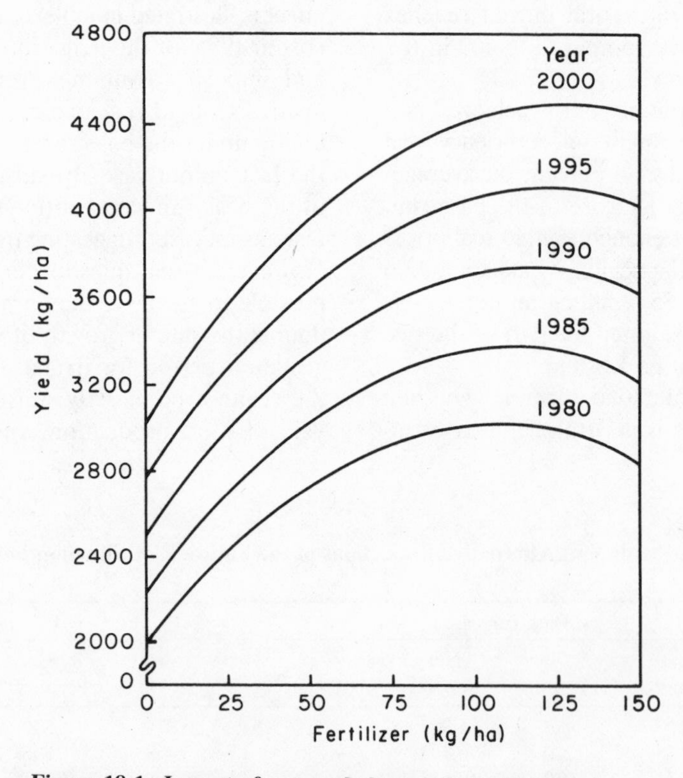

Figure 18.1. Impact of assumed changes in fertilizer and irrigation productivity on the yield response of modern variety irrigated rice in the Philippines

Table 18.7. Projected Estimates of the Impact of Increased Fertilizer and Irrigation Productivity on Production (unmilled) and Imports (milled) of Rice Needed to Maintain a Constant Real Price through the Year 2000
(million mt)

		Impact of productivity increases for					
		Fertilizer		Irrigation		Both	
Country	Base run production	Production	Imports	Production	Imports	Production	Imports
China	196.1	218.2	−5.4	231.8	−10.7	253.9	−21.4
India	99.4	114.1	6.2	120.6	3.6	135.3	3.1
Indonesia	34.1	39.2	10.0	42.9	8.5	48.0	6.2
Bangladesh	28.7	30.1	4.8	35.8	2.4	37.2	1.8
Thailand	23.8	25.8	−1.3	28.4	−2.4	30.4	−3.4
Burma	14.7	16.6	−0.2	17.4	−0.4	19.2	−1.3
Philippines	9.6	11.0	0.7	11.7	0.4	13.1	−0.3
Sri Lanka	3.1	4.9	0.0	5.5	−0.3	6.2	−0.5
Total or average	414.0	459.9	14.8	499.1	1.1	543.3	−15.8

Improvements in irrigation system management, higher levels of incentives generated by land reform, better education, extension, and new knowledge from research may all contribute to the increase in fertilizer productivity illustrated. Education and extension will provide farmers information to enable them to raise the level of potential productivity. The institutional changes of land reform may have many diverse effects, mainly upon the distribution of income, but may also contribute to higher resource productivity.

With our present knowledge, it is impossible to quantify the relationship between inputs into productivity-increasing efforts and the resulting gains, but obviously investments will be required. The few studies measuring the productivity of research and extension expenditures in the developing countries show that they have had very high payoffs,[25] but these studies may not be an adequate basis for future projections because they ignore the inputs and possible effects of land reform, general education, and other institutional changes. Thus, we do not attempt to estimate costs for productivity gains reflected in figure 18.1, but simply present the results of projections made assuming that such productivity gains can be achieved on the irrigated land in each country.

The results in table 18.7 provide dramatic evidence of the potential output gains from modest productivity improvements, if these are widely spread. Gains in either fertilizer productivity or irrigation productivity both result in substantial increases in production and reductions in imports. If productivity gains from both sources can be obtained, then the region as a whole will be more than self-sufficient, with no increase in the real price of rice.

Prospects for Rice in Asia

What then, are the prospects that Asia can be self-sufficient in rice production between 1980 and the year 2000? The answer depends very much on what resources countries are able to invest in increasing production. The technical capacity for meeting the future demand does exist. It will require substantial commitment to meet the investment requirements for irrigation; if actual performance falls far short of the rates assumed, it will be difficult for most countries to meet the needs. Fertilizer will make its biggest contribution in countries that used modest rates in 1980. In the other countries, readily achievable gains from fertilizer were already made in the 1970s. The same is true in most cases for MVs. Unless there are dramatic developments in creating drought-resistant rices or deep-water, high-yielding varieties, it is unlikely that MVs will spread rapidly in the nonirrigated rice land of eastern India, Bangladesh, and Thailand. Productivity gains in the irrigated areas and the more favorable rainfed areas will be a necessary requirement for meeting future needs. If neither the increased investment nor the needed productivity gains materialize and countries attempt to maintain constant rice prices through substantial increases in imports of rice, it is difficult to see where these might come from. Some additional wheat imports are possible, but meeting the entire additional demand is likely to place upward pressure on world wheat prices. Thus, the most likely scenario is a combination of increased output, increased price, increased productivity, and increased imports, but the precise combination is unknown.

Notes

1. For example, T. N. Barr, "The World Food Situation and Global Grain Prospects," *Science* vol. 4, no. 214 (1981) p. 4525; International Food Policy Research Institute, *Food Needs for Developing Countries*, IFPRI Research Paper No. 3 (Washington, D.C., IFPRI, 1977); and S. Handler, *World Bank Staff Working Paper No. 247* (Washington, D.C., World Bank, 1976).

2. The Economic and Social Commission for Asia and the Pacific prepared a series of useful country population monographs with the assistance of the United Nations Fund for Population Activities. These include most of the relevant population data for each of the countries of Asia. The Population and Human Resources Division of the World Bank conducts population research on most countries of the world.

3. These growth rates are reported in World Bank, *World Development Report 1981* (Washington, D.C., World Bank, 1981). They are based on UN data from the agencies mentioned in note 2.

4. United Nations Department of International Economics and Social Affairs, "World Population Prospects as Assessed in 1980," Population Studies No. 78 (New York, UN, 1981).

5. World Bank, *World Tables*, second edition (Baltimore and London, Johns Hopkins University Press, 1980).

6. World Bank, *World Development Report 1981*.

7. Food and Agriculture Organization of the United Nations, *Agricultural Commodities—Projections for 1975 and 1985* vol. 2 (Rome, FAO, 1967).

8. U. S. Department of Agriculture, *World Demand Prospects for Grain in 1980*, Foreign Agricultural Economic Report, No. 75 (Washington, D.C., USDA, 1971).

9. IFPRI, "Food Needs for Developing Countries."

10. J. W. Mellor, *The Economics of Agricultural Development* (Ithaca, New York, Cornell University Press, 1966) p. 71. Formally stated: $e_{ii} = e^*_{ii} - k_i n_i$, where e^*_{ii} is the income-compensated price effect, n_i is the income elasticity of good i, k_i is the proportion of the total budget spent on good i, and e_{ii} is the own price elasticity of good i. This relationship was used in the FAO analysis cited in note 7 and in many subsequent demand studies. For a disaggregated application see H. Bouis, "Rice Policy in the Philippines" (Ph.D. dissertation, Food Research Institute, Stanford University, 1982).

11. U. Colombo, D. G. Johnson, T. Shishido, "Reducing Malnutrition in Developing Countries: Increasing Rice Production in South and Southeast Asia," Trilateral Papers No. 16 (The Trilateral Commission, 1978) p. 4.

12. One possible exception occurs when the cost of installing irrigation exceeds its value, and farmers are forced to pay the full cost. However, such uneconomic projects are usually rejected during the project selection stage.

13. Robert Huke, *Rice Area by Type of Culture: South, Southeast Asia and East Asia* (Los Banos, Philippines, International Rice Research Institute, 1982).

14. Dana Dalrymple, *Development and Spread of High-Yielding Varieties of Wheat and Rice in the Less Developed Nations*, USDA Foreign Agricultural Economic Report No. 95 (Washington, D.C., U.S. Department of Agriculture, 1978).

15. International Rice Research Institute, *Rice Research and Production in China: An IRRI Team's View* (Los Banos, Philippines, IRRI, 1978).

16. A. M. Tang and B. Stone, *Food Production in the People's Republic of China,* IFPRI Research Report No. 15 (Washington, D.C., International Food Policy Research Institute, 1980) p. 47.

17. R. Barker, D. G. Sisler, and B. Rose, "Prospects for Growth in Grain Production," in R. Barker, R. Sinha, and B. Rose, eds., *The Chinese Agricultural Economy* (Boulder, Colo., Westview Press, 1982).

18. A. J. Nyberg and D. Prabowo, *Status and Performance of Irrigation in Indonesia and the Prospects for 1990 and 2000*, Rice Policies in Southeast Asia Project Working Paper No. 4 (Washington, D.C., International Food Policy Research Institute, 1982).

19. The 1975 data for technical, other irrigated, and rainfed sawah are from Nyberg and Prabowo, *Status and Performance of Irrigation in Indonesia*; the dryland data are from Huke, *Rice Area by Type of Culture*.

20. R. Bernsten, B. H. Siwi and H. M. Beachell, "*The Development and Diffusion of Rice Varieties in Indonesia*," IRRI Research Paper Series 71 (Los Banos, Philippines, International Rice Research Institute, 1982).

21. Dow Mongkolsmai, "*Status and Performance of Irrigation in Thailand*," Rice Policies in Southeast Asia Project, Working Paper No. 8 (Washington, D.C., IFPRI, IFDC, IRRI 1983).

22. Burma, Ministry of Planning and Finance, *Report to the Pyithu Hluttaw for 1982/83* (Rangoon, 1982).

23. For a more complete discussion of the data and their sources, see R. W. Herdt and C. Capule, *Adoption, Spread and Production Impact of Modern Rice Varieties in Asia* (Los Banos, Philippines, International Rice Research Institute, 1983).

24. Sri Lanka, Ministry of Agricultural Development and Research, *Agricultural Statistics of Sri Lanka: 1951/52–1980/81* (Colombo, Sri Lanka, 1981).

25. For a list of literature on the productivity of investment in agricultural research and extension for rice, see table 14.2.

Bibliography

Abbas, S. A. *Long-Term Projections of Supply and Demand for Selected Agricultural Products in Pakistan* (Lahore, Oxford University Press, Pakistan Branch, 1967).

Abdullah, Tahrumnesa A., and Sondra Ziendenstein. *Village Women in Bangladesh: Prospects for Change* (Oxford, Pergamon Press, 1982).

Abercrombie, K. C. "Agricultural Mechanization and Employment in Developing Countries," in Food and Agriculture Organization, *Effects of Farm Mechanization on Production and Employment* (Rome, FAO, 1975).

Adams, F. G., and J. R. Behrman. *Econometric Models of World Agricultural Commodity Markets* (Cambridge, Mass., Ballinger, 1976).

Adas, Michael. *The Burma Delta-Economic Development and Social Change on an Asian Rice Frontier, 1852–1941* (Madison, Wisc., University of Wisconsin Press, 1974).

Adiratma, I. R. "Income and Expenditure Pattern of Rice Producers in Relation to Production and Rice Marketed: A Case Study of Krawang, West Java" (Ph.D. dissertation, Institut Pertanian, Bogor, Indonesia, 1969).

AERC. *See* Agro-Economics Research Center.

Agarwal, B. "Rural Women and the High Yielding Rice Technology in India." Paper presented at Women in Rice Farming Systems Conference, Los Banos, Philippines, September 26–30, 1983.

Agrar-Und Hydrotechnik. *National Fertilizer Study—Indonesia* (Germany, Imhawen International Co., October 1972).

Agrawal, Amar N. *Indian Agriculture and its Problems* (Delhi, Ramjas College, Economics Society, 1951).

Agricultural Corporation, Agricultural Research Institute. *Fertilizer Response of Some Important Crops of Burma*, Radioisotope Laboratory Paper No. 17 (Rangoon, January 1975).

Agricultural Requisites Scheme for Asia and the Pacific. "Fertilizer Marketing, Distribution, and Use in Indonesia" (Bangkok, January 1978).

Agro-Economics Research Center (AERC), Delhi. "Evaluation of the High-Yielding Varieties Programme, Kharif, 1968" (Delhi, Ministry of Agriculture, 1970).

Agro-Economics Research Center, Jorhat. "Report on the High-Yielding Varieties Programme on Paddy in Sibsagar District, 1968/69" (Jorhat, India, Ministry of Agriculture, 1970).

Agro-Economics Research Center, Visva Bharti. "A Study of the High-Yielding Varieties Programme in the District of Cuttack, Orissa with Special Reference to Credit" (Visva Bharti, India, Ministry of Agriculture, 1967).

Ahammed, Chowdhury, S., and Robert W. Herdt. "Impacts of Farm Mechanization in a Semi-closed Input-Output Model for the Philippines," *American Journal of Agricultural Economics*, in press.

Ahmad, Mohammad Ismail, and Zainal Abidin Mohamed. "Fertilizer Marketing in Malaysia." Paper presented at the International Food Policy Research Institute, International Fertilizer Development Center, International Rice Research Institute workshop on Rice Policies in South East Asia Project, Jakarta, Indonesia, August 1982.

Ahmad, Nafis. *A New Economic Geography of Bangladesh* (New Delhi, Vikas Publishing House, 1976).

Ahmed, Kalimuddin. *Agriculture in East Pakistan* (Dacca, Ahmed Brothers Publications, 1965).

Ahmed, Raisuddin. *Foodgrain Supply, Distribution and*

Consumption Policies within a Dual Pricing Mechanism: A Case Study of Bangladesh, Research Report No. 18 (Washington, D.C., International Food Policy Research Institute, 1979).

Alam, Mahbudul, A. H. M. "The Impact of Power Tillers on Productivity, Employment, and Income Distribution: A Case Study of Bangladesh." Paper presented at ADC/IRRI workshop on the Consequences of Small Rice Farm Mechanization, Los Banos, Philippines, September 14–18, 1981.

Alamgir, M., and L.J.J.B. Berlage. "Foodgrain (Rice and Wheat) Demand, Import and Rice Policy for Bangladesh," *Bangladesh Economic Review* (January 1973) pp. 25–58.

Alderman, H., and C. Peter Timmer. "Food Policy and Food Demand in Indonesia," *Bulletin of Indonesian Economic Studies* vol. 16, no. 3 (November 1980) pp. 83–94.

Alexander, C., and C. Saleh. "The Distribution of Production Factor Inputs by Representative Farm Crops in Jati, West Java," *Agro-Economic Survey Research Notes* (Bogor, Indonesia, May 1974).

Alexander, Jennifer, and Paul Alexander. "Sugar, Rice and Irrigation in Colonial Java," *Ethnohistory* vol. 25 (1978) pp. 207–223.

Alim, A. *An Introduction to Bangladesh Agriculture* (Dacca, M. Alim, 1974).

Allen, G. C. *A Short Economic History of Modern Japan* (New York, St. Martin's Press, 1981).

Amatatsu, Katsumi. *Growing Rice in Japan* (Tokyo, Agriculture, Forestry, and Fisheries Productivity Conference, December 1959).

Amerisinghe, Nihal. "Adoption of Modern Rice Technology Under Peasant Farming Conditions in Sri Lanka," *Farm Management Notes of Asia and the Far East* no. 3 (July 1976).

———. "Economic and Social Implications of the Introduction of HYV Rice on Settlement Schemes in Ceylon" (United Nations Development Programme Global 2 Research Project, 1972).

Anden, Teresa, and Adelita Palacpac. *Data Series on Rice Statistics in the Philippines* (Los Banos, Philippines, Philippine Council for Agriculture and Resources Research, 1981).

Anderson, Kym, "South Korean Agricultural Price and Trade Policies: Their Efforts Since 1955." Paper presented at a Work-in-Progress Seminar, Korea Rural Economics Institute, Seoul, March 12, 1981.

Anderson, Robert S., Paul R. Brass, Edwin Levy, and Barrie M. Morrison, eds. *Science, Politics, and the Agricultural Revolution in Asia* (Boulder, Colo., Westview Press, 1982).

Antiporta, Donato B., and Narciso R. Deomampo. "Employment Impact of Mechanization of Rice Production Processes," discussion paper (Los Banos, Philippines, College of Development Economics and Management, n.d.) mimeo.

Anutchariya, Sarun Watt. "Demand and Supply Analysis of Rice Production in Thailand" (Ph.D. dissertation, Texas A & M University, 1978).

Apiraksirikul, Sumalee. "Rice Trade Policy as It Relates to National Objectives in the Philippines" (Master's thesis, School of Economics, University of the Philippines, n.d.).

APO. *See* Asian Productivity Organization.

Aqua, Ronald. "Local Institutions and Rural Development in South Korea," Rural Development Committee, Special Series on Rural Local Government (Ithaca, N.Y., Cornell University, November 1974).

Aquino, Rodolfo, C., and Peter R. Jennings. "Inheritance and Significance of Dwarfism in the Indica Rice Variety," *Crop Science* vol. 6 (1966) pp. 551–554.

Aquino, R.C., S. S. Virmani, and G. S. Khush. "Heterosis in Rice" (Los Banos, Philippines, International Rice Research Institute, 1981) mimeo.

Araullo, E. V., D. B. de Padua, and M. Graham, eds. *Rice Post-harvest Technology* (Ottawa, International Development Research Center, 1976).

Arndt, Thomas, M., Dana G. Dalrymple, and Vernon W. Ruttan, eds. *Resource Allocation and Productivity in National and International Agricultural Research* (Minneapolis, Minn., University of Minnesota, 1977).

Arromdee, Virach. "Economics of Rice Trade Among Countries of Southeast Asia" (Ph.D. dissertation, University of Minnesota, 1968).

Asian Productivity Organization (APO). *Farm Mechanization in Asia* (Tokyo, APO, 1983).

———. "Regional Report on Farm Mechanization Survey-1" (Tokyo, APO Symposium on Farm Mechanization, November 24–30, 1981).

———. *Fertilizer Distribution in Selected Asian Countries* (Tokyo, APO, 1979).

———. "Economics of Food Grain Distribution: The Asian Scene," Report of the Symposium on Economics of Food Grain Distribution (Tokyo, APO, 1976).

Athwal, Dilbagh, S. "Semi-dwarf Rice and Wheat in Global Food Needs," *The Quarterly Review of Biology* vol. 46 (1971) pp. 1–34.

Attanayake, D. C. "Changing Patterns of Rice Production in Ceylon, 1945–1964" (M.A. thesis, University of Wisconsin, 1969).

Aviguetero, E. F., F. V. San Antonio, I. S. Valiente, H. A. del Castillo, and E. L. San Jose. *Income and Food Consumption (Summary of 19 Economic Surveys)*, no. 78–15, Special Studies Division (Quezon City, Philippines, Department of Agriculture, 1978).

Aviguetero, E.F., F. V. San Antonio, I. G. Serrano, H. A. del Castillo, and C. K. Cabilangan. *Regional Consumption Patterns for Major Foods* no. 76–25, Special Studies Division (Quezon City, Philippines, Department of Agriculture, 1976).

Aykrod, A. "The Rice Production in India," *Indian Medical Research Memoirs* vol. 32 (1940) p. 184.

Badan Urusan Logistik. *Main Operations of BULOG* (Jakarta, September 1983).

Badruddozaa, Kaxi M. "Experience in Farmers' Fields-Fertilizer Trials and Use in Bangladesh," *First Review Meeting INPUTS* (Honolulu, East-West Food Research Institute, June 7–18, 1976).

Bagyo, Ali Sri. "The Impact of Mechanization on Production and Employment in Rice Areas of West Java" (M.S. thesis, University of the Philippines at Los Banos, 1981).

Bangladesh Agricultural Research Council and International Fertilizer Development Center. "Agricultural Production, Fertilizer Use, and Equity Considerations—Results and Analysis of Farm Survey Data, 1979/80, Bangladesh" (Muscle Shoals, Ala., International Fertilizer Development Center, 1982).

Bangladesh, Bureau of Statistics. *Monthly Statistical Bulletin of Bangladesh* (Dacca).

———. *The Yearbook of Agricultural Statistics of Bangladesh* (Dacca), formerly *The Yearbook of Agricultural Statistics*.

BARC. *See* Bangladesh Agricultural Research Council.

Bardhan, Kalpana. "Rural Employment, Wages and Labour Markets in India: A Survey of Research," *Economic and Political Weekly* vol. 12 (June–July 1977) pp. 1012–1030, 1062–1074, and 1101–1118.

———. "Price and Output Response of Marketed Surplus of Food Grains: A Cross-Sectional Study of Some North Indian Villages," *American Journal of Agricultural Economics* vol. 52 (1970) pp. 51–61.

Bardhan, Pranab K. "Size, Productivity, and Returns to Scale: An Analysis of Farm-Land Data in Indian Agriculture," *Journal of Political Economy* vol. 81 (July–December 1973) pp. 1370–1386.

Bardhan, Pranab, and Kaplana Bardhan. "Price Response to Marketed Surplus of Food Grains," *Oxford Economic Papers* vol. 23 (1971) pp. 255–267.

Bardhan, Pranab K., A. Viadyanathan, Y. Alugh, G. S. Bhalla, and A. L. Bhadem, eds. *Labour Absorption in Indian Agriculture, Some Exploratory Investigations* (Bangkok, International Labor Organization, 1978).

Barker, Randolph. "Adoption and Production Impact of New Rice Technology—the Yield Constraints Problem," in International Rice Research Institute, *Farm Level Constraints to High Rice Yields in Asia: 1974–1977* (Los Banos, Philippines, IRRI, 1979) pp. 1–26.

———. "Barriers to Efficient Capital Investment in Agriculture," in T. W. Schultz, ed., *Distortions of Agricultural Incentives* (Bloomington, Ind., Indiana University Press, 1978).

Barker, Randolph, and Violeta G. Cordova. "Labor Utilization in Rice Production," in International Rice Research Institute, *Economic Consequences of New Rice Technology* (Los Banos, Philippines, IRRI, 1978).

Barker, Randolph, Eric C. Gabler, and Donald Winkelmann. "Long-Term Consequences of Technological Change on Crop Yield Stability: The Case of Cereal Grain" in Alberto Valdez, ed., *Food Security for Developing Countries* (Boulder, Colo., Westview Press, 1981).

Barker, Randolph, and Robert W. Herdt. "Setting Priorities for Rice Research in Asia," in Robert S. Anderson, Paul R. Brass, Edwin Levy and Barrie M. Morrison, eds., *Science, Politics and the Agricultural Revolution in Asia* (Boulder, Colo., Westview Press, 1982).

———. "Rainfed Lowland Rice as a Research Priority—An Economist's View," in International Rice Research Institute, *Rainfed Lowland Rice: Selected Papers from the 1978 International Rice Research Conference* (Los Banos, Philippines, IRRI, 1979).

———. "Equity Implications of Technology Changes," in International Rice Research Institute, *Interpretative Analysis of Selected Papers from Changes in Rice Farming in Selected Areas of Asia* (Los Banos, Philippines, IRRI, 1978).

Barker, Randolph, and T. K. Pal. "Barriers to Increased Rice Production in Eastern India," IRRI Research Paper Series No. 25 (Los Banos, Philippines, International Rice Research Institute, 1979).

Barker, Randolph, Radha Sinha, and Beth Rose, eds. *The Chinese Agricultural Economy* (Boulder, Colo., Westview Press, 1982).

Barker, Randolph, D. G. Sisler, and B. Rose. "Prospects for Growth in Grain Production," in R. Barker, R. Sinha, and B. Rose, eds., *The Chinese Agricultural Economy* (Boulder, Colo., Westview Press, 1982).

Barr, T. N. "The World Food Situation and Global Grain Prospects," *Science* vol. 4, no. 214 (1981) p. 4525.

Barramela, Jose Jr. "A Case Study of Coconut Tenant-Farmers in the Bicol River Basin." Paper presented at the National Workshop on Small Farmer Credit, Albay, Philippines, FAO, 1977.

Barrau, Jacques. *Plants and the Migration of Pacific Peoples* (Honolulu, Bishop Museum Press, 1963).

Bautista, A. C., J. C. Bunoan, and R. Feuer. "Development of the Zinc Extension Component for Irrigated Rice in the Philippine Masagana 99 Production Program," *Proceedings of the Symposium on Paddy Soils* (Beijing, Science Press, 1981).

Bautista, R. C., R. M. Lantin, and O. S. Inciong. "Change in Farm Mechanization: Philippine Trends," (Tokyo, APO Symposium on Farm Mechanization, November 24–30, 1981).

Behrman, J. R. "Price Elasticity of Marketed Surplus of a Subsistence Crop," *Journal of Farm Economics* vol. 48 (1966) pp. 875–893.

Bernsten, Richard H., and Robert W. Herdt. "Toward an Understanding of Milpa Agriculture: The Belize Case," *Journal of Developing Areas* vol. 11, no. 3 (April 1977).

Bernsten, Richard H., Bernard H. Siwi, and Henry M. Beachell. "The Development and Diffusion of Rice Varieties in Indonesia," IRRI Research Paper Series

No. 71 (Los Banos, Philippines, International Rice Research Institute, 1982).

Bhati, U.N. *Some Social and Economic Aspects of the Introduction of New Varieties of Paddy in Malaysia* (Geneva, United Nations Research Institute for Social Development, 1976).

———. "Technical Knowledge as a Determinant of Farmers' Income," in Agricultural Development Council, ed., *Economic Theory and Practice in the Asian Setting* (New Delhi, Wiley Eastern Ltd., 1972).

Bhooshan. *The Development Experience of Nepal* (New Delhi, Concept Publishing Co., 1975).

Binswanger, Hans, P., ed. *The Economics of Tractors in South Asia* (New York, Agricultural Development Council, 1978).

Binswanger, Hans P., and Mark R. Rosenweig. "Contractual Arrangements, Employment and Wages in Rural Labor Markets: A Critical Review" (New York, Agricultural Development Council, 1981).

Binswanger, Hans P., Vernon W. Ruttan, et al. *Induced Innovation: Technology, Institutions, and Development* (Baltimore, Md., Johns Hopkins University Press, 1978).

Biro Pusat Statistik. *Survey Social Ekonomi Nasional October 1969–April 1970* vol. 4–5 (Jakarta, n.d.).

Blekke, Bernard H. M. *The Story of the Dutch East Indies* (Cambridge, Mass., Harvard University Press, 1945).

Blyn, George. "Price Series Correlations as a Measure of Market Integration," *Indian Journal of Agricultural Economics* vol. 28 (1973) pp. 56–58.

———. *Agricultural Trends in India 1891–1947* (Philadelphia, University of Philadelphia Press, 1966).

Bo Canh-nong (Ministry of Agriculture). "Nong-lich Viet-nam cong-hoa" (Agricultural Calendar of Vietnam) (1961).

Booth, Anne. "Irrigation in Indonesia Part I." *Bulletin of Indonesian Economic Studies* vol. 13 (March 1977) pp. 33–74.

Boserup, Esther. *Women's Role in Economic Development* (New York, St. Martin's Press, 1970).

———. *The Conditions of Agricultural Growth: The Economics of Agrarian Change Under Population Pressure* (London, George Allen and Unwin Ltd., 1965).

Bouis, Howarth. "Rice Policy in the Philippines" (Ph.D. dissertation, Food Research Institute, Stanford University, 1982).

Breeman, N. Van. "Acidity of Wetland Soils, Including Histosols, as a Constraint to Food Production," in International Rice Research Institute, *Priorities for Alleviating Soil Related Constraints to Food Production in the Tropics* (Los Banos, Philippines, IRRI and Cornell University, 1970) pp. 189–202.

Brenier, Henri. *Essai d'atlas statistique de l'Indochine française* (Hanoi-Haiphong, Imprimerie d'Extrême-Orient, 1914).

British North Borneo Company. *Handbook of the State of British North Borneo with a Supplement of Statistical and Other Useful Information* (London, n.p., n.d.).

Brohier, R. L. *Ancient Irrigation Works in Ceylon* (Colombo, Sri Lanka, Government Publications Bureau, 1977).

Brown, Gilbert T. *Korean Pricing Policies and Economic Development in the 1960s* (Baltimore, Md., Johns Hopkins University Press, 1973).

Brun, Ellen, and Jacques Hersh. *Socialist Korea: A Case Study in the Strategy of Economic Development* (New York, Monthly Review Press, 1976).

Buck, John L. *Land Utilization in China* (New York, Agricultural Development Council, 1956).

Budianto, J., and Mul Yoto. "Survey Report, Indonesia" (Tokyo, Asian Productivity Organization, Symposium on Farm Mechanization, November 24–30, 1981).

Bunge, Frederica, M., ed. *North Korea: A Country Study*, Area Handbook Series (Washington, D.C., Government Printing Office, 1981).

Burma. *Notes on Agriculture in Burma* (Rangoon, Sup't, Union Government Printing and Stationery, 1955).

Burma, Agricultural Department of Burma. *The Rice Crop in Burma*, Agricultural Survey No. 17 (Rangoon, 1932).

Burma, Central Statistical and Economics Department. *Quarterly Bulletin of Statistics* (Rangoon).

———. *Retail Prices and Consumer Price Index at Rangoon* (Rangoon).

———. *Statistical Yearbook of Burma* (Rangoon).

Burma, Department of Land Records and Agriculture. *Season and Crop Report of the Union of Burma* (Rangoon).

Burma, Ministry of Planning and Finance. *Report to the Pyithu Hluttaw for 1982/83* (Rangoon, annual).

Burma, Office of the Collector of Customs. *Annual Statement of the Seaborne Trade and Navigation of Burma* (Rangoon, Sup't, Government Prints and Stationery).

Capule, Celia, and Robert W. Herdt. "Response of Non-Irrigated Rice to Fertilizer in Farmers' Fields in Bangladesh, 1970–75," IRRI Agricultural Economics Paper 81-01 (Los Banos, Philippines, International Rice Research Institute, 1981).

Castillo, J. V. "Philippine Agricultural Data Collection: The System and Methods Used," *Philippine Agricultural Situation* vol. 3, no. 3 (September 1961) pp. 29–40.

Central Food Technological Research Institute. *Report on the Evaluation of Modern Rice Mills in Comparison with Existing Mills* (Mysore, India, n.d.).

Central Rice Research Institute. *Annual Report for 1969* (Cuttack, 1970).

Ceylon, Department of Agriculture. *The Tropical Agriculturist* (Colombo).

———. *Census of Agriculture, 1962* (Colombo, Government Press, 1965).

Ceylon, Department of Census and Statistics. "A

Report on Paddy Statistics," Monograph No. 9 (1956).
———. *Ceylon, Census of Agriculture, 1952* (Colombo, Government Press, 1956).
Ceylon, Ministry of Finance. *Economic and Social Development of Ceylon 1926-1950, A Survey* (Colombo, Government Press, 1951).
Chambers, Robert. "Water Management and Paddy Production in the Dry Zone of Sri Lanka," Occasional Publication No. 8 (Colombo, Agrarian Research and Training Institute, 1975).
Champassak, Sisouk Na. *Storm Over Laos—A Comtemporary History* (New York, Praeger, 1961).
Chand, Shin, and A. N. Kapoor. *Land and Agriculture of India* (Delhi, Metropolitan Book Co., 1959).
Chandler, Robert F. Jr. "Case History of IRRI's Research Management During the Period 1960 to 1972" (Taiwan, Asian Vegetable Research and Development Center, 1975).
Chandratat, S. "Country Report, Thailand." Paper presented at Women in Rice Farming Systems Conference, Los Banos, Philippines, September 26–30, 1983.
Chang, H.Y. "Development of Irrigation Infrastructure and Management in Taiwan, 1900–1940: Its Implications for Asian Irrigation Development," in *Economic Essays* vol. 9 (Taipei, National Taiwan University, The Graduate Institute of Economics, 1980) pp. 133–155.
Chang, Kwang-chih. *The Archaeology of Ancient China* (New Haven, Conn., Yale University Press, 1977).
Chang, L. L. "Farm Prices in Wuchin, Kiangsu, China," Booklet Series no. 19, Bureau of Foreign Trade, Ministry of Industry (Shanghai, 1932).
Chang, Te-tzu. "The Origins and Early Cultures of the Cereal Grains and Food Legumes," in David N. Keightley, ed., *The Origins of Chinese Civilization* (Berkeley, Calif., University of California Press, 1982).
———. "Hybrid Rice," in J. Sneep and A. T. Hendriksen, eds., *Plant Breeding Perspectives* (Wageningen, PUDOC, 1979) pp. 173–174.
———. "The Rice Cultures," in *The Early History of Agriculture* (London, Oxford University Press, 1977) pp. 143–155.
———. "The Origin, Evolution, Cultivation, Dissemination, and Diversification of Asian and African Rices," *Euphytica* vol. 25 (1976) pp. 425–440.
Chang, Te-tzu, and H. I. Oka. "Genetic Variousness in the Climatic Adaptation of Rice Cultivars," in *Proceedings of the Symposium on Climate and Rice* (Los Banos, Philippines, International Rice Research Institute, 1976).
Chao, Kang. *"Agricultural Production in Communist China 1949–1965* (Madison, Wisc., University of Wisconsin Press, 1970).
Chatfield, Godfrey A. *Sabah, A General Geography* (London, Eastern Universities Press SDN BHD in association with the University of London Press, Ltd., 1972).
Chaudhari, H. Ali, A. Rashid, and Q. Mohy-Ud-Din. "Gujranwala, Punjab," in International Rice Research Institute, *Changes in Rice Farming in Selected Areas of Asia* (Los Banos, Philippines, IRRI, 1975) pp. 225–240.
Chen, Hsu, and Mao. "Rice Policies in Taiwan." Paper presented at the Workshop on the Political Economy of Rice sponsored by the Food Research Institute, Stanford University (Los Banos, Philippines, International Rice Research Institute, July 1974).
Chen, Nai-ruenn. *Chinese Economic Statistics* (Chicago, Aldine Publishing Co., 1967).
Chen, Wu-hsiung, and John T. Scott, Jr. "The Simulation of Grain Supply, Producer Income and Consumer Price Policy: The Case of Rice in Taiwan," No. 181, Department of Agricultural Economics (Urbana-Champaign, The University of Illinois at Urbana-Champaign, 1982).
Cheng, Siok-hwa. *The Rice Trade of Malaya* (Singapore, University of Education Press, 1973).
———. *The Rice Industry of Burma 1852–1940* (Singapore, University of Malaya Press, 1968).
China, Ministry of Agriculture and Commerce. *Nongshang tongji biao* (Yearbook of Agriculture and Commerce Statistics) (Nanjing, 1916–1924).
China, Ministry of Industries, The National Agricultural Research Bureau, Division of Agricultural Economics. *NongQing baoguo* (Crop Reports) (Nanjing, 1930s).
China, State Statistical Bureau. *Ten Great Years*, Occasional Paper No. 5, Program in East Asian Studies (Bellingham, Wash., Western Washington State College, 1974).
China, Zhongguo Nongmin Yinhang, Jingji Yongjin Chu (Department of Economic Research). *Zhongnong jingji tongji*, (Economic and Statistical Review added English title) (Nanking).
Chinn, Dennis, L. "A Calorie-Arbitrage Model of Chinese Grain Trade," Working Paper (Stanford, Calif., Stanford Food Research Institute, Stanford University, 1979).
Chipravat, Olin, and S. Pariwat. "An Econometric Model of World Rice Markets" (Bangkok, Department of Economic Research, Bank of Thailand, 1976).
Chisholm, Anthony H., and Rodney Tyers, eds. *Food Security: Theory, Policy and Perspectives from Asia and the Pacific Rim* (Lexington, Mass., Lexington Books, 1982).
Choi, Hochin. *The Economic History of Korea from the Earliest Times to 1945* (Seoul, Freedom Library, n.d.).
Choi, Kee Ie. "Technological Diffusion in Agriculture Under the Bakuhan System," *The Journal of Asian Studies* vol. 19 (August 1971) pp. 749–759.
Chōsen Ginko (Bank of Korea). *Tokai geppo* (Bulletin of Monthly Statistics) (Seoul).
Chōsen Sōtokufu (Governor General of Korea). *Chōsen boeki nempyo* (Trade Yearbook of Korea) (annual).
———. *Chōsen tōkei nempō* (Statistical Yearbook of Korea) (Seoul).

Chung, Joseph Sang-Moon. *The North Korean Economy* (Stanford, Calif., Hoover Institution Press, Stanford University, 1974).

Clay, Edward J. "Institutional Change and Agricultural Wages in Bangladesh," *Bangladesh Development Studies* vol. 4 (1976) pp. 423–440.

Collier, William L. "Declining Labor Absorption (1878 to 1980) in Javanese Rice Production." Paper presented at the Third Biennial Meeting of the Agricultural Economics Study of Southeast Asia, Kuala Lumpur, Malaysia, November 25–29, 1979).

Collier, William L., J. Colter, Sinarhadi, and Robert d'A. Shaw. "Choice of Technique in Rice Milling in Java: A Comment," *Bulletin of Indonesian Economic Studies* vol. 10 (1974) pp. 106–120.

Collier, William L., Soentoro, K. Hidayat, and Y. Yuliati. "Labour Absorption in Javanese Rice Cultivation," in W. Gooneratne, ed., *Labour Absorption in Rice-Based Agriculture* (Bangkok, International Labor Office, 1982).

Collier, William L., Soentoro, Gunawan Wiradi, and Makali. "Agricultural Technology and Institutional Change in Java," *Food Research Institute Studies* vol. 13, no. 2 (1974) pp. 169–194.

Colombo, Umberto, D. Gale Johnson, and Toshio Shishido. "Reducing Malnutrition in Developing Countries: Increasing Rice Production in South and Southeast Asia," Trilateral Papers, No. 16 (New York, The Trilateral Commission, 1978).

Conklin, H. C. "Hanunoo Agriculture: A Report on the Systems of Shifting Cultivation in the Philippines" (Rome, Food and Agriculture Organization, 1957).

Consequences Team. "Consequences of Land Preparation Mechanization in Indonesia: South Sulawesi and West Java" (Jakarta, Regional Seminar on Appropriate Mechanization for Rural Development, January 26–30, 1981).

Contado, Tito E., and Roger A. Jaime. "Baybay, Leyte," in International Rice Research Institute, *Changes in Rice Farming in Selected Areas of Asia* (Los Banos, Philippines, IRRI, 1975) pp. 283-301.

Cordova, Violeta G., A. M. Mandac, and F. Gascon. "Some Considerations on Energy Costs of Rice Production in Central Luzon." Paper presented at the PAEDA 26th Annual Convention, Munos, Nueva Ecija, Philippines, June 1980.

Cordova, Violeta, Aida Papag, Sylvia Sardido, and Leonida Yambao. "Changes in Practices of Rice Farmers in Central Luzon: 1966–79." Paper presented to the Crop Society of the Philippines, Bacnotan, La Union, Philippines, April 1981.

Cowan, C. A. *The Economic Development of Southeast Asia: Studies in Economic History and Political Economy* (New York, Praeger, 1964).

Coward, E. Walter Jr. *Irrigation and Agricultural Development in Asia* (Ithaca, N.Y., Cornell University Press, 1980).

———. "Irrigation Development: Institutional and Organizational Issues" in E. Walter Coward, Jr., ed., *Irrigation and Agricultural Development in Asia* (Ithaca, N.Y., Cornell University Press, 1980).

Coyle, William T. *Japan's Rice Policy*, Foreign Agricultural Economic Report no. 164 (Washington, D.C., U.S. Department of Agriculture, July 1981).

Croll, Elizabeth. *Women in Rural Development—The People's Republic of China* (Geneva: ILO, 1979).

Dahil, Vidilal, ed. *A Regional Profile of Indian Agriculture* (Bombay, Vora and Co., Publishers, 1974).

Dalrymple, Dana G. *Development and Spread of High-Yielding Varieties of Wheat and Rice in the Less Developed Nations*, Foreign Agricultural Economic Report No. 95 (Washington, D.C., U.S. Department of Agriculture, 1978).

———. *Measuring the Green Revolution: The Impact of Research on Wheat and Rice Production*, Foreign Agricultural Economic Report No. 106 (Washington, D.C., U.S. Department of Agriculture, 1975).

———. *Survey of Multiple Cropping in Less Developed Nations*, Foreign Agricultural Economic Report No. 12 (Washington, D.C., U.S. Department of Agriculture, 1971).

Dandekar, V. M., and N. Rath. "Poverty in India," *Economic and Political Weekly* vol. 6, no. 1 (January 1971) pp. 25–48.

Danwood, Dayan. "Rice Importation in Indonesia, 1961–1974" (M.A. thesis, University of the Philippines, 1977).

Darrah, L. B., and F. A. Tiongson. *Agricultural Marketing in the Philippines* (Los Banos, Philippines, University of the Philippines, College of Agriculture, 1969).

Dasgupta, Buiplab. *Agrarian Change and the New Technology in India* (Geneva, United Nations Research Institute for Social Development, 1977).

David, Cristina, C. "Credit and Price Policies in Philippine Agriculture," Staff Paper Series 82-2 (Manila, Philippines, Institute for Development Studies, 1982).

———. "Analysis of Agricultural Policies in the Philippines" (Los Banos, Philippines, January 1982) mimeo.

———. "Factors Affecting Fertilizer Consumption" in International Rice Research Institute, *Interpretive Analysis of Selected Papers from Changes in Rice Farming in Selected Areas of Asia* (Los Banos, Philippines, IRRI, 1978).

———. "A Model of Fertilizer Demand of the Asian Rice Economy (A Macro-Micro Analysis)" (Ph.D. dissertation, Stanford University, 1975).

David, Cristina, and Randolph Barker. "Labor Demand in the Philippine Rice Sector" in Wibert Gooneratne, ed., *Labour Absorbtion in Rice-Based Agricuture* (Bangkok, Asian Employment Programme, International Labor Organization, 1982) pp. 119–157.

———. "Modern Rice Varieties and Fertilizer Consumption," in International Rice Research Institute, *Economic Consequences of New Rice Technology* (Los Banos, Philippines, IRRI, 1978).

Davies, David. *Thailand, The Rice Bowl of Asia* (London, England, Robert Hole Ltd., 1967).

De Datta, S. K. *Principles and Practices of Rice Production* (New York, Wiley, 1981).

De Datta, S. K., K. A. Gomez, R. W. Herdt, and R. Barker. *A Handbook on the Methodology for an Integrated Experiment-Survey on Rice Yield Constraints* (Los Banos, Philippines, International Rice Research Institute, IRRI, 1978).

De Datta, S. K., A. C. Tauro, and S. M. Balaoing. "Effect of Plant Type and Nitrogen Level in the Growth Characteristics and Grain Yield of Indian Rice in the Tropics," *Agronomy Journal* vol. 60 (November–December 1968) pp. 643-647.

de Leon, M. S. J. "Government Expenditures and Agricultural Policies in the Philippines 1955–1980," Working Paper No. 83-06 (Manila, Philippine Institute for Development Studies, 1983).

Democratic Republic of Vietnam. *Etudes Vietnamiennes* (Hanoi).

———. *So lien thong-ke 1963* (Statistical Data 1963) (Hanoi, Su That Publishing House, 1964).

Deo, G. P., and R. N. Shah. "Review of Fertilizer Investigation Work in Rice at Parwanipur (1958–1976)." Paper presented at the Fifth Rice Improvement Workshop, National Rice Improvement Programme, Department of Agriculture, Nepal, March 1978.

———. "Performance of Different Promising Lines at Different Levels of N Under Rainfed Conditions." Paper presented at the Fifth Rice Improvement Workshop, National Rice Improvement Programme, Department of Agriculture, Nepal, March 1978.

Deomampo, N. R., and R. D. Torres. "Some Economic Issues on Farm Mechanization in the Philippines" (Los Banos, Philippines, Second Agricultural Policy Conference, University of the Philippines, July 10–12, 1975).

Desai, Gunvant M. "Nitrogen Use and Foodgrain Production India, 1973–74, 1978–79, and 1983–84," Occasional Paper No. 55, USAID—Employment and Income Distribution Project (Washington, D.C., U.S. Agency for International Development, March 1973).

———. "Some Observations on the Economics of Cultivating High-Yielding Varieties of Rice in India" (Los Banos, Philippines, Rice Policy Conference, International Rice Research Institute, August 1971).

De Silva, K. M. *A History of Sri Lanka* (Berkeley, Calif., University of California Press, 1981).

Dilbagh, S. Athwal, and Mano D. Pathak. "Genetics of Resistance to Rice Insects," in International Rice Research Institute, *Rice Breeding* (Los Banos, Philippines, IRRI, 1972) pp. 375–386.

Dillon, John L. "Broad Structural Review of the Small-Farmer Technology Problems" in A. Valdez, G. Scobie, and J. Dillon, eds., *Economics and the Design of Small-Farmers Technology* (Ames, Iowa, Iowa State University Press, 1979).

Djauhari, Aman. "Present Stage of Agricultural Development, Indramayu and Lampung," *Annual Report 1975–76 Cropping Systems Research* (Bogor, Indonesia, Central Institute for Agriculture, 1977).

Djojoadinato, Desman. "Indonesia's Simple (Sederhana) Irrigation and Reclamation Program," in International Rice Research Institute, *Irrigation Policy and Management in Southeast Asia* (Los Banos, Philippines, IRRI, 1978) pp. 25–30.

Dodson, Joseph C., and Clark C. Milligan. "Japanese Crop and Livestock Statistics 1878–1950," General Headquarters Supreme Commander for the Allied Powers, Natural Resources Section Report No. 143 (Tokyo, 1951).

Donnelly, Michael W. "Setting the Price of Rice—A Study in Political Decisionmaking," in T. J. Pempel, ed., *Policymaking in Contemporary Japan* (Ithaca, N.Y., Cornell University Press, 1977).

Donner, Wolf. *The Five Faces of Thailand, An Economic Geography* (New York, St. Martin's Press, 1978).

Duff, Bart J. "Mechanization and Use of Modern Rice Varieties," in International Rice Research Institute, *Economic Consequences of the New Rice Technology* (Los Banos, Philippines, IRRI, 1978).

East Pakistan, East Pakistan Bureau of Statistics. *Abstract of Agricultural Statistics of East Pakistan* (Dacca).

East Pakistan. *Season and Crop Report of East Pakistan* (Dacca, East Bengal Government Press, annual).

Ebihara, May. "Perspectives on Sociopolitical Transformations in Cambodia/Kampuchea—A Review Article," *Journal of Asian Studies* vol. 61, no. 1 (November 1981) pp. 63–71.

Eckstein, Alexander, ed. *Quantitative Measures of China's Economic Output* (Ann Arbor, University of Michigan, 1980).

Economic and Social Commission for Asia and the Pacific. *Agro-Chemical News in Brief*, various issues.

Edmundson, Wade. "Land, Food, and Work in Three Javanese Villages" (Ph.D. dissertation, University of Hawaii, 1972).

Ekonomi dan Keucangan Indonesia (Jakarta) quarterly.

Elhance, D. N. *Economics Statistics of India Since Independence* (Delhi, Kitab Mahal Private Ltd., 1962).

Elvin, Mark. *The Pattern of the Chinese Past: A Social and Economic Interpretation* (Stanford, Calif., Stanford University Press, 1973).

Encarnacion, Jose Jr. *Philippine Economic Problems in Perspective* (Manila, Institute of Economic Development and Research, School of Economics, University of the Philippines, 1976).

Endirisinghe, Neville, and Thomas Poleman. "Implications of Government Intervention in the Rice Economy of Sri Lanka," Cornell University Agriculture Mimeograph no. 48 (Ithaca, N.Y., 1976).

Esman, Milton J. "Research and Development Organizations: A Reevaluation," in William Foote Whyte and Damon Boynton, eds., *Higher Yielding Human Systems for Agriculture* (Ithaca, N.Y., Cornell University Press, 1983).

Esmay, Merle, Soemangat, Eriyatno, and Allan Phillips. *Rice Post-Production Technology in the Tropics* (Honolulu, East-West Center, 1979).

ESSO Pakistan Fertilizer Co. "Pakistan Nitrogen Demand Forecast Study" (Karachi, 1974).

Evans, Yiyi Chit-maung. "Marketing of Export Crops in Burma and Thailand, 1845–1967" (Ph.D. dissertation, McGill University, 1972).

Evenson, Robert E. "Comparative Evidence on Returns to Investment in National and International Research Institutions," in Thomas Arndt, Dana Dalrymple, and Vernon W. Ruttan, eds., *Resource Allocation and Productivity in National and International Agricultural Research* (Minneapolis, Minn., University of Minnesota, 1977).

Evenson, Robert E., and Hans P. Binswanger, "Estimating Labor Demand Functions for Indian Agriculture," Economic Growth Center Discussion Paper No. 356 (New Haven, Conn., Yale University, August 1980).

———. "Technology Transfer and Research Allocation," in Hans P. A. Binswanger, Vernon Ruttan, et al., eds., *Technology, Institutions, and Development* (Baltimore, Md., The Johns Hopkins University Press, 1978) pp. 164–211.

Evenson, Robert E., and Pie M. Flores. "Social Returns to Rice Research," in International Rice Research Institute, ed., *Economic Consequences of the New Rice Technology* (Los Banos, Philippines, IRRI, 1978).

Evenson, Robert E., James P. Houck, Jr., and Vernon W. Ruttan. "Technical Change and Agricultural Trade: Three Examples—Sugarcane, Bananas, Rice," in Robert Vernon, ed., *The Technology Factor in International Trade* (New York, Columbia University Press, 1970).

Falcon, Walter P. "Food Self-Sufficiency: Lessons from Asia," in *International Food Policy Issues: A Proceedings*, U.S. Department of Agriculture, Foreign Agricultural Economic Report 143 (Washington, D.C., 1978).

———. "The Reliability of Punjab Agricultural Data," in Abdur Rab, ed., *Acreage, Production and Prices of Major Agricultural Crops of West Pakistan (Punjab) 1931–1959* (Karachi, Pakistan Institute of Development Economics, 1961) pp. 3–9.

Falcon, Walter P., and Eric A. Monke. "International Trade in Rice," *Food Research Institute Studies* vol. 18 (1979–80) pp. 297–306.

Falcon, Walter P., and C. Peter Timmer. "The Political Economy of Rice Production and Trade in Asia," in Lloyd Reynolds, ed., *Agriculture in Development Theory* (New Haven, Conn., Yale University Press, 1975) pp. 373–408.

Farm Machinery Industrial Research Corporation. *1982 Farm Machinery Statistics* (Tokyo, 1982).

Farruk, Muhammad O. "Structure and Performance of the Rice Marketing System in East Pakistan." Occasional Paper: USAID Employment and Income Distribution Project, No. 31 (Ithaca, N.Y., Department of Agricultural Economics, Cornell University, 1970).

Fei, John C. H., and Gustav Ranis. "Agriculture in Two Types of Open Economies," in Lloyd Reynolds, ed., *Agriculture in Development Theory* (New Haven, Conn., Yale University Press, 1975).

Fertilizer Advisory Development and Information Network for Asia and the Pacific, Economic and Social Commission for Asia and the Pacific. "Marketing, Distribution and Use of Fertilizers in Bangladesh" (Bangkok, 1980).

Fertilizer Association of India. *Fertilizer Statistics* (New Delhi, annual).

Fisher, Charles A. "Some Comments on Population Growth in Southeast Asia with Specific Reference to the Period Since 1830," in *The Economic Development of Southeast Asia—Studies in Economic History and Political Economy* (New York, Praeger, 1964).

Fitzsimmons, Thomas. *Cambodia* (New Haven, Conn., Human Resources Area Files Press, 1957).

Flores, Piedad, Robert E. Evenson, and Yujiro Hayami. "Social Returns to Rice Research in the Philippines: Domestic Benefits and Foreign Spillover," IRRI Agricultural Economics Department Paper 76-17 (Los Banos, Philippines, International Rice Research Institute, 1976).

Food and Agriculture Organization of the United Nations. *See* Food and Agriculture Organization.

Food and Agriculture Organization, *Agricultural Commodities—Projections for 1975 and 1985* vol. 2 (Rome, FAO, 1967).

———. *FAO Fertilizer Yearbook* (Rome, annual).

———. *FAO Rice Report* (Rome, 1965).

———. *Food Balance Sheets 1975–1977 and Per Capita Food Supplies, 1967 to 1977* (Rome, FAO, 1980).

———. *Food Outlook: 1981 Statistical Supplement* (Rome, FAO, January 1982).

———. *Marketing of Subsidized Fertilizers to Small Paddy Farmers*, Report to Government of Ceylon (Rome, 1968).

———. *Monthly Bulletin of Statistics* (Rome, monthly).

———. *Production Yearbook* (Rome, annual).

———. "Report on the 21st Session of the Intergovernmental Working Group on Rice" (Rome, FAO, 1978).

———. Commodities and Trade Division, General Studies Group, "Gross Domestic Product, Private Consumption Expenditure and Agricultural GDP at 1975 Constant Prices, Historical Series 1960–1975 and Projections, 1975–1990" (Rome, FAO, February 1976).

———. International Group on Rice. "Compendium of

National Rice Trade Policies" (Rome, FAO, 1973, 1977).

Frankel, Francine R. *India's Green Revolution: Economic Gains and Political Costs* (Princeton, N.J., Princeton University Press, 1971).

Freeman, J. D. "Iban Agriculture: A Report on the Shifting Cultivation of Hill Rice by the Iban of Sarawak" (London, H. M. Stationery Office, 1955).

Fukui, H. "Environmental Determinants Affecting the Potential Productivity of Rice—A Case Study of the Chao Phraya River Basin of Thailand" (Ph.D. dissertation, Kyoto University, 1973).

Gaesuwan, Yuavores, Ammar Siamwalla, and Delane Welsch. "Thai Rice Production and Consumption Data 1947–70," Thailand Rice Project Working Paper No. 2 (Bangkok, Department of Agricultural Economics, Kasetsart University, 1974).

Gavan, James, and Indrani Chandrasekera. "The Impact of Public Foodgrain Distribution on Food Consumption and Welfare in Sri Lanka," Research Report no. 13 (Washington, D.C., International Food Policy Research Institute, 1979).

Geertz, Clifford. "Organization of the Balinese Subak," in E. Walter Coward, Jr., ed., *Irrigation and Agricultural Development in Asia* (Ithaca, N.Y., Cornell University Press, 1980).

Geertz, Hildred, and Clifford Geertz. *Kinship in Bali* (Chicago, University of Chicago Press, 1975).

General Headquarters, Supreme Commander for the Allied Powers, Economic and Scientific Section. "Japanese Trade Patterns, 1930–1934" (Tokyo, 1952).

George, P. S., and V. V. Choukidar. "Production and Marketing of Paddy (A Study of Local High-Yielding Varieties in the West Godavari District)" (Ahmedabad, Indian Institute of Management, 1973).

Gill, G. J. *Farm Power in Bangladesh*, vol. 1 (Reading, England, Department of Agricultural Economics and Management, University of Reading, 1981).

Girling, John L. S. *Thailand Society and Politics* (Ithaca, N.Y., Cornell University Press, 1981).

———. "Cambodia and the Sihanouk Myths," Occasional paper no. 7 (Singapore, Institute of South East Asian Studies, 1971).

Goldman, Richard H. "Staple Food Self-Sufficiency and the Distributive Impact of Malaysian Rice Policy," *Food Research Institute Studies* vol. 14, no. 3 (1975) pp. 251–293.

———. "The Formation of Seasonal Rice Prices in Indonesia" (Ph.D. dissertation, Stanford University, 1973).

Gooneratne, W., ed. *Labour Absorption in Rice-Based Agriculture* (Bangkok, International Labor Organization, 1982).

Gotsch, Carl. "Technological Change and the Distribution of Income in Rural Areas," *American Journal of Agricultural Economics* vol. 54, no. 2 (May 1972) pp. 326–341.

Gotsch, Carl, and Gilbert Brown. *Prices, Taxes and Subsidies in Pakistan Agriculture, 1960–1976*, World Bank Staff Working Paper No. 387 (Washington, D.C., World Bank, April 1980).

Gourou, Pierre. *Land Utilization in French Indochina* (Washington, D.C., Institute of Pacific Relations, 1945).

———. *L'Utilisation du sol en Indochine Française* (Paris, Centre d'Etudes de Politique Etrangère, Travaux des Groupes d'Etudes No. 14, 1947).

Gouzhou Hua, and Yao Jianfu. "Some Aspects of and Experiences in China's Agricultural Mechanization." Paper presented at the Seminar on Mechanization of Small-scale Farming, Hangzhou, June 22–26, 1982.

Grad, Andrew J. *Land and Peasant in Japan: An Introductory Survey* (New York, International Secretariat, Institute of Pacific Relations, 1952).

Grajdanzen, Andrew J. *Statistics of Japanese Agriculture* (New York, International Secretariat, Institute of Pacific Relations, 1941).

Grant, J. W. "The Rice Crop in Burma: Its History, Cultivation, Marketing, and Improvement," Agricultural Survey no. 17 (Rangoon, Agricultural Department, 1932).

Grant, Warren R., T. Mullins, and W. F. Morrison. *World Rice Study: Disappearance, Production and Price Relationships Used to Develop the Model*, Economic Research Service No. 608 (Washington, D.C., U.S. Department of Agriculture, 1975).

Great Britian. *Colonial Reports, North Borneo* (London, His Majesty's Stationery Office).

———. *Colonial Reports, Sarawak* (London, His Majesty's Stationery Office); ceased in 1962.

Griffen, D. "Comments on Labor Utilization in Rice Production," in *Economic Consequences of the New Rice Technology* (Los Banos, Philippines, International Rice Research Institute, 1978).

Griffin, Keith. *The Political Economy of Agrarian Change* (Cambridge, Mass., Harvard University Press, 1974).

Grilches, Zvi. "Research Costs and Social Returns: Hybrid Corn and Related Innovations," *Journal of Political Economy* vol. 66 (1958) pp. 419–431.

Guangdong Provincial Academy of Agricultural Sciences. "Principal Experience in the Breeding of Dwarfed Paddy Rice," *Renmin Ribao* (Beijing, December 1964) translated in Joint Publications Research Service 28139 (January 1965) pp. 26–35.

Guino, Ricardo A. "Time Allocation Among Rice Farm Households in Central Luzon, Philippines" (M.S. thesis, University of the Philippines, 1978).

Gupta, A. P. *Marketing of Agricultural Produce in India* (Bombay, Vora and Co., Publishers, 1975).

Haessel, Walter. "The Price Elasticity of Home Consumption and Marketed Surplus of Food Grains," *American Journal of Agricultural Economics* vol. 57 (1975) pp. 111–115.

Halpern, Joel. *Economy and Society of Laos, A Brief Survey*, Monograph Series No. 5, South East Asia Studies (New Haven, Conn., Yale University, 1964).

———. "Population, Statistics, and Associated Data," Laos Project, Paper No. 3 (New Haven, Conn. March 1961) mimeo.

———. "Laotian Agricultural Statistics," Laos Project, Paper No. 9, (New Haven, Conn., April 1961).

———. "Economic and Related Statistics Dealing with Laos," Laos Project, Paper No. 11 (New Haven, Conn.) (mimeo).

Hameed, Amerasinghe, Gunasekera Panditharatna, Selvanayaham, and Selvadurai. *Rice Revolution in Sri Lanka* (Geneva, United Nations Research Institute for Social Development, 1977).

Hamid, Javed, "Agricultural Mechanization: A Case for Fractional Technology," in Tan Bock Thiam and Shao-er Ong, eds., *Readings in Asian Farm Management* (Singapore, University of Singapore Press, 1979).

Handler, S. *Developing Country Foodgrain Projections for 1985*, World Bank Staff Working Paper No. 247 (Washington, D.C., World Bank, 1976) available from the National Technical Information Service, Washington, D.C., as IBRD-WP 247.

Hansen, Gary E., ed. *Agricultural and Rural Development in Indonesia* (Boulder, Colo., Westview Press, 1981).

Hara, Yonosuke. "Labor Absorption in Asian Agriculture: The Japanese Experience," in International Labor Organization, *Labour Absorption in Agriculture: The East Asian Experience* (Geneva, ILO, 1980).

Harahap, S., H. Siregar, and B. H. Siwi. "Rice Varieties for Indonesia," in International Rice Research Institute, *Rice Breeding* (Los Banos, Philippines, IRRI, 1972).

Harahap, Z. "Breeding for Resistance to Brown Planthopper and Grassy Stunt Virus in Indonesia," in International Rice Research Institute, *Brown Planthopper: Threat to Rice Production in Asia* (Los Banos, Philippines, IRRI, 1979) pp. 201–208.

Hargrove, Thomas R. "Diffusion and Adoption of Genetic Materials Among Rice Breeding Programs in Asia," IRRI Research Paper Series No. 18 (Los Banos, Philippines, International Rice Research Institute, 1978).

———. "Genetic and Sociologic Aspects of Rice Breeding in India," IRRI Research Paper Series No. 10 (Los Banos, Philippines, International Rice Research Institute, 1977).

Hargrove, Thomas R., W. Ronnie Coffman, and Victoria L. Cabanilla. "Genetic Interrelationships of Improved Varieties in Asia," IRRI Research Paper Series No. 23 (Los Banos, Philippines, International Rice Research Institute, 1979).

Harriss, Barbara. "Allocation, Location, and Dislocation in Non-Market Rice Distribution," *Journal of Development Studies* vol. 15 (1978) pp. 87–105.

———. "Paddy and Rice Marketing in Northern Tamil Nadu," in *Studies of Marketed Surplus, Market Efficiency, Technology, and Livelihoods* (Madras, Saugram Publishers, 1977).

Hart, Gillian. "Labor Allocation Strategies in Rural Javanese Households" (Ph.D. dissertation, Cornell University, 1978).

Hasankhan, Mahmood. *The Economics of Green Revolution in Pakistan* (New York, Praeger Publishers, 1975).

Hason, Parveg. *Korea: Problems and Issues in a Rapidly Growing Economy* (Baltimore, Md., Johns Hopkins University Press, 1974).

Havens, Thomas R. *Farm and Nation in Modern Japan Agrarian Nationalism 1870–1940* (Princeton, N.J., Princeton University Press, 1974).

Hayami, Yujiro. Review of T. W. Schultz, ed., "Distortions of Agricultural Incentives," *Economic Development and Cultural Change* vol. 2, no. 29 (January 1981).

———. *Anatomy of a Peasant Economy: A Rice Village in the Philippines* (Los Banos, Philippines, International Rice Research Institute, 1978).

———. *A Century of Agricultural Growth in Japan* (Tokyo, University of Tokyo Press, 1975).

———. "Rice Policy in Japan's Economic Development," *American Journal of Agricultural Economics* vol. 54, no. 1 (February 1972) pp. 19–31.

Hayami, Yujiro, and Masakatsu Akino. "Organization and Productivity of Agricultural Research Systems in Japan," in Thomas M. Arndt, Dana Dalrymple, and Vernon W. Ruttan, eds., *Resource Allocation and Productivity in National and International Agricultural Research* (Minneapolis, Minn., University of Minnesota Press, 1977).

Hayami, Yujiro, Cristina C. David, Piedad Flores, and Masao Kikuchi. "Agricultural Growth Against a Land Resource Constraint: The Philippine Experience," *Australian Journal of Agricultural Economics* vol. 20 (1976) pp. 144–159.

Hayami, Yujiro, and Anwar Hafid. "Rice Harvesting and Welfare in Rural Java," *Bulletin of Indonesian Economic Studies* (July 1979) pp. 94–112.

Hayami, Yujiro, and Robert W. Herdt. "Market Price Effects of Technological Change on Income Distribution in Semisubsistence Agriculture," *American Journal of Agricultural Economics* vol. 59, no. 2 (May 1977) pp. 245–256.

Hayami, Yujiro, and Masao Kikuchi. *Asian Village Economy at the Crossroads, An Economic Approach to Institutional Change* (Tokyo, Tokyo University Press, 1981).

———. "Investment Inducements to Public Infrastructure: Irrigation in the Philippines," *The Review of Economics and Statistics* vol. 6, no. 1 (February 1978) pp. 70–77.

Hayami, Yujiro, and Vernon W. Ruttan. *Agricultural Development: An International Perspective* 2d ed., rev. and updated (Baltimore, Md., Johns Hopkins University Press, 1985).

Agricultural Development: An International Perspective (Baltimore, Md., Johns Hopkins University Press, 1971).

Hayami, Yujiro, Vernon W. Ruttan, and Herman M. Southworth, eds. *Agricultural Growth in Japan, Taiwan, Korea, and the Philippines* (Honolulu, University of Hawaii Press, 1979).

Hayami, Yujiro, and Saburo Yamada. "Agricultural Productivity at the Beginning of Industrialization" in Kazushi Ohkawa, Bruce Johnston, Hiromitsu Kaneda, eds., *Agricultural and Economic Growth: Japan's Experience* (Princeton, N.J., Princeton University Press, 1970).

Hazell, Peter, B. R. *Instability in Indian Foodgrain Production*, Research Report No. 30 (Washington, D.C., International Food Policy Research Institute, May 1982).

He, Kwei-ting, Amanda Te, Shi-gang Zhu, S. Lee Travers, Hsui-fang Lai, and Robert W. Herdt. "The Economics of Hybrid Rice Production in China" IRRI Department Paper No. 83-22 (Los Banos, Philippines, International Rice Research Institute, 1983).

Hedley, Douglas D. "Rice Buffer Stocks for Indonesia: A First Approximation," Working Paper No. 2, Rice Policies in Southeast Asia Project (Washington, D.C., International Food Policy Research Institute, International Fertilizer Development Center, International Rice Research Institute, 1979).

Henderson, J., et al. *Area Handbook of Burma* (Washington, D.C., Foreign Area Studies, American University, 1971).

Henry, Yves Marius. "Documents de demographic et riziculture en Indochine," *Bulletin economique de l'Indochine* (Hanoi, 1928).

Herdt, Robert W. "Perspectives, Issues, and Evidence on Rice Farm Mechanization in Developing Asian Countries," in Asian Productivity Organization, *Farm Mechanization in Asia* (Tokyo, APO, 1983) pp. 111–148.

———. "Focusing Research on Future Constraints to Rice Production," IRRI Research Paper Series No. 76 (Los Banos, Philippines, International Rice Research Institute, 1982).

———. "A Note on William L. Collier's Paper, Declining Labor Absorption (1878–1980) in Javanese Rice Production" (Los Banos, Philippines, International Rice Research Institute, 1980).

———. "An Overview of the Constraints Project Results," in International Rice Research Institute, *Farm Level Constraints to High Rice Yields in Asia: 1974–77* (Los Banos, Philippines, IRRI, 1979) pp. 397–411.

———. "Costs and Returns for Rice Production," in *Economics Consequences of the New Rice Technology* (Los Banos, Philippines, International Rice Research Institute 1978) pp. 63–80.

Herdt, Robert W., and Celia Capule. *Adoption, Spread and Production Impact of Modern Rice Varieties in Asia* (Los Banos, Philippines, International Rice Research Institute, 1983).

Herdt, Robert W., and Abraham M. Mandac. "Modern Technology and Economic Efficiency of Philippine Rice Farms," *Economic Development and Cultural Change* vol. 29 (1981) pp. 374–399.

Herdt, Robert W., and John Mellor. "The Contrasting Response of Rice to Nitrogen: India and the United States." *American Journal of Agricultural Economics* vol. 46, no. 1 (1964) pp. 150–160.

Herdt, Robert W., Amanda Te, and Randolph Barker. "The Prospects for Asian Rice Production," *Food Research Institute Studies* vol. 16 (1977–78) pp. 184–203.

Herdt, Robert W., and Thomas H. Wickham. "Exploring the Gap Between Potential and Actual Rice Yields in the Philippines," *Food Research Institute Studies* vol. 14 (1975) pp. 163–181.

Herrera, Romeo T. "Gapan, Nueva Ecija," in International Rice Research Institute *Changes in Rice Farming in Selected Areas of Asia* (Los Banos, Philippines, 1975) pp. 265–282.

Hertford, Reed, Jorge Ardlin, Andres Rocha, and Carlos Trujillo. "Productivity of Agricultural Research in Colombia," in Thomas M. Arndt, Dana Dalrymple, and Vernon W. Ruttan, eds., *Resource Allocation and Productivity in National and International Agricultural Research* (Minneapolis, Minn., University of Minnesota Press, 1977).

Hill, R. D. *Rice in Malay: A Study in Historical Geography* (Kuala Lumpur, Oxford University Press, 1977).

Hirashina, S. *Hired Labor in Rural Asia* (Tokyo, Institute of Developing Economies, 1977).

Ho, Ping-ti. "Early Ripening Rice in Chinese History," *The Economic History Review* vol. 9 (1956) pp. 200–218.

———. "Loess and the Origin of Chinese Agriculture," *The American Historical Review* vol. 75 (October 1969) pp. 1–35.

Ho, Robert. "Farmers of Central Malaya," Department of Geography Publication G/4, Research School of Pacific Studies (Canberra, The Australian National University, 1967).

Holle, K. F. *Padi-productie van Java en Madoera (met enne graphische voorstelling) (Paddy production in Java and Madura with a graph)*, Tijdschrift voor Nijverheid en Landbouw in Nederlandsch-Indie, 1982.

Hooley, Richard "An Assessment of the Macroeconomic Policy Framework for Employment Generation in the Philippines," a report submitted to U.S. Agency for International Development/Philippines, April 1981.

Hoon, K. Lee. *Land Utilization and Rural Economy in Korea* (Westport, Conn., Greenwood Press, Publishers, 1969).

Hopper, W. David. "Main Springs of Agricultural Growth in India," *Indian Journal of Agricultural*

Science vol. 35 (June 1965) pp. 3–28.

Hoq, M. N., ed. *Exploitation and the Rural Poor* (Camilla, Bangladesh, Bangladesh Academy of Rural Development, 1976).

Howard, Albert. *Crop Production in India, a Critical Survey of its Problems* (London, Oxford University Press, 1924).

Hsieh, Sam-Chung, and T. H. Lee. "Agricultural Development and its Contributions to Economic Growth in Taiwan—Input–Output and Productivity Analysis of Taiwan Agricultural Development," Economic Digest Series, No. 17 (Taipei, Joint Commission on Rural Reconstruction, 1966).

———. "An Analytical Review of Agricultural Development in Taiwan—An Input–Output and Productivity Approach," Economic Digest Series, No. 12 (Taipei, Joint Commission on Rural Reconstruction, 1958).

Hseih, Sam-Chung, and Vernon W. Ruttan. "Environmental, Technical and Institutional Factors in the Growth of Rice Production: Philippines, Thailand, and Taiwan,"*Food Research Institute Studies* vol. 7 (1967) pp. 307–341.

Huang, C. H., W. L. Chang, and Te-tzu Chang. "Ponlai Varieties and Taichung Native 1" in International Rice Research Institute, *Rice Breeding* (Los Banos, Philippines, IRRI, 1972).

Huang, C. S. "Evolution of Rice Culture in Taiwan," JCRR/PID-SC 37 (Taipei, Chinese-American Joint Commission on Rural Reconstruction, 1970) mimeo.

Hufbauer, G. C. "Cereal Consumption, Production and Prices in West Pakistan," *Pakistan Development Review* vol. 7, no. 2 (1968) pp. 292–293.

Huke, Robert E. *Rice Area by Type of Culture: South, Southeast, and East Asia* (Los Banos, Philippines, International Rice Research Institute, 1982).

———. "Geography and Climate of Rice," in *Proceedings of the Symposium on Climate and Rice* (Los Banos, Philippines, International Rice Research Institute, 1976) pp. 31–50.

———. *Shadows on the Land: An Economic Geography of the Philippines* (Manila, Bookmark Inc., 1963).

Hung, G. Nguyen Tien. *Economic Development of Socialist Vietnam 1955–80* (New York, Praeger Publishers, 1977).

Ignatius, J. G. W. "Rapport sur l'enquête par sondage sur la superficie agricole et la production du paddy au sud et au centre Viet-Nam campagne 1959–60," (Rome, Food and Agriculture Organization, n.d.).

Ihalauw, John, and Widya Utami. "Klaten, Central Java," in *Changes in Rice Farming in Selected Areas of Asia* (Los Banos, Philippines, International Rice Research Institute, 1975) pp. 149–177.

India, Central Statistical Organization. *Statistical Abstract of India* (New Delhi).

India, Department of Statistics. *Prices and Wages in India* (Calcutta, Sup't Gov't Printing, India, 1922).

India, Directorate of Economics and Statistics. *Abstracts of Agricultural Statistics* (Delhi, Manager of Publications).

———. *Agricultural Prices in India* (New Delhi).

———. *Agricultural Situation in India* (New Delhi, Controller of Publications).

———. *Agricultural Wages in India* (New Delhi).

———. *Bulletin of Food Statistics* (New Delhi).

———. *Estimates of Area and Production of Principal Crops in India* (New Delhi, Controller of Publications).

———. *Farm (Harvest) Prices of Principal Crops in India* (New Delhi).

———. *Guide to Current Agricultural Statistics* (New Delhi, 1962).

India, Directorate of Economics and Statistics and the Statistical Advisor, Institute of Agricultural Research Statistics, Indian Council of Agricultural Research. *Handbook on Methods of Collection of Agricultural Statistics in India* (New Delhi, 1959).

India, Directorate of Marketing and Inspection. *Report on the Marketing of Rice in India* (Calcutta, Government of India Press, 1954).

India, Expert Committee on Assessment and Evaluation, Ministry of Food Agriculture, Commodity Development and Cooperation. *Modernizing Indian Agriculture: Report on the Intensive Agricultural District Program 1960–68* (New Delhi, Government of India, 1969).

India, Ministry of Food and Agriculture. "India's Food Crisis and Steps to Meet It" (New Delhi, Government of India, 1959).

India, National Council of Applied Economic Research. *Fertilizer Demand Study: Interim Report* (in six volumes) (New Delhi, National Council of Applied Economic Research, 1978).

———. "Implications of Tractorization on Farm Employment, Productivity and Income: Survey and Highlights" (New Delhi, n.d.).

India, National Sample Survey Organization. *Tables on Consumer Expenditures, Twenty-eighth Round, October 1973–June 1974*, no. 240 (New Delhi, Department of Statistics, 1977).

Indochina, Inspection Générale de l'agriculture de l'élevage et des fôrets. "Riziculture en Indochine" (1931).

Indochina, Inspection Générale des Mines et d l'Industrie. *Statistique générale de l'Indochine retrospectif 1913–1929* (Hanoi, Inprimerie d' Extrême-Orient, 1931).

Indochina, Service de la Statistique Générale, *Annuaire statistique de l'Indochine* (Hanoi, Imprimerie d'Extrême-Orient).

———. *Bulletin économique de l'Indochine* (Saigon, began publication in 1898).

———. *Bulletin Statistique de l'Indochine* (Saigon).

———. *Resumé statistique relatif aux annés, 1913–1940* (Hanoi, Imprimerie d'Extrême-Orient, 1941).

Indonesia, Badan Urusan Logistik. *Pedoman pelakasanoan pengadaan dalam negeri tahun* (Jakarta).
Indonesia, Biro Pusat Statistik. *Indikator ekonomi* (Jakarta).
———. "Kompilasi data output and input usah tani padi intensifikasi per kabupaten di Java-Madura" (Jakatara).
———. *Perkembangan bulanan hargu eceran bahan makanan pokok and bohan penting lainnyadi ibukota propinsi Indonesia* (Monthly Prices Series, Retail Prices of the Essentials at Provincial Capital Cities of Indonesia) (Jakarta).
———. *Produksi tanaman bohan makanan di Indonesia* (Jakarta).
———. *Statistical Pocketbook of Indonesia* (Jakarta).
———. *Survey pertanian* (Jakarta).
Indonesia, Departement Pertanian Bogor. *Rice Bibliography of Indonesia 1842–1971* (Jakarta, underwritten by the Ford Foundation).
Ingram, James C. *Economic Change in Thailand, 1850–1970* (Stanford, Calif., Stanford University Press, 1971).
———. "Thailand's Rice Trade and Resources," in C. D. Cowan, ed., *The Economic Development of Southeast Asia Studies in Economic History and Political Economy* (New York, Praeger, 1964).
Institute of Developing Economies. "Foreign Trade Statistics of Japan 1951–1965, Time Series by Commodity," IDE Statistical Data Series No. 2 (Tokyo, 1972).
———. *One Hundred Years of Agricultural Statistics in Japan* (Tokyo, Kabushiki Kaisha Sangyo Tokei Kenkyusha, 1969).
Institute of Development Studies, University of Sussex. *Three Papers on Food and Nutrition: The Problem and the Means of Solution* (Brighton, England, 1971).
Intachaisri, Jumrush. "Survey Report, Thailand" (Tokyo, Asian Productivity Organization Symposium on Farm Mechanization, November 24–30, 1981).
Intachaisri, Jamrush, and Somnuk Pradithavanij. "Thailand," in *Impact of Fertilizer Shortage: Focus on Asia* (Asian Productivity Organization, 1975).
International Crop Research Institute for the Semi-arid Tropics. *Proceedings of the International Workshop on Socioeconomic Constraints to Development of Semi-Arid Tropical Agriculture* vol. 19, no. 23 (February 1979).
International Food Policy Research Institute. "Criteria and Approaches to the Analysis of Priorities for International Agricultural Research," IFPRI Working Paper No. 1 (Washington, D.C., 1978).
———. "Food Needs in the Developing Countries; Projections of Production and Consumption to 1990," IFPRI Research Paper No. 3 (Washington, D.C., 1977).
International Labour Organization, ed. *Labour Absorption in Agriculture: The East Asian Experience* (Geneva, ILO, 1980).
———, *Poverty and Landlessness in Rural Asia*, (Geneva, ILO, 1977).
International Rice Research Institute. *Annual Report for 1982* (Los Banos, Philippines, IRRI, 1983).
———. *Brown Planthopper: Threat to Rice Production in Asia* (Los Banos, Philippines, IRRI, 1979).
———. *Changes in Rice Farming in Selected Areas of Asia* (Los Banos, Philippines, IRRI, 1975).
———. *Climate and Rice* (Los Banos, Philippines, IRRI, 1976).
———. *Constraints to High Yields on Asian Rice Farms: An Interim Report* (Los Banos, Philippines, IRRI, 1978).
———. *Cropping Systems Research and Development for the Asian Rice Farmer* (Los Banos, Philippines, IRRI, 1977).
———. *Economic Consequences of the New Rice Technology* (Los Banos, Philippines, IRRI, 1978).
———. *Farm-Level Constraints to High Rice Yields in Asia: 1974–77* (Los Banos, Philippines, IRRI, 1979).
———. *Innovative Approaches to Rice Breeding* (Los Banos, Philippines, IRRI, 1980).
———. *International Deepwater Rice Workshop* (Los Banos, Philippines, IRRI, 1978).
———. *Interpretative Analysis of Selected Changes in Rice Farming in Selected Areas of Asia* (Los Banos, Philippines, IRRI, 1978).
———. *Interpretive Analysis of Selected Papers from Changes in Rice Farming in Selected Areas of Asia* (Los Banos, Philippines, IRRI, 1978).
———. *Irrigation Policy and Management in Southeast Asia* (Los Banos, Philippines, IRRI, 1978).
———. "Loop Survey, 1980," unpublished data.
———. *A Plan for IRRI's Third Decade* (Los Banos, Philippines, 1982).
———. *Priorities for Alleviating Soil Related Constraints to Food Production in the Tropics* (Los Banos, Philippines, IRRI and Cornell University, 1980).
———. Proceedings of the Symposium on Climate and Rice (Los Banos, Philippines, IRRI, 1976).
———. *Proceedings of the Workshop on Deepwater Rice* (Los Banos, Philippines, IRRI, 1977).
———. *Rainfed Lowland Rice* (Los Banos, Philippines, IRRI, 1979).
———. *Rice Breeding* (Los Banos, Philippines, IRRI, 1972).
———. *Rice Research and Production in China: An IRRI Team's View* (Los Banos, Philippines, IRRI, 1978).
———. *Rice Research Strategies for the Future* (Los Banos, Philippines, IRRI, 1978).
———. *Soils and Rice* (Los Banos, Philippines, IRRI, 1978).
———. "The Technical and Economic Characteristics of Rice Post Production Systems in the Bicol River Basin" (Los Banos, Philippines, IRRI and University of the Philippines at Los Banos, November 1978).
———. *Viewpoints on Rice Policy in Asia* (Los Banos, Philippines, IRRI, 1971).

International Rice Research Institute and Chinese Academy of Agricultural Sciences. *Rice Improvement in China and Other Asian Countries* (Los Banos, Philippines, IRRI, 1980).

Isamu, Baba. "Breeding of Rice Varieties Suitable for Heavy Manuring," in *Report for the Fifth Meeting of the International Rice Commission's Working Party on Rice Breeding* (Tokyo, Ministry of Agriculture and Forestry, 1954) pp. 167–185.

Ishii, Yoneo, ed. *Thailand: A Rice Growing Society*, Trans. Peter and Stephanie Hawkes (Honolulu, The University Press of Hawaii, 1978).

Ishikawa, Shigeru. *Essays on Technology, Employment, and Institutions in Economic Development*, Economic Research Series No. 19 (Tokyo, Kinokuniya Co., 1981).

———. *Labour Absorption in Asian Agriculture* (Bangkok, International Labor Organization, 1978).

———. *Economic Development in Asian Perspective* (Tokyo, Kinokuniya Co., 1967).

Isobe, Toshihiko. "Land Reform's Achievements and Limits, the Case of Japan," Symposium on Institutional Innovations and Reform (Kyoto, The Ladejinsky Legacy, Kyoto International Center, October 1977).

Isrankura, Vanrob. "A Study on Rice Production and Consumption in Thailand," 1-54 (Bangkok, Ministry of Agriculture, Division of Agricultural Economics, 1966).

Ithachat, Vichion. "Rice Premium and Its Administration" (M.A. thesis, Thammasat University, 1961).

Izumi, K., and A. S. Rantaunga. "Costs of Production of Paddy Maha 1972/73," Agrarian Research and Training Institute, Study No. 12 (Colombo, April 1974).

Jabbar, M.A., S. R. Bhuiyan, and A. K. Maksudul Bari. "Causes and Consequences of Power Tiller Utilization in Two Areas of Bangladesh" (Los Banos, Philippines, Agricultural Development Council, International Rice Research Institute Workshop on the Consequence of Small Rice Farm Mechanization, September 14–18, 1981).

Jacoby, Erich H. *Agrarian Unrest in Southeast Asia* (New York, Columbia University Press, 1949).

James, W. E. "An Economic Analysis of Land Settlement Alternatives in the Philippines," IRRI Agricultural Economics Paper 78-30 (Los Banos, Philippines, International Rice Research Institute, 1978).

Japan, Bureau of Statistics. *Annual Report on the Family Income and Expenditures Survey, 1978* (Tokyo, Prime Minister's Office, 1979).

———. *Dai Nihon teikoku tōkei nenkan* (Statistical Yearbook of the Japanese Empire) (Tokyo, Prime Minister's Office).

———. *Monthly Statistics of Japan* (Tokyo, Prime Minister's Office).

———. *Nihon tōkei nenkan* (Japan Statistical Yearbook) (Tokyo, Prime Minister's Office, 1949).

———. *Statistical Handbook of Japan* (Tokyo, Prime Minister's Office, annual).

Japan, Cabinet Impérial, Bureau de la Statistique Générale (Naikaku Tokei Kyoku). *Résumé statistique de l'empire du Japan* (Tokyo).

Japan External Trade Organization (JETRO). *White Paper on International Trade* (Tokyo, JETRO).

Japan, Food Agency. *Shokuryō kanri tōkei nenpō* (Statistical Yearbook of Food Control) (Tokyo).

Japan, General Headquarters, Supreme Commanders for the Allied Powers, Economic and Scientific Section, Research and Programs Division. "Staple Food Prices in Japan, 1930–1948" (April 1949).

Japan, Ministry of Agriculture and Forestry. *Sakumotsu tōkei* (Crop Statistics) (Tokyo).

———. *The Statistical Abstract of the Department of Agriculture and Forestry* (Tokyo).

Japan, Ministry of Agriculture, Forestry and Fisheries. *Abstract of Statistics on Agriculture, Forestry and Fisheries* (Tokyo).

———. *Norinsuisansho tōkei hyo* (Statistical Yearbook of the Ministry of Agriculture, Forestry and Fisheries) (Tokyo).

———. *Survey on Production Cost of Rice* (Tokyo, annual).

Japan, Section of Statistics, The Department of Agriculture and Commerce. *The Agricultural and Commercial Statistics* (Tokyo, annual).

Japan, Statistics and Survey Division, Ministry of Agriculture and Forestry. *Rice Statistical Compilation* (Tokyo, 1954).

Jayasuriya, S. K., Amanda Te, and Robert W. Herdt. "Mechanization and Cropping Intensification: Economic Viability of Power Tillers in the Philippines," IRRI Saturday Seminar Paper (Los Banos, Philippines, International Rice Research Institute, October 9, 1982).

Jennings, Peter R. "Plant Type as a Rice Breeding Objective," *Crop Science* vol. 4 (1964) pp. 13–15.

Johnston, Bruce F., and Peter Kilby. *Agricultural Strategies, Rural-Urban Interactions and the Expansion of Income Opportunities* (Paris, Organization for Economic Cooperation and Development, Development Center, 1973).

Joint Commission on Rural Reconstruction, Rural Economics Division. "Farm Report in Taiwan," Economic Digest Series No. 1 (Taipei, 1952).

———. "Food Administration in Taiwan," Economic Digest Series: No. 3 (Taipei, 1953).

———. "Rice Marketing in Taiwan," Economic Digest Series No. 7 (Taipei, 1955).

———. "Taiwan Agricultural Statistics, 1961–1975," Economic Digest Series No. 22 (Taipei, 1977).

———. "Taiwan Agricultural Statistics 1901–1965," Economic Digest Series No. 18 (Taipei, 1966).

Jones J. W. "Hybrid Vigor in Rice," *Journal of the American Society of Agronomy* vol. 18 (May 1926) pp. 423–428.

Jones, T. R. Frazier, and W. Henning, Jr. "An Evaluation of the Agricultural Development of Laos" (Columbus, Ohio, Ohio State University, Department of Agricultural Economics and Rural Sociology, Agricultural Finance Center, 1969).

Judd, M. Ann, James K. Boyce, and Robert E. Evenson. "Investing in Agricultural Supply" (New Haven, Conn., Economic Growth Center, Yale University, n.d.) mimeo.

Kalta, O. P. *Agricultural Policy in India* (Bombay, Popular Prakashan, 1973).

Kampuchea, Institut National de la Statistique et des Recherches Economiques, Commissariat du Plan. *Annuaire statistique* (Phnom Penh).

Kanela, Hiromitsu. "Measurement of Labor Inputs: Data and Methods," in Yujiro Hayami, Vernon W. Ruttan, and Herman M. Southworth, eds., *Agricultural Growth in Japan, Taiwan, Korea, and the Philippines* (Honolulu, University of Hawaii Press, 1979).

Karamyshev, V. P. *Sel'skoe khoziaistvo demokraticheskoi respubliki V'etnam* (Agriculture in the Democratic Republic of Vietnam) (Moscow, Gosudarstvennoe izdatel'stvo sel'skokhoziaistvennoi literatury, 1959).

Kasryno, Faisal. "Technological Progress and Its Effects on Income Distribution and Employment in Rural Areas: A Case Study in Villages in West Java, Indonesia," Agro-Economics Survey—Rural Dynamics Study (Bogor, Indonesia, 1981).

Kathirmakathamby, S. "Survey Report, Sri Lanka" (Tokyo, Asian Productivity Organization Symposium on Farm Mechanization, November 24–30, 1981).

Kato, Seizo. *Kankoku nōgyo ron* (A Discussion on Korean Agriculture) (Tokyo, 1904).

Kawano, Shigeto. *Taiwan beikoku keizai ron* (Treatise on the Rice Economy of Taiwan) (Tokyo, Yuhikaku, 1941).

Kazushi, Ohkawa, and Henry Rosovsky. "A Century of Japanese Economic Growth," in William W. Lockwood, ed., *The State of Economic Enterprise in Japan* (Princeton, N.J., Princeton University Press, 1965).

Keightley, David. *The Origins of Chinese Civilization* (Berkeley, Calif., University of California Press, 1982).

Keizo, Tsuchiya. *Productivity and Technological Progress in Japanese Agriculture* (Tokyo, University of Tokyo Press, 1976).

Kelly, William A. "Japanese Social Science Research on Irrigation Organization: A Review" (New Haven, Conn., Yale University, 1980).

Khalon, A. S., and G. Singh. "Social and Economic Implications of Large-Scale Introduction of High-Yielding Varieties of Rice in the Punjab with Special Reference to the Gurdaspur District" (Ludhania, India, Department of Economics and Sociology, Punjab Agricultural University, 1973).

———. "Social and Economic Implications of Large-Scale Introduction of High-Yielding Varieties of Wheat with Reference to Ferozepur District, Punjab" (Ludhiana, India, Punjab Agricultural University, 1973).

Khan, Mahmood Hasan. *The Economics of the Green Revolution* (New York, Praeger, 1975).

———. *The Role of Agriculture in Economic Development: A Case Study of Pakistan* (Wageningen, Centre for Agricultural Publications and Documentation, 1966).

Khush, Gurdev S., and Henry M. Beachell. "Breeding for Disease and Insect Resistance at IRRI," in International Rice Research Institute, *Rice Breeding* (Los Banos, Philippines, IRRI, 1972).

Kikuchi, Masao. "Irrigation and Rice Technology in Agricultural Development, A Comparative History of Taiwan, Korea, and the Philippines" (Ph.D. dissertation, University of Hokkaido, 1975).

Kikuchi, Masao, Geronimo Dozina, Jr., and Yujiro Hayami. "Economics of Community Work Programs: A Communal Irrigation Project in the Philippines," *Economic Development and Cultural Change* vol. 26 (January 1978) pp. 211–225.

Kikuchi, Masao, Fe B. Gascon, Luisa M. Bambo, and Robert W. Herdt. "Changes in Technology and Institutions for Rice Farming in Laguna, Philippines, 1966–1981: A Summary for Five Laguna Surveys," IRRI Agricultural Economics Department Paper 82-22 (Los Banos, Philippines, International Rice Research Institute, 1982).

Kikuchi, Masao, Keizo Mochida, and Yujiro Hayami. "Rice Statistics in Japan" (Los Banos, Philippines, International Rice Research Institute, 1975).

King, F. "Evaluation of the CB: IBRD Agricultural Credit Program in the Philippines" (Washington, D.C., World Bank, Operations Evaluation Department, 1974).

Kirsch, Thomas, and William Skinner. *Change and Persistence in Thai Culture* (Ithaca, N.Y., Cornell University Press, 1975).

Kisu, M. "Mechanization of Rice Farming in Japan" (Tokyo, Agricultural Productivity Organization Symposium on Farm Mechanization, November 24–30, 1981).

Klein, Lawrence, and Kazushi Ohkawa, eds. *Economic Growth: The Japanese Experience Since the Meiji Era* (Homewood, Ill., Richard D. Irwin, 1968).

Korea, Republic, Ministry of Agriculture and Forestry. *Food Crop Statistics* (Seoul).

———. *Report on the Results of Farm and Household Economy Survey* (Seoul).

———. *Yearbook of Agriculture and Fishery Statistics* (Seoul).

Korea, Republic, The National Agricultural Cooperative Federation. *The Agricultural Cooperatives Survey* (Seoul).

Korea, Republic, Office of Rural Development. *Success in the Green Revolution 1964–1977* (Suewon, Korea, Office of Rural Development, Ministry of Agriculture and Fisheries, Republic of Korea, n.d.).

Korten, Frances, F. *Building Rational Capacity to*

Develop Water Users Associations, World Bank Staff Working Paper No. 58 (Washington, D.C., World Bank, 1982).

Krishna, Raj. "Some Aspects of Agricultural Growth, Price Policy and Equity in Developing Countries," *Food Research Institute Studies* vol. 18 (1982) pp. 219–260.

———. "The Role of the Government in Agricultural Marketing Reform," *Review of Agricultural Economics: Malaysia* vol. 1, no. 2 (1967).

———. "The Marketed Surplus of Food Grains: Is it Inversely Related to Price?" *Economic and Political Weekly* vol. 17 (1965) pp. 325–328.

———. "A Note on the Elasticity of the Marketable Surplus of a Subsistence Crop," *Indian Journal of Agricultural Economics* vol. 17 (1962) pp. 79–84.

Kudo, Zyuro. "Implications of Farm Management Research for Government Mechanization Programs," in Herman Southworth, ed., *Farm Mechanization in East Asia* (New York, Agricultural Development Council, 1972).

Kulkarni, Vijay Ganesh. *Statistical Outline of the Indian Economy* (Bombay, Vora and Co. Publishers, 1968).

Kuo, Lesile T. C. *Agriculture in the People's Republic of China, Structural Changes and Technical Transformations* (New York, Praeger, 1976).

Kuwahara, Masanobu. *Economic Approaches to Japanese Agriculture* (Tokyo, Fuji Publishing Company Ltd., 1969).

Kyoto University, Center for Southeast Asian Studies. *Rice Culture in Malaya*, Symposium Series No. 1 (Kyoto, Kyoto University, The Center for Southeast Asian Studies, 1965).

Ladesma, Antonio J. *Landless Workers and Rice Farmers: Peasant Subclasses Under Agrarian Reform in Two Philippine Villages* (Los Banos, Philippines, International Rice Research Institute, 1982).

Lai, Weng-chieh. "Current Problems of Farm Management on Mechanized Farms," in Herman Southworth, ed., *Farm Mechanization in East Asia* (New York, Agricultural Development Council, 1972).

Lai, Wen-hui. "Trends in Agricultural Employment in Post-war Taiwan," China Council of Sino-American Cooperation in Humanities and Social Sciences, Sino-American Conference on Manpower in Taiwan (Taipei, Academica Sinica, 1972) pp. 127–134.

Lains, Alfian. "Regional Concentration of Rice Production in Indonesia" (Ph.D. dissertation, University of the Philippines, 1978).

Langham, Max R., and Ralph H. Retzlofeds, eds. *Agricultural Sector Analysis in Asia* (Bangkok, ADC, 1982).

Laos, Ministère du Plan, Service de la Statistique du Laos. *Annuaire statistique du Laos 1953–1957* (Vientiane, 1961).

Laos, Service National de la Statistique. *Bullétin de statistique du Laos* (Vientiane) quarterly, 1959–1969.

———. *Statistiques essentielles* (Vientiane).

Lardy, Nicholas. "Food Consumption in the People's Republic of China" (New Haven, Conn., Yale University, Department of Economics, 1980) mimeo.

Leaf, Murray J., "The Green Revolution and Cultural Change in a Punjab Village, 1965–1978," *Economic Development and Cultural Change* vol. 31 (1983) pp. 227–270.

Lee, B. N. "Country Report, Korea," report presented at Women in Rice Farming Systems Conference, Los Banos, Philippines, September 26–30, 1983.

Lee, C. C. "Economic and Engineering Aspects of Mechanization of Rice Harvesting in Korea," Herman Southworth, ed., *Farm Mechanization in East Asia* (New York, Agricultural Development Council, 1972).

Lee, Teng-hui. "Government Interference in the Rice Market in Taiwan," in International Rice Research Institute, *Viewpoints in Rice Policy in Asia* (Los Banos, Philippines, IRRI, 1971) chapter 4.

———. "Intersectoral Capital Flows in the Economic Development of Taiwan, 1895–1960" (Ph.D. dissertation, Cornell University, 1968).

Lee, Teng-hui, Hsi-huang Chen, and Yueh-eh Chen. "Labour Absorption in Taiwan Agriculture," in International Labor Organization, *Labour Absorption in Agriculture the East Asian Experience* (Bangkok, ILO, 1980) pp. 167–236.

Lee, Teng-hui, and Yueh-eh Chen. "Growth Rates of Taiwan Agriculture 1911–1972," *Economic Digest Series*, No. 21 (Taipei, Joint Commission on Rural Reconstruction, 1975).

Lee, Teng-hui, and T. H. Shen. "Agriculture as a Base for Socio-Economic Development," in T. H. Shen, ed., *Agriculture's Place in the Strategy of Development: The Taiwan Experience* (Taipei, Joint Commission on Rural Reconstruction, 1974).

Lele, Uma. *The Design of Rural Development* (Baltimore, Md., Johns Hopkins University Press, 1975).

———. *Food Grain Marketing in India: Private Performance and Public Policy* (Ithaca, N.Y., Cornell University Press, 1971).

Leontief, Wassily, W. "The World Economy in the Year 2000," *Scientific American* vol. 243, no. 3 (September 1980) pp. 206–231.

Levine, Gilbert. "Perspectives on Integrating Findings from Research on Irrigation Systems in Southeast Asia," no. 26 (New York, Agricultural Development Council, Teaching and Research Forum, 1982) pp. 9–15.

Li, Choh-ming. *The Statistical System of Communist China* (Berkeley, Calif., University of California Press, 1962).

Li, Dun J. *British Malaya: An Economic Analysis* (New York, The American Press, 1955).

Lim, Chong-yah. *Economic Development of Modern Malaya* (London, Oxford University Press, 1967).

Lim, Peter, and M. P. Nathan. *Basic Readings in Malayasian Economics* (Kuala Lumpur, Modern Education Publishers, 1969).

Lin, Shih-cheng, and Loung-ping Yuan. "Hybrid Rice Breeding in China," in International Rice Research

Institute, *Innovative Approaches to Rice Breeding* (Los Banos, Philippines, IRRI, 1980).

Liu, Twanmo, Yeh. *Production of Food Crops on the Chinese Mainland: Prewar and Postwar* (Santa Monica, Calif., Rand Corporation, 1964).

Lockwood, B., P. K. Mukherjee, and R. T. Shand. *The High-Yielding Varieties Programme in India, Part I* (Canberra, Planning Commission, Government of India and the Australian National University, 1971).

Lockwood, William W., ed. *The State of Economic Enterprise in Japan* (Princeton, N.J., Princeton University Press, 1965).

———. *The Economic Development of Japan: Growth and Structural Change, 1868-1938* (Princeton, N.J., Princeton University Press, 1954).

Lu, Jonathan J., and Te-tzu Chang. "Rice in its Temporal and Spatial Perspectives," in B. S. Luh, ed., *Rice: Production and Utilization* (Westport, Conn., AVI, 1980).

Luh, B. S., ed. *Rice: Production and Utilization* (Westport, Conn., AVI, 1980).

Malay, Federated Malay States, Federated Malay States Government Press. *Manual of Statistics Related to the Federated Malay States* (Kuala Lumpur, 1921).

Malay, Federation of Malay, Department of Agriculture. *Malayan Agricultural Statistics*, Economic Series Nos. 1-15 (Kuala Lumpur, Caxton Press, Ltd.).

———. *Monthly Statistical Bulletin* (Kuala Lumpur).

———. *Monthly Statistical Bulletin of Federation of Malay—Agricultural Supplement 1953* (Kuala Lumpur).

Malaysia, Department of Statistics Jabatan Perangkaan. *Buku kecil perangkaan bagi semenanjung Malaysia* (Statistical Handbook of Peninsular Malaysia) (Kuala Lumpur).

———. *Household Budget Survey of the Federation of Malaya 1957–58* (Kuala Lumpur, n.d.).

———. *Siaran perangkaan tahuman* (Annual Bulletin of Statistics) (Kuala Lumpur).

Malaysia, Sabah, Department of Agriculture. *Agricultural Statistics of Sabah* (Kota Kinabalu, Agricultural Information Division).

———. *Annual Report of the Department of Agriculture* (Kota Kinabalu).

Malaysia, Sabah, Federal Department of Information. *Sabah Annual Report* (Singapore, Eurasia Press).

Malaysia, Sabah, Jabatan Perangkaan (Department of Statistics). *Buku maklumat perangkaan* (Statistical Handbook of Sabah) (Kota Kinabalu).

———. *Perangkaan bulanan* (Monthly Statistics, Sabah) (Kota Kinabalu).

———. *Siaran perangkaan tahunan* (Annual Bulletin of Statistics, Sabah) (Kota Kinabalu).

Malaysia, Sarawak, Jabatan Perangkaan (Department of Statistics), *Buku maklumat perangkaan* (Statistical Handbook of Sarawak) (Kuching).

———. *Perangkaan perdajangan luar Sarawak* (Statistics of External Trade) (Kuching).

Malaysia, Jabatan Pertanian (Department of Agriculture). *Perangkaan pertanian Sarawak* (Agricultural Statistics of Sarawak) (Kuching).

———. *Siaran perangkaan buku pertama* (Quarterly Bulletin of Statistics) (Kuching).

———. *Siaran perangkaan tahunan* (Annual Statistical Bulletin) (Kuching).

Malcolm, John Purvis. "Evaluation and Use of Underdeveloped Agricultural Statistics: The Food Economy of Malaysia" (Ph.D. dissertation, Cornell University, 1966).

Mandac, Abraham M., and Robert W. Herdt. "Environmental and Management Constraints to High Rice Yields in Nueva Ecija, Philippines," IRRI Agricultural Economics Department Paper 79-03 (Los Banos, Philippines, International Rice Research Institute, 1979).

Mandal, G. C., and M. G. Ghosh. *Economics of the Green Revolution: A Study in East India* (London, Asia Publishers, 1976).

———. "Social and Economic Implications of the Large-Scale Introduction of High-Yielding Varieties of Foodgrains in the Eastern Region of India" (Santinikean, India, Agro-economic Research Center, 1973).

Mangahas, Mahar. "The Political Economy of Rice in the New Society," *Food Research Institute Studies* vol. 14, no. 3 (1975) pp. 295–309.

Mangahas, Mahar, and Aida R. Libero. "Study on the Social and Economic Effects of the Introduction of HYV's: Part II" (Geneva, UN Research Institute for Social Development, 1973).

———. "The High-Yielding Varieties of Rice in the Philippines: A Perspective," discussion paper No. 73-11 (Los Banos, Philippines: Institute of Economic Development and Research, School of Economics, University of the Philippines, June 1973).

Mangahas, Mahar, Virginia A. Miralao, and Romana P. de los Reyes. *Tenants, Lessees, Owners: Welfare Implications of Tenure Change* (Quezon City, Philippines, Institute of Philippine Culture, 1974).

Mangahas, Mahar, A. E. Recto, and V. W. Ruttan. *Production and Marketing Relationships for Rice and Corn in the Philippines* (Los Banos, Philippines, International Rice Research Institute, n.d.).

Manyanondh, Ruangrai. "Effect of Trade Taxes on Rice Farmers' Real Income" (M.A. thesis, Thammasat University, 1973).

Mark, P. H. "Economic Policy and Agricultural Development in Indonesia" (Ph.D. dissertation, University of California, 1977).

Marzouk, G. A. *Economic Development and Policies Case Study of Thailand* (Rotterdam, Rotterdam University Press, 1972).

Masefield, G. B. *A History of the Colonial Agricultural Service* (Oxford, Clarendon Press, 1972).

Matsuo, Takane. *Rice Culture in Japan* (Tokyo, Ministry of Agriculture and Forestry, 1954).

Maurer, Jean-Lue. "Some Consequences of Land Shortage in Four Kelurahan of the Kabupaten Bantul" (Yogyakarta, Indonesia, Gadja Mada University, 1978) mimeo.

McCalla, Alex F. *Agricultural and Food Policy Issues Analysis: Some Thoughts from an International Perspective* (Washington, D.C., International Food Policy Research Institute, 1978).

Mears, Leon. *The Rice Economy of Indonesia* (Yogyakarta, Indonesia, Gadja Mada University Press, 1982).

———. "Relationship of Rice Marketing to Rice Production in the Philippines," discussion paper no. 70-19 (Los Banos, Philippines, Institute of Economic Development and Research, School of Economics, University of the Philippines, 1970).

———. "Historical Development of Rice Marketing in the Philippines," discussion paper No. 67-14 (Los Banos, Philippines, Institute of Economic Development and Research, School of Economics, University of the Philippines, 1967).

———. *Rice Marketing in the Republic of Indonesia* (Djakarta, P. T. Pembangunan, 1961).

Mears, Leon, and Saleh Affif. "The Bimas Program and Rice Production in Indonesia Revisited," discussion paper no. 68-20, (Los Banos, Philippines, Institute of Economic Development and Research, University of the Philippines, School of Economics, June 14, 1968).

Mears, Leon, M. H. Agabin, T. L. Anden, and R. C. Marquez. *The Rice Economy of the Philippines* (Quezon City, Philippines, University of the Philippines Press, 1974).

Mears, Leon, and Teresa Anden. "Rice Prices and Rice Price Policy," discussion paper No. 71-19 (Los Banos, Philippines, Institute of Economic Development and Research, School of Economics, University of the Philippines, 1971).

———. "Rice Price Policy," discussion paper No. 71-22 (Los Banos, Philippines, Institute of Economic Development and Research, School of Economics, University of the Philippines, November 1971).

Mehra, Shakuntla. *Instability in Indian Agriculture in the Context of New Technology*, Research Report No. 25 (Washington, D.C., International Food Policy Research Institute, 1981).

———. "Some Aspects of Labour Use in Indian Agriculture," Occasional Paper No. 88 (Ithaca, N.Y., Cornell University, 1976).

Mellor, John W. *The Economics of Agricultural Development* (Ithaca, N.Y., Cornell University Press, 1966).

Mellor, John, and Ashok Dar. "Determinants and Development Implications of Foodgrain Prices in India, 1949–1964," *American Journal of Agricultural Economics* vol. 50, no. 4 (November 1968) pp. 962–974.

Mendiola, Nemesio. *Rice Culture in the Philippines* (Rizal, Philippines, The Farmer's Guide Publishing Company, 1953).

Metz, Joseph F., and Nyle C. Brady. "Foreword" to *Priorities for Alleviating Soil Related Constraints to Food Production in the Tropics* (Los Banos, Philippines, International Rice Research Institute and Cornell Unversity, 1980).

Ming, Kong Yim. "A Study into the Sources of Agricultural Growth in West Malaysia"(Los Banos, Philippines, Department of Economics, University of the Philippines, 1976/77).

Mizoguchi, Toshiyuki. *Taiwan Chōsen no keizai seicho: bukka tokei o chushin toshite* (Economic Growth in Taiwan and Korea with Special Reference to Price Statistics) (Tokyo, Iwanami, 1975).

Moermann, Michael. *Agricultural Change and Peasant Choice in a Thai Village* (Los Angeles, University of California, 1968).

Mongkolsmai, Dow. "Status and Performance of Irrigation in Thailand," Rice Policies in Southeast Asia Project, Working Paper No. 8 (Washington, D.C., International Food Policy Research Institute, International Fertilizer Development Center, International Rice Research Institute, 1983).

Monke, Eric A., Scott R. Pearson, and Narongchai Akaransansee. "Comparative Advantage, Government Policies, and International Trade in Rice," *Food Research Institute Studies* vol. 9 (1976) pp. 257–283.

Montano, Carl B., and Randolph Barker. "Economic Returns from Fertilizer Application in Tropical Rice in Relation to Solar Energy Level," *The Philippine Economic Journal* vol. 13, no. 1 (1974) pp. 27–40.

Moon, Pal Yong, and Byong Seo Yoo. "A Review of Crop Production Estimates and Key Statistics Related to Grain Policy (Korea)" (Seoul, Korea Development Institute, 1974).

Moorman, Frans R., and Nico van Breeman. *Rice: Soil, Water, Land* (Los Banos, Philippines, International Rice Research Institute, 1978).

Morgan, Dan. *Merchants of Grain* (New York, The Viking Press, 1979).

Morooka, Y., R. W. Herdt, and L. D. Haws. "An Analysis of a Labor Intensive Continuous Rice Crop Production System at IRRI," IRRI Research Paper Series 29 (Los Banos, Philippines, International Rice Research Institute, May 1979).

Motooka, Takeshi. "Agricultural Development in Thailand," 1,2,3,4, discussion Papers Nos. 26, 27, 28, 29, Center for Southeast Asian Studies (Kyoto, Kyoto University, 1971).

Moya, Piedad, Robert W. Herdt, and Shadigul I. Bhuiyan. "Returns to Irrigation Investment in Central Luzon, Philippines," Agricultural Economics Paper No. 81-23 (Los Banos, Philippines, International Rice Research Institute, 1981).

Mubyarto, and L. B. Fletcher. "The Marketable Surplus of Rice in Indonesia: A Study of Java-Madura," Mimeograph No. 4 (Ames, Iowa, Iowa State University, International Studies in Economics, 1966).

Mukherjee, P. K., and B. Lockwood. "High Yielding Varieties Programme in India—An Assessment." Paper presented at the 28th International Congress of Orientalists, Canberra, January 6–12, 1971.

Mukhopadhyay, Sudhink. "Constraints to Technological Progress in Rice Cultivation: A Study of Two Regions in India," VRF Series No. 74 (Tokyo, Institute of Developing Economies, 1980).

Mulder, Niels. *Everyday Life in Thailand: An Interpretation* (Bangkok, D. K. Books, 1979).

Muqtada, M. "The Seed Fertilizer Technology and Surplus Labor in Bangladesh Agriculture," *Bangladesh Development Studies* vol. 3, no. 4 (1975) pp. 403–423.

Murthy, A. N. Khrisna. "Shimoga Mysore," in International Rice Research Institute, *Changes in Rice Farming in Selected Areas of Asia* (Los Banos, Philippines, IRRI, 1975) pp. 117–132.

Nafis, Ahmad. "The Physiography and Crops of East Pakistan," *Agriculture Pakistan* vol. 1 no. 1 (September 1949).

Nagai, Isaburo. *Japonica Rice, Its Breeding and Culture* (Tokyo, Yokendo Ltd., 1959).

Nair, Kusum. *Blossoms in the Dust* (London, Gerald Duckworth and Co., 1961).

Nakamura, James I. *Agricultural Production and the Economic Development of Japan, 1873–1922* (Princeton, N.J., Princeton University Press, 1966).

Nam, Koon Woo. *The North Korean Communist Leadership 1945–1965: A Study of Factionalism and Political Consolidation* (University, Ala., University of Alabama Press, 1974).

Narkswasdi, Udhhis, and S. Selvadurai. "Economic Survey of Padi Production in West Malaysia," Bulletin No. 120 (Kuala Lumpur, Ministry of Agriculture and Co-operatives, 1968).

National Academy of Sciences. *National Academy of Sciences Plant Science Delegation Trip Report* (Washington, D.C., 1975).

Nazeer, Mian Mohammad. *Rice Economy of Pakistan* (1962).

Nazmul Alam, Aim. "Farm Mechanization in Bangladesh" (Tokyo, Asian Productivity Organization Symposium on Farm Mechanization, November 24–30, 1981).

Nederlandsch-Indië, Department van Economische Zaken, Centraal Kantoor voor de Statistiek. *Statistisch zakboekjivoor Nederlandsch-Indië* (Batavia).

Nederlandsch-Indië, Samengesteld door het Centraal Kantoor voor de Statistiek van het Departement van Economische Zaken. *Statistisch jarroverzicht van Nederlandsch-Indië* (Batavia, Landsdrukkerij).

Nepal, Central Bureau of Statistics. *Foreign Trade Statistics* (Kathmandu).

———. *Monthly Statistical Bulletin* (Kathmandu).

———. *Yield Data for Principal Crops* (Kathmandu).

Nepal, Ministry of Economic Planning. "Physical Production for Selected Agricultural Areas in Nepal (Kathmandu).

Nepal, National Planning Commission, Central Bureau of Statistics. *Statistical Pocketbook of Nepal* (Kathmandu).

Nguyen, van vinh. "Les réformes agraires au Viet-Nam," Université Catholique de Louvain, Faculté des Sciences Economiques et Sociales, Collection de l'Ecole des Sciences Economiques, No. 77 (Louvain, Librarie Universitaire Uystpruyst, 1961).

Nichol, Kenneth J., Somnuk Sriplung, and Earl O. Heady, eds. *Agricultural Development Planning in Thailand* (Ames, Iowa, Iowa State University Press, 1982).

Nicholls, William H. *Imperfect Competition in Agricultural Industries* (Ames, Iowa, Iowa State College Press, 1941).

Nihon, Ginko Tokukyoku (Statistics Department, The Bank of Japan), *Hundred Year Statistics of the Japanese Economy* (Tokyo, 1966).

Nitisastro, Widjojo. *Population Trends in Indonesia* (Ithaca, N.Y., Cornell University Press, 1970).

Nobufumi, Kayo. *Nihon nōgyo kiso tokei* (Basic Statistics of Japanese Agriculture) (Tokyo, Norinsuisangyo Seisansei Kojo Kuigi, 1968).

Nuttonson, M. Y. *The Physical Environment and Agriculture of Vietnam, Laos and Cambodia* (Washington, D.C., American Institute of Crop Ecology, 1963).

Nutty, Leslie. *The Green Revolution in West Pakistan: Implications for Technological Change* (New York, Praeger, 1972).

Nyberg, Albert J., and Dibyo Prabowo. *Status and Performance of Irrigation in Indonesia and the Prospects for 1990 and 2000*, Rice Policies in Southeast Asia Project, Working Paper No. 4 (Washington, D.C., International Food Policy Research Institute, International Fertilizer Development Center, International Rice Research Institute, 1982).

Nyrop, R. et al. *Area Handbook for Bangladesh* (Washington, D.C., Foreign Area Studies, The American University, 1975).

Office of Economic Research, Directorate of Intelligence, CIA, *Agricultural Acreage in Communist China, 1949–1968: A Statistical Compilation* (1969).

Ogura, Takekazu. *Can Japanese Agriculture Survive?—A Historical and Comparative Approach* (Tokyo, Agricultural Policy Research Center, 1980).

Ogura, Takekazu, ed. *Agricultural Development in Modern Japan* (Tokyo, Fuji Publishing Co., Ltd., 1963).

Ohkawa, Kazushi, Bruce F. Johnston, and Hiromitsu Kaneda, eds. *Agriculture and Economic Growth: Japan's Experience* (Princeton, N.J., Princeton University Press, 1970).

———, eds. *Agriculture and Economic Growth: Japan's Experience* (Tokyo, University of Tokyo Press, 1969).

Ohkawa, Kazushi, Miyohei Shinohara, and Mataji Umemura, eds. *Estimates of Long-term Economic Statistics of Japan Since 1868*, vol. 9, "Agriculture and Forestry" (Tokyo, Toyo Keizai Shinposha, 1965 and other years).

Okabe, Shiro. "Breeding for High-yielding Varieties in Japan," in *Rice Breeding* (Los Banos, Philippines, International Rice Research Institute, 1972).

Ongkingo, Pat S., Jose A. Galvez, and Mark Rosegrant. "Irrigation and Rice Production in the Philippines: Status and Projections," Working Paper No. 3, Rice Policies in Southeast Asia Project (Washington, D.C., International Food Policy Research Institute, International Fertilizer Development Center, International Rice Research Institute, 1982).

Oram, Peter, Juan Zapata, George Alibarubo, and Roy Shyamal. "Investment and Input Requirements for Accelerating Food Production in Low-Income Countries by 1990," Research Report No. 10 (Washington, D.C., International Food Policy Research Institute, 1979).

O'Toole, J. C., and T. T. Chang. "Drought and Rice Improvement in Perspective," IRRI Research Paper Series, no. 14 (Los Banos, Philippines, International Rice Research Institute, February 1978).

Owen, Norman, G. "The Industry of Mainland Southeast Asia 1850–1914," *Journal of Siam Society* vol. 59 (July 1971) pp. 78–142.

Owen, Wyn T. *Two Rural Sectors: Their Characteristics and Roles in the Development Process* (Ottawa, International Development Research Center, 1971).

Pakistan, Central Statistics Office. *25 Years of Pakistan in Statistics* (Karachi, 1972).

Pakistan, Department of Marketing, Intelligence and Agricultural Statistics. *Weather and Crop Report* (Karachi, Manager, Government of Pakistan Press, various dates).

Pakistan, Federal Bureau of Statistics, Statisties Division. *Monthly Statistical Bulletin* (Karachi, 1952-).

Pakistan, Ministry of Agriculture and Works, Department of Agricultural Economics and Statistics. *Land and Crop Statistics of Pakistan* (Karachi, Government of Pakistan Press, 1962).

———. *Land and Crop Statistics of Pakistan*, Fact Series III (Karachi, Government of Pakistan Press, June 1962).

Pakistan, Ministry of Finance, Planning, and Provincial Coordination, Statistics Division. *Statistical Pocketbook of Pakistan* (Karachi, Manager of Publications).

Pakistan, Ministry of Food, Agriculture, and Development, Agriculture Wing, *Yearbook of Agricultural Statistics* (Islamabad).

Pakistan, Statistics Division, *Pakistan Statistical Yearbook* (Karachi, Manager of Publications).

Pal, T. K., "Cuttack Orissa," in International Rice Research Institute, *Changes in Rice Farming in Selected Areas of Asia* (Los Banos, Philippines, IRRI, 1975) pp. 133–148.

Palacpac, Adelita, C. *World Rice Statistics* (Los Banos, Philippines, International Rice Research Institute, 1982).

Palmer, Ingrid. "The New Rice in Indonesia" (Geneva, UN Research Institute for Social Development (UNRISD), 1977.

———. *The New Rice in Asia: Conclusions from Four Country Studies* (Geneva, UN Research Institute for Social Research, 1976).

Pandey, Surya Prasad. "Review of the Rice Fertilizer Research Work at Tahara Agriculture Station." Paper presented at the Summer Crop Workshop, Department of Agriculture, Nepal, 1976/77.

Pangotra, P. N. "Survey Report, India" (Tokyo, Agricultural Productivity Organization Symposium on Farm Mechanization, November 24–30, 1981).

Panta, Sitaram R. "Nepal" in Asian Productivity Organization, *Fertilizer Distribution in Selected Asian Countries* (Tokyo, APO, 1979).

Papanek, Gustav, F. ed., *The Indonesian Economy* (New York, Praeger, 1980).

Parthasarathy, G. "West Godavari, Andhra Pradesh," International Rice Research Institute, *Changes in Rice Farming in Selected Areas of Asia* (Los Banos, Philippines, IRRI, 1975) pp. 43–70.

Parthasarathy, G., and Mohinder S. Mudahar. "Foodgrain Prices and Economic Growth," *Indian Journal of Agricultural Economics* vol. 31, no. 2 (April-June 1976) pp. 16–30.

Parthasarathy, N. "Rice Breeding in Tropical Asia up to 1960," in International Rice Research Institute, *Rice Breeding* (Los Banos, Philippines, IRRI, 1972).

Partadiredja, Ace, "Mass Extension Among Farmers in Indonesia." Paper presented at the 17th International Conference of Agricultural Economists, Banff, B.C., Canada, 1979.

Patel, S. M., and K. U. Patel. *Economics of Tubewell Irrigation* (Ahmedabad, Indian Institute of Management, 1971).

Payne, R. R. "The Nutritive Value of Asian Dietaries in Relation to the Protein and Energy Needs of Man," in Institute of Development Studies, *Three Papers on Food and Nutrition: The Problem and the Means of Its Solution* (Brighton, IDS, University of Sussex, 1971) pp. 23–24.

Peach, W. N., et al. *Basic Data on the Economy of Pakistan* (Karachi, Oxford University Press, 1959).

Pempel, T. J. *Policy and Politics in Japan* (Philadelphia, Temple University Press, 1982).

———, ed. *Policymaking in Contemporary Japan* (Ithaca, N.Y., Cornell University Press, 1977).

Peninsular Malaysia, Kementerian Pertanian (Ministry of Agriculture). *Perangkaan padi* (Paddy Statistics) (Kuala Lumpur).

———. *Rumusan perangkaan* (Statistical Digest) (Kuala Lumpur).

Penny, David A., and Masri Sigharimbun. "A Case Study of Rural Poverty," *Bulletin of Indonesian Economic Studies* vol. 8 (1972) pp. 79–88.

Perera, L. N., W. S. M. Fernando, B. V. de Mel, and T. T. Poleman. *The Effect of Income on Food Habits in Ceylon: The Findings of the Socio-Economic Survey*, Cornell Agricultural Economics Staff Paper, No. 72-25 (Ithaca, N.Y., Department of Agricultural Economics, Cornell University, 1972).

Periera, Sir C. "The Changing Pattern of Constraints on Food Production in the Third World," in T. W. Schultz, ed., *Distortion of Agricultural Incentives* (Bloomington, Ind., University of Indiana Press, 1978).

Perkins, Dwight. *Agricultural Development in China 1368–1968* (Chicago, Aldine, 1969).

Perkins, Dwight, and Yusuf. "Rural Development in the People's Republic of China," Draft prepared for World Bank, April 1980.

Petzel, Todd E., and Eric A. Monke. "The Integration of the International Rice Market," *Food Research Institute Studies* vol. 17 (1979–80) pp. 307–325.

Philippine Council of Agricultural and Resources Research. *Data Series on Rice Statistics in the Philippines* (Los Banos, Philippines, International Rice Research Institute, 1981).

Philippines, Bureau of Agricultural Economics. *Crop, Livestock, and Natural Resources Statistics* (Manila).

———. *Farm Wages* (Manila).

———. "Prices Received and Paid by Farmers" (mimeo).

Philippines, Bureau of the Census and Statistics. *Yearbook of Philippine Statistics* (Manila, Bureau of Printing) historical.

Philippines, Bureau of Commerce and Industry. *Statistical Bulletin of the Philippine Islands* (Manila, 1918–1930) historical.

Philippines, Department of Agriculture and Commerce. *Atlas of Philippine Statistics* (1939).

Philippines, Department of Agriculture and Natural Resources. *The Philippine Journal of Agriculture* (Manila, Bureau of Printing).

Philippines, Department of Agriculture and Statistics, Division of Statistics. *Statistical Handbook of the Philippine Islands* (Manila, Bureau of Printing).

Philippine National Census and Statistics Office. *Foreign Trade Statistics* (Quezon City, annual).

Philippines, National Economic and Development Authority. *Philippine Statistical Yearbook* (Manila).

Pinthong, C. "A Price Analysis of the Thai Rice Marketing System" (Ph.D. dissertation, Stanford University, 1977).

Poleman, Thomas T. "Quantifying the Nutritional Situation in Developing Countries," *Food Research Institute Studies* vol. 18, no. 1 (1981) pp. 1–58.

———. "A Reappraisal of the Extent of World Hunger," *Food Policy* vol. 6, no. 4 (November 1981) pp. 236–252.

Ponnamperuma, F. N., and A. K. Banyopudhya. "Soil Salinity as a Constraint on Food Production in the Humid Tropics," in International Rice Research Institute, *Priorities for Alleviating Soil-Related Constraints to Food Production in the Tropics* (Los Banos, Philippines, IRRI and Cornell University, 1980).

Pookkachatikul, J. S., S. Tongpan, and D. Welsch. "Thai Rice Price Data," Staff Paper No. 14 (Bangkok, Department of Agricultural Economics, Kasetsart University, 1974).

Prabowo, Dibyo, and Sajogyo. "Sidoarjo, East Java and Subang, West Java," in International Rice Research Institute, *Changes in Rice Farming in Selected Areas of Asia* (Los Banos, Philippines, IRRI, 1975) pp. 179–199.

Pradhan, B. "The Role of Women in Household Production System and Rice Farming." Paper presented at Women in Rice Farming Systems Conference, Los Banos, Philippines, September 26–30, 1983.

Prebish, Raul. *The Economic Development of Latin America and its Principal Problems* (Lake Success, N.Y., UN Department of Economic Affairs, 1950).

Price, Edwin C., and Randolph Barker. "Time Distribution of Crop Labor," *The Philippine Economic Journal* vol. 17 (1976) pp. 224–243.

Pudasaini, S. P. "Farm Mechanization, Employment and Income in Nepal: Traditional and Mechanized Farming in Bara District," IRRI Research Paper Series no. 38 (Los Banos, Philippines, International Rice Research Institute, 1979).

Purcal, John T. *Rice Economy, Employment and Income in Malaysia* (Honolulu, University Press of Hawaii, 1972).

Pyongyang Korean Central News Agency. *Choson chungang yongam* (Korean Central Yearbook).

Quasem, Md. Abul. "Factors Affecting the Use of Fertilizer in Bangladesh," *Bangladesh Development Studies* vol. 6 (1978).

Qureshi, Kamaluddin. "Survey Report, Pakistan" (Tokyo, Agricultural Productivity Organization Symposium on Farm Mechanization, November 24–30, 1981).

Rab, Abdur, *Acreage, Production and Prices of Major Agricultural Crops of West Pakistan (Punjab): 1931–1959*, Statistical Papers No. 1 (Karachi, Institute of Development Economics, June 1961).

Rachman, Anas, and Roger Montgomery. "Asian Fertilizer Demand Reconsidered: An Application to Java and Bali," *Ekonomi dan Keucangan Indonesia* vol. 28, no. 3 (September 1980).

Rahman, Abdul, Haji Yusof, and Ani bin Arope. "Rice Policy in Malaysia," in Randolph Barker, ed., *Viewpoints on Rice Policy in Asia* (Los Banos, Philippines, International Rice Research Institute, 1971).

Rajagopalan, "North Arcot, Tamil Nadu," in International Rice Research Institute, *Changes in Rice Farming in Selected Areas of Asia* (Los Banos, Philippines, IRRI, 1975) pp. 71-91.

Rajbhandany, N. K. "To Study the Response of Nitrogen in Local and Improved Rice—1977." Paper presented at the Fifth Rice Improvement Workshop, National Rice Improvement Programme, Department of Agriculture, Nepal, March 1978.

Ranade, Chandra, and Robert W. Herdt. "Shares of Farm Earnings from Rice Production," *Economic Consequences of the New Rice Technology* (Los

Banos, Philippines, International Rice Research Institute, 1978) pp. 87–104.

Ranasinghe, W., S. M. K. Milehama, and C. Gunatunga. *A Bibliography of Socioeconomic Studies in the Agrarian Sector of Sri Lanka* (Colombo, Agrarian Research and Training Institute, 1977).

Rawski, Thomas G. *Economic Growth and Employment in China* (New York, Oxford University Press, 1979).

Ray, Susanta K., Ralph W. Cummings, Jr., and Robert W. Herdt. *Policy Planning for Agricultural Development* (New Delhi, Tata McGraw-Hill, 1979).

Regmi, Mahesh. *Land Ownership in Nepal* (Berkeley, Calif., University of California Press, 1976).

Res, Alida. "Changing Labor Allocation Patterns of Women in Iloilo Rice Farm Households." Paper presented at the Conference on Women in Rice Farming Systems, International Rice Research Institute, September 26–30, 1983.

Reutlinger, S., and M. Selowsky. "Malnutrition and Poverty," World Bank Staff Occasional Paper No. 23 (Baltimore, Md., Johns Hopkins University Press, 1976).

Reynolds, Lloyd, ed. *Agriculture in Development Theory* (New Haven, Conn., Yale University Press, 1975).

Rice, Stuart, and Calvert Dedrick, "Japanese Statistical Organization, A Report to the Supreme Commander for the Allied Powers" (Washington, D.C., July 1951).

Richardson, J. Henry. "Wages in Burma," *International Labour Review* vol. 69 (1954) pp. 433–451.

Richter, Hazel. "Burma's Rice Surpluses: Accounting for the Decline," The Australian National University Development Studies Working Paper No. 3 (Canberra, 1976).

Rijik, Louis, Project Director, Intensive Public Works Program, Nepal, 1982, mimeo.

Robequain, Charles. *The Economic Development of French Indochina* (New York, Oxford University Press, 1944).

Roberts, Ivan, Robert Bain, and Eric Saxon. "Japanese Agricultural Policies: The Origin, Nature and Effects of Production and Trade" (Canberra, Bureau of Agricultural Economics, February 1981).

Robinson, Harry. *Monsoon Asia: A Geographic Survey* (New York, Praeger, 1967).

Ros, L. "Changing Labor Allocation Patterns of Women in Iloilo Rice Farm Households." Paper presented at Women in Rice Farming Systems Conference, Los Banos, Philippines, September 26–30, 1983.

Rosegrant, Mark. "The Impact of Irrigation on the Yield of Modern Varieties," IRRI Agricultural Economics Paper 76-28 (Los Banos, Philippines, International Rice Research Institute, 1976).

Rosegrant, Mark W., and Robert W. Herdt. "Simulating the Impacts of Credit Policy and Fertilizer Subsidy on Central Luzon Rice Farms, the Philippines," *American Journal of Agricultural Economics* vol. 63 no. 4 (November 1981) pp. 655–665.

Rosenberg, David A., and Jean G. Rosenberg. *Landless Peasants and Rural Poverty in Selected Asian Countries*, Rural Development Committee Special Series on Landlessness and Near Landlessness LNL3 (Ithaca, N.Y., Cornell University, 1978).

Rosenzweig, Mark T., and T. Paul Schultz. "Market Opportunities, Genetic Endowments and Intrafamily Resource Distribution: Child Survival in Rural India," *The American Economic Review* 72 (1982) pp. 803–815.

Rossiter, Fred J., L. Thelma Willahan, and William E. Cummings. *World Rice Production and Trade*, Foreign Agriculture Report No. 15 (Washington, D.C., Government Printing Office, 1946).

Roumasset, James. "Land Tenure and Labor Arrangements in Philippine Agriculture: Some Lessons from the New Institutional Economics." Paper presented at the Transition in Agricultural Organization Workshop, Los Banos, Philippines, January 1982.

———. "Fundamental Explanation of Farmer's Behavior and Agricultural Contracts" (Banff, B.C., Canada, Conference of the International Association of Agricultural Economists, 1979).

Roumassett, James A., and Arsenio M. Balisacan. "The Political Economy of Rice Policy and Trade in the Asian-Pacific Region" (Honolulu, East-West Center Resource Systems Institute, 1983).

Roy, A. C. "Fertilizer Response of Rice at BRRI Farms in Different Seasons," in Bangladesh Rice Research Institute, *Workshop on Ten Years of Modern Rice and Wheat Cultivation in Bangladesh* (Dacca, BRRI, March 7–10, 1977).

Rudner, Martin. "The Malayan Post-war Rice Crises: An Episode in Colonial Agricultural Policy," *Kajian ekonomi Malaysia* (Journal of Malaysian Economic Association) vol. 12, no. 1 (June 1975).

Rudra, Ashok. "Planning and the New Agricultural Strategy," *Economic and Political Weekly* vol. 6, no. 6 (February 1971) pp. 429–430.

Ruofang, Niu. "Does Taking Grain as the Key Link Suit Measures to Local Conditions?" *Guangming Daily (Guangdong)* December 8, 1979.

Ruttan, Vernon, W. "The International Agricultural Research Institute as a Source of Agricultural Development," *Agricultural Administration* vol. 5 (1978) pp. 1–19.

———. "Agricultural Product and Factor Markets in Southeast Asia," *Economic Development and Cultural Change* vol. 17 (1969) pp. 501–509.

Sajogyo, P. "Impact of New Farming Technology on Women's Employment." Paper presented at Women in Rice Farming Systems Conference, Los Banos, Philippines, September 26–30, 1983.

Sajogyo, P., and William L. Collier. "Adoption of New High-Yielding Rice Varieties by Java's Farmers," in R. T. Shand, ed., *Technical Change in Asian Agriculture* (Canberra, Australian National University Press, 1973) pp. 80–107.

———. "Adoption of High-Yielding Varieties by Java's

Farmers," Research Note No. 7 (Indonesia, Agro-economic Survey, May 1972).
Samson, Robert L. *The Economics of Insurgency in the Mekong Delta of Vietnam* (Cambridge, Mass., MIT Press, 1970).
Sanderson, Fred H. "Japan's Food Prospects and Policies" (Washington, D.C., The Brookings Institution, 1978).
Santos, Cynthia Lina G. "Identifying the Nutritionally Vulnerable Households in the Philippines" (Ph.D. dissertation, Cornell University, 1983).
Sarkar, Hiren. "A Simulation Model of the World Rice Economy with Special Reference to Thailand," DAE-CARD Sector Analysis Series No. 14, Center for Agricultural and Rural Development (Ames, Iowa, Iowa State University, 1978).
Sarma, J. S., Shyamal Roy, and P. S. George. *Two Analyses of Indian Foodgrain Production and Consumption Data,* IFPRI Research Report No. 12 (Washington, D.C., International Food Policy Research Institute, November 1979).
Schuh, Edward, and Helro Tolline, eds. "Costs and Benefits of Agricultural Research: State of the Art and Implications for CGIAR" (Washington, D.C., CGIAR Secretariat, World Bank, 1978) mimeo.
Schulter, M. "Differential Rates of Adoption of the New Seed Varieties in India: The Problem of the Small Farm," USAID Research Project, Occasional Paper No. 47 (Ithaca, N.Y., Cornell University, 1971).
Schultz, Theodore W. "Constraints on Agricultural Production," in T. W. Schultz, ed., *Distortion of Agricultural Incentives* (Bloomington, Ind., Indiana University Press, 1978).
———. ed. *Distortion of Agricultural Incentives* (Bloomington, Ind., Indiana University Press, 1978).
Scobie, Grant M., and Rafael T. Posada. "The Impact of Technical Change on Income and Distribution: The Case of Rice in Colombia," *American Journal of Agricultural Economics* vol. 1, no. 1 (February 1978) pp. 85–92.
———. "The Impact of High-Yielding Rice Varieties in Latin America with Special Emphasis on Colombia" (Cali, Colombia, Centro Internacional de Agricultura Tropical, 1976).
Seiichi, Tobata. "Control of the Price of Rice," Japanese Council, Institute of Pacific Relations (Tokyo, The Nippon Press, 1933).
Selvadurai, S. "Padi Farming in West Malaysia," Bulletin No. 127 (Kuala Lumpur, The Ministry of Agriculture and Fisheries, 1972).
Sen, Gita. "Paddy Production, Processing and Women Workers in India—the South Versus the Northeast." Paper presented at Women in Rice Farming Systems Conference, Los Banos, Philippines, September 26–30, 1983.
Shan, Jin-lua. "Rice Breeding in China," in International Rice Research Institute, *Rice Improvement in China and Other Asian Countries* (Los Banos, Philippines, IRRI, 1980) pp. 9–30.
Shand, R. T., ed. *Technical Change in Asian Agriculture* (Canberra, Australian National University Press, 1973).
Shanmugasundram, V. "Economic and Social Implications of High-Yielding (Paddy) Varieties Programme" (Madras, India, Department of Economics, University of Madras, 1973) mimeo.
Shapiro, K. H., and J. Muller. "Sources of Technical Efficiency: The Role of Modernization and Information," *Economic Development and Cultural Change* vol. 25 (1977) pp. 239–310.
Sharma, D. P., and V. V. Desai. *Rural Economy of India* (New Delhi, Vikas Publishing House, 1980).
Sharma, J. S. "Nainital and Varanasi, Uttar Pradesh," in International Rice Research Institute, *Changes in Rice Farming in Selected Areas of Asia* (Los Banos, Philippines, IRRI, 1975) pp. 94–116.
Shen, Lin. "The Inside Information on China's Economic Readjustment," *Zhengming*, May 1979, pp. 9–13.
Shen, T. H. *Agricultural Development on Taiwan Since World War II* (Ithaca, N.Y., Comstock Publishing Associates—A Division of Cornell University Press, 1964).
———, ed. *Agriculture's Place in the Strategy of Development: The Taiwan Experience* (Taipei, Joint Commission on Rural Reconstruction, 1974).
Shen, Tsunghan. *Agricultural Resources of China* (Ithaca, N.Y., Cornell University Press, 1951).
Shim, Young Kun. *Household Consumption Patterns of Foodgrains in Suweon* (Seoul, Department of Agricultural Economics, College of Agriculture, Seoul National University, 1968).
Shinzawa, Kagato. *Nōgyo Suiri Ron* (Treatise on Irrigation) (Tokyo, Tokyo Daigaku Shuppankai, 1955).
Short, D. E., and James C. Jackson. "The Origins of Irrigation Policy in Malaysia," *Journal of the Malaysian Branch of the Royal Asiastic Society* vol. 44 (1971) pp. 78–103.
Siamwalla, Ammar. "Farmers and Middlemen: Aspects of Agricultural Marketing in Thailand" (Bangkok, UN Asian Development Institute, Agricultural Marketing Case Study No. 2, 1975).
———. "A History of Rice Policies in Thailand," *Food Research Institute Studies* vol. 14 no. 3 (1975) pp. 233–249.
———. "Land, Labor and Capital in Three Rice Growing Deltas of Southeast Asia, 1800–1940," Economic Growth Center Discussion Paper No. 150 (New Haven, Conn., Yale University, July 1972).
Siamwalla, Ammar, and Stephen Haykin. *The World Rice Market Structure, Conduct, and Performance*, Research Report No. 39 (Washington, D.C., International Food Policy Research Institute, 1983).
Siddiqui, Akhtar, H. "Agriculture in Pakistan: A Selected Bibliography, 1947–69" (Rawalpindi, Office of Assistant Director/Agricultural Policy, USAID, 1969).

Silcock, T. H. *The Economic Development of Thai Agriculture* (Ithaca, N.Y., Cornell University Press, 1970).
———, ed. *Thailand: Social and Economic Studies in Development* (Canberra, Australian National University, 1967).
———, ed. *Readings in Malayan Economics* (Singapore, Eastern Universities Press, Ltd., 1961).
Silcock, T. H., and Fisk, E. K., eds. *The Political Economy of Independent Malaya: A Case-Study in Development* (Berkeley, Calif., University of California Press, 1963).
Sinaga, R. S. "Effects of Mechanization on Productivity: South Sulawesi, Indonesia" (Los Banos, Philippines, Agricultural Development Council, International Rice Research Institute Workshop on the Consequences of Small Rice Farm Mechanization, September 14–18, 1981).
———. "Effects of Mechanization on Productivity: West Java, Indonesia" (Los Banos, Philippines, Workshop on the Consequences of Small Rice Farm Mechanization, September 14–18, 1981).
Sinaga, R. S., and B. M. Sinaga. "Comments on Shares of Farm Earnings from Rice Production," in International Rice Research Institute, *Economic Consequences of the New Rice Technology* (Los Banos, Philippines, IRRI, 1978).
Singapore, Department of Statistics. *Monthly Digest of Statistics* (Singapore).
———. *Singapore Trade Statistics Import and Export* (Singapore).
———. *Yearbook of Statistics, Singapore* (Singapore).
Sinha, Radha. *Japan's Options for the 1980s* (New York, St. Martin's Press, 1982).
———. "Chinese Agriculture: A Quantitative Look," *The Journal of Development Studies* vol. 2, no. 3 (April 1975) pp. 201–223.
Sison, J. F., Somsak Prakongtanapan, and Y. Hayami, "Structural Changes in Rice Supply Relations: Philippines and Thailand," in International Rice Research Institute, *Economic Consequences of the New Rice Technology* (Los Banos, Philippines, IRRI, 1978).
Sison, O., E. Villanueva, E. Naveva, J. Kalaw, R. Ancheta, and R. Olan. "Country Report, Philippines." Report presented at Women in Rice Farming Systems Conference, Los Banos, Philippines, September 26–30, 1983.
Slayton, T. M., and I. G. N. Excuvirya. "The Fertilizer Situation," *Bulletin of Indonesian Economic Studies* vol. 14, no. 2 (July 1978) pp. 70–85.
Social Science Research Council. *Provincial Agricultural Statistics for Communist China* (Ithaca, N.Y., Committee on the Economy of China, Social Science Research Council, 1969).
Soe, Myint. "Economics of Production and Procurement of Paddy in Burma" (M.S. thesis, University of the Philippines, 1978).
Soejono, I. "Growth and Distributional Change of Income in Paddy Farms in Central Java, 1968–1974," *Bulletin of Indonesian Economic Studies* vol. 12, no. 2 (July 1976) pp 80–89.
Somboonsub, N. "Rice Milling Technology and Some Implications: The Case of Nakorn Pathom, Thailand, 1974" (M.S. thesis, Thammasat University, 1976).
Somboonsup, Sri-on. "The Pattern of Thai Rice Exports, 1955–1972" (M.S. thesis, Graduate School of Kasetsart University, 1975).
Sondysuwan, Prateep, ed. *Finance, Trade and Economic Development in Thailand: Essays in Honour of Khunying Suparb Yossundara* (Bangkok, Sompong Press, 1975).
Southworth, Herman M. "Some Dilemmas of Agricultural Mechanization," in Herman M. Southworth and M. Barnett, eds., *Experience in Farm Mechanization in Southeast Asia* (New York, Agricultural Development Council, 1974).
——— ed. *Farm Mechanization in East Asia* (New York, Agricultural Development Council, 1972).
Spencer, Joseph. *Shifting Cultivation in Southeast Asia*, vol. 19 (University of California Publications in Geography, 1966).
———. "The Migration of Rice from Mainland Southeast Asia into Indonesia," in Jacques Barrau, ed., *Plants and the Migration of Pacific Peoples* (Honolulu, Bishop Museum Press, 1963) pp. 84–86.
Spencer, Joseph, and William L. Thomas. *Asia East by South* (New York, Wiley, 1971).
Sri Lanka, Central Bank of Ceylon. *Central Bank of Ceylon Annual Report* (Colombo, W.R.B.K. Godakumbura, 1982).
———. *Central Bank of Ceylon Review of the Economy* (Colombo, C. L. Senanayke, 1980).
Sri Lanka, Central Bank of Ceylon, Statistics Department. *Economic and Social Statistics of Sri Lanka* (Colombo).
Sri Lanka, Department of Agriculture, Agricultural Statistical Unit. *Agricultural Statistical Information*, Agricultural Economics Publication no. 4 (1976).
Sri Lanka, Department of Census and Statistics, *Quarterly Bulletin of Statistics* (Colombo, Government Publications Bureau, 1955–).
———. *Statistical Abstract of Sri Lanka* (Colombo, Department of Government Printing, 1954–).
———. *Statistical Pocketbook of Sri Lanka* (Colombo).
———. *Sri Lanka Yearbook* (Colombo, Department of Government Printing).
Sri Lanka, Department of Census and Statistics, Ministry of Plan Implementation. *Bulletin of Selected Retail Prices 1978–80* (Colombo, 1980).
Sri Lanka, Ministry of Agricultural Development and Research. *Agricultural Statistics of Sri Lanka: 1951/52—1980/81* (Colombo, Sri Lanka, 1981).
Sriplung, Somnuk, and Koset Manowalailau. "The Demand of Rice for Domestic Consumption," Agricultural Economics Research Bulletin no. 44 (Bangkok, Division of Agricultural Economics, Ministry of Agriculture, 1972).

Sriwasdilek, Jerachone. "The Yield Performance and Economic Benefits of the High-yielding Varieties in Don Chedi, Suphanburi, Thailand" (M.S. thesis, University of the Philippines at Los Banos, 1973).

Sriwasdilek, Jerachone, Kamphol Adulavidhaya, and Sompom Isvilanonda. "Don Chedi, Suphan Buri," *Changes in Rice Farming in Selected Areas of Asia* (Los Banos, Philippines, International Rice Research Institute, 1975) pp. 243–263.

Stangel, Paul J. "World Fertilizer Sector—at a Crossroads." Paper presented at the Symposium on Food Situation in Asia and the Pacific Region, Taipei, Asian and Pacific Council, Food and Fertilizer Technology Center, April 24–29, 1980.

Steinberg, David J. *Burma's Road to Development: Growth and Ideology* (Boulder, Colo., Westview Press, 1981).

Steinberg, David J., David K. Wyatt, et al. *In Search of South East Asia: A Modern History* (London, Pall Mall, 1971).

Stone, Bruce. "The Use of Agricultural Statistics: Some National Aggregate Examples and Current State of the Art," in Randolph Barker, Radha Sinha, and Beth Rose, eds., *The Agricultural Economy of China* (Boulder, Colo., Westview Press, 1982).

Suh, Sung-chul. *Growth and Structural Changes in the Korean Economy, 1910–1940* (Cambridge, Mass., Council on East Asian Studies, Harvard University, 1978).

Suh, Wan Soo. "Factors Affecting the Rate of Adoption of Tongil Rice Varieties in Selected Locations of Korea" (M.S. thesis, University of the Philippines at Los Banos, 1976).

Sukhatme, P. V. "The Present Pattern of Production and Availability of Food in Asia," in *Three Papers on Food and Nutrition: The Problem and the Means of Its Solution* (Brighton, Institute of Development Studies, University of Sussex, 1971) pp. 1–17.

Sung, Hwan Ban. *Rural Development* (Cambridge, Mass., Council on East Asian Studies, Harvard University, 1980).

Surbakti, Pajung. "Identifying the Nutritionally Vulnerable Urban and Rural Groups in Indonesia" (Ph.D. dissertation, Cornell University, 1983).

Sutthidej, Anong. "The Rice Industry in Thailand: An Appraisal of the Various Factors Responsible for Making this Country a Big Rice Producer" (M.S. thesis, University of San Carlos, 1967).

Sweet, Norma. "Factbook—Compilation of Laotian Statistics" (Vientiane, 1967).

Swenson, C. C. "The Distribution of Benefits from Increased Rice Production in Thanjavur District, South India," *Indian Journal of Agricultural Economics* vol. 31 (January-March 1976) pp. 1–12.

Tabor, Steve. "Sources of Price Stability in Indonesian Agriculture: Implications for Growth and Equity" (M.S. thesis, Cornell University, 1983).

Tagarino, R. W., and R. D. Torres. "The Price of Irrigation Water: A Case Study of the Philippines' Upper Pampanga River Project," in International Rice Research Institute, *Irrigation Policy and Management in Southeast Asia* (Los Banos, Philippines, IRRI, 1978).

Taiwan, Bureau of Accounting and Statistics. *Report on the Survey of Family Income and Expenditures in Taiwan 1966* (Taipei, Taiwan Government, 1968).

Taiwan, Department of Agriculture and Forestry. *Taiwan Agricultural Yearbook* (Taipei).

Taiwan, Department of Budget, Accounting and Statistics. *Statistical Abstract of Taiwan* (Taipei).

———. *Statistical Yearbook of Taiwan* (Taipei).

Taiwan, Economic Planning Council Executive Yuan. *Taiwan Statistical Data Book* (Taipei).

Taiwan, Food Bureau. *Taiwan Food Statistics Book* (Taipei).

Taiwan, Governor General. *The Statistical Summary of Taiwan* (Tokyo, Japan Times Press, 1912).

———. *Taiwan beikoku yoran (Summary of Taiwan Rice Production)* (Taipei, 1939).

Taiwan, Governor General, Directorate of Statistics. *Taiwansheng wushiyinianlai tongji tiyao* (Summary of Statistics for 51 years) (Taipei, 1946).

Taiwan, Provincial Food Bureau, *Taiwan Statistical Data Book* (Taipei).

Takane, Matsuo. *Rice Culture in Japan* (Tokyo, Government of Japan, Ministry of Agriculture and Forestry, 1954).

Takase, Kunio, and Thomas Wickham. "Irrigation Management and Agricultural Development in Asia," in *Rural Asia Challenge and Opportunity: Second Asian Agricultural Survey*, Supplementary Papers, vol. 1 (Manila, Asian Development Bank, 1978).

Takaya, Yoshikazu. "Rice Cropping Patterns in Southeast Asian Deltas," *Southeast Asian Studies* vol. 13 (1975) pp. 256–281.

Takekazu, Ogura. "Agrarian Problems and Agricultural Policy in Japan: A Historical Sketch" I.A.E.A. Occasional Paper Series No. 1 (Tokyo, The Institute of Asian Economic Affairs, 1967).

Tamin, Moktar Bin, and N. Hashim Mustapha. "Kelantan, West Malaysia," *Changes in Rice Farming in Selected Areas of Asia* (Los Banos, Philippines, International Rice Research Institute, 1975) pp. 202–223.

Tan, Eva Kimpo. "Pigcawayan, Cotabato," *Changes in Rice Farming in Selected Areas of Asia* (Los Banos, Philippines, International Rice Research Institute, 1975) pp. 324–345.

Tan, Yolanda, and John A. Wicks. "Production Effects of Mechanization," Consequences of Small Farm Mechanization Working Paper 36 (Los Banos, Philippines, Department of Agricultural Engineering, International Rice Research Institute, 1981).

Tanaka, Akira. "Comparisons of Rice Growth in Different Environments," in International Rice Research

Institute, *Climate and Rice* (Los Banos, Philippines, IRRI, 1976).

Tanaka, A., K. Kawano, and J. Yamaguchi. "Photosynthesis, Respiration, and Plant Type of the Tropical Rice Plant," IRRI Technical Bulletin, No. 7 (Los Banos, Philippines, International Rice Research Institute, 1966).

Tanaka, A., J. Yamaguchi, Y. Shimazaki, and K. Shibaty. "Historical Changes in Plant Type of Rice Varieties in Hokkaido," *Soil Science* vol. 39 (1968) p. 11.

Tang, Anthony M., and Bruce Stone. *Food Production in the People's Republic of China*, IFPRI Research Report No. 15 (Washington, D.C., International Food Policy Research Institute, 1980).

Tan-Kim-Huon. *Geographie du Cambodge de l'Asie des monssons et des principales puissances* (Phnom-penh, 1963).

Taylor, Carl C., Douglas Ensminger, Helen W. Johnson, and Joyce Jean, *India's Roots of Democracy* (Calcutta, Orient Longmans, 1965).

Taylor, Donald C. *The Economics of Malaysian Paddy Production and Irrigation* (Bangkok, Agricultural Development Council, 1981).

Taylor, Donald C., Kusairi Mohd. Noh, and Mohd. Arrif Hussein. "An Economic Analysis of Irrigation Development in Malaysia," Working Paper No. 1, Rice Policies in Southeast Asia Project (Washington, D.C., International Food Policy Research Institute, International Fertilizer Development Center, International Rice Research Institute, 1981).

Te, Amanda, and Robert W. Herdt. "Fertilizer Prices, Subsidies and Rice Production." Paper presented to the 1982 Annual Convention of the Philippine Agricultural Economics Development Association, Los Banos, Philippines, June 4, 1982.

Teston, Eugene, and Maurice Percheron. *L'Indochine historique: encyclopedie administrative, touristique, artistique et économique* (Paris, Librarie de France, 1931).

Thailand, Center for Agricultural Statistics, Krasūang Kasēt (Ministry of Agriculture). *Agricultural Statistics of Thailand* (Bangkok, 1954–).

Thailand, Department of Customs. *Foreign Statistics of Thailand* (Bangkok, 1954–).

Thailand, Division of Agricultural Economics, Ministry of Agriculture. *Rice Economy of Thailand* (Bangkok, 1965).

Thailand, Krom Kān Khāo (Department of Rice), Krasūang Kasēt (Ministry of Agriculture). *Rāingan sarup phon kāntham nā* (Annual Report of Rice Production in Thailand) (Bangkok).

Thailand, Land Policy Division, Land Development Department. *Cost-Return Information for Selected Crops by Soil-Series in Ubonrajthani for 1969* (Bangkok, Ministry of National Development, 1971).

Thailand, National Statistical Office. *Household Expenditure Survey BE (1962)* (Bangkok, n.d.).

———. *Quarterly Bulletin of Statistics* (Bangkok, 1952–).

———. *Samut sathiti rāi pī khōng prathet Thai* (Statistical Yearbook of Thailand) (Bangkok, 1916–); formerly *Siam Statistical Yearbook*.

Thailand, National Statistical Office, Office of the Prime Minister. *Statistical Handbook of Thailand* (Bangkok, 1975).

Thailand, Rice Division and Planning Division, Department of Agriculture. *Thailand, Annual Research Report 1974* (Bangkok).

Thana, Khan Haeng Prathet Thai (Bank of Thailand). *Bank of Thailand Monthly Bulletin* (Bangkok, 1953–); formerly *Bank of Thailand Current Statistics*.

Thiam, Tan Bock, and Shao-er Ong, eds. *Readings in Asian Farm Management* (Singapore, University of Singapore Press, 1979).

Thorner, Daniel, and Alice Thorner. *Land and Labor in India* (Bombay, Asia Publishing House, 1962).

Timmer, C. Peter. "Food Prices and Economic Development in LDCs." Paper presented to the World Food Policy Seminar, Harvard Business School, May 13–14, 1979.

———. "The Political Economy of Rice in Asia: Indonesia," *Food Research Institute Studies* vol. 14, no. 3 (1975) pp. 197–231.

———. "The Political Economy of Rice in Asia: A Methodological Introduction," *Food Research Institute Studies* vol. 14, no. 3 (1975) pp. 191–196.

———. "Choice of Technique in Rice Milling in Java, A Reply," *Bulletin of Indonesian Economic Studies* vol. 10 (1974) pp. 121–126.

———. "A Model of Rice Marketing Margins in Indonesia," *Food Research Institute Studies* vol. 13 (1974) pp. 145–167.

———. "Choice of Technique in Rice Milling in Java," *Bulletin of Indonesian Economic Studies* vol. 9, no. 2 (1973) pp. 57–76.

Timmer, C. Peter, and Walter P. Falcon. "The Impact of Price on Rice Trade in Asia," in G. S. Tolley and P. A. Zadrogny, eds., *Trade, Agriculture and Development* (New York, Ballinger, 1975).

Timmer, C. Peter, Walter P. Falcon, and Scott R. Pearson. *Food Policy Analysis* (Baltimore, Md., The Johns Hopkins University Press, 1983).

Tolley, G. S., and P. A. Zadrogny, eds. *Trade, Agriculture and Development* (New York, Ballinger, 1975).

Tong, Wei-sen, and Sin-chaw Tu. "A Study of the Farm Economy of China Through an Analysis of Farm Accounts in Selected Districts," *Agricultural Sinica* vol. 1, no. 12 (1936) pp. 405–407.

Tongpan, Sopin. "An Economic Analysis of the Price of Thai Rice" (Ph.D. dissertation, Ohio State University, 1969).

Toquero, Z., B. Duff, A. Lacsina, and Y. Hayami. "Marketable Surplus Functions for a Subsistence Crop: Rice in the Philippines," *American Journal of Agricultural Economics* vol. 57 (1975) pp. 705–713.

Tri, Vo Nhan. *Croissance economique de la République Democratique du Vietnam 1945–65* (Hanoi, Editions en Langues Etrangères, 1967).

Tsai, Lih-yuh. "Production Costs and Returns for Rice Farms in Central Taiwan, 1895–1976: Analysis of Structural Changes" (M.A. thesis, University of the Philippines, School of Economics, 1976).

Tsuchiya, Keizo. "Mechanization and Relations Between Farm, Non-Farm and Government Sectors," in Herman Southworth, ed., *Farm Mechanization in East Asia* (New York, Agricultural Development Council, 1972).

Tsuji, Hiroshi. "A Quantitative Model of the International Rice Market and Analysis of National Rice Policies, with Special Reference to Thailand, Indonesia, Japan, and the United States," in Max R. Langham and Ralph H. Retzlafeds, eds., *Agricultural Sector Analysis in Asia* (Bangkok, Agricultural Development Council, 1982).

———. "An Economic and Institutional Analysis of the Rice Export Policy of Thailand: With Special Reference to the Rice Premium Policy," *The Developing Economies* vol. 15, no. 2 (June 1977) pp. 202–220.

———. "Rice Economy and Rice Policy in South Vietnam up to 1974: An Economic and Statistical Analysis," *Southeast Asian Studies* vol. 15, no. 3 (December 1977).

———. "An Econometric Study of the Effects of National Rice Policies and the Green Revolution on National Rice Economics and International Rice Trade Among Less Developed Countries: With Special Reference to Thailand, Indonesia, Japan and the U.S. (Ph.D. dissertation, University of Illinois, 1973).

The Tsuneta Yano Memorial Society (Yano-Buneta Kinenkai) under supervision of Ichiro Yano, ed. *Nippon: A Charted Survey of Japan* (Tokyo, Kokuseisha).

Tuan, Francis. "PRC Provincial Total Grain Production," Research Notes on Chinese Agriculture no. 2 (Washington, D.C., PRC Section, Economics and Statistics Service, U.S. Department of Agriculture, 1981).

Tubpun, Somnuk. "The Price Analysis and the Rate of Return on Holding Rice and Paddy in Thailand" (M.A. thesis, Thammasat University, 1974).

Tyers, Rodney. "Food Security in ASEAN: Potential Impacts of a Pacific Economic Community" (Canberra, Australian National University, ASEAN-Australian Economic Relations Research Project, May 1982).

United Nations. *Statistical Yearbook for Asia and the Pacific* (Bangkok, annual).

United Nations Department of International Economics and Social Affairs. "World Population Prospects as Assessed in 1980," Population Studies No. 78 (New York, 1981).

United States, Agency for International Development. "Foodgrain Technology: Agricultural Research in Nepal," AID Project Impact Evaluation no. 33 (Washington, D.C., May 1982).

United States, Department of Agriculture. "Agricultural Situation—Review of 1979 and Outlook for 1980—People's Republic of China" Supplement 6 to WAS-21 (Washington, D.C., 1979 and other years).

———. *World Demand Prospects for Grain in 1980* Foreign Agricultural Economic report No. 75 (Washington, D.C., 1971).

———. *Agricultural Statistics* (Washington, D.C., Government Printing Office, annual).

———. *Rice Situation* (Washington, D.C., Government Printing Office, annual).

———. *Yearbook of Agriculture* (Washington, D.C., Government Printing Office, annual).

United States, Department of Agriculture, Crop Reporting Board, *Agricultural Policies Annual Summary* (Washington, D.C., annual).

United States, Department of Agriculture, Economic Research Service, Economic and Statistical Analysis Division, *Agriculture in Vietnam's Economy: A System for Economic Analysis* (Washington, D.C., June 1973).

United States, Foreign Agricultural Service, Foreign Agricultural Circular, *Grains* FG-38-80 (Washington, D.C., Government Printing Office, various years).

United States, Operations Mission to Viet-nam, Division of Agriculture and Natural Resources. *Vietnamese Agricultural Statistics* (Saigon, March 1959).

Unnevehr, L. J. "The Impact of Philippine Government Intervention in Rice Markets," IRRI Agricultural Economics Paper 82-24 (Los Banos, Philippines, International Rice Research Institute, 1982).

Unnevehr, L. J., and M. L. Stanford. "Technology and the Demand for Women's Labor in Asian Rice Farming." Paper presented at Women in Rice Farming Systems Conference, Los Banos, Philippines, September 26–30, 1983.

Utami, Widya, and John Ihalauw. "Klaten, Central Java," in International Rice Research Institute, *Changes in Rice Farming in Selected Areas of Asia* (Los Banos, Philippines, IRRI, 1975) pp. 149–177.

———. "Some Consequences of Small Farm Size," *Bulletin of Indonesian Economic Studies* vol. 9 (July 1973) pp. 46–56.

Valdez, Alberto, ed. *Food Security for the Developing Countries* (Boulder, Colo., Westview Press, 1981).

Valdez, Alberto, Grant M. Scobie, and John L. Dillon, eds. *Economics and the Design of Small-farmer Technology* (Ames, Iowa, Iowa State University Press, 1979).

Valentine, R. C. "Genetic Engineering in Agriculture with Emphasis on Biological Nitrogen Fixation," in National Academy of Sciences, *Research with Recombinant DNA* (Washington, D.C., NAS, 1977).

Varca, Arlando S., and Reeshon Feurer. "The Brown Planthopper and Its Biotypes in the Philippines." Paper read before the National Conference of Farmers' Associations, Bacolod City, Philippines, April 21, 1976.

Vella, Walter. *Siam Under Rama III 1824–1851* (New York, J. J. Augustin, 1957).

———. *The Impact of the Western Government in Thailand* (Berkeley, Calif., University of California Press, 1955).

Vergara, Benito S., Roberto Lilis, and Akira Tanaka. "Studies of Internode Elongation of the Rice Plant: In Relationship Between Growth Duration and Internode Elongation," *Soil Science and Plant Nutrition* vol. 11 (1965) pp. 26–30.

Vernon, Robert E., ed. *The Technology Factor in International Trade* (New York, Columbia University Press, 1970).

Viadyanathan, A. "Labor Use in Indian Agriculture: An Analysis Based on Farm Management Survey Data," in P. K. Bardhan, A. Viadyanathan, Y. Alugh, G. S. Bhalla, and A. L. Bhadem, eds., *Labour Absorption in Indian Agriculture, Some Exploratory Investigations* (Bangkok, International Labor Organization, 1978) pp. 33–118.

Viêt-nam Công-Hòa, Bô Kê-hoach và Phát-triên Quôc-gia (Ministry of National Planning and Development), Viên Quôc-gia Thông-kê (National Institute of Statistics). *Viêt-nam niên giám thông-kê* (Statistical Yearbook of Vietnam) (Saigon, 1949/50–1972/73).

Viêt-nam Công-hòa, So Thông-kê và kinh-tê Nông-nghiêp. (Agricultural Economics and Statistics Service). *Viêt-nam thông-kê canh-nông năm 1959, 1960* (Vietnamese Agricultural Statistics, 1959, 1960) (Saigon).

Viêt-nam, Công-hòa, So Thông-kê và kinh-tê Nông-nghiêp. *Nièn-giám thông-kê nông-nghiêp* (Agricultural Statistical Yearbook) (Saigon, 1949/50–1972/73).

Viêt-nam Công-hòa, So Thông-kê và kinh-tê Nông-nghiêp. *Canh-nông thông-kê nguyêt-san* (Monthly Bulletin of Statistics) (Saigon).

Vietnam, Democratic Republic of Vietnam, Central Statistical Office. *Nam nam xay dung kinh te va van hòa* (Five Years of Economic and Cultural Building) (Hanoi, Central Statistical Office, 1960).

Wackernagel, Frederick W. III. "Rice for the Highlands: Cold Tolerant Varieties and Other Strategies for Increasing Rice Production in the Mountains of Southeast Asia" (Ph.D. dissertation, Cornell University, 1984).

Wai, U. Tun. "The Economic Development of Burma from 1800–1940" (Rangoon, Department of Economics, University of Rangoon, 1961).

Walinsky, Louis J. *Economic Development in Burma 1951–1960* (New York, The Twentieth Century Fund, 1962).

Walker, Kenneth. "Provincial Grain Output in China 1952–57: A Statistical Compilation," Research Notes and Studies No. 3 (London, Contemporary China Institute, School of Oriental and African Studies, University of London, 1977).

Watabe, Tadayo. "The Development of Rice Cultivation," in Ishii Yoneo, ed., *Thailand: A Rice Growing Society* (Honolulu, University of Hawaii Press, 1978) pp. 6–10.

Wattananukit, Atchana. "Comparative Advantage of Rice Production in Thailand: A Domestic Resource Cost Study" (M.A. thesis, Thammasat University, 1975).

Weitz-Hettelsater Engineers. *Rice Storage, Handling, and Marketing, The Republic of Indonesia* sponsored by United States Agency for International Development (USAID) (1972).

West Pakistan, Bureau of Statistics, Planning and Development. *Statistical Handbook of West Pakistan* (Lahore, 1963–1968).

Whitaker, A., et. al. *Area Handbook for the Khmer Republic* (Washington, D.C., Foreign Area Studies, American University, 1973).

White, Benjamin. "Women and the Modernization of Rice Agriculture: Some General Issues and a Javanese Case Study." Paper presented at Women in Rice Farming Systems Conference, Los Banos, Philippines, September 26–30, 1983.

———. "Population, Involution, and Employment in Rural Java," *Development and Change* vol. 7 (1976) pp. 267–290.

Whyte, R. O. *Rural Nutrition in Monsoon Asia* (Kuala Lumpur, Malaysia, Oxford University Press, 1974).

Whyte, William Foote, and Damon Boynton, eds. *Higher Yielding Human Systems for Agriculture* (Ithaca, N.Y., Cornell University Press, 1983).

Wickham, T. H. "Predicting Yield Benefits in Lowland Rice Through a Water Balance Model," in International Rice Research Institute, *Water Management in Philippine Irrigation Systems: Research and Operation* (Los Banos, Philippines, IRRI, 1973).

Wickham, T. H., R. Barker, and M. W. Rosegrant. "Complementarities Among Irrigation, Fertilizers, and Modern Rice Varieties," in International Rice Research Institute, *Economic Consequences of New Rice Technology* (Los Banos, Philippines, IRRI, 1978).

Wickizer, V., and M. Bennett. *The Rice Economy of Monsoon Asia* (Stanford, Calif., Stanford University Press, 1941).

Wiens, Thomas B. "The Limits of Agricultural Intensification: The Suzhou Experience," in Beth Rose and Randolph Barker, eds., *Agricultural and Rural Development in China Today* (Ithaca, N.Y., Cornell University, 1983) pp. 54–77.

———. "Agricultural Statistics in the People's Republic of China," in Alexander Eckstein, ed., *Quantitative Measures of China's Economic Output* (Ann Arbor, Mich., University of Michigan Press, 1980) pp. 104–107.

———. "The Evolution of Policy Capabilities in China's

Agricultural Technology," in *The Chinese Economy Post-Mao, Policy and Performance*, vol. 1. U.S. Congress, Joint Economic Committee (1978) pp. 671–703.

Wilkinson, Endymion Porter. *Studies in Chinese Price History* (New York, Garland Publishing Inc., 1980).

Win, Khin, and Nyi Nyi. "Factors Contributing to Increased Rice Production in Burma" (Rangoon, Agriculture Corporation, 1980).

Wittfogel, Karl. *Oriental Despotism: A Comparative Study of Total Power* (New Haven, Conn., Yale University Press, 1957).

Wong, Chung Ming. "A Model of the Rice Economy of Thailand" (Ph.D. dissertation, University of Chicago, 1976).

Wongsangaroonsri, Anuwat. "Effects of Mechanization on Employment and Intensity of Labor Use" (Los Banos, Philippines, Agricultural Development Council/International Rice Research Institute, Workshop on the Consequences of Small Rice Farm Mechanization, September 14–18, 1981).

Wood, G. "Class Differentiation and Power in Bandakgram: The Minifundist Case," in M. N. Hoq, ed., *Exploitation and the Rural Poor* (Camilla, Bangladesh, Bangladesh Academy of Rural Development, 1976).

World Bank. *Agricultural Sector Survey Indonesia* Annex 13, Report No. 183 (Washington, D.C.)

———. *Commodity Trade and Price Trends* (Washington, D.C., various dates).

———. "Nepal Agricultural Sector Review," Report No. 2205-NEP (Washington, D.C., 1979).

———. "Nepal Agricultural Sector Review," Report No. 519a-NEP (Washington, D.C., 1974).

———. "Nepal Development Performance and Prospects" (Washington, D.C., December 1979).

———. *The Philippines, Priorities and Prospects for Development* (Washington, D.C., 1976).

———. "Philippines Sector Study: Grain Production Policy Review," Report No. 2192a-Ph (Washington, D.C., January 22, 1979).

———. *Thailand Toward a Development Strategy of Full Participation* (Washington, D.C., East Asia and Pacific Regional Office, World Bank, 1980).

———. *World Development Report* (Washington, D.C., The World Bank, various years).

———. *World Tables*, second edition (Baltimore, Md., Johns Hopkins University Press, 1980).

Wortman, Sterling, and Ralph W. Cumings, Jr. *To Feed This World: The Challenge and the Strategy* (Baltimore, Md., Johns Hopkins University Press, 1978).

Wu, Kung-Hsien, Carson. "Analysis of Machinery–Labor Relationship in Farm Mechanization," in Herman Southworth ed., *Farm Mechanization in East Asia* (New York, Agricultural Development Council, 1972).

Wu, Trong-chuang. "Government Policies Promoting Farm Mechanization," in Herman Southworth, ed., *Farm Mechanization in East Asia* (New York, Agricultural Development Council, 1972).

Xue-Bin, X. "Half-Sky Role of China's Women in Rice Farming System." Paper presented at Women in Rice Farming Systems Conference, Los Banos, Philippines, September 26–30, 1983.

Yap, Kim Lian. "The Role of Women in Paddy Production and Processing in Malaysia—An Economic Analysis and Perspective Trend." Paper presented at the Workshop on Women's Participation in Paddy Production and Processing, Kata Bharu, Malaysia, October 21–28, 1981.

Yeh, S. M. "Rice Marketing in Taiwan," *Economic Digest Series*, no. 7 (Taipei, Joint Commission on Rural Reconstruction, 1955).

Yoneo, Ishii, ed. *Thailand: A Rice Growing Society* (Honolulu, University of Hawaii Press, 1978).

Yoshida, K. "Country Report, Japan." Report presented at Women in Rice Farming Systems Conference, Los Banos, Philippines, September 26–30, 1983.

Yoshida, Shouichi. *Fundamentals of Rice Crop Science* (Los Banos, Philippines, International Rice Research Institute, 1971).

Young, Ralph. "An Economic Analysis of Uncertainty in Production" (Ph.D. dissertation, Cornell University, 1981).

Zaman, M. Raquibuz, and M. Asaduzzaman. "An Analysis of Rice Prices in Bangladesh 1952/53–1967/68," Research Report Series no. 2 (Dacca, Bangladesh Institute of Development Economics, 1972).

Zandstra, H. G., E. C. Price, J. A. Litsinger, and R. A. Morris. *A Methodology for On-Farm Cropping Systems Research* (Los Banos, Philippines, International Rice Research Institute, 1981).

Zhongguo jingji nianjian 1982 (Chinese Economic Yearbook of 1982) (Beijing, Jingji Guanli Zazhishe, 1982) (annual).

Zhongguo Kexue Yuan, Dili Yanjiu Suo, Jingji Dili Yanjiushi. *Zhongguo nongye dili zonghu* (General Treatise on the Agricultural Geography of China) (Beijing, Kexue Chubanshe, 1980).

Zhongguo nongye nianjian 1980 (Agricultural Yearbook of China 1980) (Beijing, Nongye Chubanshe, 1981) (annual).

Zongguo tongji nianjian 1981 (Statistical Yearbook of China) (Beijing, Zongguo Tongji Chubanshe, 1981).

INDEX

Page numbers in boldface indicate tables.

The book was set in Clearface display and Times Roman text type by Electronic Publishing Services (EPS) Group Inc., Baltimore, Maryland. It was printed by Optic Graphics Inc., Glen Burnie, Maryland.